MicroRNAs: From Bench to Bedside

MicroRNAs: From Bench to Bedside

Edited by Daniel Fletcher

www.statesacademicpress.com

States Academic Press,
109 South 5th Street,
Brooklyn, NY 11249, USA

Visit us on the World Wide Web at:
www.statesacademicpress.com

ISBN: 978-1-63989-719-3

Cataloging-in-Publication Data

MicroRNAs : from bench to bedside / edited by Daniel Fletcher.
   p. cm.
Includes bibliographical references and index.
ISBN 978-1-63989-719-3
1. MicroRNA. 2. Small interfering RNA. 3. RNA. 4. Microbiology. I. Fletcher, Daniel.
QP623.5.S63 M53 2023
572.88--dc23

Table of Contents

Preface

This book has been a concerted effort by a group of academicians, researchers and scientists, who have contributed their research works for the realization of the book. This book has materialized in the wake of emerging advancements and innovations in this field. Therefore, the need of the hour was to compile all the required researches and disseminate the knowledge to a broad spectrum of people comprising of students, researchers and specialists of the field.

MicroRNAs (miRNAs) are a category of non-coding RNAs which play a crucial role in the regulation of gene expression. A vast range of miRNAs are transcribed from DNA sequences in primary miRNAs, which are then processed into precursor miRNAs and then into mature miRNAs. The interaction of miRNAs with their target genes is dynamic and depends on a variety of factors including the abundance of miRNAs and target mRNAs, the affinity of miRNA-mRNA interactions, and miRNA subcellular location. They can be secreted into extracellular fluids and transferred to target cells by vesicles like exosomes or through a bind to proteins such as argonautes. Furthermore, miRNAs regulate the activities of host cell and also secrete and transport those host cells to recipient cells. This book provides significant information to help develop a good understanding of microRNAs. The readers would gain knowledge that would broaden their perspective in this area of study.

At the end of the preface, I would like to thank the authors for their brilliant chapters and the publisher for guiding us all-through the making of the book till its final stage. Also, I would like to thank my family for providing the support and encouragement throughout my academic career and research projects.

Editor

Integration of miRNA and mRNA Co-Expression Reveals Potential Regulatory Roles of miRNAs in Developmental and Immunological Processes in Calf Ileum during Early Growth

Duy N. Do [1,2], Pier-Luc Dudemaine [1], Bridget E. Fomenky [1,3] and Eveline M. Ibeagha-Awemu [1,*]

[1] Agriculture and Agri-Food Canada, Sherbrooke Research and Development Centre, Sherbrooke, QC J1M 0C8, Canada; DuyNgoc.Do@AGR.GC.CA (D.N.D.); Pier-Luc.Dudemaine@AGR.GC.CA (P.-L.D.); bridget.fomenky.1@ulaval.ca (B.E.F.)

[2] Department of Animal Science, McGill University, Ste-Anne-de-Bellevue, QC H9X 3V9, Canada

[3] Département de Sciences Animale, Université Laval, Quebec, QC G1V 0A6, Canada

* Correspondence: eveline.ibeagha-awemu@agr.gc.ca;

Abstract: This study aimed to investigate the potential regulatory roles of miRNAs in calf ileum developmental transition from the pre- to the post-weaning period. For this purpose, ileum tissues were collected from eight calves at the pre-weaning period and another eight calves at the post-weaning period and miRNA expression characterized by miRNA sequencing, followed by functional analyses. A total of 388 miRNAs, including 81 novel miRNAs, were identified. A total of 220 miRNAs were differentially expressed (DE) between the two periods. The potential functions of DE miRNAs in ileum development were supported by significant enrichment of their target genes in gene ontology terms related to metabolic processes and transcription factor activities or pathways related to metabolism (peroxisomes), vitamin digestion and absorption, lipid and protein metabolism, as well as intracellular signaling. Integration of DE miRNAs and DE mRNAs revealed several DE miRNA-mRNA pairs with crucial roles in ileum development (bta-miR-374a—*FBXO18*, bta-miR-374a—*GTPBP3*, bta-miR-374a—*GNB2*) and immune function (bta-miR-15b—*IKBKB*). This is the first integrated miRNA-mRNA analysis exploring the potential roles of miRNAs in calf ileum growth and development during early life.

Keywords: calf; Ileum; miRNA-mRNA integration; miRNA sequencing; growth; development

1. Introduction

MicroRNAs (miRNAs) are small (~22 nucleotides) endogenous RNA molecules that regulate gene expression post-transcriptionally by targeting principally the 3' untranslated region (3'UTR) of genes and to some extend the 5'UTR, introns and coding region of mRNAs [1]. They play key roles in a wide range of biological processes [2]. In bovine, miRNAs have been shown to play important roles in embryonic development [3–5], mammary gland [6,7] and adipose tissue [8] functions and also in the regulation of production traits such as milk yield [6,9,10], milk quality [11] and diseases like mastitis [12,13] and Johne's disease [14]. The importance of miRNAs in gut development and disease (mostly inflammatory bowel disease) has been extensively studied in humans [15–18]. For instance, several miRNAs have been reported to be relevant for different aspects of inflammatory bowel disease (IBD), including miRNAs important for intestinal fibrosis (miR-29, miR-200b, miR-21, miR-192), epithelial barrier and immune function in IBD pathogenesis (miR-192, miR-21, miR-126, miR-155, miR-106a) [19].

Data from a few studies suggest important roles for miRNAs in calf's early life [20,21]. For instance, Liang et al. (2016a) [21] identified several dominantly expressed miRNAs (miR-143 (30% of read counts), miR-192 (15%), miR-10a (12%) and miR-10b (8%)) in ileum tissues of dairy calves collected at 30 min after birth and at 7, 21 and 42 days old. Furthermore, several temporally expressed miRNAs (miR-146, miR-191, miR-33, miR-7, miR-99/100, miR-486, miR-145, miR-196 and miR-211), regional specific miRNAs (miR-192/215, miR-194, miR-196, miR-205 and miR-31) and miRNAs (miR-15/16, miR-29 and miR-196) linked to bacterial abundance in the jejunum and ileum were also reported [21]. Moreover, several ileum miRNAs are reported to play important roles in host responses to *Mycobacterium avium* subspecies paratuberculosis infection, such as the role of bta-miR-196b in the proliferation of endothelial cells and bta-miR-146b in bacteria recognition and regulation of the inflammatory response [22].

During early life, calves undergo major physiological and digestive changes, including adaptation to diet changes from pre- to post-weaning. The interactions between transcriptional and post-transcriptional mechanisms are known to coordinate these developmental transitions via regulation of gene expression. Recently, we characterized the long non-coding RNA (lncRNA) expression in ileum tissues of calves and functional inference of identified lncRNA (623 known and 1505 novel) *cis* target genes revealed potential roles in growth and development as well as in posttranscriptional gene silencing by RNA or miRNA processing processes and in disease resistance mechanisms [23]. Moreover, we also observed that 122 miRNAs were significantly differentially expressed between the pre- and post-weaning periods in the rumen, suggesting important roles of miRNAs in calf gut during early life [24]. Therefore, we hypothesize that miRNAs might play important roles in ileum development during the pre- and post-weaning periods. Thus, in the current study, we performed integrated miRNA-mRNA co-expression analyses to uncover the potential roles of miRNAs in ileum development at the pre- and post-weaning periods.

2. Materials and Methods

2.1. Animals and Management

Procedures for animal management were conducted according to the Canadian national codes of practice for the care and handling of farm animals (http://www.nfacc.ca/codes-of-practice) and approved by the animal care and ethics committee of Agriculture and Agri-Food Canada (CIPA #442).

Experimental details have been described in our previous studies [23,24]. Briefly, sixteen 2–7 days-old Holstein calves were raised for a period of 96 days in individual pens. In the first week of the experiment, calves were fed milk replacer (6 L/day for the first four days and 9 L/day thereafter, Goliath XLR 27-16, La Coop, Montreal, QC, Canada), and then starter feed (Munerie Sawyerville Inc., Cookshire-Eaton, QC, Canada) was introduced (ad libitum) from the second week. After weaning, calves were fed with starter feed and hay ad libitum. The calves were weighed weekly until euthanization. At experiment D33 (pre-weaning), eight calves were humanely euthanized and another eight calves on D96 (post-weaning), for the collection of ileum tissue samples. Tissues were aseptically collected, snap frozen in liquid nitrogen, and stored at −80 °C until used.

2.2. Total RNA Purification

Ileum tissue (30 mg/sample) was used for total RNA isolation using miRNeasy Kit (Qiagen Inc., Toronto, ON, Canada). Potentially contaminating genomic DNA was removed by treating 10 μg of purified RNA (10 μg) with DNase (Turbo DNA-free™ Kit, Ambion Inc., Foster City, CA, USA). The RNA concentration (before and after digestion) and its quality (integrity) after DNase treatment were assessed with Nanodrop ND-1000 (NanoDrop Technologies, Wilmington, DE, USA) and Agilent 2100 Bioanalyzer (Agilent Technologies, Santa Clara, CA, USA), respectively. The RNA integrity number (RIN) of all samples was greater than 8 and a small RNA peak area was visible on the electropherogram [25].

2.3. miRNA Library Preparation and Sequencing

Libraries (n = 16) were prepared and barcoded for sequencing according to Do et al. [10]. Briefly, polyacrylamide gel electrophoresis was used to size separate miRNA libraries from other RNA species. An elution buffer (10 mM Tris-HCl pH 7.5; 50 mM NaCl, 1 mM EDTA) was used to elute the libraries from the gel. Eluted library was concentrated using DNA clean and concentrator-5 (Zymo Research, Irvine, CA, USA). The concentration of purified libraries was evaluated using Picogreen assay (Life Technologies, Waltham, MA, USA) and a Nanodrop 3300 fluorescent spectrophotometer (NanoDrop Technologies), and further confirmed by qPCR using the Kapa Library Quantification Kit for Illumina platforms (KAPA Biosystems, Wilmington, MA, USA). Libraries were multiplexed in equimolar concentrations and sequenced in one lane on an Illumina HiSeq 2500 platform following Illumina's recommended protocol to generate single end data of 50-bases by The Centre for Applied Genomics, The Hospital for Sick Children, Toronto, Canada (http://www.tcag.ca/).

2.4. Small RNA Sequence Data Analysis

Bioinformatics processing of generated small RNA sequences was done as previously described [10,24]. Briefly, the raw sequence data (16 fastq files) was checked for sequencing quality with FastQC program v0.11.3 (http://www.bioinformatics.babraham.ac.uk/projects/fastqc/). Trimming of 3′ and 5′ adaptor sequences, contaminants and repeats was accomplished with Cutadapt v1.2.2 (https://cutadapt.readthedocs.org/). Then, FASTQ Quality Filter tool of FASTX-toolkit (http://hannonlab.cshl.edu/fastx_toolkit/) was used to remove reads having a Phred score <20 for at least 50% of the bases and reads shorter than 18 nucleotides or longer than 30 nucleotides. Clean reads that passed all filtering criteria from the 16 files were parsed into one file and mapped to the bovine genome (UMD3.1) using bowtie 1.0.0 (http://bowtie-bio.sourceforge.net/index.shtml) [26]. Reads that mapped to other RNA species (rRNA, tRNA, snRNA and snoRNA) in the Rfam RNA family database (http://rfam.xfam.org/) or to more than five positions of the genome were removed.

2.5. Identification of Known miRNA and Novel miRNA Discovery

The identification of known miRNAs was performed with miRBase v21 http://www.mirbase.org/) (Kozomara and Griffiths-Jones, 2014), while novel miRNA discovery was achieved with miRDeep2 v2.0.0.8 (https://github.com/rajewsky-lab/mirdeep2) [27]. MiRDeep2 was designed to detect miRNAs from deep sequence reads using a probabilistic algorithm based on the miRNA biogenesis model. The core and quantifier modules of miRDeep2 were applied to discover novel miRNAs in the pooled dataset of all the libraries while the quantifier module was used to profile the detected miRNAs in each library. MiRDeep2 score higher than five was used as cuff point for the identification of novel miRNAs. Subsequently, a threshold of 10 counts per million and present in ≥2 libraries was applied to remove lowly expressed miRNAs. MiRNAs meeting these criteria were further used in downstream analyses including differential expression (DE) analysis.

2.6. Differential miRNA Expression

DeSeq2 (v1.14.1) (https://bioconductor.org/packages/release/bioc/html/DESeq2.html) [28], which implements a negative binomial model, was used to perform differential miRNA expression analysis. Following normalization, normalized counts of miRNAs at D96 were compared with corresponding values on D33. Significant differential miRNA expression between D33 (pre-weaning) and D96 (post-weaning) was defined as having a Benjamini and Hochberg [29] false discovery rate (FDR) or corrected p-value < 0.05.

2.7. Predicted Target Genes of miRNAs

In order to investigate the functions of the most highly expressed miRNAs and differently expressed (DE) miRNAs, we firstly predicted their target mRNAs. Perl scripts (targetscan_60.pl and targetscan_61_context_scores.pl) (http://targetscan.org) were used to predict target mRNAs and to calculate their context scores, respectively. Predicted target mRNAs with context + scores above 95th percentile were further used [9,10,30]. The predicted target mRNAs were then filtered against the mRNA transcriptome obtained from ileum tissues of the same animals. Only predicted target genes that were expressed in the mRNA transcriptome of the ileum tissues of the animals [23] were retained for further analysis.

2.8. miRNA–mRNA Co-Expression Analysis and Target Gene Enrichment

For miRNA–mRNA co-expression, the Pearson correlation coefficient between target mRNAs (retained above) and DE miRNAs were calculated. A miRNA-mRNA pair was considered co-expressed if it had a negative and significant correlation value at FDR < 0.05. The mRNAs significantly correlated with miRNAs were then used for downstream target gene ontology and KEGG pathways enrichment using ClueGO (http://apps.cytoscape.org/apps/cluego) [31]. For ClueGO analysis, a hypergeometric test was used for enrichment analyses and Benjamini–Hochberg [29] correction was used for multiple testing correction (FDR < 0.05). Since KEGG pathways enrichment relied on the human database (due to lack of information in bovine), we used a less stringent threshold (uncorrected p-value < 0.05) to declare if a pathway was significantly enriched. Interactions between miRNAs and mRNAs were visualized with Cytoscape (http://www.cytoscape.org/) [32].

2.9. Real-Time Quantitative PCR

The method of real-time quantitative PCR was used to validate the expression of four DE (bta-miR-142-5p, miR-146a, miR-24-3p and miR-374b) and two non-DE (bta-miR-486-5p and miR-193b) miRNAs. The same total RNA used in miRNA-sequencing was used. Total RNA was reverse transcribed with Universal cDNA Synthesis Kit II from Exiqon (Exiqon Inc., Woburn, MA, USA), following the manufacturer's instructions. ExiLENT SYBR® Green Master Mix Kit (Exiqon, Woburn, MA, USA) and the miRCURY LNA™ Assay (Exiqon, Woburn, MA, USA) specific for each miRNA listed above were used to perform Quantitative qPCR on a StepOne Plus System (Applied Biosystems, Foster City, CA, USA) according to the manufacturer's instructions. Bta-miR-126-3p was used as endogenous control to assess the expression level of miRNAs using the comparative Ct (ΔΔCt) method. Bta-miR-126-3p was selected as an endogenous control based on its consistent expression throughout all the analyzed samples on D33 and D96.

3. Results

3.1. Identification and Characterization of Ileum miRNAs

MiRNA sequencing of 16 libraries generated a total of 185,458,022 reads. After adaptor trimming, size selection and quality filtering, 150,999,506 (81.4%) reads with length ranging from 18 to 30 nucleotides and having a phred score >20 were retained for analysis (Table S1). Out of this number, 133,698,161 reads (88.5%) mapped to unique positions on the bovine genome (University of Maryland assembly of *B. taurus*, release 3.1; UMD.3.1), 10,661,520 (7.1%) were unmapped, while 1,150,263 (0.8%) mapped to more than five positions and were discarded (Table S1). Mapped reads belonging to other RNA species, tRNA (3,153,316 (2.1%)), rRNA (480,099 (0.3%)), snRNA (236,118 (0.2%)) and snoRNA (1,620,029 (1.1%)) were discarded. The majority of miRNAs retained for further analyses were 22 nucleotides long (Table S2).

Novel miRNAs were considered to have a minimum MiRDeep2 score of five, as shown in Table S3. After removing lowly expressed reads, a total of 307 known and 81 novel miRNAs satisfying the conditions of having at least 10 read counts per million and present in a minimum of two libraries were used for DE analysis (Table S4a,b).

Abundantly expressed miRNAs having >3% of the total read counts on D33 and D96 were bta-miR-143, bta-miR-192, bta-miR-26a and bta-miR-21-5p, while bta-miR-191, bta-miR-10b, bta-miR-148a and bta-miR-10a were highly expressed with >3% of total read counts on D96 (post-weaning) only (Table 1). The 20 commonly highly expressed miRNAs (>1% of total read counts) targeted 2609 unique genes (Tables 1 and S5a). The target genes were significantly enriched in 459 biological processes (BP), 53 cellular components (CC) and 43 molecular function (MF) gene ontology (GO) terms (Table S5b), as well as in 14 KEGG pathways (Table S5c). Single-organism developmental process (FDR = 1.13×10^{-10}), intracellular (FDR = 4.63×10^{-17}), and protein binding (FDR = 1.10×10^{-5}) were the most significantly enriched BP, CC and MF GO terms, respectively (Table 2), while MAPK signaling pathway was the most significantly enriched KEGG pathway (Table 3). Moreover, a novel miRNA, bta-miR-22-24033, was the most highly expressed among novel miRNAs (accounted for 0.3% of total read counts) (Table S4b).

Table 1. The 20 most abundantly expressed miRNAs in ileum tissue of calves.

miRNA	Pre-Weaning (D33)		Post-Weaning (D96)		Both Periods	
	Read Counts	% of Total	Read Counts	% of Total	Read Counts	% of Total
bta-miR-143	16,742,092	24.51	7,468,034	13.65	24,210,126	19.68
bta-miR-192	4,435,941	6.50	5,383,934	9.84	9,819,875	7.98
bta-miR-26a	2,861,953	4.19	4,861,644	8.89	7,723,597	6.28
bta-miR-191	1,691,133	2.48	4,106,915	7.51	5,798,048	4.71
bta-miR-10b	1,693,467	2.48	2,031,077	3.71	3,724,544	3.03
bta-miR-148a	1,702,572	2.49	1,754,282	3.21	3,456,854	2.81
bta-miR-10a	1,413,716	2.07	1,734,580	3.17	3,148,296	2.56
bta-miR-21-5p	4,233,604	6.20	1,673,008	3.06	5,906,612	4.80
bta-miR-99a-5p	1,282,425	1.88	1,452,116	2.65	2,734,541	2.22
bta-miR-215	1,320,560	1.93	1,373,575	2.51	2,694,135	2.19
bta-miR-27b	1,557,472	2.28	1,310,625	2.40	2,868,097	2.33
bta-let-7a-5p	1,729,397	2.53	1,231,252	2.25	2,960,649	2.41
bta-let-7f	1,668,877	2.44	1,226,489	2.24	2,895,366	2.35
bta-miR-125b	923,543	1.35	987,775	1.81	1,911,318	1.55
bta-miR-145	1,152,495	1.69	798,715	1.46	1,951,210	1.59
bta-miR-30e-5p	699,838	1.02	735,220	1.34	1,435,058	1.17
bta-let-7g	1,083,909	1.59	685,389	1.25	1,769,298	1.44
bta-miR-194	1,498,650	2.19	571,343	1.04	2,069,993	1.68
bta-miR-30d	747,419	1.09	560,435	1.02	1,307,854	1.06

Table 2. Enriched gene ontology (GO) terms for target genes of 20 most abundantly expressed miRNAs.

GO Class	GOID	GO Term	p-Value	FDR
Biological process	GO:0044767	Single-organism developmental process	4.40×10^{-13}	1.13×10^{-10}
	GO:0044260	Cellular macromolecule metabolic process	7.96×10^{-13}	1.79×10^{-10}
	GO:0044237	Cellular metabolic process	3.29×10^{-12}	6.57×10^{-10}
	GO:0007275	Multicellular organismal development	2.62×10^{-11}	4.71×10^{-9}
	GO:0048731	System development	7.45×10^{-11}	1.22×10^{-8}
	GO:0009888	Tissue development	9.21×10^{-11}	1.38×10^{-8}
	GO:0048856	Anatomical structure development	1.24×10^{-10}	1.72×10^{-8}
	GO:0036211	Protein modification process	8.55×10^{-10}	9.61×10^{-8}
	GO:0030154	Cell differentiation	1.74×10^{-9}	1.74×10^{-7}
Cellular component	GO:0043412	Macromolecule modification	3.53×10^{-9}	3.34×10^{-7}
	GO:0005622	Intracellular	2.57×10^{-20}	4.63×10^{-17}
	GO:0044424	Intracellular part	1.35×10^{-19}	1.22×10^{-16}
	GO:0043227	Membrane-bounded organelle	3.43×10^{-17}	2.06×10^{-14}
	GO:0043231	Intracellular membrane-bounded organelle	1.95×10^{-15}	8.79×10^{-13}
	GO:0043229	Intracellular organelle	3.08×10^{-15}	1.11×10^{-12}
	GO:0005737	Cytoplasm	4.55×10^{-15}	1.36×10^{-12}
	GO:0044422	Organelle part	2.39×10^{-10}	3.07×10^{-8}
	GO:0044446	Intracellular organelle part	3.80×10^{-10}	4.56×10^{-8}
	GO:0044444	Cytoplasmic part	9.28×10^{-10}	9.83×10^{-8}
	GO:0005634	Nucleus	7.71×10^{-8}	4.34×10^{-6}
Molecular function	GO:0005515	Protein binding	3.18×10^{-7}	1.10×10^{-5}
	GO:0019207	Kinase regulator activity	4.91×10^{-6}	1.08×10^{-4}
	GO:0019887	Protein kinase regulator activity	3.01×10^{-5}	4.79×10^{-4}
	GO:0003723	RNA binding	3.24×10^{-5}	5.06×10^{-4}
	GO:0019210	Kinase inhibitor activity	3.78×10^{-5}	5.76×10^{-4}
	GO:0004702	Receptor signaling protein serine/threonine kinase activity	8.23×10^{-5}	1.06×10^{-3}
	GO:0005057	Receptor signaling protein activity	1.94×10^{-4}	2.02×10^{-3}
	GO:0061650	Ubiquitin-like protein conjugating enzyme activity	3.20×10^{-4}	2.96×10^{-3}
	GO:0003700	Transcription factor activity, sequence-specific DNA binding	4.99×10^{-4}	4.17×10^{-3}

Table 3. Enriched KEGG pathways for target genes of 20 most abundantly expressed miRNAs.

KEGG Pathway	p-Value	FDR
Cysteine and methionine metabolism	2.56×10^{-3}	4.39×10^{-2}
Amino sugar and nucleotide sugar metabolism	1.73×10^{-3}	3.47×10^{-2}
TGF-beta signaling pathway	9.58×10^{-4}	2.30×10^{-2}
Signaling pathways regulating pluripotency of stem cells	6.67×10^{-4}	2.00×10^{-2}
Pathways in cancer	3.81×10^{-5}	4.57×10^{-3}
Transcriptional misregulation in cancer	1.76×10^{-3}	3.26×10^{-2}
Proteoglycans in cancer	1.39×10^{-4}	8.37×10^{-3}
MAPK signaling pathway	1.22×10^{-5}	2.94×10^{-3}
Cell cycle	3.48×10^{-4}	1.67×10^{-2}
p53 signaling pathway	9.32×10^{-4}	2.49×10^{-2}
Protein processing in endoplasmic reticulum	1.29×10^{-3}	2.81×10^{-2}
ErbB signaling pathway	5.55×10^{-4}	1.90×10^{-2}
FoxO signaling pathway	1.04×10^{-4}	8.35×10^{-3}
Chronic myeloid leukemia	5.21×10^{-4}	2.08×10^{-2}

3.2. Differentially Expressed miRNAs and Downstream Target Gene Enrichment Analyses

A total of 220 miRNAs (104 up-regulated and 116 down-regulated) were significantly DE between D33 (pre-weaning) and D96 (post-weaning) (Figure 1, Table S6a). Bta-miR-374a (FDR = 5.00×10^{29}), bta-miR-15b (FDR = 7.96×10^{24}) and bta-miR-26a (FDR = 1.30×10^{20}) were the most significantly down-regulated miRNAs, while bta-miR-455-5p (FDR = 1.01×10^{23}), bta-miR-210 (FDR = 4.23×10^{20}) and bta-miR-497 (FDR = 9.95×10^{20}) were the most significantly up-regulated miRNAs (Table 4).

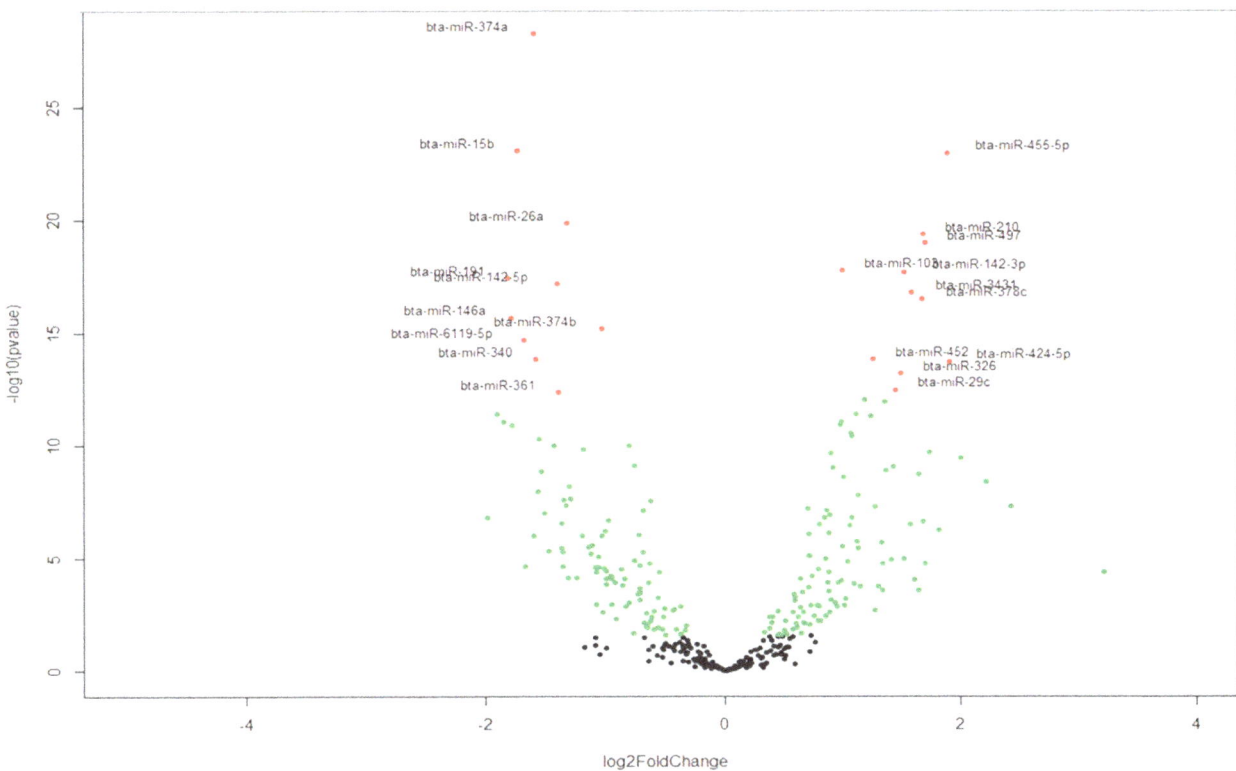

Figure 1. Volcano plot depicting miRNA differential expression results. Each dot represents a miRNA. Green and red dots represent miRNAs significantly differentially expressed at FDR < 0.05 and FDR < 1×10^{11}, respectively. Black dots represent miRNAs that were not differentially expressed. Differential expression analysis was accomplished with DeSeq2.

Table 4. The ten most up- and down-regulated miRNAs between D33 (pre-weaning) and D96 (post-weaning).

miRNA	Base Mean	Log2fold Change	Fold Change	p-Value	FDR
bta-miR-374a	8851.83	−1.59	−3.01	5.00×10^{-29}	1.94×10^{-26}
bta-miR-15b	17,755.36	−1.73	−3.32	7.96×10^{-24}	1.31×10^{-21}
bta-miR-26a	494,446.71	−1.32	−2.50	1.30×10^{-20}	1.26×10^{-18}
bta-miR-191	367,869.82	−1.81	−3.51	3.59×10^{-18}	1.55×10^{-16}
bta-miR-142-5p	92,116.08	−1.40	−2.64	6.64×10^{-18}	2.58×10^{-16}
bta-miR-146a	73,690.12	−1.79	−3.45	2.04×10^{-16}	6.08×10^{-15}
bta-miR-374b	4629.32	−1.03	−2.04	6.14×10^{-16}	1.70×10^{-14}
bta-miR-6119-5p	1796.83	−1.68	−3.19	2.08×10^{-15}	5.38×10^{-14}
bta-miR-340	360.85	−1.58	−2.99	1.40×10^{-14}	3.31×10^{-13}
bta-miR-361	9880.75	−1.39	−2.62	4.33×10^{-13}	8.00×10^{-12}
bta-miR-455-5p	282.27	1.89	3.70	1.01×10^{-23}	1.31×10^{-21}
bta-miR-210	486.48	1.67	3.19	4.23×10^{-20}	3.28×10^{-18}
bta-miR-497	692.69	1.70	3.24	9.95×10^{-20}	6.43×10^{-18}
bta-miR-103	39,682.96	1.00	2.00	1.69×10^{-18}	9.35×10^{-17}
bta-miR-142-3p	1742.08	1.52	2.87	1.96×10^{-18}	9.52×10^{-17}
bta-miR-3431	614.77	1.58	2.98	1.58×10^{-17}	5.57×10^{-16}
bta-miR-378c	1268.42	1.67	3.18	3.21×10^{-17}	1.04×10^{-15}
bta-miR-452	393.30	1.26	2.39	1.45×10^{-14}	3.31×10^{-13}
bta-miR-424-5p	301.96	1.90	3.73	1.84×10^{-14}	3.97×10^{-13}
bta-miR-326	378.11	1.49	2.82	6.07×10^{-14}	1.24×10^{-12}

The DE miRNAs (220 miRNAs) were predicted to target 11,691 mRNAs (Table S6b). Using mRNA transcriptome data of the same samples, 1560 mRNAs out of the predicted 11,691 mRNAs, were significantly and negatively correlated with their targeting miRNAs (Table S6c). Bta-miR-2285f had the highest number of target genes (172), while *AGO2* gene was the most popular target for DE miRNAs (targeted by 25 DE miRNAs) (Table S6c). Other common target genes for DE miRNAs were *SLC25A46*, *KCTD13* and *PAXIP1*, each targeted by 9 DE miRNAs (Table S6c). The GO enrichment analyses of the 1560 target genes (significantly and negatively correlated with miRNA) indicated that 158, 26 and 28 of them were significantly enriched in BP-, CC- and MF-GO terms, respectively (Table S7a–c). The most enriched BP-, CC- and MF-GO terms were cellular macromolecule metabolic process (FDR = 9.38×10^{10}), intracellular (FDR = 3.37×10^{19}) and organic cyclic compound binding (FDR = 1.19×10^4), respectively (Table 5, Figures 2–4). Moreover, 16 KEGG pathways were significantly enriched for the target genes (1560) of 220 DE miRNAs, and peroxisome ($p = 0.004$) and Hedgehog signaling pathways ($p = 0.006$) were the most significantly enriched (Table S7d, Figure 5). Moreover, among the 1560 target genes negatively correlated with miRNAs, 278 were also significantly DE between D33 and D96 in our previous study (Table S8) [23]. The 278 genes were the targets for 64 DE miRNAs. SOX4 was the most common target, since it was targeted by 6 different miRNAs (bta-miR-191, bta-miR-30e-5p, bta-miR-15-11508, bta-miR-2285f, bta-miR-92b and bta-miR-2285q). Meanwhile, bta-miR-2285f and bta-miR-874 had the highest number of target genes (37 and 28, respectively) (Figure 6).

Table 5. Most significantly enriched gene ontology (GO) terms for target genes of differentially expressed miRNAs.

GO Class	GO ID	GO Term	*p*-Value	FDR
Biological process	GO:0044260	Cellular macromolecule metabolic process	9.38×10^{-10}	5.11×10^{-7}
	GO:0043170	Macromolecule metabolic process	6.49×10^{-10}	7.07×10^{-7}
	GO:0019222	Regulation of metabolic process	9.86×10^{-9}	3.58×10^{-6}
	GO:0048518	Positive regulation of biological process	1.67×10^{-8}	4.55×10^{-6}
	GO:0071704	Organic substance metabolic process	4.25×10^{-8}	7.72×10^{-6}
	GO:0090304	Nucleic acid metabolic process	3.66×10^{-8}	7.97×10^{-6}
	GO:c0060255	Regulation of macromolecule metabolic process	9.84×10^{-8}	1.34×10^{-5}
	GO:0044237	Cellular metabolic process	8.67×10^{-8}	1.35×10^{-5}
	GO:0009059	Macromolecule biosynthetic process	1.76×10^{-7}	1.60×10^{-5}
	GO:0048522	Positive regulation of cellular process	1.92×10^{-7}	1.61×10^{-5}
Cellular component	GO:0005622	Intracellular	2.76×10^{-21}	3.37×10^{-19}
	GO:0043229	Intracellular organelle	2.39×10^{-21}	5.85×10^{-19}
	GO:0044424	Intracellular part	1.21×10^{-20}	9.85×10^{-19}
	GO:0043231	Intracellular membrane-bounded organelle	5.54×10^{-18}	3.39×10^{-16}
	GO:0043227	Membrane-bounded organelle	1.80×10^{-17}	8.80×10^{-16}
	GO:0005634	Nucleus	1.41×10^{-12}	5.76×10^{-11}
	GO:0005737	Cytoplasm	1.69×10^{-10}	5.90×10^{-9}
	GO:0005654	Nucleoplasm	1.80×10^{-9}	5.52×10^{-8}
	GO:0044428	Nuclear part	4.96×10^{-9}	1.35×10^{-7}
Molecular function	GO:0044446	Intracellular organelle part	1.06×10^{-8}	2.59×10^{-7}
	GO:0097159	Organic cyclic compound binding	1.19×10^{-6}	1.19×10^{-4}
	GO:1901363	Heterocyclic compound binding	6.87×10^{-7}	1.37×10^{-4}
	GO:0043167	Ion binding	2.32×10^{-5}	1.16×10^{-3}
	GO:0003676	Nucleic acid binding	2.03×10^{-5}	1.36×10^{-3}
	GO:0005515	Protein binding	3.90×10^{-5}	1.56×10^{-3}
	GO:0019207	Kinase regulator activity	5.59×10^{-5}	1.60×10^{-3}
	GO:0019887	Protein kinase regulator activity	5.31×10^{-5}	1.77×10^{-3}
	GO:0043169	Cation binding	1.39×10^{-4}	3.47×10^{-3}
	GO:0003700	Transcription factor activity, sequence-specific DNA binding	1.63×10^{-4}	3.63×10^{-3}
	GO:0019899	Enzyme binding	5.60×10^{-4}	8.61×10^{-3}

Figure 2. The ClueGO results for biological processes gene ontology terms enrichment for target genes (mRNAs) of differentially expressed miRNAs and relationships between them. The nodes (round shape) represent gene ontology terms, node color represents the level of significance as indicated in the legend, while node size reflects the number of genes enriched in each gene ontology term.

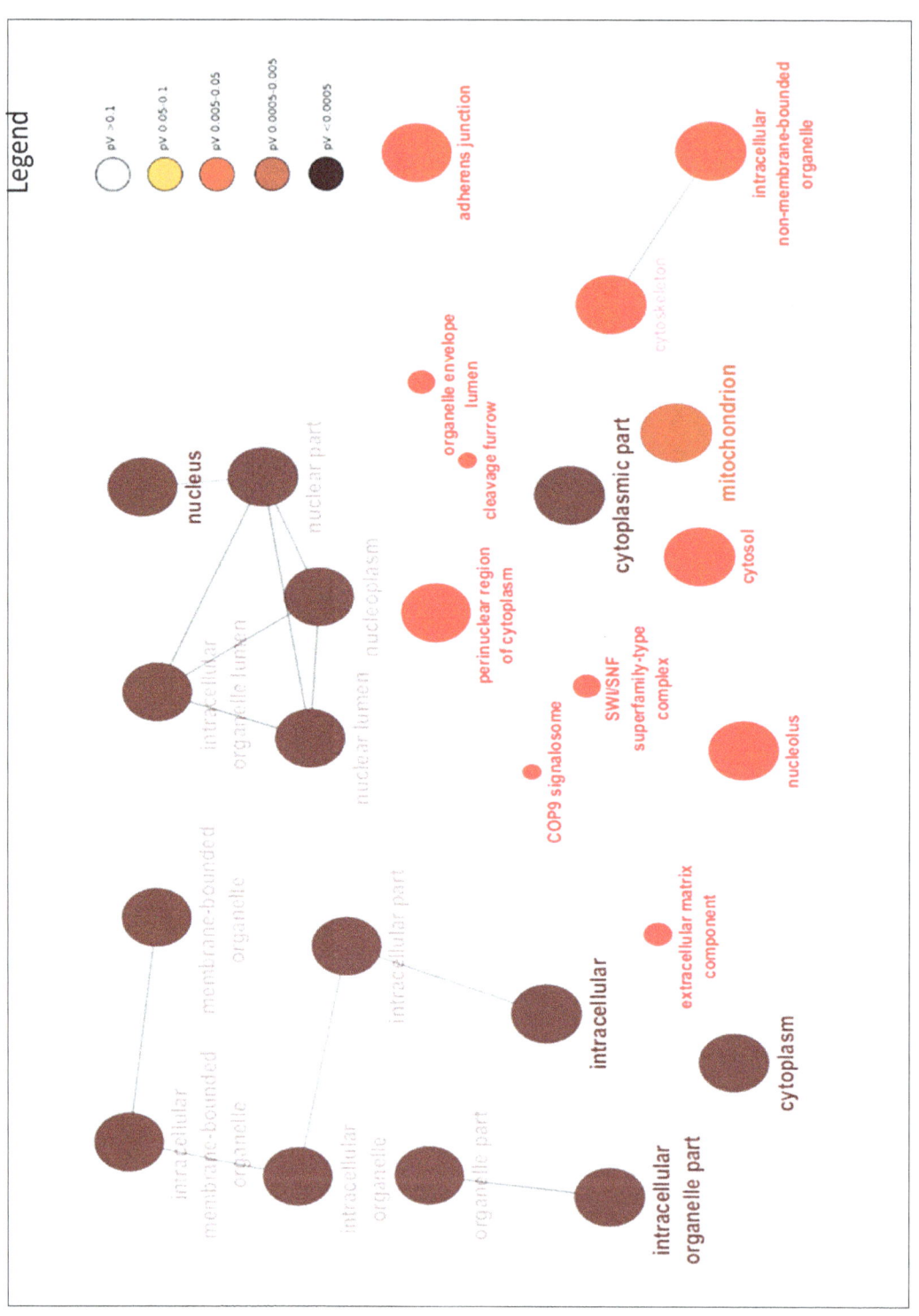

Figure 3. The ClueGO results for cellular processes gene ontology terms enrichment for target genes of differentially expressed miRNAs and relationships between them. The nodes (round shape) represent gene ontology terms, node color represents the level of significance as indicated in the legend, while node size reflects the number of genes enriched in each gene ontology term.

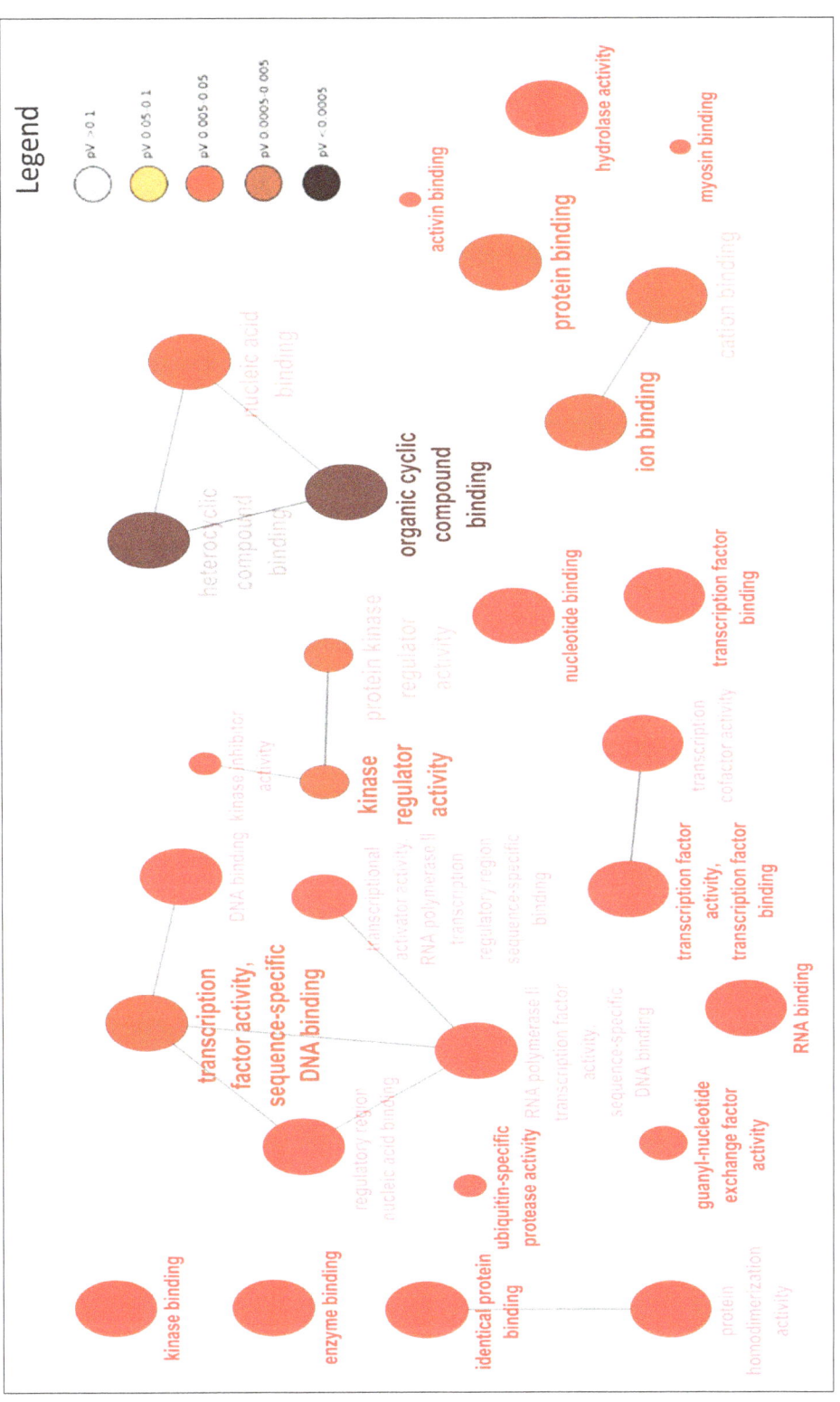

Figure 4. The ClueGO results for molecular functions gene ontology terms enrichment for target genes of differentially expressed miRNAs and relationships between them. The nodes (round shape) represent gene ontology terms, node color represents the level of significance as indicated in the legend, while node size reflects the number of genes enriched in each gene ontology term.

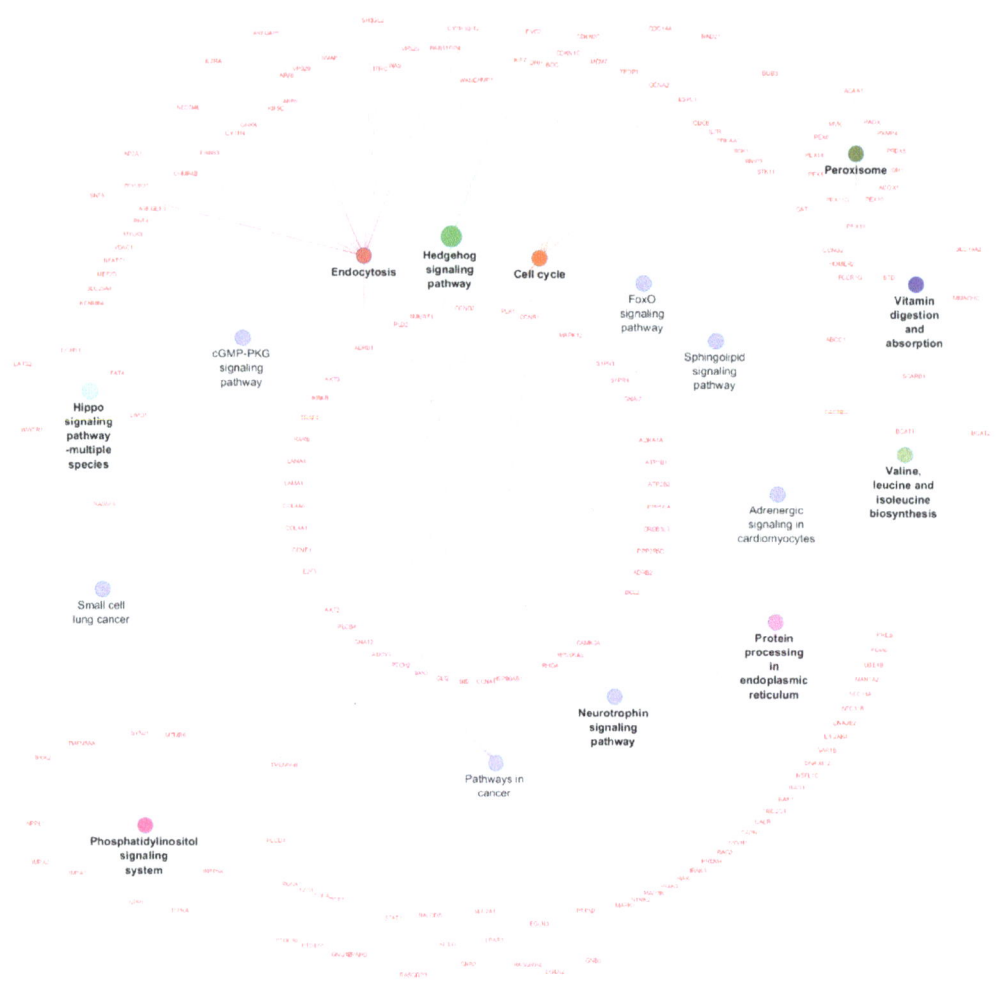

Figure 5. The ClueGO results for KEGG pathways enrichment for target genes of differentially expressed miRNAs and relationships between them. The nodes (round shapes) represent KEGG pathways or genes enriched in the pathways.

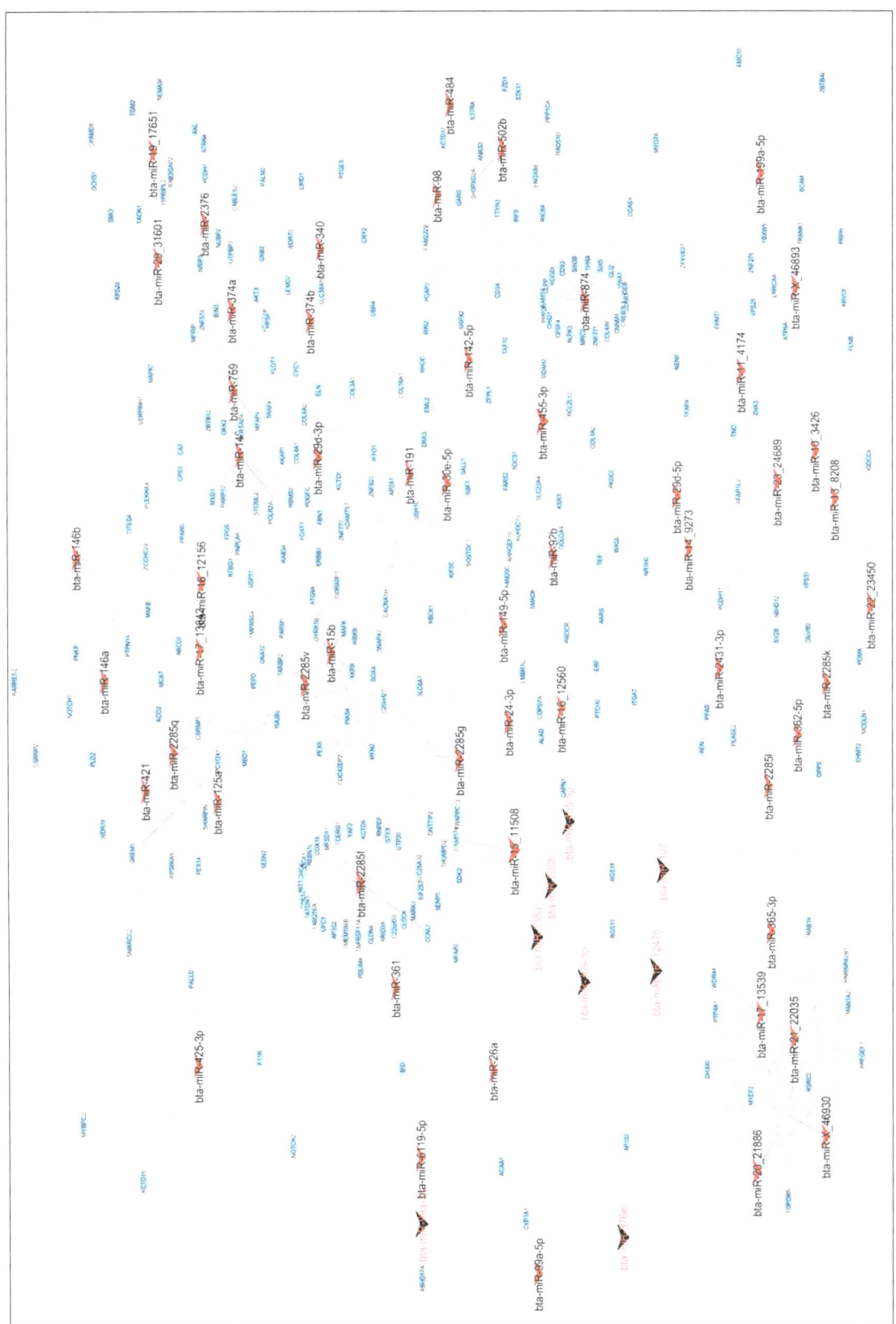

Figure 6. The Cytoscape visualization of the relationships between differentially expressed miRNAs and their target mRNAs. The nodes present either genes (round shape) or miRNAs (V shape). The up- and down-regulated miRNAs are colored red and black, respectively.

3.3. Real Time Quantitative PCR Validation

The RNA-Seq expression of 6 miRNAs was validated by qPCR. Two of them (bta-miR-486-5p and bta-miR-193b) were non DE, while four of them (bta-miR-142-5p, miR-146a, miR-24-3p and miR-374b) were DE between D33 and D96 by RNA-Seq. Observed fold changes for DE miRNAs between both methods were similar, except for the non-DE miRNAs, where an opposite trend was observed (Figure 7).

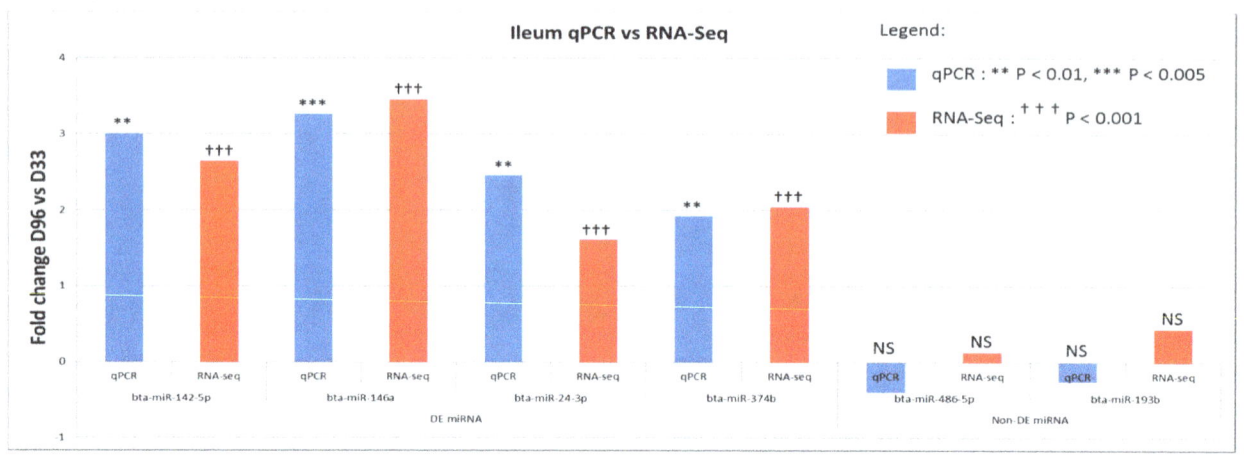

Figure 7. Result of qPCR validation of the expression of miRNAs between day 33 (pre-weaning) and day 96 (post-weaning), and compared with results obtained by miRNA sequencing.

4. Discussion

Physiologically, major metabolic changes that take place in calf gastrointestinal tract following the transition from liquid to solid food are accompanied by rapid changes in gene expression controlled by the signal-mediated coordination of transcriptional and post-transcriptional mechanisms [33]. Previously, we reported that ~20% of bovine genes were significantly DE between the pre-weaning (day 33) and the post-weaning (day 96) period, and enrichment analysis revealed the importance of DE genes in biological processes necessary for the switch in nutrition and developmental stage from the pre-weaning to the post-weaning period [23]. While it is well known that miRNAs are important for the regulation of these processes, little is known about how they participate in the regulation of ileum functions from the pre-weaning to the post-weaning period.

Highly expressed and DE miRNAs identified in this study suggest potential roles in ileum developmental processes during the transition from the pre-weaning to the post-weaning period. However, some highly expressed miRNAs such as bta-miR-21-5p, bta-miR-26a, bta-miR-148a and bta-let-7a-5p (Table 1) were also highly expressed in other tissues such as milk fat [10], milk whey/somatic cells [25], and mammary gland epithelial cells [13]. Interestingly, bta-miR-143, the most highly expressed miRNA, was also among the most abundant miRNAs in bovine testis [34] and reported as the most highly expressed miRNA in ileum tissue of calves at the pre- and post-weaning periods [35]. It was suggested that bta-miR-143 might regulate key genes involved in differentiation of connective tissue cells, the major components of the gut; hence, its high abundance might be important for the regulation of rapid development and growth of the gastrointestinal tract during early life. Indeed, enrichment analysis of target genes of the top 20 abundant miRNAs indicated enrichment in many biological processes and molecular function GO terms (Tables 2 and S5) involved in developmental processes, therefore supporting important roles for highly abundant miRNAs, including bta-miR-143, in these processes. Further supporting evidence was derived from enriched KEGG pathways crucial for cellular development processes such as FoxO signaling pathway, cell cycle, MAPK signaling pathway, and p53 signaling pathway (Table 3). Moreover, we also detected 81 novel miRNAs in this study that were lowly expressed. The most abundant novel miRNA, bta-miR-22_24033, accounted for only 0.3% of the total read counts. Forty-four novel miRNAs were significantly DE between D33 and D96, therefore suggesting roles in the regulation of ileum gene expression during the early period of growth. Nevertheless, novel miRNAs identified in this study will enrich the bovine miRNome as well as enhance knowledge of the potential roles of miRNAs in calf GIT. However, further functional validations to clarify the roles of identified miRNAs in the development of calf gut during the early period of growth are needed.

Bta-miR-374a was the most significantly DE miRNA in this study (Table 4). Bta-miR-374a was found to be differentially expressed between lactating and non-lactating cows [36]. Bta-miR-374a potentially targeted 36 different genes (Table S6b) and some of them might be important for ileum functions such as *EIF2AK4*, *FBXO18*, *GTPBP3* and *GNB2*, etc. *EIF2AK4* is an important transcription factor in host response to infection with pathogenic bacteria associated with Crohn's disease [37]. Moreover, *FBXO18*, *GTPBP3* and *GNB2* have been reported to be significantly DE in calf gastrointestinal tract between the pre- and post-weaning periods [23]. *FBXO18* encodes a member of the F-box protein family with function in phosphorylation-dependent ubiquitination [38]. *FBXO18* is implicated in the regulation of stress-induced apoptosis processes and homologous recombination in familial and sporadic breast cancer [39]. Cells deficient in *FBXO18* were unable to activate the cytotoxic-stress-induced cascade, resulting in increased cell survival [40]. *GTPBP3* is an important gene for mitochondrial functions and a mutation in this gene resulted in defective mitochondrial energy production through oxidative phosphorylation [41]. *GNB2* is important for neuronal apoptosis and was induced by lidocaine in the rat [42]. Nevertheless, the functions of these genes (*FBXO18*, *GTPBP3* and *GNB2*) in the ileum are unknown. Bta-miR-15b, the second most significantly up-regulated miRNA, belongs to the miR-15b family cluster. This miRNA cluster can target cell cycle proteins and the anti-apoptotic *Bcl-2* gene to control cell proliferation and apoptosis [43]. Bta-miR-15b might also play a role in mastitis disease development in cows [44]. Moreover, Liang et al. [35] also reported that bta-miR-15b was significantly DE between 0-day-old and 7-day-old calves and its expression correlated with bacterial population, thus suggesting roles in the regulation of gut development, immune, and digestive functions. Furthermore, bta-miR-15b potentially targeted *IKBKB* (Figure 6), a gene known as an essential molecule for NF-κB signaling pathway [45] with important roles in both innate and acquired immunity [46].

Interestingly, bta-miR-26a, one of the most significantly DE miRNA (Table 4), was also one of the most highly expressed miRNA (Table 1). The human homologue of this miRNA plays important roles in Crohn's disease [15]. In cattle, this miRNA regulates the expression of *PCK1* gene, which is important for semen quality and longevity of Holstein bulls [47]. Bta-miR-26a potentially targeted *BID* gene and their expressions were significantly correlated in this study (Figure 6). *BID* is a pro-apoptotic member of the Bcl-2 protein family with roles in the regulation of apoptosis processes [46]. Therefore, bta-miR-26a might have roles in ileum development via targeting the *BID* gene. Bta-miR-455-5p was the most down-regulated miRNA between D33 and D96. Bta-miR-455-5p was reported to be important for the function of granulosa cells of subordinate and dominant follicles during the early luteal phase of the bovine estrous cycle [48]. In humans, this miRNA homologue down regulated *RAB18* gene in gastric cancer [49]. In fact, bta-miR-455-5p also potentially targeted *RAB18* gene in this study (Figure 6).

Among the top 20 DE miRNAs in this study, bta-miR-142-3p, bta-miR-142-5p, bta-miR-191, bta-miR-146a, bta-miR-210 and bta-miR-424-5p were also found to be DE in calf ileum at the pre-weaning and weaning periods [35]. Some of these miRNAs have been reported to play important roles in immune functions. For example, bta-miR-146a inhibited the mRNA and protein expression levels of *TRAF6* gene and acted as a negative feedback regulator of bovine inflammation and innate immunity through down regulation of the TLR4/TRAF6/NF-κB pathway in bovine mammary epithelial cells [50]. Furthermore, bta-miR-142-5p was important for bovine alveolar macrophage response to *Mycobacterium bovis* infection [51].

As expected, the target genes of DE miRNAs were enriched in important biological process GO-terms related to metabolic processes (such as cellular macromolecule metabolic process, macromolecule metabolic process and regulation of metabolic process) (Table 5 and Figure 2) and molecular function GO-terms related to metabolism of the macromolecule compound (such as organic cyclic compound binding, heterocyclic compound binding, nucleic acid binding and protein binding) (Table 5 and Figure 4), thus suggesting roles in the regulation of these processes. Interestingly, the target genes of DE miRNAs were also enriched in GO-terms like transcription factor activity and sequence-specific DNA binding (FDR = 3.63×10^{-3}) (Table 5), thus suggesting their importance in transcription factor activity. The interaction between miRNAs and transcription factors to regulate gene expression in biological processes is well documented [52,53]. MiRNAs might inhibit transcription factor activities by either directly inhibiting the expression of their encoding genes or by inhibiting other gene(s) that have impact on their activities [53,54]. Previously, we observed that some transcription factors might play important roles in mediating miRNA regulatory functions in cow milk yield and milk component traits [9]. In humans, several miRNAs have been reported to participate in the regulation of intestinal transcription factors; miR-196b inhibited the *GATA6* intestinal transcription factor to control intestinal cell homeostasis and tumorigenesis in colon cancer patients [55], while miR-30 family controlled intestinal epithelial cell proliferation and differentiation by targeting SOX9 (transcription factor) and other genes in ubiquitin ligase pathway [56]. In fact, as mentioned above, the most DE miRNA in this study (bta-miR-374) also potentially targeted a transcription factor (*EIF2AK4*) known to be important for human Crohn's disease [37]. The most important pathway enriched for DE miRNAs target genes was Hedgehog signaling pathway ($p = 0.003$, Table S7, Figure 5). Hedgehog signaling pathway is important for cell growth, survival and fate, as well as for normal embryonic development [57,58]. This pathway also has multiple patterning functions during mammalian gut development [59]; therefore, it may be important for ileum functions during the early part of life. Another important pathway enriched for target genes of DE miRNAs was peroxisomes pathway ($p = 0.006$, Table S7, Figure 5). The peroxisomes pathway is crucial for metabolic processes such as fatty acid oxidation, biosynthesis of ether lipids, and free radical detoxification [60]. Since one of the main functions of the ileum is to absorb bile salts, one of the products of fatty acid oxidation, enrichment of the peroxisome pathway supports its role in normal ileum function. Another important role of the ileum is vitamin absorption and the enriched pathway, vitamin digestion and absorption (Table S7 and Figure 5), might reflect the changes in gene expression for different vitamin requirements between the pre- and post-weaning periods. Other enriched pathways also reflect the importance of miRNAs in the regulation of genes involved in lipid metabolism (phosphatidylinositol signaling system and sphingolipid signaling pathway), protein metabolism (valine, leucine and isoleucine biosynthesis and protein processing in endoplasmic reticulum) and intracellular signaling (cGMP-PKG signaling, FoxO signaling and neurotrophin signaling pathway) in the development of the ileum during the early part of life.

5. Conclusions

This is the first integrated miRNA-mRNA analysis characterizing the function of miRNAs in calf ileum during early life. Eighty-one novel miRNAs were identified that will enrich the bovine miRNome repertoire and contribute to the understanding of regulatory processes in calf ileum. This study highlighted potential roles of bta-miR-143, bta-miR-192, bta-miR-26a and bta-miR-21-5p in growth and developmental processes during the transition from the pre-weaning to the post-weaning period. This study also suggested roles for DE miRNAs in metabolic processes, metabolism of the macromolecule compound, transcription factor activities, as well as involvement in pathways related to metabolism (peroxisomes), vitamin digestion and absorption, lipid and protein metabolism, and intracellular signaling (Hedgehog signaling, GMP-PKG signaling, FoxO signaling, neurotrophin signaling pathway). Moreover, several DE miRNAs—DE mRNAs pairs such as bta-miR-374a—*FBXO18*, bta-miR-374a—*GTPBP3* and bta-miR-374a—*GNB2* with potential roles in tissue development, and bta-miR-15b—*IKBKB* with vital roles in immune functions were revealed. This study, therefore, provided insights on miRNA expression and their potential functions in calf ileum development during early life, which might facilitate identification of miRNA biomarkers for growth, nutritional and disease challenges during the pre- and post-weaning periods.

Supplementary Materials:
Table S1: Mapping statistics of miRNA sequencing reads; Table S2: Length distribution (nt) of miRNA reads; Table S3: Novel miRNAs identified by miRDeep2; Table S4: Novel and known miRNAs and their read count summary; Table S5; The 20 most highly expressed miRNAs and their predicted target genes (a); enriched gene ontology terms (b) and enriched KEGG pathways (c); Table S6; Differentially expressed miRNAs between day 33 (pre-weaning) and day 96 (post-weaning) (a); predicted target genes of DE miRNAs (b) and differentially expressed genes which are both negatively correlated and predicted targets of miRNAs (c); Table S7; Gene ontology and KEGG pathways enriched for target genes of differential expressed miRNAs; Table S8; Differentially expressed genes that were targets of DE miRNAs and also negatively correlated.

Author Contributions: Conceived and designed the experiments, E.M.I.-A.; Performed the experiments, B.E.F. and P.-L.D.; Data analysis/curation, P.-L.D. and D.N.D.; Writing-Original Draft Preparation, D.N.D.; Writing-Review & Editing, E.M.I.-A.; Visualization, D.N.D.; Supervision and Project Administration, E.M.I.-A.; Funding Acquisition, E.M.I.-A.; all authors revised and approved the final draft.

Acknowledgments: Authors thank farm staff of Agriculture and Agri-Food Canada's Sherbrooke Research and Development Center for assistance in animal management.

References

1. Ambros, V. The functions of animal micrornas. *Nature* **2004**, *431*, 350–355. [CrossRef] [PubMed]

2. Carthew, R.W.; Sontheimer, E.J. Origins and mechanisms of miRNAs and siRNAs. *Cell* **2009**, *136*, 642–655. [CrossRef] [PubMed]

3. Coutinho, L.L.; Matukumalli, L.K.; Sonstegard, T.S.; Van Tassell, C.P.; Gasbarre, L.C.; Capuco, A.V.; Smith, T.P. Discovery and profiling of bovine micrornas from immune-related and embryonic tissues. *Physiol. Genet.* **2007**, *29*, 35–43. [CrossRef] [PubMed]

4. Ponsuksili, S.; Tesfaye, D.; Schellander, K.; Hoelker, M.; Hadlich, F.; Schwerin, M.; Wimmers, K. Differential expression of mirnas and their target mRNAs in endometria prior to maternal recognition of pregnancy associates with endometrial receptivity for in vivo-and in vitro-produced bovine embryos. *Biol. Reprod.* **2014**, *91*, 1–12. [CrossRef] [PubMed]

5. Hossain, M.; Salilew-Wondim, D.; Schellander, K.; Tesfaye, D. The role of micrornas in mammalian oocytes and embryos. *Anim. Reprod. Sci.* **2012**, *134*, 36–44. [CrossRef] [PubMed]

6. Do, D.N.; Ibeagha-Awemu, E.M. Non-coding RNA roles in ruminant mammary gland development and lactation. In *Current Topics in Lactation*; InTech: Rijeka, Croatia, 2017.

7. Gu, Z.; Eleswarapu, S.; Jiang, H. Identification and characterization of micrornas from the bovine adipose tissue and mammary gland. *FEBS Lett.* **2007**, *581*, 981–988. [CrossRef] [PubMed]

8. Jin, W.; Dodson, M.V.; Moore, S.S.; Basarab, J.A. Characterization of microrna expression in bovine adipose tissues: A potential regulatory mechanism of subcutaneous adipose tissue development. *BMC Mol. Biol.* **2010**, *11*, 29. [CrossRef] [PubMed]

9. Do, D.N.; Dudemaine, P.-L.; Li, R.; Ibeagha-Awemu, E.M. Co-expression network and pathway analyses reveal important modules of mirnas regulating milk yield and component traits. *Int. J. Mol. Sci.* **2017**, *18*, 1560. [CrossRef] [PubMed]

10. Do, D.N.; Li, R.; Dudemaine, P.-L.; Ibeagha-Awemu, E.M. Microrna roles in signalling during lactation: An insight from differential expression, time course and pathway analyses of deep sequence data. *Sci. Rep.* **2017**, *7*. [CrossRef] [PubMed]

11. Jabed, A.; Wagner, S.; McCracken, J.; Wells, D.N.; Laible, G. Targeted microrna expression in dairy cattle directs production of β-lactoglobulin-free, high-casein milk. *Proc. Natl. Acad. Sci. USA* **2012**, *109*, 16811–16816. [CrossRef] [PubMed]

12. Naeem, A.; Zhong, K.; Moisá, S.; Drackley, J.; Moyes, K.; Loor, J. Bioinformatics analysis of microRNA and putative target genes in bovine mammary tissue infected with streptococcus uberis. *J. Dairy Sci.* **2012**, *95*, 6397–6408. [CrossRef] [PubMed]

13. Jin, W.; Ibeagha-Awemu, E.M.; Liang, G.; Beaudoin, F.; Zhao, X.; Guan, L.L. Transcriptome microrna profiling of bovine mammary epithelial cells challenged with *Escherichia coli* or staphylococcus aureusbacteria reveals pathogen directed microRNA expression profiles. *BMC Genet.* **2014**, *15*, 181.

14. Farrell, D.; Shaughnessy, R.G.; Britton, L.; MacHugh, D.E.; Markey, B.; Gordon, S.V. The identification of circulating mirna in bovine serum and their potential as novel biomarkers of early mycobacterium avium subsp paratuberculosis infection. *PLoS ONE* **2015**, *10*, e0134310. [CrossRef] [PubMed]

15. Wu, F.; Zhang, S.; Dassopoulos, T.; Harris, M.L.; Bayless, T.M.; Meltzer, S.J.; Brant, S.R.; Kwon, J.H. Identification of microRNAs associated with ileal and colonic crohn's disease. *Inflamm. Bowel Dis.* **2010**, *16*, 1729–1738. [CrossRef] [PubMed]

16. Paraskevi, A.; Theodoropoulos, G.; Papaconstantinou, I.; Mantzaris, G.; Nikiteas, N.; Gazouli, M. Circulating microRNA in inflammatory bowel disease. *J. Crohn's Colitis* **2012**, *6*, 900–904. [CrossRef] [PubMed]

17. De Souza, H.S.; Fiocchi, C. Immunopathogenesis of ibd: Current state of the art. *Nat. Rev. Gastroenterol. Hepatol.* **2016**, *13*, 13–27. [CrossRef] [PubMed]

18. Dalal, S.R.; Kwon, J.H. The role of microrna in inflammatory bowel disease. *Gastroenterol. Hepatol.* **2010**, *6*, 714.

19. Chapman, C.G.; Pekow, J. The emerging role of miRNAs in inflammatory bowel disease: A review. *Ther. Adv. Gastroenterol.* **2015**, *8*, 4–22. [CrossRef] [PubMed]

20. Liang, G.; Malmuthuge, N.; Griebel, P. Model systems to analyze the role of mirnas and commensal microflora in bovine mucosal immune system development. *Mol. Immunol.* **2015**, *66*, 57–67. [CrossRef] [PubMed]

21. Liang, G.; Malmuthuge, N.; Bao, H.; Stothard, P.; Griebel, P.J. Transcriptome analysis reveals regional and temporal differences in mucosal immune system development in the small intestine of neonatal calves. *BMC Genet.* **2016**, *17*, 602. [CrossRef] [PubMed]

22. Liang, G.; Malmuthuge, N.; Guan, Y.; Ren, Y.; Griebel, P.J.; Guan, L.L. Altered microrna expression and pre-mRNA splicing events reveal new mechanisms associated with early stage mycobacterium avium subspecies paratuberculosis infection. *Sci. Rep.* **2016**, *6*. [CrossRef] [PubMed]

23. Ibeagha-Awemu, E.M.; Do, D.N.; Dudemaine, P.-L.; Fomenky, B.E.; Bissonnette, N. Integration of lncRNA and mrna transcriptome analyses reveals genes and pathways potentially involved in calves' intestinal growth and development during the early weeks of life. *Genes* **2017**. submitted.

24. Do, D.N.; Dudemaine, P.-L.; Fomenky, B.E.; Ibeagha-Awemu, E.M. Integration of miRNA weighted gene co-expression network and miRNA-mRNA co-expression analyses reveals potential regulatory functions of mirnas in calf rumen development. *Genomics* **2018**. [CrossRef] [PubMed]

25. Li, R.; Dudemaine, P.-L.; Zhao, X.; Lei, C.; Ibeagha-Awemu, E.M. Comparative analysis of the mirnome of bovine milk fat, whey and cells. *PLoS ONE* **2016**, *11*, e0154129. [CrossRef] [PubMed]

26. Langmead, B.; Trapnell, C.; Pop, M.; Salzberg, S.L. Ultrafast and memory-efficient alignment of short DNA sequences to the human genome. *Genet. Biol.* **2009**, *10*, R25. [CrossRef] [PubMed]

27. Friedlander, M.R.; Chen, W.; Adamidi, C.; Maaskola, J.; Einspanier, R.; Knespel, S.; Rajewsky, N. Discovering micrornas from deep sequencing data using mirdeep. *Nat. Biotechnol.* **2008**, *26*, 407–415. [CrossRef] [PubMed]

28. Love, M.I.; Huber, W.; Anders, S. Moderated estimation of fold change and dispersion for RNA-seq data with DESeq2. *Genet. Biol.* **2014**, *15*, 550. [CrossRef] [PubMed]

29. Benjamini, Y.; Hochberg, Y. Controlling the false discovery rate: A practical and powerful approach to multiple testing. *J. R. Stat. Soc. Ser. B (Methodol.)* **1995**, *57*, 289–300.

30. Li, R.; Beaudoin, F.; Ammah, A.A.; Bissonnette, N.; Benchaar, C.; Zhao, X.; Lei, C.; Ibeagha-Awemu, E.M. Deep sequencing shows microrna involvement in bovine mammary gland adaptation to diets supplemented with linseed oil or safflower oil. *BMC Genet.* **2015**, *16*, 884. [CrossRef] [PubMed]

31. Bindea, G.; Mlecnik, B.; Hackl, H.; Charoentong, P.; Tosolini, M.; Kirilovsky, A.; Fridman, W.-H.; Pagès, F.; Trajanoski, Z.; Galon, J. Cluego: A cytoscape plug-in to decipher functionally grouped gene ontology and pathway annotation networks. *Bioinformatics* **2009**, *25*, 1091–1093. [CrossRef] [PubMed]

32. Shannon, P.; Markiel, A.; Ozier, O.; Baliga, N.S.; Wang, J.T.; Ramage, D.; Amin, N.; Schwikowski, B.; Ideker, T. Cytoscape: A software environment for integrated models of biomolecular interaction networks. *Genet. Res.* **2003**, *13*, 2498–2504. [CrossRef] [PubMed]

33. Turner, M.; Galloway, A.; Vigorito, E. Noncoding RNA and its associated proteins as regulatory elements of the immune system. *Nat. Immunol.* **2014**, *15*, 484–491. [CrossRef] [PubMed]

34. Huang, J.; Ju, Z.; Li, Q.; Hou, Q.; Wang, C.; Li, J.; Li, R.; Wang, L.; Sun, T.; Hang, S.; et al. Solexa sequencing of novel and differentially expressed micrornas in testicular and ovarian tissues in holstein cattle. *Int. J. Biol. Sci.* **2011**, *7*, 1016–1026. [CrossRef] [PubMed]

35. Liang, G.; Malmuthuge, N.; McFadden, T.B.; Bao, H.; Griebel, P.J.; Stothard, P. Potential regulatory role of micrornas in the development of bovine gastrointestinal tract during early life. *PLoS ONE* **2014**, *9*, e92592. [CrossRef] [PubMed]

36. Li, Z.; Liu, H.; Jin, X.; Lo, L.; Liu, J. Expression profiles of micrornas from lactating and non-lactating bovine mammary glands and identification of mirna related to lactation. *BMC Genet.* **2012**, *13*, 731. [CrossRef] [PubMed]

37. Bretin, A.; Carriere, J.; Dalmasso, G.; Bergougnoux, A.; B'Chir, W.; Maurin, A.C.; Muller, S.; Seibold, F.; Barnich, N.; Bruhat, A.; et al. Activation of the EIF2AK4-EIF2A/eIF2α-ATF4 pathway triggers autophagy response to crohn disease-associated adherent-invasive *Escherichia coli* infection. *Autophagy* **2016**, *12*, 770–783. [CrossRef] [PubMed]

38. Kim, J.; Kim, J.-H.; Lee, S.-H.; Kim, D.-H.; Kang, H.-Y.; Bae, S.-H.; Pan, Z.-Q.; Seo, Y.-S. The novel human DNA helicase hFBH1 is an F-box protein. *J..Biol. Chem.* **2002**, *277*, 24530–24537. [CrossRef] [PubMed]

39. Heyn, H.; Sayols, S.; Moutinho, C.; Vidal, E.; Sanchez-Mut, J.V.; Stefansson Olafur, A.; Nadal, E.; Moran, S.; Eyfjord Jorunn, E.; Gonzalez-Suarez, E.; et al. Linkage of DNA methylation quantitative trait loci to human cancer risk. *Cell Rep.* **2014**, *7*, 331–338. [CrossRef] [PubMed]

40. Laulier, C.; Cheng, A.; Huang, N.; Stark, J.M. Mammalian fbh1 is important to restore normal mitotic progression following decatenation stress. *DNA Repair* **2010**, *9*, 708–717. [CrossRef] [PubMed]

41. Kopajtich, R.; Nicholls Thomas, J.; Rorbach, J.; Metodiev Metodi, D.; Freisinger, P.; Mandel, H.; Vanlander, A.; Ghezzi, D.; Carrozzo, R.; Taylor Robert, W.; et al. Mutations in gtpbp3 cause a mitochondrial translation defect associated with hypertrophic cardiomyopathy, lactic acidosis, and encephalopathy. *Am. J. Hum. Genet.* **2014**, *95*, 708–720. [CrossRef] [PubMed]

42. Tan, Y.; Wang, Q.; Zhao, B.; She, Y.; Bi, X. GNB2 is a mediator of lidocaine-induced apoptosis in rat pheochromocytoma PC12 cells. *NeuroToxicol.* **2016**, *54*, 53–64. [CrossRef] [PubMed]

43. Yue, J.; Tigyi, G. Conservation of miR-15a/16-1 and miR-15b/16-2 clusters. *Mamm. Genome* **2010**, *21*, 88–94. [CrossRef] [PubMed]

44. Chen, L.; Liu, X.; Li, Z.; Wang, H.; Liu, Y.; He, H.; Yang, J.; Niu, F.; Wang, L.; Guo, J. Expression differences of mirnas and genes on NF-κB pathway between the healthy and the mastitis chinese holstein cows. *Gene* **2014**, *545*, 117–125. [CrossRef] [PubMed]

45. Schmid, J.A.; Birbach, A. Iκb kinase β (ikkβ/ikk2/ikbkb)—A key molecule in signaling to the transcription factor NF-κB. *Cytokine Growth Factor Rev.* **2008**, *19*, 157–165. [CrossRef] [PubMed]

46. Pannicke, U.; Baumann, B.; Fuchs, S.; Henneke, P.; Rensing-Ehl, A.; Rizzi, M.; Janda, A.; Hese, K.; Schlesier, M.; Holzmann, K.; et al. Deficiency of innate and acquired immunity caused by an IKBKB mutation. *New Eng. J. Med.* **2013**, *369*, 2504–2514. [CrossRef] [PubMed]

47. Huang, J.; Guo, F.; Zhang, Z.; Zhang, Y.; Wang, X.; Ju, Z.; Yang, C.; Wang, C.; Hou, M.; Zhong, J. PCK1 is negatively regulated by bta-miR-26a, and a single-nucleotide polymorphism in the 3′ untranslated region is involved in semen quality and longevity of holstein bulls. *Mol. Reprod. Dev.* **2016**, *83*, 217–225. [CrossRef] [PubMed]

48. Salilew-Wondim, D.; Ahmad, I.; Gebremedhn, S.; Sahadevan, S.; Hossain, M.M.; Rings, F.; Hoelker, M.; Tholen, E.; Neuhoff, C.; Looft, C. The expression pattern of micrornas in granulosa cells of subordinate and dominant follicles during the early luteal phase of the bovine estrous cycle. *PLoS ONE* **2014**, *9*, e106795. [CrossRef] [PubMed]

49. Liu, J.; Zhang, J.; Li, Y.; Wang, L.; Sui, B.; Dai, D. MiR-455-5p acts as a novel tumor suppressor in gastric cancer by down-regulating rab18. *Gene* **2016**, *592*, 308–315. [CrossRef] [PubMed]

50. Wang, X.P.; Luoreng, Z.M.; Zan, L.S.; Li, F.; Li, N. Bovine miR-146a regulates inflammatory cytokines of bovine mammary epithelial cells via targeting the TRAF6 gene. *J. Dairy Sci.* **2017**, *100*, 7648–7658. [CrossRef] [PubMed]

51. Vegh, P.; Magee, D.A.; Nalpas, N.C.; Bryan, K.; McCabe, M.S.; Browne, J.A.; Conlon, K.M.; Gordon, S.V.; Bradley, D.G.; MacHugh, D.E. Microrna profiling of the bovine alveolar macrophage response to

mycobacterium bovis infection suggests pathogen survival is enhanced by microrna regulation of endocytosis and lysosome trafficking. *Tuberculosis* **2015**, *95*, 60–67. [CrossRef] [PubMed]

52. Martinez, N.J.; Walhout, A.J. The interplay between transcription factors and micrornas in genome-scale regulatory networks. *Bioessays* **2009**, *31*, 435–445. [CrossRef] [PubMed]

53. Hobert, O. Gene regulation by transcription factors and micrornas. *Science* **2008**, *319*, 1785–1786. [CrossRef] [PubMed]

54. Shalgi, R.; Lieber, D.; Oren, M.; Pilpel, Y. Global and local architecture of the mammalian microrna–transcription factor regulatory network. *PLoS Comput. Biol.* **2007**, *3*, e131. [CrossRef] [PubMed]

55. Fantini, S.; Salsi, V.; Reggiani, L.; Maiorana, A.; Zappavigna, V. The miR-196b mirna inhibits the gata6 intestinal transcription factor and is upregulated in colon cancer patients. *Oncotarget* **2017**, *8*, 4747. [CrossRef] [PubMed]

56. Peck, B.C.E.; Sincavage, J.; Feinstein, S.; Mah, A.T.; Simmons, J.G.; Lund, P.K.; Sethupathy, P. MiR-30 family controls proliferation and differentiation of intestinal epithelial cell models by directing a broad gene expression program that includes sox9 and the ubiquitin ligase pathway. *J. Biol. Chem.* **2016**, *291*, 15975–15984. [CrossRef] [PubMed]

57. Mazumdar, T.; DeVecchio, J.; Shi, T.; Jones, J.; Agyeman, A.; Houghton, J.A. Hedgehog signaling drives cellular survival in human colon carcinoma cells. *Cancer Res.* **2011**, *71*, 1092–1102. [CrossRef] [PubMed]

58. Varjosalo, M.; Taipale, J. Hedgehog: Functions and mechanisms. *Genes Dev.* **2008**, *22*, 2454–2472. [CrossRef] [PubMed]

59. Zacharias, W.J.; Madison, B.B.; Kretovich, K.E.; Walton, K.D.; Richards, N.; Udager, A.M.; Li, X.; Gumucio, D.L. Hedgehog signaling controls homeostasis of adult intestinal smooth muscle. *Dev. Biol.* **2011**, *355*, 152–162. [CrossRef] [PubMed]

60. Kim, P.; Hettema, E. Multiple pathways for protein transport to peroxisomes. *J. Mol. Biol.* **2015**, *427*, 1176–1190. [CrossRef] [PubMed]

2

MicroRNAs as Biomarkers in Amyotrophic Lateral Sclerosis

Claudia Ricci *, Carlotta Marzocchi and Stefania Battistini

Department of Medical, Surgical and Neurological Sciences, University of Siena, 53100 Siena, Italy; carlottamarzocchi@libero.it (C.M.), stefania.battistini@unisi.it (S.B.)
* Correspondence: claudia.ricci@unisi.it;

Abstract: Amyotrophic lateral sclerosis (ALS) is an incurable and fatal disorder characterized by the progressive loss of motor neurons in the cerebral cortex, brain stem, and spinal cord. Sporadic ALS form accounts for the majority of patients, but in 1–13.5% of cases the disease is inherited. The diagnosis of ALS is mainly based on clinical assessment and electrophysiological examinations with a history of symptom progression and is then made with a significant delay from symptom onset. Thus, the identification of biomarkers specific for ALS could be of a fundamental importance in the clinical practice. An ideal biomarker should display high specificity and sensitivity for discriminating ALS from control subjects and from ALS-mimics and other neurological diseases, and should then monitor disease progression within individual patients. microRNAs (miRNAs) are considered promising biomarkers for neurodegenerative diseases, since they are remarkably stable in human body fluids and can reflect physiological and pathological processes relevant for ALS. Here, we review the state of the art of miRNA biomarker identification for ALS in cerebrospinal fluid (CSF), blood and muscle tissue; we discuss advantages and disadvantages of different approaches, and underline the limits but also the great potential of this research for future practical applications.

Keywords: amyotrophic lateral sclerosis (ALS); biomarker; microRNA; cerebrospinal fluid (CSF); muscle biopsy; circulating miRNAs

1. Introduction

Amyotrophic lateral sclerosis (ALS), the most common adult-onset neurodegenerative disorder, is an incurable and invariably fatal condition characterized by the progressive loss of motor neurons in the motor cortex, brain stem, and spinal cord [1]. Motor neurons are selectively affected by degeneration and death, however the collective evidence is that ALS is non-cell autonomous, but rather pathogenesis and disease progression depend on the active participation of non-neuronal neighboring cells such as microglia, astrocytes, muscle and T cells [2,3]. Motor neuron degeneration causes progressive weakness of limb, thoracic, abdominal, and bulbar muscles.

During the early stages of the disease symptoms may vary depending on dysfunction of upper motor neurons (UMN) in the motor cortex (resulting in hyperreflexia, extensor plantar response, and increased muscle tone), or lower motor neuron (LMN) in the brainstem and spinal cord (leading to generalized weakness, muscle atrophy, hyporeflexia, fasciculations, and muscle cramps) [1]. Patients with bulbar onset ALS usually develop slurred and nasal speech and difficulty chewing or swallowing. Bulbar onset occurs less frequently than limb involvement, and accounts for about 25% of ALS cases. During the disease course, most cases show the presence of both LMN and UMN signs affecting spinal and brainstem regions [4]. Death, mainly due to bulbar dysfunction and respiratory insufficiency, occurs within 2–4 years of first symptoms; however, a small group of patients with ALS may survive for 10 or more years [5].

1.1. Epidemiology and Genetic Factors

The incidence of ALS is 2.1 per 100,000 persons per year, with an estimated prevalence of 5.4 cases per 100,000 population [6]. Based on data collected by population-based registers, the incidence of ALS increases after the age of 40, shows a peak in the late 60s or early 70s, and then displays a fast decline [7]. The reported male to female ratio varies widely with the age: a sex ratio of 2 or higher is observed for younger patients, while it appears to decrease towards 1 when the proportion of older patients increases [8]. Over the years, several environmental and lifestyle risk factors have been suggested as potential contributors to the cause of ALS. Nevertheless, no conclusive data are yet available, and further studies are required to identify exogenous risk factors of ALS [7,9].

Most cases (around 90%) are classified as sporadic ALS (SALS), since they are not associated with a documented family history. In 1–13% of patients the disease is inherited and defined as familial ALS (FALS), most frequently with a Mendelian dominant inheritance and high penetrance, even though pedigrees with recessive inheritance or incomplete penetrance have been described [10]. The mean age of onset for FALS is 46 years and for SALS is 56 years. In familial ALS, age of onset displays a Gaussian distribution, whereas an age-dependent incidence characterizes sporadic ALS [4]. Disease with an onset prior to 25 years of age is defined as "juvenile ALS" [11]. Apart from the mean age of onset, sporadic and familial forms are clinically indistinguishable suggesting a common pathogenesis.

Several genes have been associated with pathogenesis of ALS. The most common ALS causative genes include chromosome 9 open reading frame 72 (C9orf72), Cu2+/Zn2+ superoxide dismutase (SOD1), TAR DNA-binding protein 43 (TARDBP), and RNA binding protein FUS (FUS) [12–14], but a lot of other genes have been associated with the disease [15]. Notably, the mutated genes in ALS encode for proteins with very distinct functions in the cell. However, interestingly many ALS-linked genes, particularly TARDBP and FUS, are involved in RNA metabolism, including microRNA (miRNA) processing [16,17].

1.2. Diagnosis and Treatment

There is no objective laboratory test able to provide the diagnosis of ALS, which remains mainly based on clinical assessment, electrophysiological examinations, and exclusion of conditions that can mimic ALS. The certainty level of the diagnosis of ALS may be classified into different categories by clinical and laboratory assessments based on El Escorial criteria [18].

Currently, riluzole and edaravone represent the only drugs approved by the FDA for ALS, providing however a limited improvement in survival [5]. The most significant benefit of riluzole is observed after intervention in the early stages of the disease [19]. Thus, an early diagnosis of ALS could provide the most effective results. Since diagnosis of ALS relies on clinical symptoms, and the time from the first symptoms to diagnosis is about 12 months, there is a delay hindering a successful therapy [5]. This phenomenon underlies the importance of the development of screening tests able to detect the disease in early stages.

2. Role of Biomarkers in ALS

In the last years, research has been focused on the identification of potential biological markers to use in diagnostic procedure and clinical practice.

According to the National Institutes of Health Biomarkers Definitions Working Group, a biomarker is defined as "a characteristic that is objectively measured and evaluated as an indicator of normal biological processes, pathogenic processes, or pharmacologic responses to a therapeutic intervention" [20]. Biomarkers can be classified into three general categories: (1) diagnostic biomarkers, which are used for differential diagnosis; (2) prognostic biomarkers, which can differentiate a good or a bad outcome of the disease; and (3) predictive biomarkers, which are utilized for assessing whether a treatment may be effective for a specific patient or not.

In the case of ALS, biomarkers would allow an earlier and more accurate diagnosis, with the opportunity to start an earlier treatment able to modify the disease course. They could help the classification/stratification of ALS patients, monitor the disease progression and identify patients who will respond better to a particular drug. Biomarkers can also provide a valuable tool for the identification of new therapeutic approaches and drive patients' enrollment in clinical trials. Furthermore, they may represent a link between the results obtained in animal models and the human patients, providing insight on potential therapeutic targets.

Over the last two decades, intensive work has been carried out to find consistent biomarkers for ALS. Several candidates involved in excitotoxicity, oxidative stress, neuroinflammation, metabolic dysfunction, and neurodegeneration processes have been explored [21], but, unfortunately, none of these biomarkers has been currently translated into a practical diagnostic tool.

3. miRNAs as Biomarkers

Recently, among the different categories of potential biomarkers, miRNAs have aroused great interest in several fields of research. miRNAs are short (about 22 nucleotides in length) non-coding RNA molecules that play an important role as endogenous regulators of gene expression acting at the post-transcriptional level. miRNAs are synthesized from primary miRNAs, which are transcribed in the nucleus. Primary miRNAs are processed into pre-miRNAs by Drosha and then exported to the cytoplasm. Pre-miRNAs are eventually processed by the Dicer complex, resulting in mature miRNAs, which form RNA-induced silencing complexes [22]. miRNAs have a tissue-specific expression and this knowledge can help to better understand a normal and a disease development of the respective tissue [23]. miRNAs are known to play important roles in many physiological and pathological processes, including tumorigenesis [24], metabolism [25], immune function [26], and several neurodegenerative disorders [27], such as Parkinson's disease, Alzheimer's disease, Huntington's disease [28] and also ALS [29].

miRNAs have several intrinsic characteristics that make them promising as biomarkers. An ideal biomarker should display high sensitivity, specificity, and predictive power. miRNAs have been shown to have high specificity, and, in particular in cancer research, where a plethora of publications has been generated, it has been demonstrated that miRNA expression profiles differ among cancer types according to diagnosis and developmental stage of the tumor, with a better resolution than traditional gene expression analysis [30]. Moreover, unlike other RNA classes, miRNAs are remarkably stable and therefore can be robustly measured in many biological body fluids including plasma, tears, saliva and cerebrospinal fluid [31]. Indeed, miRNAs appear resistant to boiling, repeated freeze-thawing cycles, pH changes, and fragmentation by chemical or enzymes [32–34]. Furthermore, recent evidence indicates that miRNAs can be detected in biological fluids and can be used to "capture" changes in the cells of origin, including neurons [35].

In addition to these general considerations, several findings suggest a specific involvement of miRNAs in ALS. For example, the loss of Dicer is sufficient to cause progressive degeneration of spinal motor neurons [36]; in addition, a global down-regulation of miRNAs is a frequent molecular denominator for multiple forms of human ALS [37]. Moreover, a common theme for several ALS-related genes is a role in RNA processing pathways [38]. FUS facilitates co-transcriptional Drosha recruitment to specific miRNA loci [39] and TARDBP participate to miRNA biogenesis as a component of both Drosha and Dicer complexes [16].

miRNA Detection

During the last decade, the development of methods for detecting miRNAs has risen to become a very attractive area of research. Although miRNAs have characteristics that made them suitable biomarkers, the detection of these molecules is challenging due to their intrinsic characteristics including small size, sequence similarity among various members, low level and tissue-specific or developmental stage-specific expression. Two approaches commonly used in the research of miRNAs as biomarkers, including studies in the area of neurodegenerative diseases and in particular in ALS, are reported below.

(1) Measurement of hundreds of miRNAs in specimens from patients with a pathology of interest and from control subjects using profiling methods, such as microarray, quantitative Real-Time Polymerase Chain Reaction (qRT-PCR)-based array, quantitative nCounter or Next Generation Sequencing (NGS), with subsequent validation of identified miRNAs by qRT-PCR;

(2) Analysis of selected miRNA(s) already known as related to specific tissues, cell types, or gene expression pathways. In this case, the number of miRNA(s) to be tested is limited, which makes the use of individual qRT-PCR appropriate, increasing sensitivity and reproducibility of the analysis.

Among the profiling methods, microarray is a powerful high-throughput widely used tool that screens large numbers of miRNAs analyzing simultaneously several samples processed in parallel in a single experiment [40]. An alternative method is deep-sequencing, which relays on NGS machines that can process millions of sequence reads in parallel in just a few days [41,42]. Sequence reads are processed by bioinformatics analysis, which identifies both known and novel miRNAs in the data sets, and perform a relative quantification using a digital approach [43]. Finally, qRT-PCR arrays can also be used to detect multiple miRNAs at the same time [44]. This approach is able to detect miRNAs in very low copy number [45]. This is an important aspect, since large amounts of RNA from clinical samples can be difficult to obtain. Other advantages of qRT-PCR-based techniques used in routine diagnostic are sensitivity, specificity, speed and simplicity [46]. Of note, potential biomarkers selected by array-based analysis need to be confirmed by qRT-PCR, due to high variability and low reproducibility of results obtained from these techniques [47].

A critical issue in qRT-PCR analysis is the data normalization approach. Normalization refers to adjusting for variations in data that are due to known factors (usually technical factors) and not related to the biological differences that are being investigated, and that could otherwise lead to inaccurate quantification. For this reason, stable normalizers are needed, but identifying such molecules is challenging, and it is often necessary to select them on a case-by-case basis [48]. Normalization to reference invariant miRNAs [49] is effective in many cases, but this approach requires that the reference miRNA is not influenced by the condition being studied. Exogenous spike-in controls added to samples during the miRNAs extraction may be used to compensate the variability caused by extraction efficiency and possible presence of inhibitors [50]. The combined use of two or more normalizers usually allows reducing experimental variability and improving reliability of the analysis.

4. miRNAs as Biomarkers for ALS

The first paper about miRNAs as biomarkers for ALS in human samples was published by De Felice and colleagues in 2012 [51]. Since then, a large number of studies have been performed on cerebrospinal fluid (CSF), blood and muscle biopsies from ALS patients. See Figure 1 for a schematic workflow of identification of miRNA-based biomarkers.

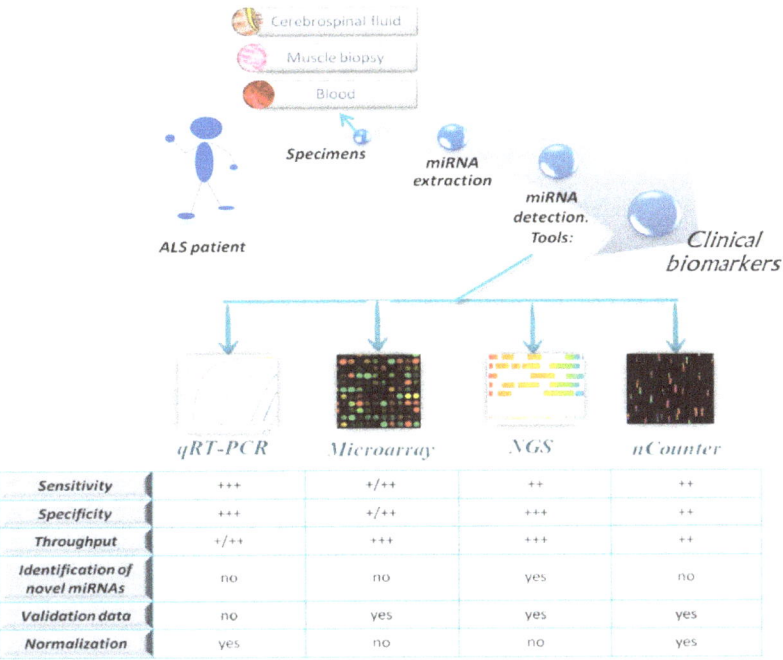

Figure 1. MicroRNA (miRNA)-based biomarkers in amyotrophic lateral sclerosis (ALS) patients. Schematic workflow to identify possible miRNAs as biomarkers starting from ALS patients' sample using different quantitative approaches. The comparison among the common characteristics of miRNA detection platforms is summarized in the figure. Sensibility, specificity and throughput are classified as follows: +++ (very high), ++ (moderate), +/++ (moderate to low) and + (low). Abbreviations: qRT-PCR, quantitative Real-Time Polymerase Chain Reaction; NGS, Next Generation Sequencing.

4.1. miRNAs in Cerebrospinal Fluid

Cerebrospinal fluid (CSF) is the fluid that bathes the central nervous system (CNS) and, due to this direct interaction, represents a potentially ideal source for identifying biomarkers for ALS. miRNAs present in CSF can mirror CNS physiological and pathological conditions representing more sensitive biomarkers of brain changes than those present in other biofluids [35]. The presence of miRNAs in CSF was first demonstrated by Cogswell and colleagues [52]. The authors reported that the amount of miRNAs secreted or excreted from other organs to CSF is very limited and that the major source of miRNAs detected in CSF are immune cells present in this biofluid. In addition, other studies showed that the miRNAs present in CSF derived also from neurons [53].

CSF samples are obtained by lumbar puncture, a procedure used for diagnostic purposes to confirm ALS diagnosis and exclude other pathologies, as inflammatory nerve conditions. Lumbar puncture, however, represents an invasive procedure, that cannot be repeated during the disease course for ethical implications. Thus, analysis of miRNAs in CSF is not suitable to identify biomarkers to follow disease progression.

Up to date, five studies have been published about the identification of miRNAs as biomarkers in CSF from ALS patients. Results are shown in Table 1.

The first three studies in Table 1 selected a limited set of miRNAs to analyze: 43 miRNAs found up-regulated in SOD1 spinal cord CD39+ microglia and splenic Ly6Chi monocytes [54], a group of *TARDBP* binding miRNAs [55], or one selected miRNA, over-expressed in ALS blood leucocytes [56], respectively. The other two studies performed a miRNA expression profiling, using qRT-PCR [57] or small RNA sequencing (NGS) [58]. In both profiling studies, results were validated by qRT-PCR for each single miRNA. While Benigni and colleagues found eight out of fourteen miRNAs as significantly deregulated, Waller and coworkers failed to confirm statistically significant differences in miRNA expression [57,58].

Table 1. Deregulated microRNAs (miRNAs) in cerebrospinal fluid (CSF) of amyotrophic lateral sclerosis (ALS) patients compared to healthy controls.

miRNA (Hsa-miR)	miRNA Expression Change	No. of Specimens	miRNA Detection Approach	Ref.
150, 99b, 146a	↑ in SALS	SALS: 10 FALS: 5 HCs: 10	qRT-PCR	[54]
27b, 328, 532-3p	↑ in SALS and FALS			
132-5p, 132-3p, 143-3p	↓	SALS: 22 HCs: 24	qRT-PCR	[55]
143-5p, 574-5p	↑			
338-3p	↑	SALS: 10 HCs: 10	qRT-PCR	[56]
181a-5p	↑	SALS: 24 HCs: 24	qRT-PCR	[57]
21-5p, 195-5p, 148-3p, 15b-5p, let7a-5p, let7b-5p, let7f-5p	↓			
124-3p, 127-3p, 143-3p, 125b-2-3p, 9-5p, 27b-3p	↑	SALS: 32 HCs: 10 NCs: 6	NGS	[58]
486-5p, let7f-5p, 16-5p, 28-3p, 146a-3p, 150-5p, 378a-3p, 142-5p, 92a-5p	↓			

Abbreviations: Ref., Reference; ↑/↓, up-regulated/down-regulated; SALS, sporadic amyotrophic lateral sclerosis patients; FALS, familial amyotrophic lateral sclerosis patients; HCs, healthy controls; NCs, neurological disease control subjects (multiple sclerosis); qRT-PCR, quantitative Real-Time Polymerase Chain Reaction; NGS, Next Generation Sequencing.

A common feature observed by the authors is an overall down-regulation of miRNAs in CSF samples from ALS patients [55,57,58], in agreement with other studies showing that the majority of deregulated miRNAs in tissues from ALS models and ALS patients are down-regulated [59]. This could suggest a general default in RNA metabolism in ALS [38].

In general, however, these studies highlight a wide heterogeneity among miRNAs significantly deregulated. A possible explanation could be the variability in terms of experimental approach and technical procedures and the reduced number of CSF samples analyzed in each study.

Some authors evaluated the correlation between CSF and serum miRNA expression levels. A significant positive correlation between expression levels in CSF and serum from ALS patients was found for miR-338-3p [56] and miR-143-3p [55]. However, the amount of most miRNAs was independently regulated between the two biofluids at individual level. This suggests that CSF miRNAs do not simply reflect the usually more abundant serum miRNAs, and changes in the serum do not necessarily reproduce alterations of CSF levels [55].

It should be noted that, among the different body fluids, the lowest abundance of miRNAs appears in CSF. Thus, it is possible that some potentially promising and informative miRNAs, identified both in vivo and in vitro ALS models, are below the limit of detection of the available methods of analysis. For example, the miR-218, a motor neurons-enriched miRNA, has been found increased in CSF of ALS rodent models: its expression correlates with the number of remaining spinal motor neurons and is responsive to motor neuron sparing therapy [60]. miR-218 could thus represent a potential biomarker to assess drug effects on motor neurons during clinical trials in ALS patients. However, at the present time, this miRNA is detectable only in some CSF samples, and thus a comparison between ALS patients and controls is not possible.

An approach to overcome the technical limits due to low abundance of several miRNAs in CSF could be to focus on those miRNAs found up-regulated in ALS patients. Among these, miR-338-3p seems to be very promising, since it has been reported as consistently upregulated in CSF, serum and leukocytes from ALS patients [56]. miR-338-3p is involved in several molecular pathways and could contribute to ALS pathogenesis through different modalities, such as neurodegeneration and apoptosis. Recent evidence suggests that miR-338 participates in the control of neuroblast apoptosis and in neuroblastoma pathogenesis [61] and it is able to suppress neuroblastoma proliferation, invasion and migration [62]. Interestingly, also another miRNA found up-regulated in CSF from ALS patients, miR181a-5p, has been proposed as an anti-oncomir, which acts as a tumor suppressor in normal tissues, promoting growth inhibition and apoptosis [63]. These findings suggest that these up-regulated miRNAs are involved in ALS pathogenetic process through apoptotic mechanisms responsible for cell death.

In order to increase diagnostic accuracy, up-regulated miRNAs can be used in combination with other miRNAs, identified as down-regulated. Benigni and colleagues reported that the ratios of miR-181a-5p/miR-15b-5p and miR-181a-5p/miR-21-5p considerably increased the specificity with a slight decrease in sensitivity compared with each individual miRNA [57]. A wider use of this strategy could allow improvements in the performance of identified biomarkers and should be taken into account for future studies, as further discussed in the following paragraphs.

4.2. Circulating miRNAs

The use of blood samples in the diagnostic routine presents several advantages. Blood specimens are easy to obtain, process and store, and the samples required for the analysis can be collected without using invasive procedures for the patients. The lack of ethical implications as compared with CSF and muscle biopsy makes it possible to repeat the blood draw during the disease progression. Since miRNAs circulate in the blood in a highly stable form, this may facilitate the procedure of storage and conservation and increase the flexibility of the analysis.

Blood-based biomarkers may originate from the CNS through a transfer between the blood and CSF at the blood–CSF barrier [64,65], suggesting that the same biomarkers could be present in both biofluids. They may be generated also by other organs and tissues affected during ALS, such as degenerating muscles or peripheral blood cells. Therefore, blood can represent an excellent biofluid for discovery and validation of biomarkers for ALS [66]. On the other hand, miRNAs present in blood can reflect other pathophysiological conditions concurrent but not directly related to ALS disease (e.g. inflammatory status, response to pharmacological treatments, etc.), which may represent confounding factors.

Several studies on circulating miRNAs as potential biomarkers for ALS have been published. The findings from such studies are summarized in Table 2.

Table 2. Deregulated circulating microRNAs (miRNAs) in amyotrophic lateral sclerosis (ALS) patients compared to healthy controls.

miRNA (Has-miR) Expression Change	Source	miRNA Detection Approach	No. of Specimens for miRNAs Validation	Ref.
↑: 338-3p	Leukocytes	Microarray→ miRNAs validation with qRT-PCR	SALS: 14 HCs: 14	[51]
↑: 27a, 155, 142-5p, 223, 30b, 532-3p	Monocytes (CD14+ CD16-)	Nanostring nCounter [1] → miRNAs validation with qRT-PCR	SALS: 22 FALS: 4 HCs: 24	[54]
↓: 132-3p, 132-5p, 143-3p, 143-5p, let-7b	Serum	Nine *TARDBP* binding miRNAs and miR-9-5p → qRT-PCR	SALS: 22 HCs: 24	[55]

Table 2. *Cont.*

miRNA (Has-miR) Expression Change	Source	miRNA Detection Approach	No. of Specimens for miRNAs Validation	Ref.
↑: 206, 106b	Serum	Microarray [1] → miRNAs validation with qRT-PCR	SALS: 12 HCs: 12	[67]
↑: 338-3p	Leukocytes and serum	miR-338-3p → qRT-PCR	SALS: 10 HCs: 10	[56]
↓ in FALS/SALS: 4745-5p, 3665, 4530 ↓ in FALS: 1915-3p	Serum	Microarray → miRNAs validation with qRT-PCR	FALS: 23 HCs: 24 SALS: 14 HCs: 14	[68]
↓ in FALS/SALS: 1825 ↓ in SALS: 1234-3p	Serum	Microarray→ miRNAs validation with qRT-PCR	SALS: 20 HCs: 20 FALS: 13 HCs: 13	[69]
↑: 4649-5p ↓: 4299	Plasma	Microarray → miRNAs validation with qRT-PCR	SALS: 48 HCs: 47	[70]
↓: 183, 193b, 451, 3935	Leukocytes	Microarray → miRNAs validation with qRT-PCR	SALS: 83 HCs: 61	[71]
↑: 424, 206	Plasma	Microarray [2] → miRNAs validation with qRT-PCR	SALS: 39 HCs: 39	[72]
↑: 206, 133a,133b ↓: 146a, 149*, 27a	Serum	Preselected myo-miRNAs, inflammatory and angiogenic miRNA → qRT-PCR	SALS: 14 HCs: 8	[73]
↑: 206 Deregulated MicroRN pairs: 206/338-3p 9*/129-3p 335-5p/338-3p	Plasma	Thirty seven brain-enriched and inflammation-associated microRNAs → qRT-PCR	ALS: 50 HCs: 50	[74]
↑ [†]: 1, 133a-3p, 133b, 144-5p, 192-3p, 195-5p, 19a-3p ↓ [†]: let-7d-3p, 320a, 320b, 320c, 425-5p, 139-5p	Serum	qRT-PCR array	SALS: 20 FALS: 3 HCs: 30 NCs: 103	[75]
↑: 206, 143-3p ↓: 374b-5p	Serum	qRT-PCR array → miRNAs validation with qRT-PCR	SALS: 23 CRL: 22	[76]
↑: 9, 338, 638, 663a, 124a, 451a, 132, 206, let-7b	Leukocytes	Preselected 10 miRNAs → miRNAs validation with qRT-PCR	SALS: 84 HCs: 27	[77]
↑: 142-3p ↓: 1249-3p	Serum	Microarray [1] → miRNAs validation with qRT-PCR	SALS: 20 HCs: 20	[78]
↓: 27a-3p	Serum exosomes	miR-27a-3p → qRT-PCR	ALS: 10 HCs: 20	[79]
↓: let-7a-5p, let-7d-5p, let-7f-5p, let-7g-5p, let-7i-5p, 103a-3p, 106b-3p, 128-3p, 130a-3p, 130b-3p, 144-5p, 148a-3p, 148b-3p, 15a-5p, 15b-5p, 151a-5p, 151b, 16-5p, 182-5p, 183-5p, 186-5p, 22-3p, 221-3p, 223-3p, 23a-3p, 26a-5p, 26b-5p, 27b-3p, 28-3p, 30b-5p, 30c-5p, 342-3p, 425-5p, 451a, 532-5p, 550a-3p, 584-5p, 93-5p	Whole blood	NGS → qRT-PCR	SALS: 50 HCs: 15	[80]

[1] analysis carried out on samples from transgenic mice; [2] analysis carried out on samples of ALS patients' skeletal muscle biopsies; [†], miRNAs deregulated in ALS patients compared to healthy controls and neurological controls (including multiple sclerosis and Alzheimer's disease patients). Abbreviations: Ref., Reference; ↑/↓, up-regulated/down-regulated; SALS, sporadic amyotrophic lateral sclerosis patients; FALS, familial amyotrophic lateral sclerosis patients; HCs, healthy controls; NCs, neurological controls; qRT-PCR, quantitative Real-Time Polymerase Chain Reaction; NGS, Next Generation Sequencing.

As reported in Table 2, several studies have identified numerous potential miRNA biomarkers in peripheral blood from ALS patients, however their results rarely overlap with each other. This high discrepancy in the identified miRNAs is probably associated with the variability of quantification methods, miRNA normalizers used, number of samples included, clinical features of patients, and also with the differences in selected source of miRNAs (serum, plasma, leukocytes, and whole blood). Another possible reason for the poor reproducibility of results may be the high level of heterogeneity in miRNA profiles of SALS patients in comparison to FALS patients. Freischmidt and colleagues initially reported a signature of 22 miRNAs significantly down-regulated in FALS and presymptomatic mutation carriers [68]. Subsequently, the same authors replicated the analysis of these miRNAs in a larger SALS sample group using identical technical procedures, and found only 2 miRNAs significantly down-regulated in all SALS patients. A more accurate analysis of results revealed that around 60% of SALS patients shared a serum miRNA fingerprint with genetic cases, while the remaining around 40% of patients were evenly distributed among control samples. The absence of FALS-like miRNA patterns in these patients may mirror a higher impact of exogenous factors and possibly a lower and/or different genetic influence in a subgroup of SALS patients [69].

Interestingly, the miRNA expression profiles derived from the study performed by Freischmidt and colleagues [68] were re-elaborated applying principal component analysis (PCA)-based unsupervised feature extraction (FE), another analysis approach [81]. The authors identified a total of 51 deregulated miRNAs, 27 down-regulated and 24 up-regulated in ALS patients in comparison with healthy controls. Applying the linear discriminant analysis (LDA) to these selected miRNAs, overall accuracy was 0.66 including healthy controls, ALS mutation carriers, FALS and SALS patients. Of note, excluding SALS patients, LDA was able to successfully discriminate healthy controls, ALS mutation carriers and FALS patients, with an accuracy rising up to 0.84, confirming as the heterogeneity of SALS group can introduce a wider variability in circulating miRNA profiles.

Among the studies published until now, a largely used approach is miRNA profiling on blood samples from ALS patients and controls, carried out by microarray [51,68–71], PCR-array [75,76] and NGS [80]. Other studies performed analysis on specific miRNAs, selected from data previously reported in the literature [55,56,73,74,77]. In other cases, the first step of the research was a microarray analysis on samples from transgenic mice [54,67,78] or skeletal muscle biopsies from ALS patients [68], followed by validation of miRNAs found deregulated in the first step of analysis.

Only one study analyzed miRNA expression specifically in serum exosomes [79]. Exosomes are double lipid vesicles secreted by a variety of cells and widespread in the peripheral body fluid. They can reflect physiological and pathological changes of the cells of origin, representing potential new biomarkers for disease diagnosis [82]. miRNAs are enriched in exosomes, and the exosome membrane structure can protect them from degradation by RNA enzymes. The authors investigated the expression of only miR-27a-3p, previously reported as present in myoblast-derived exosomes [83], and found a down-regulation of this miRNA in ALS patients, suggesting that miRNA exosome analysis could represent a future perspective for ALS biomarker identification.

Despite a poor overlapping among the miRNAs identified as deregulated in ALS, some circulating miRNAs seem to be particularly promising as potential biomarkers in ALS patients. Table 3 summarizes these miRNAs, reported as de-regulated in two or more papers.

As shown in the Table 3, some common pathways emerge: some miRNAs are involved in neurodegeneration and apoptosis (miR-338, miR-142, miR-183 and let-7d), other miRNAs act at muscle level (miR-206, miR-133a, miR-133b and miR-27a). In particular, miR-206, miR-133a and miR-133b are myo-miRNAs, molecules specifically expressed in striated muscle and involved in muscle proliferation, repair and regeneration. Their expression levels change during the process of myogenesis, development, atrophy, degeneration, and myopathies [84]. The more recurrent result is an up-regulation of circulating miR-206 in ALS patients. miR-206 is a human skeletal muscle-specific miRNA that promotes the formation of new neuromuscular junctions following nerve injury, and therefore plays a crucial role in the reinnervation process [85]. In miR-206 knock-out mice, delayed and incomplete muscular reinnervation was observed in comparison to those animals that expressed miR-206. In addition, high expression levels of miR-206 were found in a mouse model of ALS, and its under-expression was associated with a faster progression of the disease [86]. A consensus for higher expression levels of this miRNA in ALS patients compared to controls was reported by several authors [67,72–74,76,77]. Although miR-206 seems to represent a valid circulating biomarker for ALS, it is still to define whether the elevated expression of this miRNA is the result of the disease or its cause.

Table 3. The most promising circulating microRNAs (miRNAs) detected as potential biomarkers in amyotrophic lateral sclerosis (ALS) patients.

miRNAs (Has-miR)	miRNA Change	Role in ALS	Ref.
206	↑	Myo-miRNA: muscle proliferation, repair and regeneration. It promotes neuromuscular connectivity and enhances reinnervation	[67,72–74,76,77]
338	↑	Involvement in different pathways such as apoptosis, neurodegeneration, and/or glutamate clearance	[51,67,74,77]
133a	↑	Myo-miRNA: muscle proliferation, repair and regeneration	[73,75]
133b	↑	Myo-miRNA: muscle proliferation, repair and regeneration	[73,75]
142	↑	miRNA predicted to target a specific set of genes associated to the pathophysiology of ALS, including *TARDBP* and *C9orf72*.	[54,78]
183	↓	miRNA involved in neurodegenerative signaling pathway, including PI3K-Akt and MAPK pathway. miR-183/mTOR pathway contributes to spinal muscular atrophy pathology	[71,80]
27a	↓	miRNA involved in muscle growth, myoblast proliferation acting on myostatin. It is present in myoblast-derived exosomes	[73,79]
let-7d	↓	Involvement in apoptosis by the Hippo signaling pathway	[75,80]

Abbreviations: Ref., Reference; ↑/↓, up-regulated/down-regulated.

While all the works performed a comparison between samples from ALS patients and healthy controls, only a subset of them included also samples from patients affected by other neurological disorders. Neurological controls comprised Parkinson's disease [56,70,71,74], Alzeihmer's Disease [56,69,74,75], Huntington's disease [56,69,71], Multiple Sclerosis [54,75] and ALS-mimic conditions [76]. The use of neurological controls can help to discriminate whether identified miRNAs are really specific for ALS or are common features linked to neurodegenerative processes. For example, the comparison of miRNA expression between ALS and Parkinson's disease patients suggested that miR-183 might be specific for SALS, whereas miR-451 and miR-3935 might be more general biomarkers linked to neurodegenerative disorders [71]. In addition, the inclusion of an ALS-mimic patients' group may contribute to identify miRNA biomarkers to use in the differential diagnosis in the early stages of the disease.

Only a part of the studies performed until now investigated the potential correlations among miRNA expression levels and ALS clinical features, sometimes in longitudinal studies, measuring miRNA levels in the same ALS patient over time [70,72,75,76,80]. In some case this analysis failed to find any association [67,69,77], in other cases specific correlations were reported. Some authors described associations of miRNA expression levels with ALS site of onset [70,74,76,80], ALS Functional Rating Scale-revised (ALSFRS-R) and/or vital capacity (VC) [70,75,78,80], Medical Research Council (MRC) sumscore [72] and with the disease progression rate [72,80]. Only two studies investigated the possible associations of specific serum miRNAs with riluzole treatment, failing to identify any correlation [75,76]. Such results must be anyway considered with caution, since the number of subjects included in every group is limited. They need to be confirmed in larger cohorts of ALS patients, to really define the role of miRNA expression in ALS clinical presentation and progression. From this perspective, it would be very important that, after the identification of potential miRNA biomarkers, more longitudinal studies were performed, to evaluate if these miRNAs could be used as prognostic indicators.

As already mentioned for CSF studies, also in serum the analysis of combinations of several miRNAs has shown a higher accuracy than single miRNAs in discriminating ALS from healthy controls or other neurological disorders [71,74,75]. A very interesting approach is reported by Sheinerman and colleagues, who developed a strategy based on miRNA pairs, consisting of one miRNA enriched in synapses of a brain region affected by the disease and another miRNA enriched in a different brain region or cell type. The use of the pair of miRNA derived from the same organ allowed decreasing potential overlap with pathologies of other organs and reducing also inter-individual variability. The authors demonstrated that, combining two or three effective miRNA pairs into a single miRNA classifier, they could achieve a greater accuracy in discriminating ALS both from healthy controls and patients affected by other neurological disorders [74]. Thus, in the future studies it should be considered that, while the deregulation of a single miRNA can be a feature common to several neurological diseases, panels of deregulated miRNAs, or combinations of them, may result highly specific for ALS and represent a signature for this disease.

Finally, a relevant aspect of the use of miRNAs as ALS biomarkers is their potential of identifying the disease in very early stages, also before any clinical manifestation. In their work, Freischmidt and colleagues showed that a specific subset of miRNAs, reduced in the serum of patients with familial and sporadic ALS, was reduced also in presymptomatic carriers of pathogenic ALS mutations. Moreover, the down-regulation was largely independent of the underlying disease gene and was stronger in patients with familial ALS than in pre-manifest mutation carriers, suggesting that alterations of miRNA profiles could be progressive when comparing the pre-manifest and manifest phase of the disease [68]. If confirmed, these findings may be of fundamental importance for the development of screening tests able to detect ALS in early asymptomatic stages and for future preventive therapeutic strategies before the occurrence of clinically evaluable symptoms.

4.3. miRNAs in Muscle Biopsies

Skeletal muscle is another potential source for the identification of candidate miRNA biomarkers. In the last years, it has become evident that ALS does not affect only motor neurons but also other cell types, including striated muscle, which play an active role in the disease pathogenesis. Before the clinical onset and during the disease progression, the affected skeletal muscle of ALS patients attempts to restore function by futile cycles of reinnervation and denervation [87]. Eventually, persistent muscle wasting exceeds the ability to repair and consequently the atrophy process starts. Due to the crucial role of the skeletal muscle in ALS pathogenesis, recent studies have focused their research on the identification of specific muscle miRNAs in ALS tissues, which could potentially be use as prognostic biomarkers of disease. Moreover, miRNAs identified in skeletal muscle of ALS patients could be used as biomarkers also in plasma or serum, where they can be released by the affected tissues. This strategy seems to be particularly interesting, since muscle biopsy is unfortunately an invasive practice and cannot be proposed for longitudinal studies to follow disease progression.

Several studies focused on analysis of myo-miRNAs, including miR-1, miR-133a, miR-133b, miR-206, miR-208a, miR-208b, miR-499, and miR-486 [88]. Most of them explored the role of these miRNAs in mouse models (for a review see [89]), but only few studies investigated the role of these molecules as possible markers in muscle biopsies of patients with ALS, due to the rarity and difficulty to obtain this kind of samples. miRNAs found deregulated in muscle biopsies from ALS patients compared to healthy control subjects are shown in Table 4.

Most studies focused on the expression of myo-miRNAs [90–93]; only in some cases also other miRNAs were included, for example miRNAs related to inflammation/angiogenesis [93] or selected by microarray [72] or NGS approaches [94,95]. Overall, results are sometimes contrasting and poorly reproducible. These non-concordant finding could be attributed to different types of muscle used for biopsy, discordance among the samples in terms of inclusion criteria of patients (age, gender, evolution of disease, onset) and different techniques and internal control molecules used to assess miRNA expression levels.

Table 4. Deregulated miRNAs in skeletal muscle biopsies of amyotrophic lateral sclerosis (ALS) patients compared to healthy controls.

miRNA (Hsa-miR)	miRNA Expression Change	Type of Muscle	No. of Muscle Biopsies	miRNA Detection Approach	Ref.
206	↑	Deltoid, anconeus	FALS: 1 SALS: 10 HCs: 6	mir-206 → qRT-PCR	[90]
23a, 29b, 206, 455, 31	↑	Vastus lateralis	ALS: 14 HCs: 10	Myo-miRNAs and miRNAs dysregulated in human muscle disease → qRT-PCR	[91]
1, 26a, 133a, 455	↓	Vastus lateralis	ALS: 5 HCs: 7	Myo-miRNAs → qRT-PCR	[92]
424, 214, 206	↑	Biceps brachii	ALS: 5 HCs: 5	Microarray → miRNAs validation with qRT-PCR	[72]
1, 206, 133a, 133b, 27a, 155, 146a, 221	↑	Quadriceps femoris	SALS: 13 HCs: 5	Inflammatory/angiogenic miRNAs and myo-miRNAs → qRT-PCR	[93]
1, 10b-5p, 100-5p, 133a-3p, 133b-3p	↓	Biceps, deltoid, tibialis anterior, vastus lateralis	ALS: 19 HCs: 9	NGS [1] and qRT-PCR [1] → qRT-PCR	[94]
100-5p, 10a, 125a-5p, 133a-1/-2-3p, 362, 500a-3p, 542-5p, 99a-5p	↑	Vastus lateralis	FALS: 2 SALS: 9 HCs: 11	NGS	[95]
1303-3p, 150-5p, 26a-1/-2-5p, 486-1/-2-5p,	↓				

[1] analysis carried out on samples from transgenic mice.   Abbreviations: Ref.,   Reference;   ↑/↓, up-regulated/down-regulated; SALS, sporadic amyotrophic lateral sclerosis patients; FALS, familial amyotrophic lateral sclerosis patients; HCs, healthy controls; qRT-PCR, quantitative Real-Time Polymerase Chain Reaction; NGS, Next Generation Sequencing.

Some authors performed also a correlation analysis among miRNA expression and ALS clinical features. Table 5 reports miRNAs altered in tissue of specific stratified ALS patients' groups analyzed in comparison to control subjects.

In addition, in other papers the associations with clinical variables were analyzed comparing groups of patients to each other. Stratifying ALS patients, an up-regulation of myo-miRNAs (miR-206, miR-133a, miR-133b and miR-27a) and of inflammatory miRNAs (miR-155, miR-146a and miR-221) was discovered in ALS patients with earlier age at onset (<55 years) and longer disease duration [93]. Moreover, significantly higher expression levels of the same myo-miRNAs and inflammatory miRNAs were detected in male than in female. This gender difference has been hypothesized to be related to a difference in hormonal regulation, implying a slower disease progression in women [93]. In another paper, miR-29c, miR-208b and miR-499 were reported as increased in patients with slow disease course [96]. Expression data were analyzed in patients categorized into "early" and "late" based on disease duration at the moment of biopsy (more or less one year). miR-9 and miR-206 significantly increased in the early patients' group and, of note, miR-206 inversely correlated with the time from symptoms onset to muscle biopsy, indicating an early response to denervation in skeletal muscle [96].

Table 5. Deregulated miRNAs in skeletal muscle biopsies of specific amyotrophic lateral sclerosis (ALS) patients' groups analyzed in comparison to healthy controls.

miRNA (Hsa-miR)	miRNA Expression Change in Specific ALS Patients' Group	Type of Muscle	No. of Muscle Biopsies	miRNA Detection Approach	Ref.
133a, 29c, 9, 208b	↑in ALS slow group [1]	Deltoid and quadriceps	FALS: 3 SALS: 11 HCs: 24	Eleven skeletal muscle related miRNAs → qRT-PCR	[96]
1, 208b	↓ in ALS rapid group [2]		Slow group [1]: 6 Rapid group [2]: 5 Early group [3]: 4 Late group [4]:9		
133a, 133b, 206, 29c, 9, 155, 23a	↑ in early stage group [3]				
100-5p, 199a-1/-2, 199b-3p, 27a-5p, 3607-3p, 424-5p, 450a-1/-2-5p, 450b-5p, 501-3p, 502-3p, 542-5p, 660-5p	↑ in higher disease severity [5]	Vastus lateralis	Higher disease group [5]: 7 HCs: 11	NGS	[95]
1303-3p, 133a-1/-2-3p, 150-5p, 378, 486-1/-2-5p, 502-3p, 855-3p	↓ in higher disease severity [5]				

[1], ALS slow group (≥4 years of disease progression without requiring respiratory supports); [2], ALS rapid group (<4 years of disease progression without respiratory supports or death occurring <4 years from symptoms onset); [3], early stage group (less than one year from symptom onset to muscle biopsy); [4], late stage group (more than one year from symptom onset to muscle biopsy); [5], group of patients with higher disease severity. Abbreviations: Ref., Reference; ↑/↓, up-regulated/down-regulated; SALS, sporadic amyotrophic lateral sclerosis patients; FALS, familial amyotrophic lateral sclerosis patients; HCs, healthy controls; qRT-PCR, quantitative Real-Time Polymerase Chain Reaction; NGS, Next Generation Sequencing.

Although the results are often inconsistent among different studies, some trends in miRNAs deregulation seem to emerge. One of the most interesting miRNA is miR-133a, which was found to be up-regulated in human ALS tissues [93,95,96], particularly in patients with slow disease progression and in biopsies obtained before one year from the symptom onset [96]. At the same time, a significant reduction of this miRNA was present in a specific ALS patients' group with higher disease severity [95], suggesting changes in its expression during the disease progression. In contrast, however, other studies detected a down-regulation of miR-133a in human biopsies, as reported also in mice [92,94]. At the moment, the strongest data are those concerning miR-206. Indeed, the mechanisms responsible for the increase of this miRNA seem to be conserved in the skeletal muscle of mouse models and in that from ALS patients, and the up-regulation described in both cases is an ALS-specific response to the denervation. miR-206 was found significantly up-regulated in muscle samples from ALS patients compared to control subjects [72,90,91,93], similarly to what observed in blood samples, strengthening the role of this miRNA as potential biomarker for ALS. Of note, miR-206 showed an increased trend in muscle biopsies from long-term survivor patients, even though below the statistically significance [90]. De Andrade and colleagues [72] reported that this miRNA was over-expressed both in plasma and skeletal muscle of patients with ALS, but the over-expression was not progressive during the follow-up. They supposed that miR-206 expression increased early in the disease course, reaches a plateau and then begins to fall. In agreement with this hypothesis, an up-regulation of miR-206 was described in muscle biopsies from ALS patients within one year from the clinical onset, becoming less evident as the disease progresses to a later stage [96]. Finally, Si and collaborators reported a non-significant upward trend in miR-206 in muscle samples from ALS patients compared to controls. This result, however, was correlated to a high standard error for this miRNA due to the variability among samples. Moreover, the authors reported a significant inverse correlation between this miRNA and the muscle power of the biopsied muscle, hypothesizing that it could be a marker of disease activity. This finding highlights the importance to associate miR-206 levels with muscle-specific clinical assessment rather than overall clinical status [94].

Although a concordant miRNA signature have not been identified yet in ALS patient muscle biopsies, these findings show that miRNAs could be useful prognostic markers to better understand the course of disease. In particular, the identification of a specific muscular miRNA profile through multicenter studies, able to increase the statistical power of the analysis, could lead to a stratification of the patients in order to identify prognostic biomarkers to use as indicator of disease progression, facilitating the clinical management of patients.

5. Conclusions and Future Perspectives

Despite the intense research activity of the last years, the use of miRNAs as biomarkers for diagnosis of ALS and clinical management of patients is still in an early stage of development. Several interesting data have been obtained so far, with important insights into the disease processes. However, results achieved in different studies are most of the time conflicting and poorly reproducible, making it difficult to unequivocally identify which miRNA(s) may be selected as biomarker in clinical practice. In order to overcome these limits, some improvements in the research approach should be taken into account.

First of all, one factor strongly complicating the comparison among data reported by different research groups is the wide range of methods used for the identification of potential miRNA biomarkers and the different techniques for miRNA measurement and data normalization. A common acceptance of certain guidelines, standard research protocols, and strong methods of statistical analysis of miRNAs will be important in the future to achieve reliable biomarkers.

Another critical issue is the relatively small number of patients included in the studies performed until now. Results are often interesting, but they need to be verified in larger cohorts of ALS patients. It would be really important that those miRNAs, which have shown initial promise, were validated in independent laboratories and/or in multicenter collaborations. In addition, since ALS is a highly heterogeneous disease, replication studies should increase the number of patients stratifying them based on clinical and genetic features, in order to obtain a better assessment of the potential associations among miRNAs and these variables.

Further, in several studies miRNA levels of ALS patients have been compared only to those of control subjects not affected by neurological disorders. This approach may bring to the identification of miRNAs able to successfully differentiate patients from healthy control subjects, but these miRNAs are often associated with common pathologic processes of neurodegeneration and are not specific for ALS. It will be of fundamental importance to extend the comparison to patients affected by other neurodegenerative diseases, in particular ALS-mimic disorders, to evaluate the specificity of deregulated miRNAs for ALS.

One of the more interesting approaches to miRNA biomarker identification is the use of a complex set of biomarkers, or combinations or ratios of biomarkers from different pathogenic pathways, rather than the employ of a single marker. This strategy has been shown to increase the sensitivity and/or specificity of potential ALS biomarkers and to contain more exhaustive diagnostic information, and should be more widely used in future researches.

At the same time, when possible, future studies should try to combine data obtained from multiple source of sample (blood, CSF, muscle) of the same patient. Up to date, only few studies have performed this kind of analysis, and their results are quite conflicting. However, an extensive analysis of correlations among different samples could be helpful to obtain more informative data and improve patients' stratification. In addition, for circulating miRNAs, it would be important to perform longitudinal studies on a large number of patients, in order to identify potential biomarkers of disease progression, and evaluate their role as prognostic indicators.

In conclusion, miRNAs constitute very promising biomarkers for ALS, but there is still much work to be done to validate and use them in clinical routine. The ultimate objective is to include these biomarkers in all phases of ALS management, from the diagnosis to the clinical trials, and, in perspective, to the identification of future therapeutic approaches.

Author Contributions: Writing—review and editing, C.R. and C.M.; supervision, S.B.

References

1. Rowland, L.P.; Shneider, N.A. Amyotrophic lateral sclerosis. *N. Engl. J. Med.* **2001**, *344*, 1688–1700. [CrossRef] [PubMed]
2. Rothstein, J.D. Current hypotheses for the underlying biology of amyotrophic lateral sclerosis. *Ann. Neurol.* **2009**, *65*, S3–S9. [CrossRef] [PubMed]
3. Ilieva, H.; Polymenidou, M.; Cleveland, D.W. Non-cell autonomous toxicity in neurodegenerative disorders: ALS and beyond. *J. Cell Biol.* **2009**, *187*, 761–772. [CrossRef] [PubMed]
4. Wijesekera, L.C.; Leigh, P.N. Amyotrophic lateral sclerosis. *Orphanet J. Rare Dis.* **2009**, *4*, 3. [CrossRef] [PubMed]
5. Brown, R.H., Jr.; Al-Chalabi, A. Amyotrophic lateral sclerosis. *N. Engl. J. Med.* **2017**, *377*, 162–172. [CrossRef] [PubMed]
6. Chiò, A.; Logroscino, G.; Traynor, B.J.; Collins, J.; Simeone, J.C.; Goldstein, L.A.; White, L.A. Global epidemiology of amyotrophic lateral sclerosis: A systematic review of the published literature. *Neuroepidemiology* **2013**, *41*, 118–130. [CrossRef] [PubMed]
7. Logroscino, G.; Traynor, B.J.; Hardiman, O.; Chio', A.; Couratier, P.; Mitchell, J.D.; Swingler, R.J.; Beghi, E. Descriptive epidemiology of amyotrophic lateral sclerosis: New evidence and unsolved issues. *J. Neurol. Neurosurg. Psychiatry* **2008**, *79*, 6–11. [CrossRef] [PubMed]
8. Manjaly, Z.R.; Scott, K.M.; Abhinav, K.; Wijesekera, L.; Ganesalingam, J.; Goldstein, L.H.; Janssen, A.; Dougherty, A.; Willey, E.; Stanton, B.R.; et al. The sex ratio in amyotrophic lateral sclerosis: A population based study. *Amyotroph. Lateral Scler.* **2010**, *11*, 439–442. [CrossRef] [PubMed]
9. Sutedja, N.A.; Fischer, K.; Veldink, J.H.; van der Heijden, G.J.; Kromhout, H.; Heederik, D.; Huisman, M.H.; Wokke, J.J.; van den Berg, L.H. What we truly know about occupation as a risk factor for ALS: A critical and systematic review. *Amyotroph. Lateral Scler.* **2009**, *10*, 295–301. [CrossRef] [PubMed]
10. Andersen, P.M. Amyotrophic lateral sclerosis associated with mutations in the CuZn superoxide dismutase gene. *Curr. Neurol. Neurosci. Rep.* **2006**, *6*, 37–46. [CrossRef] [PubMed]
11. Ben Hamida, M.; Hentati, F.; Ben Hamida, C. Hereditary motor system diseases (chronic juvenile amyotrophic lateral sclerosis). Conditions combining a bilateral pyramidal syndrome with limb and bulbar amyotrophy. *Brain* **1990**, *113*, 347–363. [CrossRef] [PubMed]
12. Andersen, P.M.; Al-Chalabi, A. Clinical genetics of amyotrophic lateral sclerosis: What do we really know? *Nat. Rev. Neurol.* **2011**, *7*, 603–615. [CrossRef] [PubMed]
13. Bigio, E.H.; Weintraub, S.; Rademakers, R.; Baker, M.; Ahmadian, S.S.; Rademaker, A.; Weitner, B.B.; Mao, Q.; Lee, K.H.; Mishra, M.; Ganti, R.A.; Mesulam, M.M. Frontotemporal lobar degeneration with TDP-43 proteinopathy and chromosome 9p repeat expansion in C9ORF72: Clinicopathologic correlation. *Neuropathology* **2013**, *33*, 122–133. [CrossRef] [PubMed]
14. Rizzo, F.; Riboldi, G.; Salani, S.; Nizzardo, M.; Simone, C.; Corti, S.; Hedlund, E. Cellular therapy to target neuroinflammation in amyotrophic lateral sclerosis. *Cell. Mol. Life Sci.* **2014**, *71*, 999–1015. [CrossRef] [PubMed]
15. Volk, A.E.; Weishaupt, J.H.; Andersen, P.M.; Ludolph, A.C.; Kubisch, C. Current knowledge and recent insights into the genetic basis of amyotrophic lateral sclerosis. *Med. Genet.* **2018**, *30*, 252–258. [CrossRef] [PubMed]
16. Kawahara, Y.; Mieda-Sato, A. TDP-43 promotes microRNA biogenesis as a component of the drosha and dicer complexes. *Proc. Natl. Acad. Sci. USA* **2012**, *109*, 3347–3352. [CrossRef] [PubMed]

17. Lagier-Tourenne, C.; Polymenidou, M.; Cleveland, D.W. TDP-43 and FUS/TLS: Emerging roles in RNA processing and neurodegeneration. *Hum. Mol. Genet.* **2010**, *19*, R46–R64. [CrossRef] [PubMed]

18. Brooks, B.R.; Miller, R.G.; Swash, M.; Munsat, T.L.; World Federation of Neurology Research Group on Motor Neuron Diseases. El escorial revisited: Revised criteria for the diagnosis of amyotrophic lateral sclerosis. *Amyotroph. Lateral Scler. Other Motor. Neuron Disord.* **2000**, *1*, 293–299. [CrossRef] [PubMed]

19. Zoing, M.C.; Burke, D.; Pamphlett, R.; Kiernan, M.C. Riluzole therapy for motor neurone disease: An early Australian experience (1996–2002). *J. Clin. Neurosci.* **2006**, *13*, 78–83. [CrossRef] [PubMed]

20. Biomarkers Definitions Working Group. Biomarkers and surrogate endpoints: Preferred definitions and conceptual framework. *Clin. Pharmacol. Ther.* **2001**, *69*, 89–95. [CrossRef] [PubMed]

21. Robelin, L.; Gonzalez De Aguilar, J.L. Blood biomarkers for amyotrophic lateral sclerosis: Myth or reality? *Biomed. Res. Int.* **2014**, *2014*, 10. [CrossRef] [PubMed]

22. O'Brien, J.; Hayder, H.; Zayed, Y.; Peng, C. Overview of microrna biogenesis, mechanisms of actions, and circulation. *Front. Endocrinol. Lausanne* **2018**, *9*, 402. [CrossRef] [PubMed]

23. Ludwig, N.; Leidinger, P.; Becker, K.; Backes, C.; Fehlmann, T.; Pallasch, C.; Rheinheimer, S.; Meder, B.; Stähler, C.; Meese, A.; et al. Distribution of miRNA expression across human tissues. *Nucleic Acids Res.* **2016**, *44*, 3865–3877. [CrossRef] [PubMed]

24. Mocellin, S.; Pasquali, S.; Pilati, P. Oncomirs: From tumor biology to molecularly targeted anticancer strategies. *Mini Rev. Med. Chem.* **2009**, *9*, 70–80. [CrossRef] [PubMed]

25. Aumiller, V.; Förstemann, K. Roles of microRNAs beyond development—metabolism and neural plasticity. *Biochim. Biophys. Acta.* **2008**, *1779*, 692–696. [CrossRef] [PubMed]

26. Carissimi, C.; Fulci, V.; Macino, G. MicroRNAs: Novel regulators of immunity. *Autoimmun. Rev.* **2009**, *8*, 520–524. [CrossRef] [PubMed]

27. Bushati, N.; Cohen, S.M. MicroRNAs in neurodegeneration. *Curr. Opin. Neurobiol.* **2008**, *18*, 292–296. [CrossRef] [PubMed]

28. Rajgor, D. Macro roles for microRNAs in neurodegenerative diseases. *Noncoding RNA Res.* **2018**, *3*, 154–159. [CrossRef] [PubMed]

29. Rinchetti, P.; Rizzuti, M.; Faravelli, I.; Corti, S. MicroRNA Metabolism and dysregulation in amyotrophic lateral sclerosis. *Mol. Neurobiol.* **2018**, *55*, 2617–2630. [CrossRef] [PubMed]

30. Lu, J.; Getz, G.; Miska, E.A.; Alvarez-Saavedra, E.; Lamb, J.; Peck, D.; Sweet-Cordero, A.; Ebert, B.L.; Mak, R.H.; Ferrando, A.A.; et al. MicroRNA expression profiles classify human cancers. *Nature.* **2005**, *435*, 834–838. [CrossRef] [PubMed]

31. Weber, J.A.; Baxter, D.H.; Zhang, S.; Huang, D.Y.; Huang, K.H.; Lee, M.J.; Galas, D.J.; Wang, K. The microRNA spectrum in 12 body fluids. *Clin. Chem.* **2010**, *56*, 1733–1741. [CrossRef] [PubMed]

32. Mitchell, P.S.; Parkin, R.K.; Kroh, E.M.; Fritz, B.R.; Wyman, S.K.; Pogosova-Agadjanyan, E.L.; Peterson, A.; Noteboom, J.; O'Briant, K.C.; Allen, A.; et al. Circulating microRNAs as stable blood-based markers for cancer detection. *Proc. Natl. Acad. Sci. USA* **2008**, *105*, 10513–10518. [CrossRef] [PubMed]

33. Cortez, M.A.; Bueso-Ramos, C.; Ferdin, J.; Lopez-Berestein, G.; Sood, A.K.; Calin, G.A. MicroRNAs in body fluids-the mix of hormones and biomarkers. *Nat. Rev. Clin. Oncol.* **2011**, *8*, 467–477. [CrossRef] [PubMed]

34. Mo, M.H.; Chen, L.; Fu, Y.; Wang, W.; Fu, S.W. Cell-free circulating miRNA biomarkers in cancer. *J. Cancer* **2012**, *3*, 432–448. [CrossRef] [PubMed]

35. Rao, P.; Benito, E.; Fischer, A. MicroRNAs as biomarkers for CNS disease. *Front. Mol. Neurosci.* **2013**, *6*, 39. [CrossRef] [PubMed]

36. Haramati, S.; Chapnik, E.; Sztainberg, Y.; Eilam, R.; Zwang, R.; Gershoni, N.; McGlinn, E.; Heiser, P.W.; Wills, A.M.; Wirguin, I.; et al. MiRNA malfunction causes spinal motor neuron disease. *Proc. Natl. Acad. Sci. USA* **2010**, *107*, 13111–13116. [CrossRef] [PubMed]

37. Emde, A.; Eitan, C.; Liou, L.L.; Libby, R.T.; Rivkin, N.; Magen, I.; Reichenstein, I.; Oppenheim, H.; Eilam, R.; Silvestroni, A.; et al. Dysregulated miRNA biogenesis downstream of cellular stress and als-causing mutations: A new mechanism for ALS. *EMBO J.* **2015**, *34*, 2633–2651. [CrossRef] [PubMed]

38. Strong, M.J. The evidence for altered RNA metabolism in amyotrophic lateral sclerosis (ALS). *J. Neurol. Sci.* **2010**, *288*, 1–12. [CrossRef] [PubMed]

39. Morlando, M.; Dini Modigliani, S.; Torrelli, G.; Rosa, A.; Di Carlo, V.; Caffarelli, E.; Bozzoni, I. FUS stimulates microRNA biogenesis by facilitating co-transcriptional drosha recruitment. *EMBO J.* **2012**, *31*, 4502–4510.

[CrossRef] [PubMed]

40. Li, W.; Ruan, K. MicroRNA detection by microarray. *Anal. Bioanal. Chem.* **2009**, *394*, 1117–1124. [CrossRef] [PubMed]

41. Motameny, S.; Wolters, S.; Nürnberg, P.; Schumacher, B. Next generation sequencing of miRNAs–strategies, resources and methods. *Genes* **2010**, *1*, 70–84. [CrossRef] [PubMed]

42. Friedländer, M.R.; Chen, W.; Adamidi, C.; Maaskola, J.; Einspanier, R.; Knespel, S.; Rajewsky, N. Discovering MicroRNAs from deep sequencing data using miRDeep. *Nat. Biotechnol.* **2008**, *26*, 407–415. [CrossRef] [PubMed]

43. Creighton, C.J.; Reid, J.G.; Gunaratne, P.H. Expression profiling of microRNAs by deep sequencing. *Brief Bioinform.* **2009**, *10*, 490–497. [CrossRef] [PubMed]

44. Mestdagh, P.; Feys, T.; Bernard, N.; Guenther, S.; Chen, C.; Speleman, F.; Vandesompele, J. High-throughput stem-loop RT-qPCR miRNA expression profiling using minute amounts of input RNA. *Nucleic Acids Res.* **2008**, *36*, e143. [CrossRef] [PubMed]

45. Schmittgen, T.D.; Jiang, J.; Liu, Q.; Yang, L. A high-throughput method to monitor the expression of microRNA precursors. *Nucleic Acids Res.* **2004**, *32*, e43. [CrossRef] [PubMed]

46. Murphy, J.; Bustin, S.A. Reliability of real-time reverse-transcription PCR in clinical diagnostics: Gold standard or substandard? *Expert. Rev. Mol. Diagn.* **2009**, *9*, 187–197. [CrossRef] [PubMed]

47. De Planell-Saguer, M.; Rodicio, M.C. Detection methods for microRNAs in clinic practice. *Clin. Biochem.* **2013**, *46*, 869–878. [CrossRef] [PubMed]

48. Schwarzenbach, H.; da Silva, A.M.; Calin, G.; Pantel, K. Data normalization strategies for microRNA quantification. *Clin. Chem.* **2015**, *61*, 1333–1342. [CrossRef] [PubMed]

49. Peltier, H.J.; Latham, G.J. Normalization of microRNA expression levels in quantitative RT-PCR assays: Identification of suitable reference RNA targets in normal and cancerous human solid tissues. *RNA* **2008**, *14*, 844–852. [CrossRef] [PubMed]

50. Kroh, E.M.; Parkin, R.K.; Mitchell, P.S.; Tewari, M. Analysis of circulating microRNA biomarkers in plasma and serum using quantitative reverse transcription-PCR (qRT-PCR). *Methods* **2010**, *50*, 298–301. [CrossRef] [PubMed]

51. De Felice, B.; Guida, M.; Guida, M.; Coppola, C.; De Mieri, G.; Cotrufo, R. A miRNA signature in leukocytes from sporadic amyotrophic lateral sclerosis. *Gene* **2012**, *508*, 35–40. [CrossRef] [PubMed]

52. Cogswell, J.P.; Ward, J.; Taylor, I.A.; Waters, M.; Shi, Y.; Cannon, B.; Kelnar, K.; Kemppainen, J.; Brown, D.; Chen, C.; et al. Identification of miRNA changes in Alzheimer's disease brain and CSF yields putative biomarkers and insights into disease pathways. *J. Alzheimers Dis.* **2008**, *14*, 27–41. [CrossRef] [PubMed]

53. Alexandrov, P.N.; Dua, P.; Hill, J.M.; Bhattacharjee, S.; Zhao, Y.; Lukiw, W.J. microRNA (miRNA) speciation in Alzheimer's disease (AD) cerebrospinal fluid (CSF) and extracellular fluid (ECF). *Int. J. Biochem. Mol. Biol.* **2012**, *3*, 365–373. [PubMed]

54. Butovsky, O.; Siddiqui, S.; Gabriely, G.; Lanser, A.J.; Dake, B.; Murugaiyan, G.; Doykan, C.E.; Wu, P.M.; Gali, R.R.; Iyer, L.K.; et al. Modulating inflammatory monocytes with a unique microRNA gene signature ameliorates murine ALS. *J. Clin. Invest.* **2012**, *122*, 3063–3087. [CrossRef] [PubMed]

55. Freischmidt, A.; Müller, K.; Ludolph, A.C.; Weishaupt, J.H. Systemic dysregulation of TDP-43 binding microRNAs in amyotrophic lateral sclerosis. *Acta Neuropathol. Commun.* **2013**, *1*, 42. [CrossRef] [PubMed]

56. De Felice, B.; Annunziata, A.; Fiorentino, G.; Borra, M.; Biffali, E.; Coppola, C.; Cotrufo, R.; Brettschneider, J.; Giordana, M.L.; Dalmay, T.; et al. MiR-338-3p is over-expressed in blood, CFS, serum and spinal cord from sporadic amyotrophic lateral sclerosis patients. *Neurogenetics* **2014**, *15*, 243–253. [CrossRef] [PubMed]

57. Benigni, M.; Ricci, C.; Jones, A.R.; Giannini, F.; Al-Chalabi, A.; Battistini, S. Identification of miRNAs as potential biomarkers in cerebrospinal fluid from amyotrophic lateral sclerosis patients. *Neuromolecular Med.* **2016**, *18*, 551–560. [CrossRef] [PubMed]

58. Waller, R.; Wyles, M.; Heath, P.R.; Kazoka, M.; Wollff, H.; Shaw, P.J.; Kirby, J. Small RNA sequencing of sporadic amyotrophic lateral sclerosis cerebrospinal fluid reveals differentially expressed miRNAs related to neural and glial activity. *Front. Neurosci.* **2018**, *11*, 731. [CrossRef] [PubMed]

59. Paez-Colasante, X.; Figueroa-Romero, C.; Sakowski, S.A.; Goutman, S.A.; Feldman, E.L. Amyotrophic lateral sclerosis: mechanisms and therapeutics in the epigenomic era. *Nat. Rev. Neurol.* **2015**, *11*, 266–279. [CrossRef] [PubMed]

60. Hoye, M.L.; Koval, E.D.; Wegener, A.J.; Hyman, T.S.; Yang, C.; O'Brien, D.R.; Miller, R.L.; Cole, T.;

Schoch, K.M.; Shen, T.; et al. MicroRNA profiling reveals marker of motor neuron disease in ALS models. *J. Neurosci.* **2017**, *37*, 5574–5586. [CrossRef] [PubMed]

61. Kos, A.; Olde Loohuis, N.F.; Wieczorek, M.L.; Glennon, J.C.; Martens, G.J.; Kolk, S.M.; Aschrafi, A. A potential regulatory role for intronic microRNA-338-3p for its host gene encoding apoptosis-associated tyrosine kinase. *PLoS ONE* **2012**, *7*, e31022. [CrossRef] [PubMed]

62. Chen, X.; Pan, M.; Han, L.; Lu, H.; Hao, X.; Dong, Q. miR-338-3p suppresses neuroblastoma proliferation, invasion and migration through targeting PREX2a. *FEBS Lett.* **2013**, *587*, 3729–3737. [CrossRef] [PubMed]

63. Conti, A.; Aguennouz, M.; La Torre, D.; Tomasello, C.; Cardali, S.; Angileri, F.F.; Maio, F.; Cama, A.; Germanò, A.; Vita, G.; et al. miR-21 and 221 upregulation and miR-181b downregulation in human grade II-IV astrocytic tumors. *J. Neurooncol.* **2009**, *93*, 325–332. [CrossRef] [PubMed]

64. Johanson, C.E.; Stopa, E.G.; McMillan, P.N. The blood-cerebrospinal fluid barrier: Structure and functional significance. *Methods Mol. Biol.* **2011**, *686*, 101–131. [CrossRef] [PubMed]

65. Spector, R.; Robert Snodgrass, S.; Johanson, C.E. A balanced view of the cerebrospinal fluid composition and functions: Focus on adult humans. *Exp. Neurol.* **2015**, *273*, 57–68. [CrossRef] [PubMed]

66. Vu, L.T.; Bowser, R. Fluid-based biomarkers for amyotrophic lateral sclerosis. *Neurotherapeutics* **2017**, *14*, 119–134. [CrossRef] [PubMed]

67. Toivonen, J.M.; Manzano, R.; Oliván, S.; Zaragoza, P.; García-Redondo, A.; Osta, R. MicroRNA-206: A potential circulating biomarker candidate for amyotrophic lateral sclerosis. *PLoS ONE.* **2014**, *9*, e89065. [CrossRef] [PubMed]

68. Freischmidt, A.; Müller, K.; Zondler, L.; Weydt, P.; Volk, A.E.; Božič, A.L.; Walter, M.; Bonin, M.; Mayer, B.; von Arnim, C.A.; et al. Serum microRNAs in patients with genetic amyotrophic lateral sclerosis and pre-manifest mutation carriers. *Brain* **2014**, *137*, 2938–2950. [CrossRef] [PubMed]

69. Freischmidt, A.; Müller, K.; Zondler, L.; Weydt, P.; Mayer, B.; von Arnim, C.A.; Hübers, A.; Dorst, J.; Otto, M.; Holzmann, K.; et al. Serum microRNAs in sporadic amyotrophic lateral sclerosis. *Neurobiol. Aging* **2015**, *36*, 2660.e15–2660.e20. [CrossRef] [PubMed]

70. Takahashi, I.; Hama, Y.; Matsushima, M.; Hirotani, M.; Kano, T.; Hohzen, H.; Yabe, I.; Utsumi, J.; Sasaki, H. Identification of plasma microRNAs as a biomarker of sporadic amyotrophic lateral Sclerosis. *Mol. Brain* **2015**, *8*, 67. [CrossRef] [PubMed]

71. Chen, Y.; Wei, Q.; Chen, X.; Li, C.; Cao, B.; Ou, R.; Hadano, S.; Shang, H.F. Aberration of miRNAs expression in leukocytes from sporadic amyotrophic lateral sclerosis. *Front. Mol. Neurosci.* **2016**, *9*, 69. [CrossRef] [PubMed]

72. De Andrade, H.M.; de Albuquerque, M.; Avansini, S.H.; de S Rocha, C.; Dogini, D.B.; Nucci, A.; Carvalho, B.; Lopes-Cendes, I.; França, M.C., Jr. MicroRNAs-424 and 206 are potential prognostic markers in spinal onset amyotrophic lateral sclerosis. *J. Neurol. Sci.* **2016**, *368*, 19–24. [CrossRef] [PubMed]

73. Tasca, E.; Pegoraro, V.; Merico, A.; Angelini, C. circulating microRNAs as biomarkers of muscle differentiation and atrophy in ALS. *Clin. Neuropathol.* **2016**, *35*, 22–30. [CrossRef] [PubMed]

74. Sheinerman, K.S.; Toledo, J.B.; Tsivinsky, V.G.; Irwin, D.; Grossman, M.; Weintraub, D.; Hurtig, H.I.; Chen-Plotkin, A.; Wolk, D.A.; McCluskey, L.F.; et al. Circulating brain-enriched microRNAs as novel biomarkers for detection and differentiation of neurodegenerative diseases. *Alzheimers Res. Ther.* **2017**, *9*, 89. [CrossRef] [PubMed]

75. Raheja, R.; Regev, K.; Healy, B.C.; Mazzola, M.A.; Beynon, V.; Von Glehn, F.; Paul, A.; Diaz-Cruz, C.; Gholipour, T.; Glanz, B.I.; et al. Correlating serum micrornas and clinical parameters in amyotrophic lateral sclerosis. *Muscle Nerve* **2018**, *58*, 261–269. [CrossRef] [PubMed]

76. Waller, R.; Goodall, E.F.; Milo, M.; Cooper-Knock, J.; Da Costa, M.; Hobson, E.; Kazoka, M.; Wollff, H.; Heath, P.R.; Shaw, P.J. Serum miRNAs miR-206, 143–3p and 374b-5p as potential biomarkers for amyotrophic lateral sclerosis (ALS). *Neurobiol. Aging* **2017**, *55*, 123–131. [CrossRef] [PubMed]

77. Vrabec, K.; Boštjančič, E.; Koritnik, B.; Leonardis, L.; Dolenc Grošelj, L.; Zidar, J.; Rogelj, B.; Glavač, D.; Ravnik-Glavač, M. Differential expression of several miRNAs and the host genes AATK and DNM2 in leukocytes of sporadic ALS patients. *Front. Mol. Neurosci.* **2018**, *11*, 106. [CrossRef] [PubMed]

78. Matamala, J.M.; Arias-Carrasco, R.; Sanchez, C.; Uhrig, M.; Bargsted, L.; Matus, S.; Maracaja-Coutinho, V.; Abarzua, S.; van Zundert, B.; Verdugo, R. Genome-wide circulating microRNA expression profiling reveals potential biomarkers for amyotrophic lateral sclerosis. *Neurobiol. Aging* **2018**, *64*, 123–138. [CrossRef] [PubMed]

79. Xu, Q.; Zhao, Y.; Zhou, X.; Luan, J.; Cui, Y.; Han, J. Comparison of the extraction and determination of serum exosome and miRNA in serum and the detection of miR-27a-3p in serum exosome of ALS patients. *Intractable Rare Dis. Res.* **2018**, *7*, 13–18. [CrossRef] [PubMed]

80. Liguori, M.; Nuzziello, N.; Introna, A.; Consiglio, A.; Licciulli, F.; D'Errico, E.; Scarafino, A.; Distaso, E.; Simone, I.L. Dysregulation of MicroRNAs and target genes networks in peripheral blood of patients with sporadic amyotrophic lateral sclerosis. *Front. Mol. Neurosci.* **2018**, *11*, 288. [CrossRef] [PubMed]

81. Taguchi, Y.H.; Wang, H. Exploring microRNA biomarker for amyotrophic lateral sclerosis. *Int. J. Mol. Sci.* **2018**, *19*, 1318. [CrossRef] [PubMed]

82. Bang, C.; Thum, T. Exosomes: New players in cell-cell communication. *Int. J. Biochem. Cell. Biol.* **2012**, *44*, 2060–2064. [CrossRef] [PubMed]

83. Cui, Y.; Luan, J.; Li, H.; Zhou, X.; Han, J. Exosomes derived from mineralizing osteoblasts promote ST2 cell osteogenic differentiation by alteration of microRNA expression. *FEBS Lett.* **2016**, *590*, 185–192. [CrossRef] [PubMed]

84. Sharma, M.; Juvvuna, P.K.; Kukreti, H.; McFarlane, C. Mega roles of microRNAs in regulation of skeletal muscle health and disease. *Front. Physiol.* **2014**, *5*, 239. [CrossRef] [PubMed]

85. Ma, G.; Wang, Y.; Li, Y.; Cui, L.; Zhao, Y.; Zhao, B.; Li, K. MiR-206, a key modulator of skeletal muscle development and disease. *Int. J. Biol. Sci.* **2015**, *11*, 345–352. [CrossRef] [PubMed]

86. Williams, A.H.; Valdez, G.; Moresi, V.; Qi, X.; McAnally, J.; Elliott, J.L.; Bassel-Duby, R.; Sanes, J.R.; Olson, E.N. MicroRNA-206 delays ALS progression and promotes regeneration of neuromuscular synapses in mice. *Science* **2009**, *326*, 1549–1554. [CrossRef] [PubMed]

87. Loeffler, J.P.; Picchiarelli, G.; Dupuis, L.; Gonzalez De Aguilar, J.L. The role of skeletal muscle in amyotrophic lateral sclerosis. *Brain Pathol.* **2016**, *26*, 227–236. [CrossRef] [PubMed]

88. Horak, M.; Novak, J.; Bienertova-Vasku, J. Muscle-specific microRNAs in skeletal muscle development. *Dev. Biol.* **2016**, *410*, 1–13. [CrossRef] [PubMed]

89. Di Pietro, L.; Lattanzi, W.; Bernardini, C. Skeletal muscle MicroRNAs as key players in the pathogenesis of amyotrophic lateral sclerosis. *Int. J. Mol. Sci.* **2018**, *19*, 1534. [CrossRef] [PubMed]

90. Bruneteau, G.; Simonet, T.; Bauché, S.; Mandjee, N.; Malfatti, E.; Girard, E.; Tanguy, M.L.; Behin, A.; Khiami, F.; Sariali, E.; et al. Muscle histone deacetylase 4 upregulation in amyotrophic lateral sclerosis: potential role in reinnervation ability and disease progression. *Brain* **2013**, *136*, 2359–2368. [CrossRef] [PubMed]

91. Russell, A.P.; Wada, S.; Vergani, L.; Hock, M.B.; Lamon, S.; Léger, B.; Ushida, T.; Cartoni, R.; Wadley, G.D.; Hespel, P.; et al. Disruption of skeletal muscle mitochondrial network genes and miRNAs in amyotrophic lateral sclerosis. *Neurobiol. Dis.* **2013**, *49*, 107–117. [CrossRef] [PubMed]

92. Jensen, L.; Jørgensen, L.H.; Bech, R.D.; Frandsen, U.; Schrøder, H.D. Skeletal muscle remodelling as a function of disease progression in amyotrophic lateral sclerosis. *Biomed. Res. Int.* **2016**, *2016*, 5930621. [CrossRef] [PubMed]

93. Pegoraro, V.; Merico, A.; Angelini, C. Micro-RNAs in ALS muscle: Differences in gender, age at onset and disease duration. *J. Neurol. Sci.* **2017**, *380*, 58–63. [CrossRef] [PubMed]

94. Si, Y.; Cui, X.; Crossman, D.K.; Hao, J.; Kazamel, M.; Kwon, Y.; King, PH. Muscle microRNA signatures as biomarkers of disease progression in amyotrophic lateral sclerosis. *Neurobiol. Dis.* **2018**, *114*, 85–94. [CrossRef] [PubMed]

95. Kovanda, A.; Leonardis, L.; Zidar, J.; Koritnik, B.; Dolenc-Groselj, L.; Ristic Kovacic, S.; Curk, T.; Rogelj, B. Differential expression of microRNAs and other small RNAs in muscle tissue of patients with ALS and healthy age-matched controls. *Sci. Rep.* **2018**, *8*, 5609. [CrossRef] [PubMed]

96. Di Pietro, L.; Baranzini, M.; Berardinelli, M.G.; Lattanzi, W.; Monforte, M.; Tasca, G.; Conte, A.; Logroscino, G.; Michetti, F.; Ricci, E.; et al. Potential therapeutic targets for ALS: MIR206, MIR208b and MIR499 are modulated during disease progression in the skeletal muscle of patients. *Sci. Rep.* **2017**, *7*, 9538. [CrossRef] [PubMed]

3

Expression of miR159 is Altered in Tomato Plants Undergoing Drought Stress

María José López-Galiano [1], Inmaculada García-Robles [1], Ana I. González-Hernández [2], Gemma Camañes [2], Begonya Vicedo [2], M. Dolores Real [1] and Carolina Rausell [1,*]

[1] Department of Genetics, University of Valencia, Burjassot, 46100 Valencia, Spain
[2] Plant Physiology Area, Biochemistry and Biotechnology Group, Department CAMN, University Jaume I, 12071 Castellón, Spain
*   Correspondence: carolina.rausell@uv.es;

Abstract: In a scenario of global climate change, water scarcity is a major threat for agriculture, severely limiting crop yields. Therefore, alternatives are urgently needed for improving plant adaptation to drought stress. Among them, gene expression reprogramming by microRNAs (miRNAs) might offer a biotechnologically sound strategy. Drought-responsive miRNAs have been reported in many plant species, and some of them are known to participate in complex regulatory networks via their regulation of transcription factors involved in water stress signaling. We explored the role of miR159 in the response of *Solanum lycopersicum* Mill. plants to drought stress by analyzing the expression of sly-miR159 and its target SlMYB transcription factor genes in tomato plants of cv. Ailsa Craig grown in deprived water conditions or in response to mechanical damage caused by the Colorado potato beetle, a devastating insect pest of Solanaceae plants. Results showed that sly-miR159 regulatory function in the tomato plants response to distinct stresses might be mediated by differential stress-specific MYB transcription factor targeting. sly-miR159 targeting of SlMYB33 transcription factor transcript correlated with accumulation of the osmoprotective compounds proline and putrescine, which promote drought tolerance. This highlights the potential role of sly-miR159 in tomato plants' adaptation to water deficit conditions.

Keywords: *Solanum lycopersicum*; drought; Colorado potato beetle; miR159; MYB transcription factors; *P5CS*; proline; putrescine

1. Introduction

Climate change due to increasing concentration of CO_2 in the atmosphere is leading to rising temperatures, altered rainfall patterns, and more frequent and severe drought episodes [1], which negatively impact crop production. Therefore, gaining knowledge about how plants regulate their adaptation to stress is critical to find ways to enhance plant performance in eventually drier environments.

To cope with drought, plants activate a complex cascade of events at the cellular level that include extensive metabolic and gene transcriptional reprogramming to protect cells from osmotic stress, and limit water loss. The response of plants to drought stress involves genes related to diverse functional categories such as genes encoding proteins participating in the direct protection of essential proteins and membranes (osmoprotectants, free radical scavengers, etc.), genes encoding membrane transporters and ion channels that promote water uptake, and genes encoding stress related regulatory proteins such as kinases and transcription factors belonging to the V-myb myeloblastosis viral oncogene homolog (MYB), basic-helix-loop-helix (bHLH), basic region/leucine zipper (bZIP), NAM, ATAF1/2, and CUC (NAC), and APETALA2/ethylene-responsive element binding protein (AP2/EREBP) families [2].

The phytohormone Abscisic acid (ABA) coordinates the plant's response to reduced water availability by modulating the expression of some of the drought responsive genes [3]. Interestingly, microRNAs (miRNAs) have been recently reported to mediate drought tolerance by post-transcriptionally regulating drought-responsive genes, some of which are known to be controlled by ABA signaling pathways [4]. An example of such intricate regulatory network is provided by miR159, which in *Arabidopsis* germinating seeds, has been reported to be induced by ABA and drought treatments, and promote transcript cleavage of the ABA positive regulators MYB33 and MYB101 transcription factors, thereby playing a key role in ABA response [5].

The miR159 family is highly conserved among monocot and dicot plants, but in plants undergoing drought, miR159's relative abundance varies in a tissue- and species-specific manner. For instance, miR159 was reported to be up-regulated by drought stress in *Arabidopsis* [6], and maize [7], but down-regulated in cotton [8], and potato [9], whereas in barley and alfalfa, miR159 was down-regulated in roots and up-regulated in leaves in response to drought stress [10,11]. Pegler et al. [12] proposed that the differential miRNA abundance across species following drought or salt stress exposure might be in part due to differential distribution of regulatory transcription factor binding sites within the putative promoter region of the miRNA gene, which encodes the highly conserved, stress-responsive miRNA.

To expand our knowledge on the miR159 regulatory network involved in tomato plants' response to drought stress, in the present work we analyzed the expression of miR159 and its predicted target genes in tomato plants of *Solanum lycopersicum* Mill. cv. Ailsa Craig undergoing drought stress, in which we previously reported that ABA hormone is accumulated after water deprivation [13].

2. Results and Discussion

2.1. Expression of miR159 in Tomato Plants Undergoing Drought Stress

To assess miR159 expression in tomato plants of *Solanum lycopersicum* Mill. cv. Ailsa Craig following a seven-day water deprivation, we analyzed sly-miR159 (GenBank: 102464332) transcript levels by RT-qPCR in control tomato plants and plants undergoing drought stress. Results showed significantly reduced expression of sly-miR159 in response to stress (Figure 1A). However, in recent high-throughput sequencing studies performed by Liu et al. [14,15], miR159 was not found among the miRNAs differentially expressed after 10 days of drought stress in a sensitive and a tolerant tomato cultivar. This apparent discrepancy with our results might be due to the differential experimental conditions or techniques used to measure miRNA expression, but is most probably due to the fact that the tomato cultivars were different, since it has been reported that miRNAs respond to environmental stresses in a genotype-dependent manner [16]. As in plants, most miRNAs negatively regulate their target genes, we hypothesized that sly-miR159 gene targets that are upregulated in tomato plants grown in water-limited conditions in our experimental conditions may play beneficial roles in the adaptive responses to drought stress.

In *Arabidopsis*, a clade of seven closely related *GAMYB*-like genes (*MYB33, MYB101, MYB65, MYB81, MYB97, MYB104*, and *MYB120*) share a conserved putative miR159-binding site [17]. The *GAMYB*-like genes encode a highly conserved family of R2R3-type MYB domain transcription factors that are regulated by Gibberellic acid (GA) and ABA and participate in the GA signaling pathway [18]. Recent studies in potato plants highlight the involvement of miR159 and its targets *GAMYB*-like genes in the response of this species to water stress [9]. Using psRNATarget software [19] we identified the following putative *GAMYB*-like transcription factor genes that are sly-miR159 targets in tomato: *SlMYB33* (Solyc01g009070.2.1), *SlMYB65* (Solyc06g073640.2.1), *SlMYB104* (Solyc11g072060.1.1), *SlMYB97* (Solyc10g019260.1.1), and *SlMYB120* (Solyc01g090530.1.1). Figure 1B shows the nucleotide sequence of the sly-miR159-binding sites in the tomato *SlMYB* transcripts identified, which strongly resemble those found in *AtMYB* transcripts targeted by miR159 in *Arabidopsis* [20].

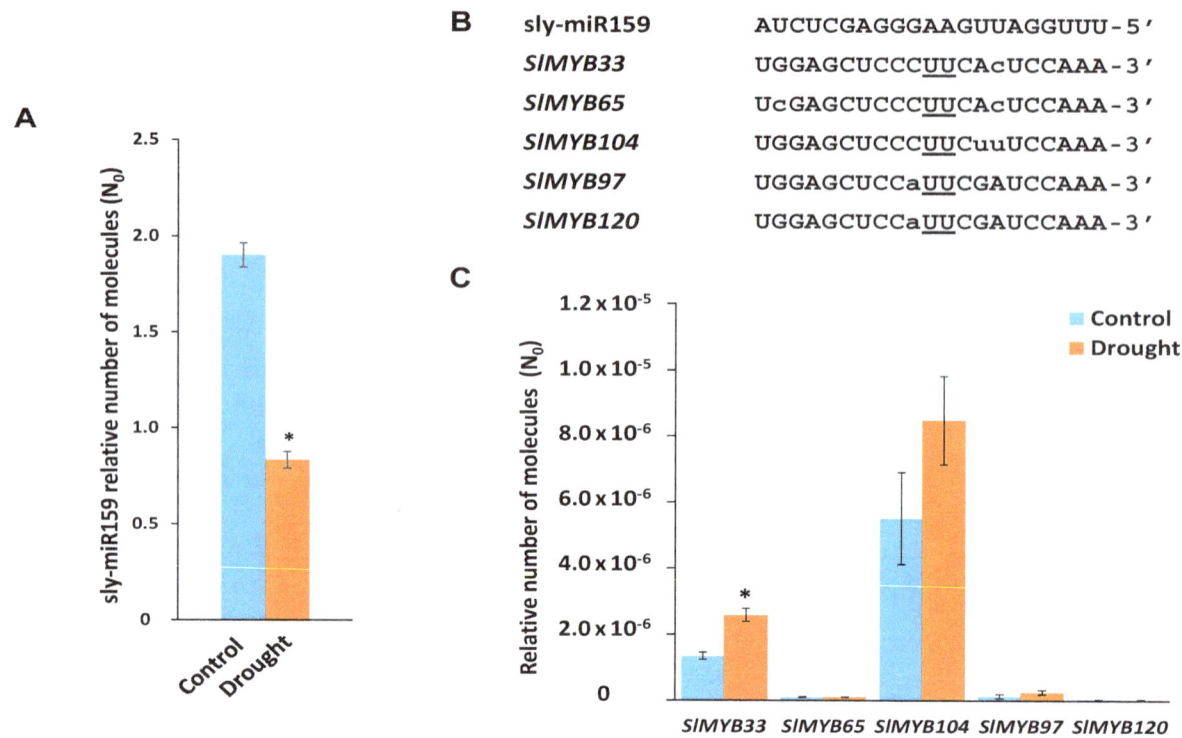

Figure 1. RT-qPCR analysis of sly-miR159 expression and its *MYB* predicted targets in tomato plants undergoing drought stress. (**A**) RT-qPCR analysis of sly-miR159 expression in control tomato plants and tomato plants following 7-day water deprivation. (**B**) Nucleotide sequence of sly-miR159-binding sites in tomato GAMYB-like transcripts. Nucleotides in the cleavage site are underlined, lower-case red letters indicate mismatches to sly-miR159, and G:U pairing is shown in uppercase green letters. (**C**) RT-qPCR analysis of *SlMYB33, SlMYB65, SlMYB104, SlMYB97*, and *SlMYB120* genes expression in control tomato plants and tomato plants following 7-day water deprivation. In panels (A) and (C), data shown are the mean of three independent experiments ± standard error (SE). Asterisk indicates that differences between means of control and undergoing drought stress tomato plants were statistically significant (Student's *t*-test, $p < 0.05$).

Li et al. [21] identified 127 *MYB* genes in the tomato genome and classified the corresponding proteins into 18 subgroups based on domain similarity and phylogenetic topology, and suggested that conserved motifs outside the MYB domain might reflect their functional conservation. SlMYB33, SlMYB65, and SlMYB104 proteins cluster in subgroup 12, in which the three of them are the only ones (out of the thirteen subgroup members) sharing the conserved motifs 14 and 15 outside the MYB domain. SlMYB97 and SlMYB120 proteins constitute subgroup 15, which is composed only by these two MYB proteins that have no conserved motifs outside the MYB domain.

We analyzed the expression of sly-miR159 *MYB* predicted targets in control tomato plants and plants undergoing drought stress by RT-qPCR (Figure 1C). Only *SlMYB33* gene showed statistically significant induction in water-stressed tomato plants, exhibiting an opposite pattern of expression relative to that of sly-miR159, which suggests that this *MYB* gene may be regulated by sly-miR159 in tomato plants in response to drought stress. In line with this hypothesis, in potato plants in which the *CBP80* gene encoding a protein involved in RNA processing was silenced, improved tolerance to water stress was correlated with decreased levels of miR159 and enhanced *MYB33* gene expression [22].

To further assess the involvement of sly-miR159 in the regulation of *SlMYB33* gene expression under drought stress, we aimed at analyzing *SlMYB33* cleavage fragments. We designed two pairs of primers to amplify *SlMYB33* mRNA fragments in small RNA samples isolated from total RNA of control tomato plants and tomato plants following a seven-day water deprivation (Materials and Methods, Section 4.3). Figure 2A shows the annealing positions of both pairs of PCR primers. The primer pair

O_{Fw} and O_{Rv} anneals to sequences within a *SlMYB33* mRNA region downstream of the predicted sly-miR159-binding site, yielding a 199 bp *SlMYB33* amplification product. The primer pair F_{Fw} and F_{Rv} anneals to sequences flanking the putative cleavage site in the predicted sly-miR159-binding region, yielding a 200 bp *SlMYB33* amplification product only when the SlMYB33 mRNA is not cleaved at the sly-miR159 cleavage site. Therefore, we hypothesized that if sly-miR159 is not involved in the regulation of *SlMYB33* gene expression of the same amplification patterns of control vs. drought, then small RNA samples with both primer pairs would be expected. Figure 2B shows the results obtained in the RT-PCR amplifications using the two pairs of primers. Lower amounts of amplification products were obtained using primers O_{Fw} and O_{Rv} in tomato plants grown under water scarcity compared to control plants. In contrast, higher amount of PCR amplified product was observed in drought-stressed tomato plants than in control tomato plants using primers F_{Fw} and F_{Rv}. Collectively, these results support targeted cleavage of *SlMYB33* transcripts by sly-miR159 that might participate in the transcriptional regulation of the tomato plants' response to drought stress.

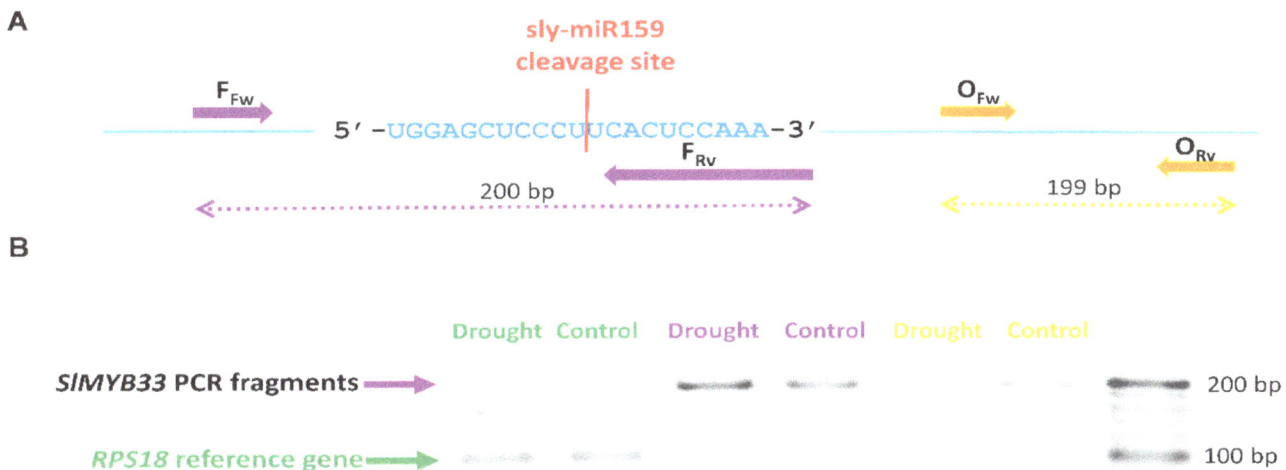

Figure 2. RT-PCR amplification of SlMYB33 mRNA fragments. (**A**) Nucleotide sequence of sly-miR159-binding sites in SlMYB33 transcripts. Bar in red depicts the putative cleavage site and arrows indicate the annealing positions of primer pair OFw and ORv, outside the sly-miR159-binding region, and primer pair FFw and FRv, flanking the putative cleavage site in sly-miR159-binding region. (**B**) RT-PCR analysis of SlMYB33 small RNA fragments in control tomato plants and tomato plants following 7-day water deprivation using primers OFw and ORv, or FFw and FRv. RPS18 gene expression was used as normalization control. For each sample, three biological replicates were pooled and analyzed.

Interestingly, Qin et al. [23] proposed that MYB33 transcription factor may enhance drought tolerance by means of promoting osmotic pressure balance reconstruction and reactive oxidative species (ROS) scavenging, since ectopic over-expression of wheat *MYB33* gene in *Arabidopsis* induced the expression of *AtP5CS* and *AtZAT12* genes involved in proline synthesis and ascorbate peroxidase synthesis, respectively. Accordingly, we observed an induction of *SlP5CS* gene expression and a remarkable increase in proline levels relative to other amino acids in tomato plants grown in water-shortage conditions compared to irrigated control plants (Figure 3A,B), suggesting that sly-miR159 might participate in the tomato plants' adaptive response to drought stress via induction of *SlMYB33* transcription factor gene expression. Nevertheless, further research is needed to demonstrate whether the sly-miR159-SlMYB33 pathway is necessary for drought tolerance in the tomato cultivar Ailsa Craig.

Tonon et al. [24] proposed a strong metabolic coordination between polyamines and proline pathways in response to osmotic stresses. Therefore, we analyzed polyamine levels in tomato plants undergoing drought stress and non-stressed control plants (Figure 3C), and results showed increased accumulation of putrescine, a polyamine reported to have a role in protecting plants during water-deficient conditions, as well as oxidative stress [25]. In wheat, Pál et al. [26] recently described that ABA pre-treatments induced the expression of *P5CS* gene and enhanced the accumulation of putrescine. Authors suggested that the connection between polyamine metabolism and ABA signaling may control the regulation and maintenance of polyamine and proline levels under osmotic stress conditions in wheat seedlings.

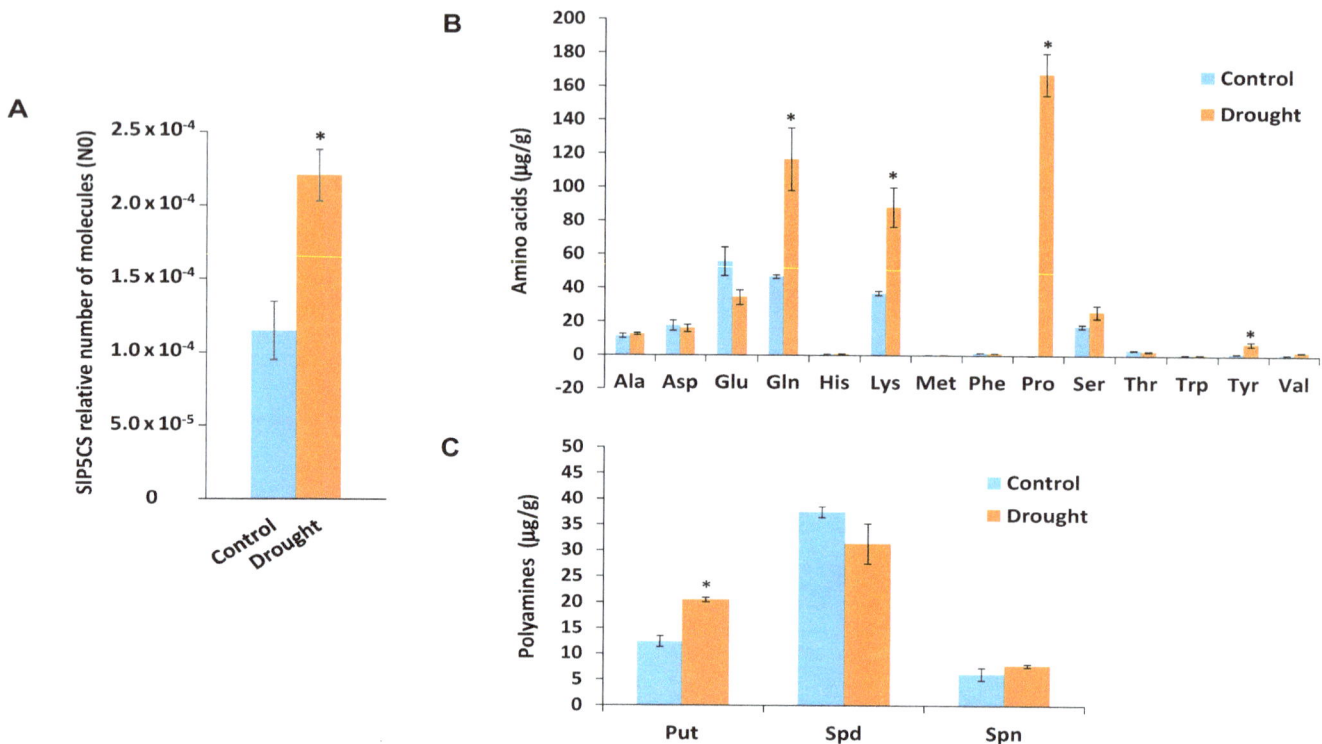

Figure 3. Analysis of *SlP5CS* gene expression, and amino acid and polyamines in tomato plants undergoing drought stress. (**A**) RT-qPCR analysis of *SlP5CS* expression in control tomato plants and tomato plants following 7-day water deprivation. (**B**) Amino acids levels upon drought treatment. Amino acids levels are expressed in µg/g DW. (**C**) Polyamines levels upon drought treatment. Polyamines levels are expressed in µg/g DW. Put (putrescine), Spd (Spermidine), Spn (Spermine). Tomato leaves were collected from plants that were properly irrigated (Control) or deprived of water 1 week (Drought). Data shown are the mean of three independent experiments ± standard error (SE). Asterisk indicates that differences between means of control and undergoing drought stress tomato plants were statistically significant (Student's *t*-test, $p < 0.05$).

2.2. Assessment of sly-miR159 Stress-Specific Targeting of SlMYB33

To ascertain whether *SlMYB33* targeting by sly-miR159 is stress-specific, we analyzed the expression of sly-miR159 and its predicted MYB target genes in tomato plants attacked by the coleopteran insect pest Colorado potato beetle (CPB), in which we previously reported that, as opposed to tomato plants undergoing drought stress, ABA was not accumulated [13]. In the present work, neither *SlP5CS* gene expression were induced, nor were increased proline and putrescine levels observed in infested tomato plants compared to tomato control plants (Figure 4), corroborating that the plants' response to this biotic stress is different from the plant response to water stress.

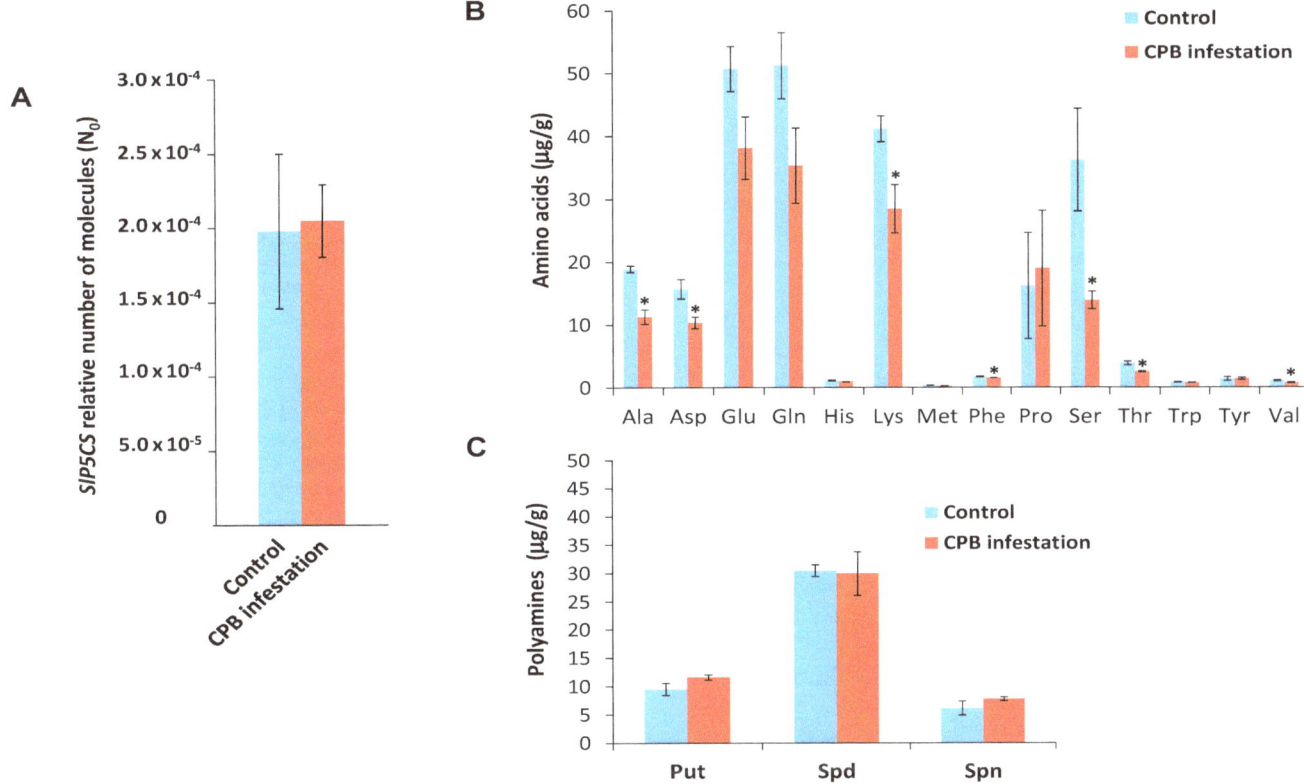

Figure 4. Analysis of *SlP5CS* gene expression, and amino acid and polyamines in tomato plants infested by Colorado potato beetle (CPB) larvae. (**A**) RT-qPCR analysis of *SlP5CS* expression in control tomato plants and tomato plants infested by CPB larvae. (**B**) Amino acids levels upon CPB infestation. Tomato leaves were collected from non-infested plants (Control) or plants infested by CPB. Amino acids levels are expressed in μg/g DW. (**C**) Polyamines levels upon CPB larvae infestation. Tomato leaves were collected from non-infested plants (Control) or plants infested by CPB. Polyamines levels are expressed in μg/g DW. Put (putrescine), Spd (Spermidine), Spn (Spermine). Data shown are the mean of three independent experiments ± standard error (SE). Asterisk indicates that differences between means of control and undergoing drought stress tomato plants were statistically significant (Student´s *t*-test, $p < 0.05$).

Intriguingly, as it was observed in plants deprived of water, in infested tomato plants, sly-miR159 was significantly down-regulated compared to non-infested control plants (Figure 5A). However, in plants attacked by CPB, among sly-miR159 putative *MYB* targets, only the *SlMYB104* transcript factor gene was significantly up-regulated (Figure 5B), suggesting that sly-miR159 might be regulating this specific MYB transcription factor in response to CPB damage.

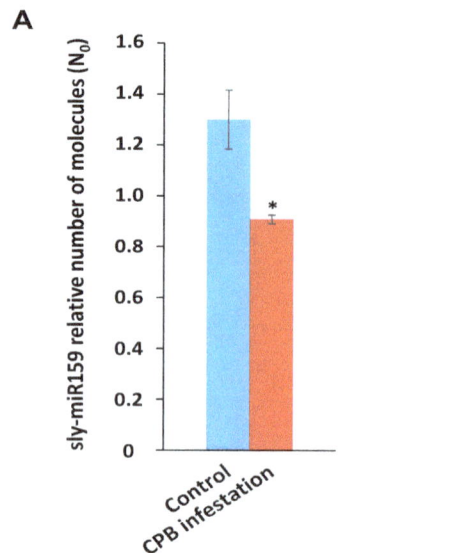
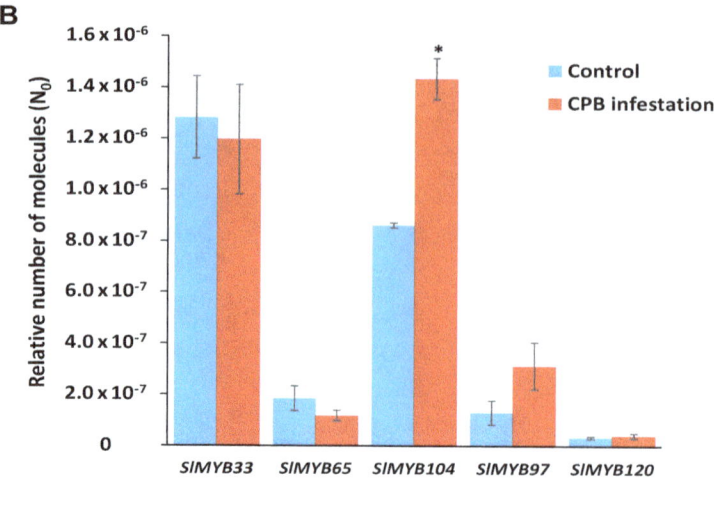

Figure 5. RT-qPCR analysis of sly-miR159 expression and its *MYB* predicted targets in tomato plants infested by CPB larvae. (**A**) RT-qPCR analysis of sly-miR159 expression in control tomato plants and tomato plants infested by CPB larvae. (**B**) RT-qPCR analysis of *SlMYB33, SlMYB65, SlMYB104, SlMYB97,* and *SlMYB120* genes expression in control tomato plants and tomato plants upon CPB larvae infestation. In panels (A) and (B), data shown are the mean of three independent experiments ± standard error (SE). Asterisk indicates that differences between means of control tomato plants and tomato plants infested by CPB larvae were statistically significant (Student´s *t*-test, $p < 0.05$).

In contrast, correlating with the lack of proline and putrescine accumulation, no variation was detected in *SlMYB33* transcription factor gene expression. This suggests that the specificity of the stress response regulated by sly-miR159 might, at least in part, rely on the distinct *MYB* transcription factor transcript that the sly-miR159 sRNA specifically regulates under each stress condition. It has been proposed that additional factors other than complementarity and cleavage, such as target accessibility and secondary structure, RNA binding proteins, and target site context may modulate silencing efficiency [27], which might lie at the root of the stress specific miR159 regulation of MYB transcription factors, and deserve further research.

3. Conclusions

Overall, the results obtained in this work show the potential involvement of sly-miR159 in the tomato plants' response to different stresses through stress-specific *MYB* transcription factor targeting. Under drought-stress, sly-miR159 targeting of *SlMYB33* correlates with induction of *SlP5CS* gene expression and accumulation of the osmoprotective compounds proline and putrescine, pointing to the possible participation of this miR in the regulation of drought stress tolerance. Understanding the regulatory network underlying drought stress response may provide new biotechnological approaches to generate plants better adapted to dry environments. Our results support that in addition to using *SlMYB33* transcription factor as a biotechnological target for metabolic engineering by ectopic expression, *SlMYB33* gene expression reprogramming by sly-miR159 might develop into a useful system to improve plant drought tolerance in tomato plants.

4. Materials and Methods

4.1. Plants

Thirty-day-old tomato plants of *Solanum lycopersicum* Mill. cv. Ailsa Craig (four-week-old) were grown from germinated seeds in a growth chamber under the following environmental conditions: 16/8 h light/night cycle, 26/18 °C day/night temperature cycle, and 60% relative humidity (RH). Seeds were irrigated twice a week with distilled water during the first week, and with Hoagland solution thereafter [28].

For drought stress experiments, thirty-day-old tomato plants were deprived of water for 7 days, and leaf tissue from 3rd and 4th leaves was collected, frozen in liquid nitrogen, and stored at −80 °C. Leaf tissue from 3rd and 4th leaves of irrigated plants was also collected as control.

For Colorado potato beetle (CPB) infestation, 15 CPB larvae of different developmental stages were placed on the 3rd and 4th leaves of thirty-day-old tomato plants. When necessary, non-cooperative larvae (molting or not eating) were removed and substituted. Leaf tissue left after 3 h of CPB feeding and that of the non-infested control plants were harvested, frozen in liquid nitrogen, and stored at −80 °C.

4.2. Total RNA Isolation and RT-qPCR Analysis

Total RNA was isolated from leaves of control tomato plants and plants undergoing drought stress or CPB infestation using RiboPure Kit (Ambion, Cat. No. AM1924), following the manufacturer's protocol. TURBO DNA-free kit (Ambion, Cat. No. AM1907) was used to remove contaminating genomic DNA from RNA preparations and RNA quality was evaluated by 1% agarose gel electrophoresis and quantified spectrophotometrically (NanoDrop 2000, Thermo Scientific, Waltham, MA, USA).

RT-qPCR amplification was performed using SYBR Premix Ex Taq II (Takara).

For sly-miR159 amplification 1 µg of RNA was polyadenylated in a final volume of 10 µL, including 1 µL of 10x poly(A) polymerase buffer, 1 mM of ATP, and 1 unit of poly(A) polymerase (New England Biolabs, Ipswich, MA, USA), and incubated at 37 °C for 15 min and then at 65 °C for 20 min. Polyadenylated RNA was reverse transcribed to complementary DNA (cDNA) using the Universal RT-primer (Integrated DNA Technologies, Coralville, IA, USA) described in Balcells et al. [29] (5′-CAGGTCCAGTTTTTTTTTTTTTTTVN-3′, where V is A, C, and G, and N is A, C, G, and T). Reverse transcription reaction was performed using PrimeScript™ RT reagent Kit (Takara) in a final volume of 10 µL, including 2 µL of 5X PrimeScript™ Buffer, 0.5 µL of PrimeScript™ RT Enzyme Mix I, and 1 µM of Universal RT-primer, and it was incubated at 37 °C for 15 min followed by enzyme inactivation at 85 °C for 5 s. Forward and reverse primers for miRNA RT-qPCR amplification were designed according to Balcells et al. [29] (Table 1).

Table 1. Primers used to analyze by RT-qPCR sly-miR159, *SlMYB*, and *SlP5CS* gene expression in tomato plants.

Gene	Forward Primer (5′-3′)	Reverse Primer (5′-3′)	Product Size (bp)
sly-miR159	CGCAGTTTGGATTGAAGGGAG	CAGGTCCAGTTTTTTTTTTTTTTTTAGAG	50
SlMYB33	TATGGGCATCCAGTCTCTCC	TGGGACTGGAAAAGATCGTC	199
SlMYB65	TCTGCTGCATCGGTGTTTAG	TCTGGCCTGGGACAGATAAG	164
SlMYB104	TTTCGGAATTGTTTGGAAGC	TGAAGAAGTTGCCGACAATG	110
SlMYB97	CATGTCCCCTTGGAAGATTTAG	CTAGTGGCAAAGCAAAGTCATC	181
SlMYB120	CACATTCCAGTCCAAACCAAC	CCTAGGTCGGAAGCACTGAG	116
SlP5CS	TGCTCAACAGGCCGGATATG	AAAGTGTGACCAAGGGGCTC	126
U6 snRNA	GGGGACATCCGATAAAATTGGAAC	TGGACCATTTCTCGATTTGTGC	88
RPS18	GGGCATTCGTATTTCATAGTCAGAG	CGGTTCTTGATTAATGAAAACATCCT	105

For *SlMYB33, SlMYB65, SlMYB104, SlMYB97, SlMYB120,* and *SlP5CS* transcript amplification, the PrimeScript™ RT reagent kit (Takara) was used for cDNA synthesis according to the manufacturer's protocol using 50 ng/µL oligo(dT) (Promega), and 2.5 µM random hexamers (Applied Biosystems). Ten ng cDNA, and gene specific forward (F) and reverse (R) primers (Table 1), designed with PRIMER3PLUS software [30], were used.

A StepOnePlus Real-Time PCR system (Applied Biosystems) was used, under the conditions recommended by the manufacturer, and the cycling parameters were: Initial polymerase activation step at 95 °C for 30 s, 40 cycles of denaturation at 95 °C for 5 s, annealing, and elongation at 60 °C for 30 s. For each sample, three biological replicates (with 3 technical replicates each) were analyzed. Relative-fold calculations were made using *RPS18* (ribosomal protein S18, GeneBank: 3950409) gene to normalize gene expression, and *U6* snRNA gene (GenBank: X51447.1) to normalize sly-miR159 expression (Table 1). LingReg software [31] was employed for the analysis of RT-qPCR experiments and data were analyzed by Student's *t*-test for statistically significant differences ($p < 0.05$).

Each biological sample from the 3rd and 4th leaves of plants undergoing drought stress and their corresponding controls consisted of a pool of total RNA from 25 plants. Biological samples in CPB infestation experiments and their corresponding controls also consisted of a pool of total RNA from 25 plants.

4.3. Small RNA Isolation and RT-PCR Analysis

The small RNA fraction in total RNA samples of control tomato plants and tomato plants following 7-day water deprivation was isolated using Nucleospin® miRNA (Macherey-Nagel, Bethlehem, PA, USA) following the manufacturer's instructions.

For *SlMYB33* small mRNA amplification, the PrimeScript™ RT reagent kit (Takara, Shiga, Japan) was used for cDNA synthesis according to the manufacturer's protocol using 50 ng/μL oligo(dT) (Promega, Madison, WI, USA) and 2.5 μM random hexamers (Applied Biosystems, Waltham, MA, USA), 10 ng cDNA, and gene specific forward (F) and reverse (R) primers (Table 2), designed with PRIMER3PLUS software [30]. *RPS18* (ribosomal protein S18, GeneBank: 3950409) was used as a reference gene.

Table 2. Primers used to analyze *SlMYB33* small transcript fragments by RT-PCR in tomato plants annealing to a region outside the predicted sly-miR159 binding site in *SlMYB33* mRNA (O_{Fw}, O_{Rv}) or flanking the putative cleavage site within the predicted sly-miR159 binding site in *SlMYB33* mRNA (F_{Fw}, F_{Rv}).

Primer Pair	Forward Primer (5′-3′)	Reverse Primer (5′-3′)	Product Size (bp)
O_{Fw}, O_{Rv}	TATGGGCATCCAGTCTCTCC	TGGGACTGGAAAAGATCGTC	199
F_{Fw}, F_{Rv}	ATGACGGTTCTTTGCTTGCT	CTGTCTGGTTTTGGAGTGAAGG	200
$RPS18_{FW}$, $RPS18_{RV}$	GGGCATTCGTATTTCATAGTCAGAG	CGGTTCTTGATTAATGAAAACATCCT	105

The cycling parameters were as follows: Initial polymerase activation step at 95 °C for 30 s, 40 cycles of denaturation at 95 °C for 5 s, annealing, and elongation at 60 °C for 30 s. For each sample, three biological replicates were pooled and analyzed. Five microliters of the reaction volume were separated in a 3% agarose gel.

4.4. Amino Acids and Polyamines Quantification

Leaves were recollected after stress condition and frozen in liquid N_2, ground, and lyophilized.

For amino acids analysis, dry tissue (0.1 g) was homogenized with 800 μL of extraction solution: 400 μL of distilled water, 200 μL of chloroform, and 200 μL of methanol per sample. Moreover, a mixture of internal standards was added prior to extraction (100 ng of Phe $^{13}C_9^{15}N$ and 100 ng of Thr $^{13}C_4^{15}N$). Samples were filtered, and a final concentration of 1 mM perfluoroheptanoic acid as ion-pairing reagent was added to each sample. A 20 μL aliquot was injected into a high-performance liquid chromatography system (HPLC) with an XSelect HSS C18 column (5 μm 2.1 × 100 mm) which was interfaced with a triple quadrupole mass spectrometer (TQD, Waters, Manchester, UK).

Polyamine analysis was conducted according to the method described by Sánchez-López et al. [32], using as internal standards a mixture of $[^{13}C_4]$-putrescine and 1,7-diaminoheptane. To analyze each condition, ten independent biological replicates per sample were generated and three independent experiments were conducted.

Author Contributions: Conceptualization, I.G.-R., M.D.R. and C.R.; Funding acquisition, I.G.-R., G.C., B.V., M.D.R. and C.R.; investigation, M.J.L.-G., I.G.-R., A.I.G.-H. and G.C.; writing—original draft, M.D.R. and C.R.; writing—review and editing, M.J.L.-G., I.G.-R., G.C. and B.V.

Acknowledgments: We thank the Genomics, Proteomics and Greenhouse Facilities from the SCSIE of the University of Valencia and the SCIC from the University Jaume I.

References

1. Swann, A.L.S. Plants and drought in a changing climate. *Curr. Clim. Chang. Rep.* **2018**, *4*, 192–201. [CrossRef]

2. Hossain, M.A.; Wani, S.H.; Bhattacharjee, S.; Burritt, D.J.; Tran, L.-S.P. *Drought Stress Tolerance in Plants*, 1st ed.; Springer International Publishing: Basel, Switzerland, 2016; Volume 2, ISBN 978-3-319-32421-0.

3. Cutler, S.R.; Rodriguez, P.R.; Finkelstein, R.R.; Abrams, S.R. Abscisic acid: Emergence of a core signaling network. *Annu. Rev. Plant Biol.* **2010**, *61*, 651–679. [CrossRef] [PubMed]

4. Ding, Y.; Tao, Y.; Zhu, C. Emerging roles of microRNAs in the mediation of drought stress response in plants. *J. Exp. Bot.* **2013**, *64*, 3077–3086. [CrossRef] [PubMed]

5. Reyes, J.L.; Chua, N.-H. ABA induction of miR159 controls transcript levels of two MYB factors during *Arabidopsis* seed germination. *Plant J.* **2007**, *49*, 592–606. [CrossRef] [PubMed]

6. Liu, H.-H.; Tian, X.; Li, Y.-J.; Wu, C.-A.; Zheng, C.-C. Microarray-based analysis of stress-regulated microRNAs in *Arabidopsis thaliana*. *RNA* **2008**, *14*, 836–843. [CrossRef]

7. Wei, L.; Zhang, D.; Xiang, F.; Zhang, Z. Differentially expressed miRNAs potentially involved in the regulation of defense mechanism to drought stress in maize seedlings. *Int. J. Plant Sci.* **2009**, *170*, 979–989. [CrossRef]

8. Xie, F.; Wang, Q.; Sun, R.; Zhang, B. Deep sequencing reveals important roles of microRNAs in response to drought and salinity stress in cotton. *J. Exp. Bot.* **2015**, *66*, 789–804. [CrossRef]

9. Yang, J.; Zhang, N.; Mi, X.; Wu, L.; Ma, R.; Zhu, X.; Yao, L.; Jin, X.; Si, H.; Wang, D. Identification of miR159s and their target genes and expression analysis under drought stress in potato. *Comput. Biol. Chem.* **2014**, *53*, 204–213. [CrossRef]

10. Hackenberg, M.; Gustafson, P.; Langridge, P.; Shi, B.-J. Differential expression of microRNAs and other small RNAs in barley between water and drought conditions. *Plant Biotechnol. J.* **2015**, *13*, 2–13. [CrossRef]

11. Li, Y.; Wan, L.; Bi, S.; Wan, X.; Li, Z.; Cao, J.; Tong, Z.; Xu, H.; He, F.; Li, X. Identification of drought-responsive microRNAs from roots and leaves of alfalfa by high-throughput sequencing. *Genes* **2017**, *8*, 119. [CrossRef]

12. Pegler, J.L.; Grof, C.P.L.; Eamens, A.L. Profiling of the differential abundance of drought and salt stress-responsive microRNAs across grass crop and genetic model plant species. *Agronomy* **2018**, *8*, 118. [CrossRef]

13. López-Galiano, M.J.; González-Hernández, A.I.; Crespo-Salvador, O.; Rausell, C.; Real, M.D.; Escamilla, M.; Camañes, G.; García-Agustín, P.; González-Bosch, C.; García-Robles, I. Epigenetic regulation of the expression of WRKY75 transcription factor in response to biotic and abiotic stresses in Solanaceae plants. *Plant Cell Rep.* **2018**, *37*, 167–176. [CrossRef] [PubMed]

14. Liu, M.; Yu, H.; Zhao, G.; Huang, Q.; Lu, Y.; Ouyang, B. Profiling of drought-responsive microRNA and mRNA in tomato using high-throughput sequencing. *BMC Genom.* **2017**, *18*, 481. [CrossRef]

15. Liu, M.; Yu, H.; Zhao, G.; Huang, Q.; Lu, Y.; Ouyang, B. Identification of drought-responsive microRNAs in tomato using high-throughput sequencing. *Funct. Integr. Genom.* **2018**, *18*, 67–78. [CrossRef] [PubMed]

16. Zhang, B. MicroRNA: A new target for improving plant tolerance to abiotic stress. *J. Exp. Bot.* **2015**, *66*, 1749–1761. [CrossRef]

17. Allen, R.S.; Li, J.; Stahle, M.I.; Dubroué, A.; Gubler, F.; Millar, A.A. Genetic analysis reveals functional redundancy and the major target genes of the *Arabidopsis* miR159 family. *Proc. Natl. Acad. Sci. USA* **2007**, *104*, 16371–16376. [CrossRef]

18. Woodger, F.J.; Millar, A.; Murray, F.; Jacobsen, J.V.; Gubler, F. The role of GAMYB transcription factors in GA-regulated gene expression. *J. Plant Growth Regul.* **2003**, *22*, 176–184. [CrossRef]

19. Dai, X.; Zhao, P.X. psRNATarget: A plant small RNA target analysis server. *Nucleic Acids Res.* **2011**, *39*, W155–W159. [CrossRef]

20. Zheng, Z.; Reichel, M.; Deveson, I.; Wong, G.; Li, J.; Millar, A.A. Target RNA secondary structure is a major determinant of miR159 efficacy. *Plant Physiol.* **2017**, *174*, 1764–1778. [CrossRef]

21. Li, Z.; Peng, R.; Tian, Y.; Han, H.; Xu, J.; Yao, Q. Genome-wide identification and analysis of the MYB transcription factor superfamily in *Solanum lycopersicum*. *Plant Cell Physiol.* **2016**, *57*, 1657–1677. [CrossRef]

22. Pieczynski, M.; Marczewski, W.; Hennig, J.; Dolata, J.; Bielewicz, D.; Piontek, P.; Wyrzykowska, A.; Krusiewicz, D.; Strzelczyk-Zyta, D.; Konopka-Postupolska, D.; et al. Down-regulation of CBP80 gene expression as a strategy to engineer a drought-tolerant potato. *Plant Biotechnol. J.* **2013**, *11*, 459–469. [CrossRef] [PubMed]

23. Qin, Y.; Wang, M.; Tian, Y.; He, W.; Han, L.; Xia, G. Over-expression of TaMYB33 encoding a novel wheat MYB transcription factor increases salt and drought tolerance in *Arabidopsis*. *Mol. Biol. Rep.* **2012**, *39*, 7183–7192. [CrossRef] [PubMed]

24. Tonon, G.; Kevers, C.; Faivre-Rampant, O.; Graziani, M.; Gaspar, T. Effect of NaCl and mannitol iso-osmotic stresses on proline and free polyamine levels in embryogenic *Fraxinus angustifolia* callus. *J. Plant Physiol.* **2004**, *161*, 701–708. [CrossRef] [PubMed]

25. Alcázar, R.; Cuevas, J.C.; Patron, M.; Altabella, T.; Tiburcio, A.F. Abscisic acid modulates polyamine metabolism under water stress in *Arabidopsis thaliana*. *Physiol. Plant* **2006**, *128*, 448–455. [CrossRef]

26. Pál, M.; Tajti, J.; Szalai, G.; Peeva, V.; Végh, B.; Janda, T. Interaction of polyamines, abscisic acid and proline under osmotic stress in the leaves of wheat plants. *Sci. Rep.* **2018**, *8*, 12839. [CrossRef] [PubMed]

27. Li, J.; Reichel, M.; Li, Y.; Millar, A.A. The functional scope of plant microRNA-mediated silencing. *Trends Plant Sci.* **2014**, *19*, 750–756. [CrossRef] [PubMed]

28. Hoagland, D.R.; Arnon, D.I. The water-culture method for growing plants without soil. *Circ. Calif. Agric. Exp. Sta.* **1950**, *347*, 1–32.

29. Balcells, I.; Cirera, S.; Busk, P.K. Specific and sensitive quantitative RT-PCR of miRNAs with DNA primers. *BMC Biotechnol.* **2011**, *11*, 70. [CrossRef] [PubMed]

30. Untergasser, A.; Nijveen, H.; Rao, X.; Bisseling, T.; Geurts, R.; Leunissen, J.A.M. Primer3Plus, an enhanced web interface to Primer3. *Nucleic Acids Res.* **2007**, *35*, W71–W74. [CrossRef]

31. Ruijter, J.M.; Ramakers, C.; Hoogaars, W.M.H.; Karlen, Y.; Bakker, O.; van den Hoff, M.J.B.; Moorman, A.F.M. Amplification efficiency: Linking baseline and bias in the analysis of quantitative PCR data. *Nucleic Acids Res.* **2009**, *37*, e45. [CrossRef]

32. Sánchez-López, J.; Camañes, G.; Flors, V.; Vicent, C.; Pastor, V.; Vicedo, B.; Cerezo, M.; García-Agustín, P. Underivatized polyamine analysis in plant samples by ion pair LC coupled with electrospray tandem mass spectrometry. *Plant Physiol. Biochem.* **2009**, *47*, 592–598. [CrossRef]

miR-338-3p is Regulated by Estrogens through GPER in Breast Cancer Cells and Cancer-Associated Fibroblasts (CAFs)

Adele Vivacqua [1,*], Anna Sebastiani [1], Anna Maria Miglietta [2], Damiano Cosimo Rigiracciolo [1], Francesca Cirillo [1], Giulia Raffaella Galli [1], Marianna Talia [1], Maria Francesca Santolla [1], Rosamaria Lappano [1], Francesca Giordano [1], Maria Luisa Panno [1] and Marcello Maggiolini [1,*]

[1] Department of Pharmacy, Health and Nutritional Sciences, University of Calabria, 87036 Rende, Italy; annasebastiani86@gmail.com (A.S.); damianorigiracciolo@yahoo.it (D.C.R.); francesca89cirillo@libero.it (F.C.); giulia.r.galli@gmail.com (G.R.G.); mariannatalia11@gmail.com (M.T.); m.f.s@hotmail.it (M.F.S.); lappanorosamaria@yahoo.it (R.L.); francesca.giordano@unical.it (F.G.); mamissina@yahoo.it (M.L.P.)

[2] Regional HospitalCosenza, 87100 Cosenza, Italy; annamariamiglietta@virgilio.it

[*] Correspondence: adele.vivacqua@unical.it (A.V.); marcellomaggiolini@yahoo.it (M.M.);

Abstract: Estrogens acting through the classic estrogen receptors (ERs) and the G protein estrogen receptor (GPER) regulate the expression of diverse miRNAs, small sequences of non-coding RNA involved in several pathophysiological conditions, including breast cancer. In order to provide novel insights on miRNAs regulation by estrogens in breast tumor, we evaluated the expression of 754 miRNAs by TaqMan Array in ER-negative and GPER-positive SkBr3 breast cancer cells and cancer-associated fibroblasts (CAFs) upon 17β-estradiol (E2) treatment. Various miRNAs were regulated by E2 in a peculiar manner in SkBr3 cancer cells and CAFs, while miR-338-3p displayed a similar regulation in both cell types. By METABRIC database analysis we ascertained that miR-338-3p positively correlates with overall survival in breast cancer patients, according to previous studies showing that miR-338-3p may suppress the growth and invasion of different cancer cells. Well-fitting with these data, a miR-338-3p mimic sequence decreased and a miR-338-3p inhibitor sequence rescued the expression of genes and the proliferative effects induced by E2 through GPER in SkBr3 cancer cells and CAFs. Altogether, our results provide novel evidence on the molecular mechanisms by which E2 may regulate miR-338-3p toward breast cancer progression.

Keywords: breast cancer; CAFs; estrogens; GPER; miR-338-3p; c-Fos; Cyclin D1

1. Introduction

Estrogens play a crucial role in diverse pathophysiological conditions, including cancer [1]. The action of estrogens are mainly mediated by the classic estrogen receptors (ERs) [2], however several data have also indicated that the G protein estrogen receptor (GPER) may trigger a network of transduction pathways toward the progression of several types of tumors [3–8]. Among numerous biological targets, estrogens may modulate the expression of diverse microRNAs (miRNAs) [6], which are small non-coding RNA molecules of 22–25 nucleotides [9]. In particular, miRNAs inhibit the expression of certain genes at both transcriptional and post-transcriptional levels binding to complementary sequences located within the 3′ untranslated region (UTR) of target mRNAs [10,11]. Therefore, miRNAs may be involved in important biological processes, including cancer development [12–20]. The involvement of ERs in miRNA regulation by estrogens has been established [6]. Likewise, it has been also reported that GPER may regulate the expression of certain miRNAs in normal and cancer cell contexts characterized by the presence or absence of ERs [21–25].

MiR-338-3p is a highly conserved gene located on the chromosome 17q25 and precisely on the 7th intron of the apoptosis-associated tyrosine kinase (AATK) [26,27]. MiR-338-3p, initially identified as a brain specifically expressed miRNA, has been involved in the formation of basolateral polarity and regulation of axonal respiration [28,29]. Various studies have also shown that miR-338-3p is downregulated in many types of malignancies, hence suggesting its potential role in tumor progression [30–34]. Nevertheless, the biological function of miR-338-3p and its prognostic significance remains to be fully understood.

In this present study we provide novel insights into the ability of estrogens to regulate miR-338-3p expression and function through GPER in ER-negative breast cancer cells and cancer associated fibroblasts (CAFs), which are main components of the tumor microenvironment [35,36]. On the basis of our findings miR-338-3p may be included among the miRNAs involved in breast tumor development.

2. Materials and Methods

2.1. Reagents

17β-estradiol (E2) was purchased from Sigma-Aldrich Corp. (Milan, Italy); rel-1-[4-(6-bromo-1,3-benzodioxol-5-yl)-3aR,4S,5,9bS-tetrahydro-3H-cyclopenta[c]quinolin-8-yl]-ethanone (G-1) was obtained from Tocris Bioscience (Space, Milan, Italy). All compounds were solubilized in dimethyl sulfoxide (DMSO).

2.2. Cell Cultures

Breast cancer cell line SkBr3 (ER-negative and GPER-positive) was obtained from ATCC (Manassas, VA, USA), used less than six months after revival and routinely tested and authenticated according to the ATCC suggestions. CAFs (ER-negative and GPER-positive) were extracted from invasive mammary ductal carcinomas obtained from mastectomies. Briefly, samples were cut into smaller pieces (1–2 mm diameter), placed in digestion solution (400 IU collagenase I, 100 IU hyaluronidase, and 10% FBS, containing antibiotic and antimycotic solution) and incubated overnight at 37 °C. The cells were then separated by differential centrifugation at $90 \times g$ for 2 min. Supernatant containing fibroblasts was centrifuged at $485 \times g$ for 8 min; the pellet obtained was suspended in fibroblasts growth medium (Medium 199 and Ham's F12 mixed 1:1 and supplemented with 10% FBS) and cultured at 37 °C in 5% CO_2. Primary cells cultures of breast fibroblasts were characterized by immunofluorescence. Briefly cells were incubated with human anti-vimentin (V9, sc-6260) and human anti-cytokeratin 14 (LL001 sc-53253), both from Santa Cruz Biotechnology (DBA, Milan, Italy) (data not shown). To characterize fibroblasts activation, we used anti-fibroblast activated protein α (FAPα) antibody (SS-13, sc-100528; Santa Cruz Biotechnology, DBA, Milan, Italy) (data not shown). Signed informed consent from all the patients was obtained and samples were collected, identified and used in accordance with approval by the Institutional Ethical Committee Board (Regional Hospital, Cosenza, Italy). Cell types were grown in a 37 °C incubator with 5% CO_2. SkBr3 breast cancer cells were maintained in RPMI-1640 without phenol red supplemented with 10% fetal bovine serum (FBS) and 100 μg/mL of penicillin/streptomycin (Gibco, Life Technologies, Milan, Italy). CAFs were cultured in a mixture of MEDIUM 199 and HAM'S F-12 (1:1) supplemented with 10% FBS and 100 μg/mL of penicillin/streptomycin (Gibco, Life Technologies, Milan, Italy). Cells were switched to medium without serum the day before experimental analysis.

2.3. RNA Extraction

Cells were maintained in regular growth medium and then switched to medium lacking serum before performing the indicated assays. Total RNA was extracted from cultured cells using miRVana Isolation Kit (Ambion, Life Technologies, Milan, Italy) according to the manufacturer's recommendations. The RNA concentrations were determined using Gene5 2.01 Software in Synergy H1 Hybrid Multi-Mode Microplate Reader (BioTek, AHSI, Milan, Italy).

2.4. miRNA Expression Profiling

TaqMan™ Array Human MicroRNA A+B Cards Set v3.0 was used for global miRNA profiling. The panel includes two 384-well microfluidic cards (human miRNA pool A and pool B) that contain primers and probes for 754 different miRNAs in addition to small nucleolar RNAs that function as endogenous controls for data normalization. Equal quantity (100 ng) of RNA extracted from SkBr3 breast cancer cells and CAFs treated with vehicle or 100 nM E2 for 4 h was reverse-transcribed for cDNA synthesis using the Megaplex RT Primer Pool A or B and the TaqMan MicroRNA Reverse Transcription kit (Applied Biosystems).in a final volume of 7.5 μL (Applied Biosystems, Milan, Italy). The reverse transcription reaction was incubated for 2 min at 16 °C, 1 min at 42 °C and 1 s at 50 °C for 40 cycles, followed by 5 min at 85 °C to deactivate the enzyme. The cDNA obtained was pre-amplified using Megaplex Preamp primer pool A or B and TaqMan PreAmp Master Mix 2X in a final volume of 25 μL using the same temperature conditions above described. The product was diluted 1:4 in TE 0.1X, to which were added TaqMan Universal Master Mix no UNG 2X and nuclease free water. 100 μL of the sample/master mix for each multiplex pool were loaded into fill reservoirs on the microfluidic card. The array was then centrifuged, mechanically sealed with the Applied Biosystems sealer device and run on QuantStudio 6&7 Flex Real Time PCR System (Applied Biosystems, Life Technologies, Milan, Italy). The raw array data were analysed by DataAssist™. The baseline was set automatically, while the threshold was set manually at 0.2. Samples that had Ct values>32 were removed from the analysis. Each miRNA was normalized against the mean of the four RNU6B and its expression was then assessed in the E2 treated cells against the vehicle treated cells using the $2^{-\Delta\Delta CT}$ method [37]. miRNAs showing an increased value of 2-fold expression and a 50% reduction respect to vehicle-treated cells were selected. Venn diagram was obtained by http://bioinformatics.psb.ugent.be/cgibin/liste/Venn/calculate_venn.htpl.

2.5. Analysis of Public Data Set from METABRIC and Kaplan-Meier Plotter

Prognostic values of miR-338-3p levels, using METABRIC data set, were analyzed by Kaplan–Meier survival curves of breast cancer patients, using Kaplan-Meier Plotter (www.kmplot. com/analysis) [38]. Log-rank test was used for statistical analysis.

2.6. Real Time-PCR

cDNA for miRNA expression was synthesized from 100 ng of total RNA using the TaqMan microRNA Reverse Transcription Kit (Applied Biosystems, Life Technologies, Milan, Italy). The expression levels of miR-338-3p were quantified by TaqMan microRNA Assay Kit (Applied Biosystems, Milan, Italy), using the primers for the internal control RNU6B (assay ID 001093) and miR-338-3p (assay ID 002252). In order to measure the mRNA levels of c-Fos and Cyclin D1, 3μg of total RNA were reversely transcribed using the murine leukemia virus reverse transcriptase (Life Technologies, Milan, Italy), as indicated by the manufacturer. The quantitative PCR was performed using SYBR Green PCR Master Mix (Applied Biosystems, Life Technologies, Milan, Italy). Specific primers for Actin, which was used as internal control, c-Fos and Cyclin D1 genes were designed using Primer Express version 2.0 software (Applied Biosystems Inc, Milano, Italy). The sequences were as follows: Actin Fwd: 5′-AAGCCAACCCCACTTCTCTCTAA-3′ and Rev: 5′-CACCTCCCCTGTGTGGACTT-3′; c-Fos Fwd: 5′-CGAGCCCTTTGATGACTTCCT-3′ and Rev: 5′-GGAGCGGGCTGTCTCAGA-3′; Cyclin D1 Fwd: 5′-CCGTCCATGCGGAAGATC-3′ and Rev: 5′-ATGGCCAGCGGGAAGAC-3′. All experiments were performed in triplicate using QuantStudio 6&7 Flex Real Time PCR System (Applied Biosystems, Life Technologies, Milan, Italy). Data were normalized to the geometric mean of housekeeping gene to control the variability into expression levels and fold changes were calculated by relative quantification compared to respective scrambled controls [32].

2.7. Bioinformatic Tools

The sites miRNAbase (http://www.miRNAbase.org), Targetscan (http://www.targetscan.org) and miRDip (http://ophid.utoronto.ca/mirDIP/) were used to identified miR-338-3p target genes.

2.8. Constructs and Transfections

The negative control (miR-Ctrl), the miR-338-3p mimic (miR-338-3p m) (ID MC10716) and miR-338-3p inhibitor (miR-338-3p i) (ID MH10716) sequences were purchased from Ambion (Life Technologies, Milan, Italy) and transfected into the cells 48 h before the treatments, using X-treme GENE 9 DNA Transfection Reagent (Roche Diagnostics, Sigma-Adrich, Milan, Italy). Silencing of GPER expression was obtained by using the construct previously described [39]. The plasmid DN-Fos, which encodes a c-Fos mutant that heterodimerizes with c-Fos dimerization partners but does not allow DNA binding, was a kind gift from Dr. C. Vinson (NIH, Bethesda, MD, USA).

2.9. Western Blotting

Cells were maintained in complete medium before the transfection assays, which are performed in medium without serum for 48 h and then treated as indicated. Cells were lysed in RIPA buffer containing a mixture of protease inhibitors. Equal amounts of protein extract were resolved on SDS-polyacrylamide gel, transferred to a nitrocellulose membrane (Amersham Biosciences, Italy), probed overnight at 4 °C with antibodies against: c-Fos (E-8, sc-166940) and β-Actin (AC-15, sc-69879) (Santa Cruz Biotechnology, DBA, Italy), GPER (AB137479) (Abcam, Euroclone, Milan, Italy) and Cyclin D1 (Origene, DBA, Milan, Italy). Proteins were detected by horseradish peroxidase-linked secondary antibodies (Biorad, Milan, Italy) and revealed using the chemiluminescent substrate for western blotting Westar Nova 2.0 (Cyanagen, Biogenerica, Catania, Italy).

2.10. Luciferase Assays

Cells were seeded in regular growth medium into 24-well plates. The next day the growth medium was replaced with medium lacking serum and the transfection was performed using X-tremeGene9 reagent, as recommended by the manufacturer (Roche Diagnostics), with a mixture containing Cyclin-D1-luc, the internal control pRL-TK and miR-Ctrl, miR-338-3p m, alone or in presence of miR-338-3p i, shGPER, DN-Fos as indicated. The cells were treated overnight with 100 nM of E2 or G1. Luciferase activity was measured using the Dual Luciferase kit (Promega, Milan, Italy) according to the manufacturer's instructions. Firefly luciferase values were normalized to the internal transfection control provided by the Renilla luciferase activity. The normalized relative light unit (RLU) values obtained from cells transfected with respective scrambled controls were set as 1-fold induction upon which the activity induced by the treatment was calculated.

2.11. Cell Proliferation Assays

For quantitative proliferation assay, cells (1×10^4) were seeded in 24-well plates in regular growth medium. Cells were washed, once they had attached, and then incubated in medium containing 2.5% charcoal stripped fetal bovine serum, before the transfection with 25 nM miR-338-p m and 50 nM miR-338-3p i, as indicated. Transfection was renewed every 2 day, while the cells were treated every day. Evaluation of cell growth was performed on day 6 using automatic counter (Countess™-Invitrogen).

2.12. Cell Cycle Analysis

To analyze cell cycle distribution, CAFs were cultured in regular medium and shifted in medium containing 2.5% charcoal-stripped FBS at the 70% confluence. Next, miRNA sequences as indicated were added to cells using X-treamGene9 reagent (Roche Diagnostics, Milan, Italy). After 24 h, 100 nM E2 or 100 nM G-1 were put in the medium for additional 24 h. Cells were pelleted, once washed with phosphate buffered saline and stained with a solution containing 50 μg/mL propidium iodide in 1 x PBS (PI), 20 U/mL RNAse-A and 0.1% Triton (Sigma-Aldrich, Milan, Italy). The DNA content was measured using a FACScan flow cytometer (Becton Dickinson, Mountain View, CA, USA) and the data acquired using CellQuest software. Cell cycle profiles were determined using ModFit LT. The proportion of the cells in G0/G1, S and G2/M phases was each estimated as a percentage of the total events (10,000 cells).

2.13. Statistical Analysis

Data were analyzed by one-way ANOVA with Dunnett's multiple comparisons where applicable, using GraphPad Prism version 6.01 (GraphPad Software, Inc., San Diego, CA, USA). $p < 0.05$ (*) was considered statistically significant.

3. Results

3.1. miRNAs Expression by E2 in SkBr3 Cancer Cells and CAFs

In order to provide novel insights on the action of estrogens toward miRNAs modulation in breast cancer, the ER-negative SkBr3 breast cancer cells and CAFs were treated with 100 nM E2 for 4 h and then analyzed by TaqMan™ Array Human MicroRNA. A total amount of 754 miRNAs involved in diverse pathophysiological conditions (www.thermofisher.com/order/catalog/product/4444913) were evaluated, thereafter we focused our attention on miRNAs displaying a Ct< 32 along with at least 2 fold increase or 50% reduction upon E2 exposure respect to vehicle-treated cells. On the basis of these criteria, we identified 25 and 29 E2-regulated miRNAs in SkBr3 cancer cells (Figure 1A) and CAFs (Figure 2A), respectively. In particular, in SkBr3 cancer cells 23 miRNAs were up-regulated and 2 miRNAs were down-regulated by E2 treatment (Figure 1B). As it concerns CAFs, among the 29 E2-regulated miRNAs, 7 showed an increase and 22 a reduction upon E2 stimulation (Figure 2B). To identify unique and shared E2-regulated miRNAs in both cell types, we then calculated a Venn diagram. SkBr3s cancer cells and CAFs shared only the expression of 2 miRNAs (Figure 3A), namely miR-144 and miR-338-3p, which exhibited a similar response (Figure 3B). Considering that in our previous studies we evaluated the role of miR-144 in tumor cell growth [25], in the present investigation we aimed to determine the mechanisms leading to the estrogen regulation of miR-338-3p and its action in breast cancer. Hence, we began our study ascertaining that miR-338-3p expression correlates positively with the overall survival in 1283 breast tumor patients, as reported in the Molecular Taxonomy of Breast Cancer International Consortium (METABRIC) database [40] (Figure 3C). Nicely fitting with these findings, previous evidence has suggested that miR-338-3p may function as a tumor suppressor in certain malignancies including breast cancer [30–34].

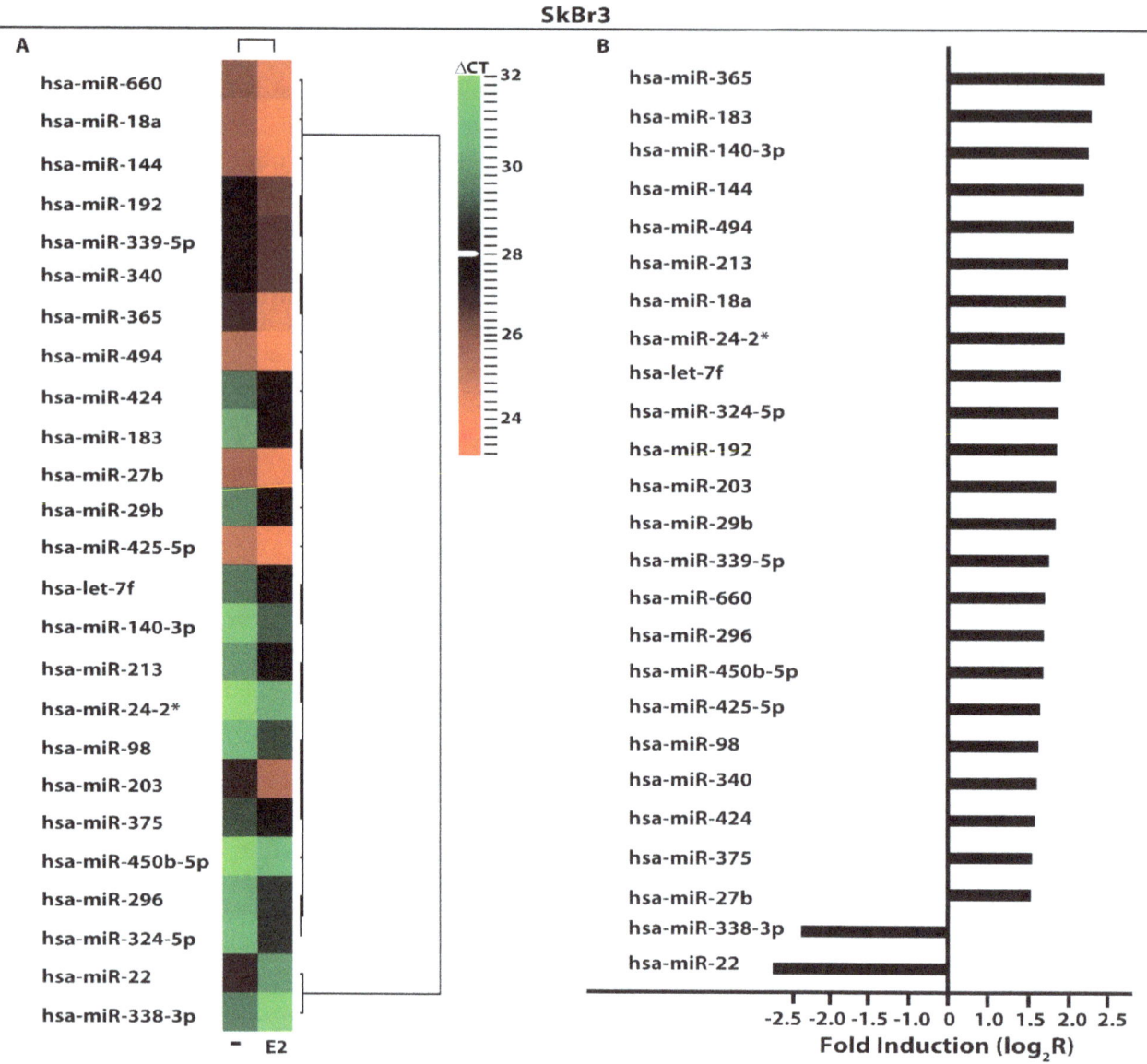

Figure 1. E2-modulated miRNAs expression in SkBr3 breast cancer cells. (**A**) Heat Map representation of E2-modulated miRNAs in SkBr3 cancer cells treated with 100 nM E2 for 4 h and analyzed by TaqMan Low-Density Array Human miRNA. Row represents a miRNA and column represents the treatment used. Each column is illustrated according to a color scale from green (low expression) to red (high expression). The distance measured is Euclidean Distance and the clustering method is complete linkage. Dendrograms of clustering analysis for miRNAs and samples are displayed on the top and right, respectively. (**B**) Up- and down-regulated miRNAs in SkBr3 breast cancer cells upon E2 stimulation. The values are indicated as log2 fold change (R) calculated respect to vehicle (-).

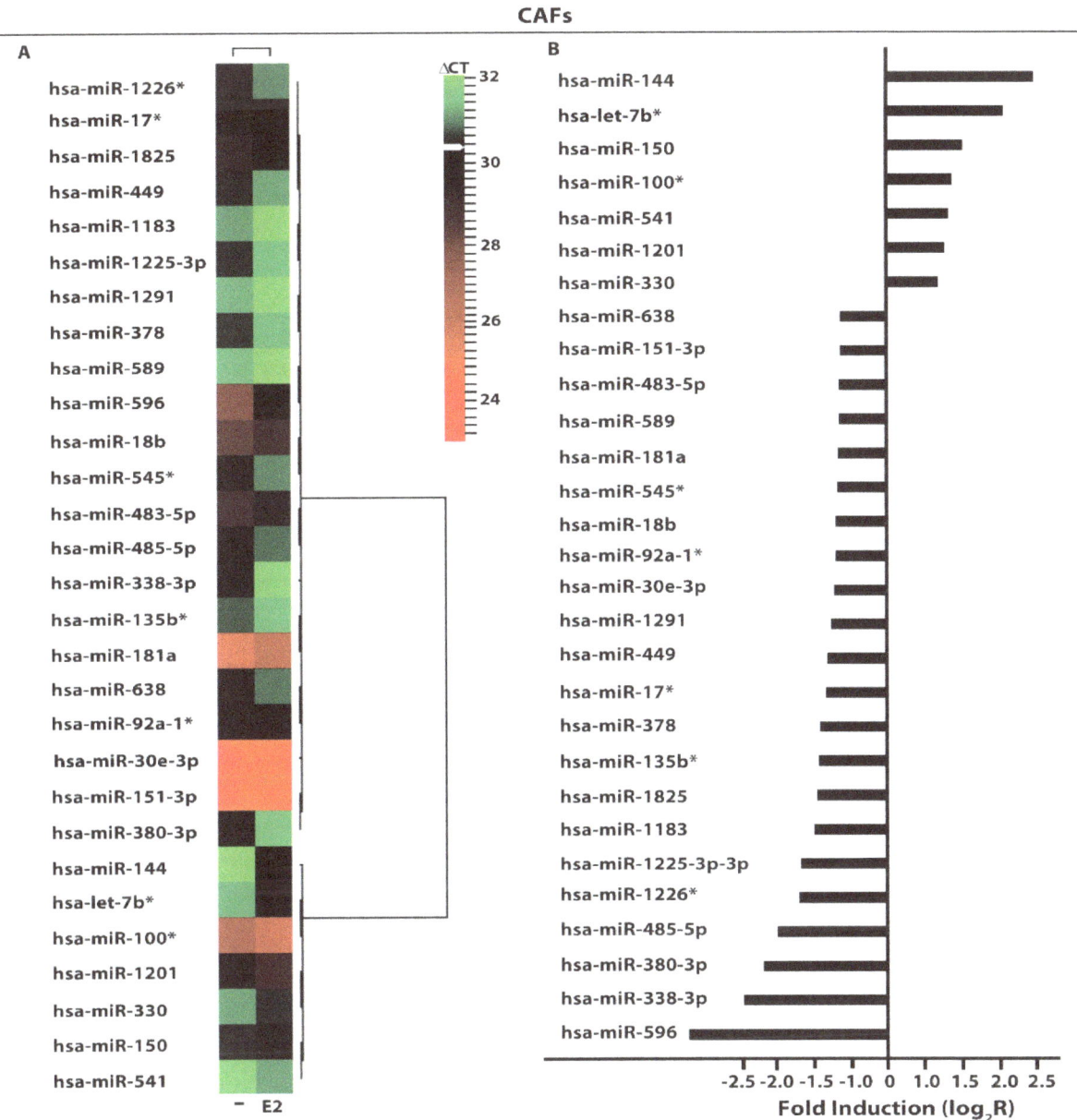

Figure 2. E2-modulated miRNAs expression in CAFs. (**A**) Heat Map representation of E2-modulated miRNAs in CAFs treated with 100 nM E2 for 4 h and analyzed by TaqMan Low-Density Array Human miRNA. Row represents a miRNA and column represents the treatment used. Each column is illustrated according to a color scale from green (low expression) to red (high expression). The distance measured is Euclidean Distance and the clustering method is complete linkage. Dendrograms of clustering analysis for miRNAs and samples are displayed on the top and right, respectively. (**B**) Up- and down-regulated miRNAs in CAFs upon E2 stimulation. The values are indicated as log2 fold change (R) calculated respect to vehicle (-).

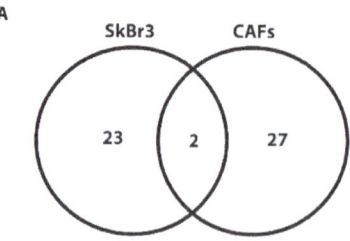

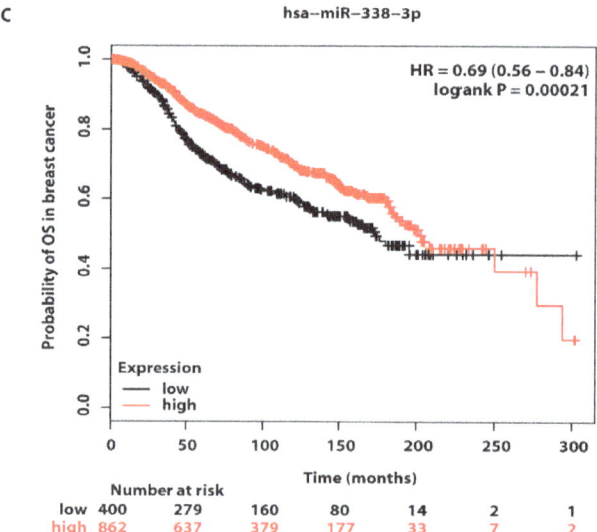

Figure 3. Exclusive and shared expression of miRNAs between SkBr3 and CAFs. (**A**) Venn Diagram of E2-modulated miRNAs in SkBr3 cancer cells and CAFs. (**B**) Up and down-regulated miRNAs by 100 nM E2 treatment for 4 h in SkBr3 cancer cells and CAFs. (**C**) The expression of miR-338-3p is associated with higher overall survival in breast cancer patients. The evaluation was performed by Kaplan–Meier Plotter (http://www.kmplot.com). Statistical analysis was made using the log-rank test.

3.2. GPER Is Involved in the Regulation of miR-338-3p by E2 and G-1 in SkBr3 Cancer Cells and CAFs

On the basis of the aforementioned results, we then attempted to define the molecular mechanisms involved in the estrogenic regulation of miR-338-3p performing a time-course study upon 100 nM of E2 and 100 nM of the selective GPER ligand G-1. Worthy, the inhibitory effects of E2 and G-1 on miR-338-3p expression were no longer evident silencing GPER in SkBr3 cancer cells (Figure 4A–C) and in CAFs (Figure 4D–F). Thereafter, we aimed to identify putative target genes of miR-338-3p by a bioinformatic analysis of available algorithms (http://ophid.utoronto.ca/mirDIP; http://www.microrna.org;http://www.targetscan.org). Among others, two putative target sequences of miR-338-3p located within the 3'-UTR of the oncogene c-Fos were found (Figure 5A). According to our previous studies showing that estrogens regulate c-Fos levels in diverse cancer cell types [41–44], the induction of c-Fos mRNA and protein expression upon a 4 h treatment with 100 nM E2 and 100 nM G-1 was abolished silencing GPER in SkBr3 cancer cells (Figure 5B,C) and CAFs (Figure 5D,E). Next, we found that in SkBr3 cells and CAFs transfected for 48 h with a miR-338-3p mimic sequence, the treatment for 4 h with 100 nM E2 and 100 nM G-1 is no longer able to induce c-Fos mRNA and protein levels, a response rescued transfecting the miR-338-3p mimic sequence in combination with a miR-338-3p inhibitor sequence (Figure 6A–F).

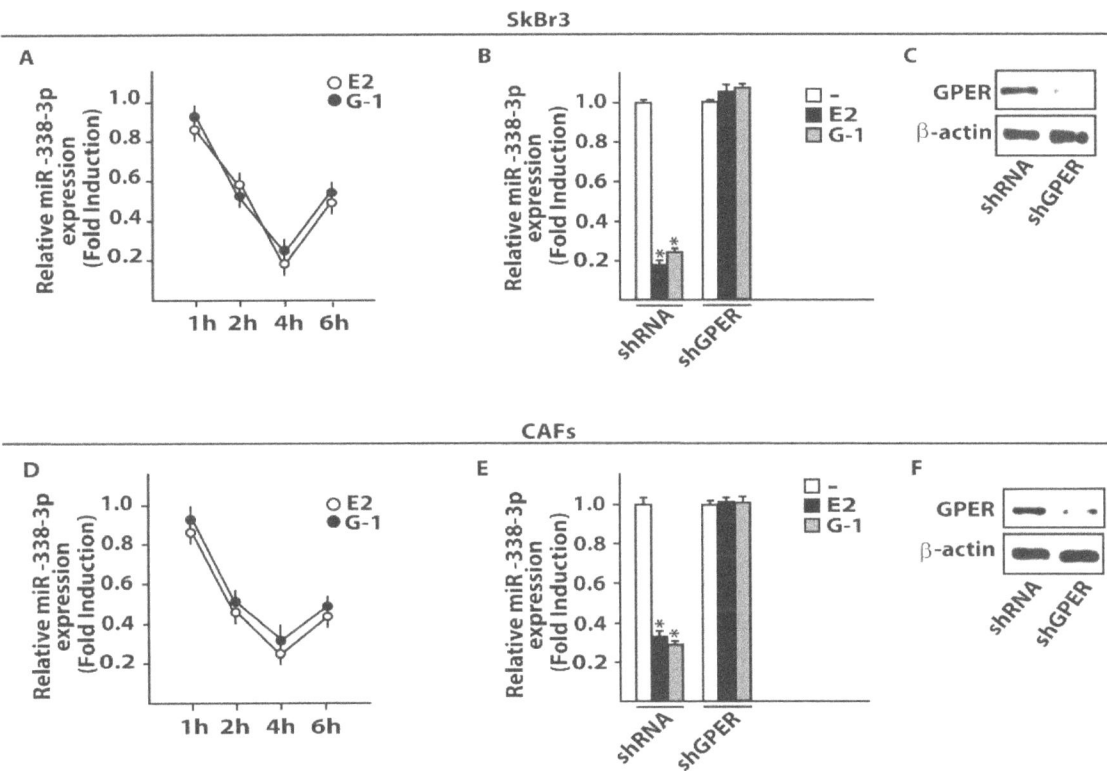

Figure 4. E2 and G-1 down-regulate miR-338-3p levels in SkBr3 cancer cells and CAFs. SkBr3 breast cancer cells (**A**) and CAFs (**D**) were stimulated with 100 nM E2 or 100 nM G-1 as indicated and analyzed by RT-PCR. Each point is plotted as fold changes of cells receiving treatments respect to cells treated with vehicle (-) and represents the mean ± SD of three independent experiments performed in triplicate. MiR-338-3p expression upon a 4 h treatment with 100 nM E2 or 100 nM G-1 in SkBr3 cells (**B**) and CAFs (**E**) previously transfected with shRNA or shGPER for 48 h. Each column represents the mean ± SD of three independent experiments performed in triplicate. Efficacy of GPER silencing in SkBr3 cells (**C**) and CAFs (**F**). β-actin serves as a loading control. (*) indicates $p < 0.05$, for cells receiving treatments vs cells treated with vehicle.

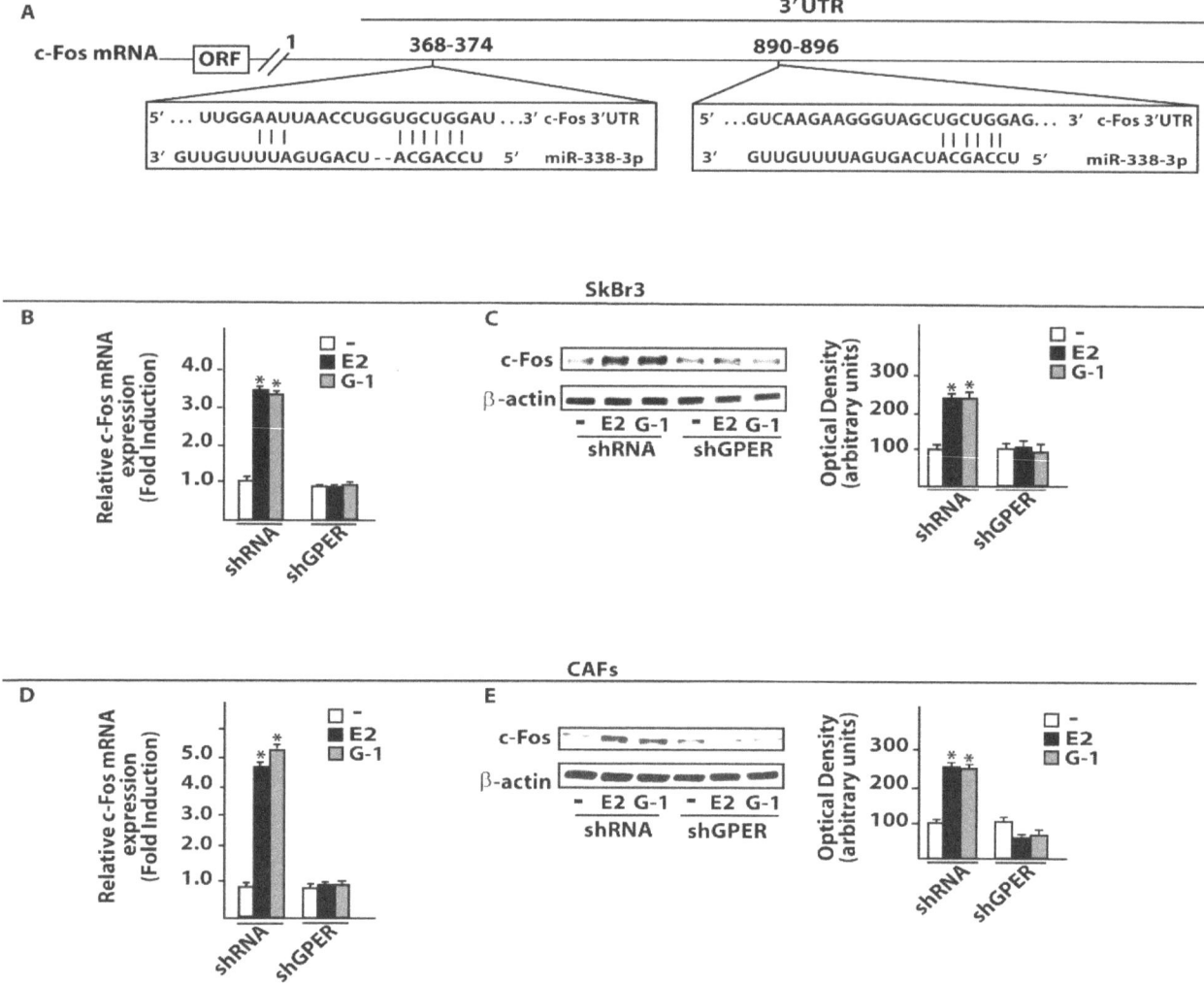

Figure 5. c-Fos is a target gene of miR-338-3p. (**A**) Schematic alignment between the miR-338-3p sequence and the 3′-UTR mRNA region of c-Fos. mRNA expression of c-Fos in SkBr3 cancer cells (**B**) and CAFs (**D**) transfected with shRNA or shGPER for 48 h and then treated for 4 h with 100 nM E2 or 100 nM G-1. Each column represents the mean ± SD of three independent experiments performed in triplicate. c-Fos protein expression in SkBr3 cancer cells (**C**) and CAFs (**E**) transfected with shRNA or shGPER for 48 h and then treated for 4 h with 100 nM E2 or 100 nM G-1. Side panels show densitometry analysis of the blots normalized to the loading control β-actin.

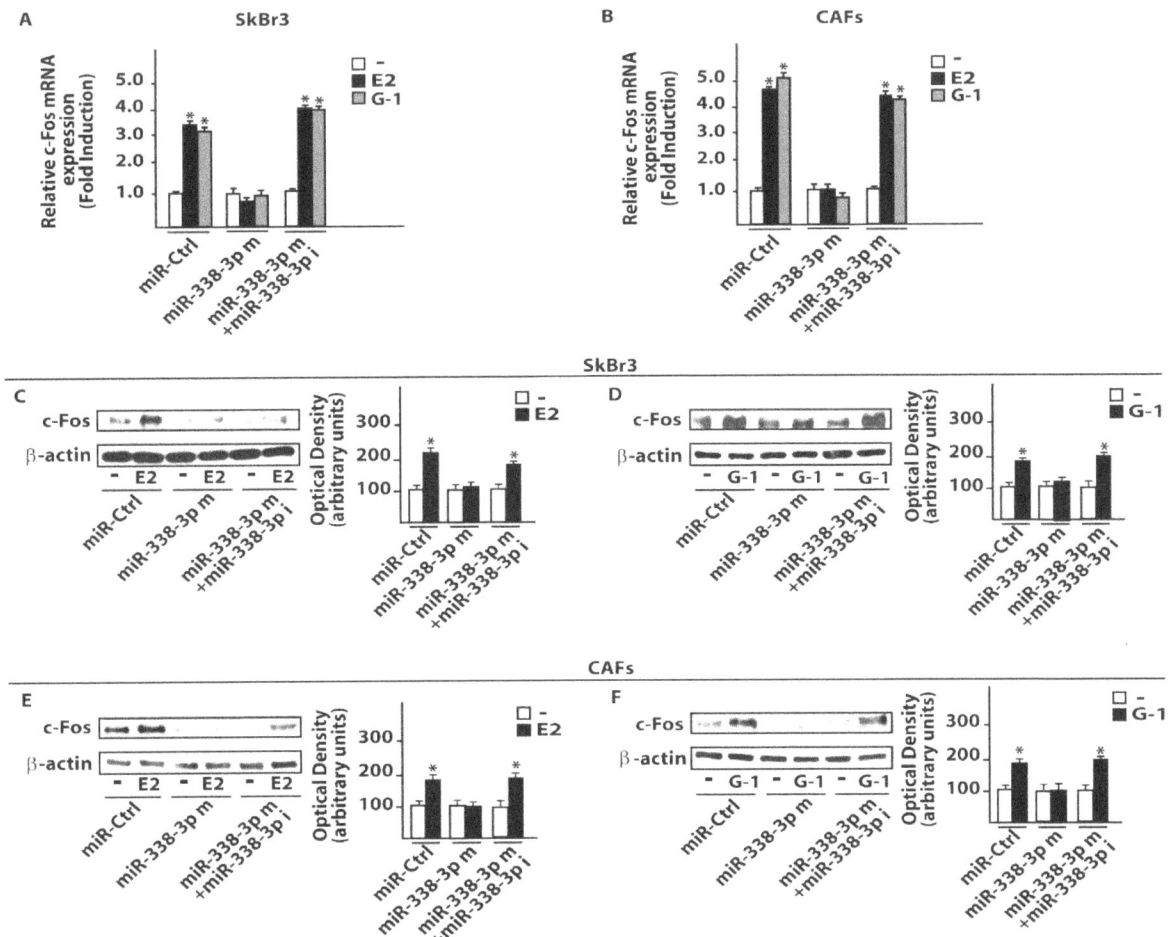

Figure 6. miR-338-3p prevents c-fos induction by E2 and G-1 in SkBr3 cancer cells and CAFs. mRNA levels of c-Fos in SkBr3 cancer cells (**A**) and CAFs (**B**) transfected for 48 h with 25 nM miR-Ctrl or miR-338-3p mimic (miR-338-3p m) in combination or not with 50 nM miR-338-3p inhibitor (miR-338-3p i) and then treated for 4 h with 100 nM E2 or 100 nM G-1. Each column represents the mean ± SD of three independent experiments performed in triplicate. c-Fos protein levels in SkBr3 cancer cells (**C, D**) and CAFs (**E, F**) transfected for 48 h with 25 nM miR-Ctrl or miR-338-3p mimic (miR-338-3p m) in combination or not with 50 nM miR-338-3p inhibitor (miR-338-3p i) and then stimulated for 4 h with 100 nM E2 or 100 nM G-1. Side panels show densitometry analysis of the blots normalized to the loading control β-actin. (*) indicates $p < 0.05$, for cells receiving treatments vs cells treated with vehicle (-).

3.3. miR-338-3p Triggers Inhibitory Effects on the Proliferation Induced by E2 and G-1

As in our previous investigations c-Fos was involved in the regulation of cyclins [43,45], we assessed that the transactivation of the Cyclin D1 promoter sequence by 100 nM E2 and 100 nM G-1 was prevented co-transfecting a dominant negative c-Fos expression construct (DN-Fos) in SkBr3 and CAFs (Figure 7A,B). Nicely recapitulating the aforementioned results, the Cyclin D1 promoter luciferase activity induced by 100 nM E2 and 100 nM G-1 was inhibited using the miR-338-3p mimic, an effect rescued by the miR-338-3p inhibitor sequence (Figure 7C,D). In addition, similar findings were observed evaluating the regulation of Cyclin D1 at both mRNA (Figure 7E,F) and protein levels (Figure 8A–D). As biological counterpart, the proliferative responses elicited by 100 nM E2 and 100 nM G-1 in SkBr3 cancer cells and CAFs were prevented silencing GPER or transfecting the DN-Fos construct (Figure 9A,B). Furthermore, the miR-338-3p mimic sequence decreased the proliferation induced by 100 nM E2 and 100 nM G-1 (Figure 9A,B), however this effect was rescued co-transfecting the miR-338-3p inhibitor (Figure 9A,B). Further supporting the aforementioned findings, the treatment for 24 h with 100 nM E2 and 100 nM G-1 triggered inhibitory effects on cell cycle progression transfecting CAFs with the miR-338-3p mimic sequence, however this response was rescued in the presence of the miR-338-3p inhibitor sequence (Figure 9C). Overall, these results suggest that estrogenic GPER signaling regulates miR-338-3p expression and function in SkBr3 cancer cells and CAFs.

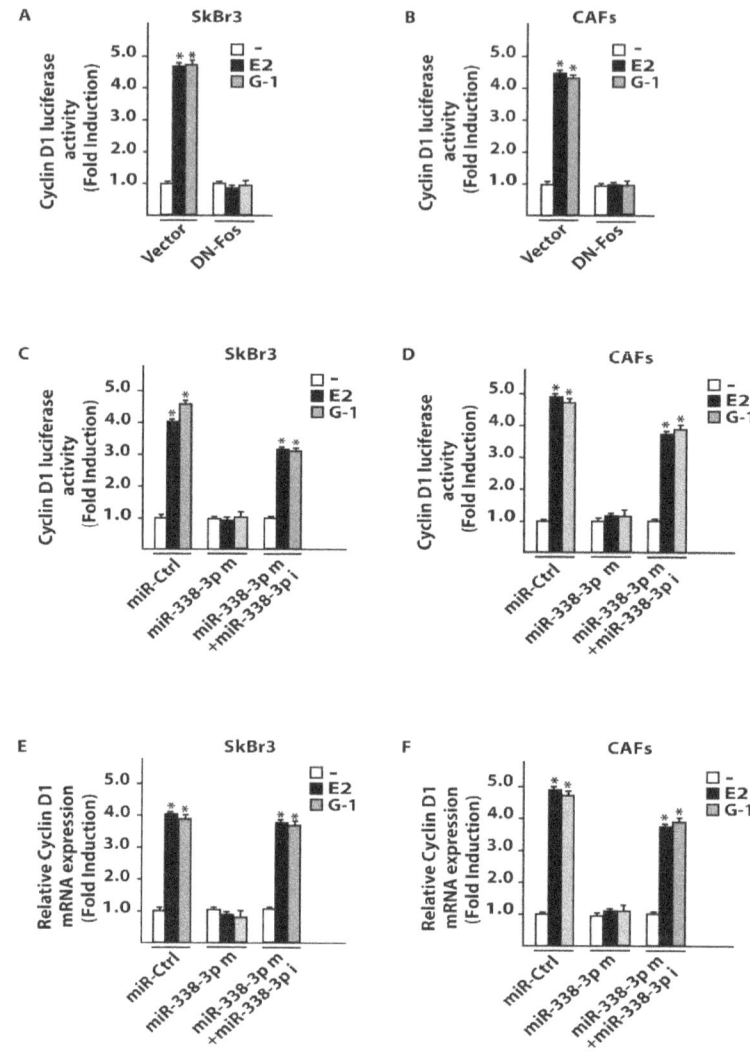

Figure 7. c-Fos and miR-338-3p are involved in Cyclin D1 regulation in SkBr3 cancer cells and CAFs. Luciferase activity of Cyclin D1 reporter gene in SkBr3 cancer cells (**A**) and CAFs (**B**) transfected for 8 h with a vector or a dominant-negative c-Fos construct (DN-Fos) before treatment with 100 nM of E2 and 100 nM G-1 for 18 h. Luciferase activity of Cyclin D1 reporter gene in SkBr3 cancer cells (**C**) and CAFs (**D**) transfected for 24 h with 25 nM miR-Ctrl or miR-338-3p mimic (miR-338-3p m) in combination or not with 50 nM miR-338-3p inhibitor (miR-338-3p i) before treatment for 18 h with 100 nM E2 or 100 nM G-1. The luciferase activity was normalized to the internal transfection control, values of cells receiving vehicle (-) were set as 1-fold induction upon which the activity obtained upon the indicated treatments was calculated. mRNA expression of Cyclin D1 in SkBr3 cells (**E**) and CAFs (**F**) transfected for 48 h with 25 nM miR-Ctrl or miR-338-3p mimic (miR-338-3p m) in combination or not with 50 nM miR-338-3p inhibitor (miR-338-3p i) before treatment for 8 h with 100 nM E2 or 100 nM G-1. Each column represents the mean ± SD of three independent experiments performed in triplicate. (*) indicates $p < 0.05$ for cells receiving treatments vs cells treated with vehicle (-).

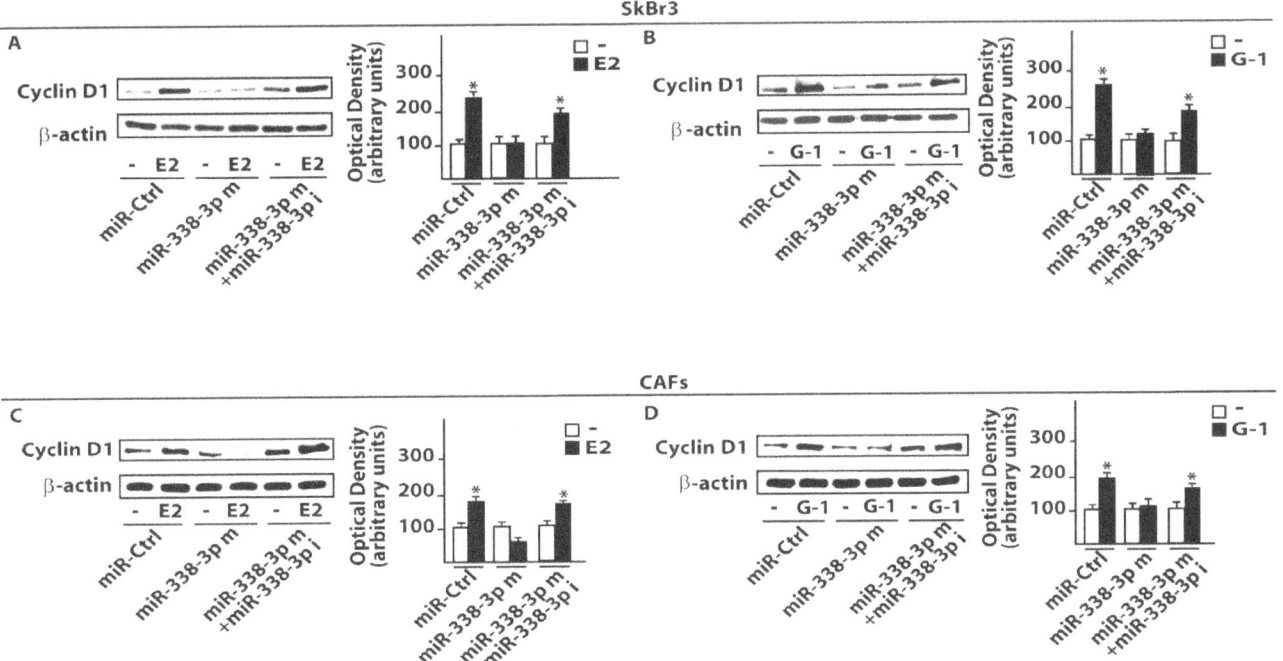

Figure 8. miR-338-3p prevents Cyclin D1 protein induction by E2 and G1 in SkBr3 cancer cells and CAFs. Cyclin D1 protein expression in SkBr3 cancer cells (**A,B**) and CAFs (**C,D**) transfected for 48 h with 25 nM miR-Ctrl or miR-338-3p mimic (miR-338-3p m) in combination or not with 50 nM miR-338-3p inhibitor (miR-338-3p i) before treatment for 12h with 100 nM E2 or 100 nM G-1. Side panels show densitometry analysis of the blots normalized to the loading control β-actin. (*) indicates $p < 0.05$ for cells receiving treatments vs cells treated with vehicle (-).

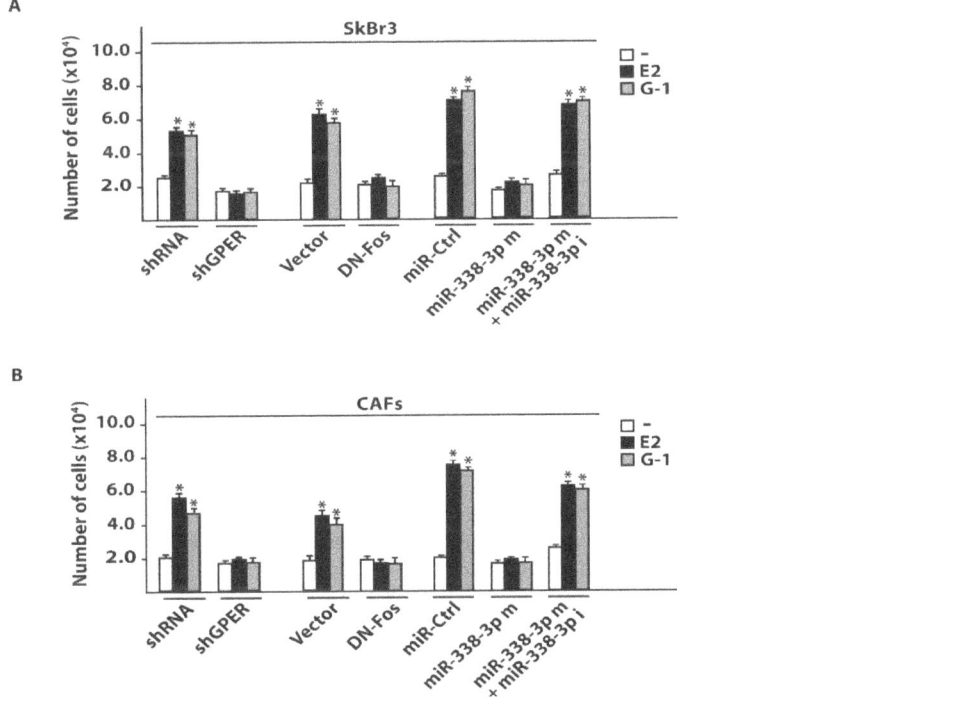

Figure 9. *Cont.*

C

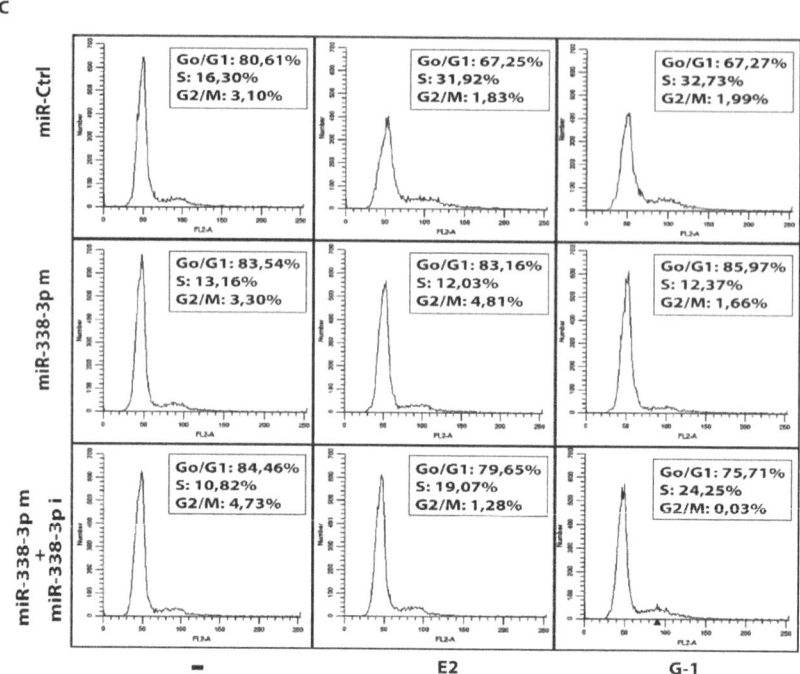

Figure 9. miR-338-3p decreases the proliferation of SkBr3 cancer cells and CAFs induced by E2 and G-1. Cell proliferation in SkBr3 cancer cells (**A**) and CAFs (**B**) transfected every 2 days with 100ng shRNA or shGPER, 100ng vector or a dominant-negative c-Fos construct (DN-Fos) and 25 nM miR-Ctrl or miR-338-3p mimic (miR-338-3p m) in combination or not with 50 nM miR-338-3p inhibitor (miR-338-3p i). Cells were treated every day with 100 nM E2 or 100 nM G-1 and counted on day 6. Each column represents the mean ± SD of three independent experiments performed in triplicate. (*) indicates $p <$ 0.05 for cells receiving treatments vs cells treated with vehicle (-). (**C**) Representative pictures of cell cycle analysis in CAFs transfected for 48 h with 25 nM miR-Ctrl or miR-338-3p mimic (miR-338-3p m) in combination or not with 50 nM miR-338-3p inhibitor (miR-338-3p i) before the treatment for 24 h with 100 nM E2 and 100 nM G-1. In each panel, the percentages of cells in G0/G1, S and G2/M phases of the cell cycle are indicated. Values represent the mean ± SD of three independent experiments.

4. Discussion

Performing a microarray analysis of 754 miRNAs involved in diverse diseases, in the present study we determined that diverse miRNAs are regulated by E2 in both SkBr3 breast cancer cells and CAFs. In particular, we assessed that E2 increases 23 miRNAs and lowers 2 miRNAs in SkBr3 cells, while E2 triggers the up-regulation of 7 miRNAs and the down-regulation of 22 miRNAs in CAFs. In addition, in both cell types E2 induced the expression of miR-144 and repressed the levels of miR-338-3p, which is known as an inhibitor of cancer progression [30–34]. Considering that miR-144 was investigated in our previous study [25], we attempted to provide novel insights into the estrogen regulation of miR-338-3p. First, we performed a METABRIC analysis that revealed a positive correlation of miR-338-3p with the overall survival in breast cancer patients. Then, we evidenced that a miR-338-3p mimic sequence prevents the expression of c-Fos, Cyclin-D1 and the growth effects induced by E2 and G-1 through GPER in SkBr3 cells and CAFs. Worthy, these effects triggered by E2 and G-1 were rescued using a miR-338-3p inhibitor sequence. Altogether, the aforementioned results provide new insights on the molecular mechanisms involved in the expression and function of certain miRNAs upon estrogen exposure in both breast cancer cells and CAFs.

Breast tumor is the most common malignancy in females and its incidence is increasing worldwide [46]. Several studies are ongoing in order to identify novels biological targets that may be considered toward innovative therapeutic approaches. To date, few markers like the estrogen receptor (ER), the progesterone receptor (PR) and the human epidermal growth factor receptor 2 (HER2), have been identified as predictors of clinical responses to breast cancer treatments [47]. None of these markers, however, well evaluates tumor invasion or provides early detection of cancer progression [48]. In this context, GPER has been suggested as a further predictor of breast cancer aggressiveness as its expression was found positively associated with clinic-pathological features of cancer progression and poor survival rates [49,50]. Moreover, GPER has been also indicated as an independent factor to predict a reduced disease-free survival in patients treated with tamoxifen [49]. The lack of GPER in the plasma membrane was also related to excellent long-term prognosis in ER-positive breast tumors treated with tamoxifen, an observation that highlighted the potential importance of GPER expression in different cancer cell types [51].

Despite the stimulatory effects elicited by GPER on the growth of diverse cancer cells [3–6], high doses of the GPER agonist G-1 ($\geq$1 μM) have been shown to exert an inhibitory action on the proliferation of certain cancer cell lines [52–56]. Therefore, the different biological responses mediated by GPER in distinct tumor cell contexts may depend on the receptor expression repertoire, the signaling pathways activated and other factors that remain to be fully elucidated.

The involvement of diverse miRNAs in breast cancer progression has been well established [6]. For instance, it has been reported that let-7d, miR-210 and miR-221 are down-regulated in the breast ductal carcinoma in situ and up-regulated following the invasive transition. Moreover, miR-9, miR-10b, miR-21, miR-29a, miR-155 and miR-373-520 family were found to promote the metastatic tumor dissemination [57]. Next, member of the let-7, miR-200, miR-34 and miR-125b families, were able to regulate the epithelial-mesenchymal transition in breast cancer [57]. According to the results obtained in the present investigation, previous studies have indicated that in diverse pathophysiological conditions, including breast cancer, the regulation of certain miRNAs by E2 may involve GPER activation [21,25,58,59]. It has been shown that GPER activation by estrogens stimulates a network of transduction pathways, which triggers key factors involved in cell growth, differentiation and transformation, like c-Fos [5,44,60,61]. The proto-oncogene c-Fos represents a prototypical "immediate early" gene since its expression is rapidly induced by different extracellular stimuli through the activation of the serine-threonine kinases of mitogen-activated protein kinase (MAPK) family [62,63]. The nuclear protein encoded by c-Fos interacts with Jun family members to form the heterodimeric activating protein-1 transcription factor complex (AP-1), which binds to TGAC/GTC/AA sequences (AP-1 responsive elements) located within the promoter sequences of target genes [62,64]. Many studies focusing on the oncogenic functions of c-Fos have demonstrated its involvement in tumor growth through the modulation of Cyclin D1, which is a nuclear regulatory subunit of the cyclin-dependent kinases (CDK)-4 and CDK-6 [65–67]. Nicely fitting with these data, we determined that in SkBr3 cancer cells and CAFs E2 and G-1 induce c-Fos and Cyclin D1 expression toward cell proliferation. According to the inhibitory function of miR-338-3p in certain cancer types [30–34], we also found that miR-338-3p abrogates the abovementioned effects triggered by E2 and G-1 in SkBr3 cells and in important components of the tumor microenvironment as CAFs [35,36]. In this regard, our data highlight additional mechanisms by which tumor cells and CAFs cooperate toward worse cancer features. Well-fitting with the present findings, it has been established that cancer development involves the functional interaction of malignant cells with the tumor microenvironment [68,69]. For instance, stromal cells like CAFs generate a dynamic signaling network through the secretion of growth factors and cytokines that stimulate the proliferation and dissemination of cancer cells [70,71]. In this context, the regulation of miR-338-3p shared by breast cancer cells and CAFs may be a further mechanism linking the estrogen stimulation of both the tumor microenvironment and tumor cells.

5. Conclusions

miRNAs target numerous genes involved in the cell growth and survival of diverse types of tumors, including breast cancer [72]. Therefore, changes in miRNAs expression may have a prognostic role along with a therapeutic perspective in cancer patients. Here, we have provided novel insights on the molecular mechanisms through which estrogenic GPER signaling in both breast cancer cells and

CAFs lowers the expression of miR-338-3p, which has been reported to act as an inhibitor of cancer cell growth and invasion [30–34]. Further studies are needed to better define the functions of miR-338-3p and its usefulness in innovative therapeutic approaches in breast cancer patients.

Author Contributions: A.V., M.L.P. and M.M. conceived and designed the study. A.V., A.S., D.C.R., F.C., G.R.G., M.T., M.F.S., R.L., F.G. performed the experiments. A.M.M. provided breast tumor samples. A.V., M.L.P. and M.M. analyzed and interpreted the data. A.V. and M.M. wrote the manuscript. M.M. acquired the funding. All authors have read and approved the final manuscript.

References

1. Burns, K.A.; Korach, K.S. Estrogen receptors and human disease: An update. *Arch. Toxicol.* **2012**, *86*, 1491–1504. [CrossRef] [PubMed]
2. Nilsson, S.; Gustafsson, J. Estrogen receptors: Therapies targeted to receptor subtypes. *Clin. Pharmacol. Ther.* **2011**, *89*, 44–55. [CrossRef] [PubMed]
3. Cirillo, F.; Pellegrino, M.; Malivindi, R.; Rago, V.; Avino, S.; Muto, L.; Dolce, V.; Vivacqua, A.; Rigiracciolo, D.C.; De Marco, P.; et al. GPER is involved in the regulation of the estrogen-metabolizing CYP1B1 enzyme in breast cancer. *Oncotarget* **2017**, *8*, 106608–106624. [CrossRef] [PubMed]
4. Santolla, M.F.; Avino, S.; Pellegrino, M.; De Francesco, E.M.; De Marco, P.; Lappano, R.; Vivacqua, A.; Cirillo, F.; Rigiracciolo, D.C.; Scarpelli, A.; et al. SIRT1 is involved in oncogenic signaling mediated by GPER in breast cancer. *Cell Death Dis.* **2015**, *6*, e1834. [CrossRef] [PubMed]
5. Prossnitz, E.R.; Barton, M. Estrogen biology: New insights into GPER function and clinical opportunities. *Mol. Cell. Endocrinol.* **2014**, *389*, 71–83. [CrossRef] [PubMed]
6. Vrtačnik, P.; Ostanek, B.; Mencej-Bedrač, S.; Marc, J. The many faces of estrogen signaling. *Biochem. Med.* **2014**, *24*, 329–342. [CrossRef] [PubMed]
7. Santolla, M.F.; Lappano, R.; De Marco, P.; Pupo, M.; Vivacqua, A.; Sisci, D.; Abonante, S.; Iacopetta, D.; Cappello, A.R.; Dolce, V.; et al. G protein-coupled estrogen receptor mediates the up-regulation of fatty acid synthase induced by 17β-estradiol in cancer cells and cancer-associated fibroblasts. *J. Biol. Chem.* **2012**, *287*, 43234–43245. [CrossRef] [PubMed]
8. Vivacqua, A.; Romeo, E.; De Marco, P.; De Francesco, E.M.; Abonante, S.; Maggiolini, M. GPER mediates the Egr-1 expression induced by 17β-estradiol and 4-hydroxitamoxifen in breast and endometrial cancer cells. *Breast Cancer Res. Treat.* **2012**, *133*, 1025–1035. [CrossRef] [PubMed]
9. Bartel, D.P. MicroRNAs: Genomics, biogenesis, mechanism, and function. *Cell* **2004**, *116*, 281–297. [CrossRef]
10. Bartel, D.P. MicroRNAs: Target recognition and regulatory functions. *Cell* **2009**, *136*, 215–233. [PubMed]
11. Fabian, M.R.; Sonenberg, N.; Filipowicz, W. Regulation of mRNA translation and stability by microRNAs. *Annu. Rev. Biochem.* **2010**, *79*, 351–379. [CrossRef] [PubMed]
12. Ben-Hamo, R.; Efroni, S. MicroRNA regulation of molecular pathways as a generic mechanism and as a core disease phenotype. *Oncotarget* **2015**, *6*, 1594–1604. [CrossRef] [PubMed]
13. Gross, N.; Kropp, J.; Khatib, H. MicroRNA signaling in embryo development. *Biology* **2017**, *6*, 34. [CrossRef] [PubMed]
14. Montagner, S.; Dehó, L.; Monticelli, S. MicroRNAs in hematopoietic development. *BMC Immunol.* **2014**, *15*, 14. [CrossRef] [PubMed]
15. Singh, R.P.; Massachi, I.; Manickavel, S.; Singh, S.; Rao, N.P.; Hasan, S.; Mc Curdy, D.K.; Sharma, S.; Wong, D.; Hahn, B.H.; et al. The role of miRNA in inflammation and autoimmunity. *Autoimmun. Rev.* **2013**, *12*, 1160–1165. [CrossRef] [PubMed]
16. Malemud, C.J. MicroRNAs and osteoarthritis. *Cells* **2018**, *7*, 92. [CrossRef] [PubMed]
17. Vannini, I.; Fanini, F.; Fabbri, M. Emerging roles of microRNAs in cancer. *Curr. Opin. Genet. Dev.* **2018**, *48*, 128–133. [CrossRef] [PubMed]
18. Santolla, M.F.; Lappano, R.; Cirillo, F.; Rigiracciolo, D.C.; Sebastiani, A.; Abonante, S.; Tassone, P.; Tagliaferri, P.; Di Martino, M.T.; Maggiolini, M.; et al. miR-221 stimulates breast cancer cells and cancer-associated fibroblasts (CAFs) through selective interference with the A20/c-Rel/CTGF signaling. *J. Exp. Clin. Cancer Res.* **2018** *37*, 94. [CrossRef] [PubMed]

19. Inamura, K. Major tumor suppressor and oncogenic non-coding RNAs: Clinical relevance in lung cancer. *Cells* **2017**, *6*, 12. [CrossRef] [PubMed]

20. Vivacqua, A.; De Marco, P.; Belfiore, A.; Maggiolini, M. Recent advances on the role of microRNAs in both insulin resistance and cancer. *Curr. Pharm. Des.* **2017**, *23*, 3658–3666. [CrossRef] [PubMed]

21. Jacovetti, C.; Abderrahmani, A.; Parnaud, G.; Jonas, J.C.; Peyot, M.L.; Cornu, M.; Laybutt, R.; Meugnier, E.; Rome, S.; Thorens, B.; et al. MicroRNAs contribute to compensatory β cell expansion during pregnancy and obesity. *J. Clin. Investig.* **2012**, *122*, 3541–3551. [CrossRef] [PubMed]

22. Wang, Y.; Liu, Z.; Shen, J. MicroRNA-421-targeted PDCD4 regulates breast cancer cell proliferation. *Int. J. Mol. Med.* **2018**. [CrossRef] [PubMed]

23. Tao, S.; He, H.; Chen, Q.; Yue, W. GPER mediated estradiol reduces miR-148a to promote HLA-G expression in breast cancer. *Biochem. Biophys. Res. Commun.* **2014**, *451*, 74–78. [CrossRef] [PubMed]

24. Zhang, Y.; Fang, J.; Zhao, H.; Yu, Y.; Cao, X.; Zhang, B. Downregulation of microRNA-1469 promotes the development of breast cancer via targeting HOXA1 and activating PTEN/PI3K/AKT and Wnt/β-catenin pathways. *J. Cell. Biochem.* **2018**. [CrossRef] [PubMed]

25. Vivacqua, A.; De Marco, P.; Santolla, M.F.; Cirillo, F.; Pellegrino, M.; Panno, M.L.; Abonante, S.; Maggiolini, M. Estrogenic gper signaling regulates mir144 expression in cancer cells and cancer-associated fibroblasts (cafs). *Oncotarget* **2015**, *6*, 16573–16587. [CrossRef] [PubMed]

26. Rodriguez, A.; Griffiths-Jones, S.; Ashurst, J.L.; Bradley, A. Identification of mammalian microRNA host genes and transcription units. *Genome Res.* **2004**, *14*, 1902–1910. [CrossRef] [PubMed]

27. Raghunath, M.; Patti, R.; Bannerman, P.; Lee, C.M.; Baker, S.; Sutton, L.N.; Phillips, P.C.; Damodar Reddya, C. A novel kinase, AATYK induces and promotes neuronal differentiation in a human neuroblastoma (SH-SY5Y) cell line. *Mol. Brain Res.* **2000**, *77*, 151–162. [CrossRef]

28. Tsuchiya, S.; Oku, M.; Imanaka, Y.; Kunimoto, R.; Okuno, Y.; Terasawa, K.; Sato, F.; Tsujimoto, G.; Shimizu, K. MicroRNA-338-3p and microRNA-451 contribute to the formation of basolateral polarity in epithelial cells. *Nucleic Acids Res.* **2009**, *37*, 3821–3827. [CrossRef] [PubMed]

29. Aschrafi, A.; Schwechter, A.D.; Mameza, M.G.; Natera-Naranjo, O.; Gioio, A.E.; Kaplan, B.B. MicroRNA-338 regulates local cytochrome c oxidase IV mRNA levels and oxidative phosphorylation in the axons of sympathetic neurons. *J. Neurosci.* **2008**, *28*, 12581–12590. [CrossRef] [PubMed]

30. Cao, R.; Shao, J.; Hu, Y.; Wang, L.; Li, Z.; Sun, G.; Gao, X. microRNA-338-3p inhibits proliferation, migration, invasion, and EMT in osteosarcoma cells by targeting activator of 90 kDa heat shock protein ATPase homolog 1. *Cancer Cell Int.* **2018**, *18*, 49. [CrossRef] [PubMed]

31. Li, Y.; Chen, P.; Zu, L.; Liu, B.; Wang, M.; Zhou, Q. MicroRNA-338-3p suppresses metastasis of lung cancer cells by targeting the EMT regulator Sox4. *Am. J. Cancer Res.* **2016**, *6*, 127–140. [PubMed]

32. Jin, Y.; Zhao, M.; Xie, Q.; Zhang, H.; Wang, Q.; Ma, Q. MicroRNA-338-3p functions as tumor suppressor in breast cancer by targeting SOX4. *Int. J. Oncol.* **2015**, *47*, 1594–1602. [CrossRef] [PubMed]

33. Wen, C.; Liu, X.; Ma, H.; Zhang, W.; Li, H. miR-338-3p suppresses tumor growth of ovarian epithelial carcinoma by targeting Runx2. *Int. J. Oncol.* **2015**, *46*, 2277–2285. [CrossRef] [PubMed]

34. Huang, X.H.; Chen, J.S.; Wang, Q.; Chen, X.L.; Wen, L.; Chen, L.Z.; Bi, J.; Zhang, L.J.; Su, Q.; Zeng, W.T. miR-338-3p suppresses invasion of liver cancer cell by targeting smoothened. *J. Pathol.* **2011**, *225*, 463–472. [CrossRef] [PubMed]

35. Farhood, B.; Najafi, M.; Mortezaee, K. Cancer-associated fibroblasts: Secretions, interactions, and therapy. *J. Cell. Biochem.* **2018**. [CrossRef] [PubMed]

36. Kalluri, R.; Zeisberg, M. Fibroblasts in cancer. *Nat. Rev. Cancer* **2006**, *6*, 392–401. [CrossRef] [PubMed]

37. Pfaffl, M.W. A new mathematical model for relative quantification in real-time RT-PCR. *Nucleic Acids Res.* **2001**, *29*, e45. [CrossRef] [PubMed]

38. Lanczky, A.; Nagy, A.; Bottai, G.; Munkacsy, G.; Paladini, L.; Szabo, A.; Santarpia, L.; Gyorffy, B. miRpower: A web-tool to validate survival-associated miRNAs utilizing expression data from 2178 breast cancer patients. *Breast Cancer Res Treat.* **2016**, *160*, 439–446. [CrossRef] [PubMed]

39. Vivacqua, A.; Lappano, R.; De Marco, P.; Sisci, D.; Aquila, S.; De Amicis, F.; Fuqua, S.A.; Andò, S.; Maggiolini, M. G protein-coupled receptor 30 expression is up-regulated by EGF and TGF alpha in estrogen receptor alpha-positive cancer cells. *Mol. Endocrinol.* **2009**, *23*, 1815–1826. [CrossRef] [PubMed]

40. Curtis, C.; Shah, S.P.; Chin, S.F.; Turashvili, G.; Rueda, O.M.; Dunning, M.J.; Speed, D.; Lynch, A.G.;

Samarajiwa, S.; Yuan, Y.; et al. The genomic and transcriptomic architecture of 2,000 breast tumours reveals novel subgroups. *Nature* **2012**, *486*, 346–352. [CrossRef] [PubMed]

41. Albanito, L.; Madeo, A.; Lappano, R.; Vivacqua, A.; Rago, V.; Carpino, A.; Oprea, T.I.; Prossnitz, E.R.; Musti, A.M.; Andò, S.; et al. G protein-coupled receptor 30 (GPR30) mediates gene expression changes and growth response to 17β-estradiol and selective GPR30 ligand G-1 in ovarian cancer cells. *Cancer Res.* **2007**, *67*, 1859–1866. [CrossRef] [PubMed]

42. Vivacqua, A.; Bonofiglio, D.; Recchia, A.G.; Musti, A.M.; Picard, D.; Andò, S.; Maggiolini, M. The G protein-coupled receptor GPR30 mediates the proliferative effects induced by 17β-estradiol and hydroxytamoxifen in endometrial cancer cells. *Mol. Endocrinol.* **2006**, *20*, 631–646. [CrossRef] [PubMed]

43. Vivacqua, A.; Bonofiglio, D.; Albanito, L.; Madeo, A.; Rago, V.; Carpino, A.; Musti, A.M.; Picard, D.; Andò, S.; Maggiolini, M. 17β-estradiol, genistein, and 4-hydroxytamoxifen induce the proliferation of thyroid cancer cells through the G protein-coupled receptor GPR30. *Mol. Pharmacol.* **2006**, *70*, 1414–1423. [CrossRef] [PubMed]

44. Maggiolini, M.; Vivacqua, A.; Fasanella, G.; Recchia, A.G.; Sisci, D.; Pezzi, V.; Montanaro, D.; Musti, A.M.; Picard, D.; Andò, S. The G protein-coupled receptor GPR30 mediates c-fos up-regulation by 17β-estradiol and phytoestrogens in breast cancer cells. *J. Biol. Chem.* **2004**, *279*, 27008–27016. [CrossRef] [PubMed]

45. Madeo, A.; Maggiolini, M. Nuclear alternate estrogen receptor GPR30 mediates 17β-estradiol-induced gene expression and migration in breast cancer-associated fibroblasts. *Cancer Res.* **2010**, *70*, 6036–6046. [CrossRef] [PubMed]

46. Siegel, R.; Naishadham, D.; Jemal, A. Cancer statistics, 2013. *Cancer J. Clin.* **2013**, *63*, 11–30. [CrossRef] [PubMed]

47. Schettini, F.; Buono, G.; Cardalesi, C.; Desideri, I.; De Placido, S.; Del Mastro, L. Hormone receptor/human epidermal growth factor receptor 2-positive breast cancer: Where we are now and where we are going. *Cancer Treat. Rev.* **2016**, *46*, 20–26. [CrossRef] [PubMed]

48. Jiang, W.G.; Sanders, A.J.; Katoh, M.; Ungefroren, H.; Gieseler, F.; Prince, M.; Thompson, S.K.; Zollo, M.; Spano, D.; Dhawan, P.; et al. Tissue invasion and metastasis: Molecular, biological and clinical perspectives. *Semin. Cancer Biol.* **2015**, *35*, S244–S275. [CrossRef] [PubMed]

49. Molina, L.; Figueroa, C.D.; Bhoola, K.D.; Ehrenfeld, P. GPER-1/GPR30 a novel estrogen receptor sited in the cell membrane: Therapeutic coupling to breast cancer. *Expert Opin. Ther. Targets* **2017**, *21*, 755–766. [CrossRef] [PubMed]

50. Filardo, E.J. A role for G-protein coupled estrogen receptor (GPER) in estrogen-induced carcinogenesis: Dysregulated glandular homeostasis, survival and metastasis. *J. Steroid Biochem. Mol. Biol.* **2018**, *176*, 38–48. [CrossRef] [PubMed]

51. Sjöström, M.; Hartman, L.; Grabau, D.; Fornander, T.; Malmström, P.; Nordenskjöld, B.; Sgroi, D.C.; Skoog, L.; Stål, O.; Fredrik Leeb-Lundberg, L.M.; et al. Lack of G protein-coupled estrogen receptor (GPER) in the plasma membrane is associated with excellent long- term prognosis in breast cancer. *Breast Cancer Res. Treat.* **2014**, *145*, 61–71. [CrossRef] [PubMed]

52. Ariazi, E.A.; Brailoiu, E.; Yerrum, S.; Shupp, H.A.; Slifker, M.J.; Cunliffe, H.E.; Black, M.A.; Donato, A.L.; Arterburn, J.B.; Oprea, T.I.; et al. The G protein-coupled receptor GPR30 inhibits proliferation of estrogen receptor-positive breast cancer cells. *Cancer Res.* **2014**, *70*, 1184–1194. [CrossRef] [PubMed]

53. Chimento, A.; Casaburi, I.; Rosano, C.; Avena, P.; De Luca, A.; Campana, C.; Martire, E.; Santolla, M.F.; Maggiolini, M.; Pezzi, V.; et al. Oleuropein and hydroxytyrosol activate GPER/GPR30-dependent pathways leading to apoptosis of ERnegative SKBR3 breast cancer cells. *Mol. Nutr. Food Res.* **2014**, *58*, 478–489. [CrossRef] [PubMed]

54. Weißenborn, C.; Ignatov, T.; Ochel, H.J.; Costa, S.D.; Zenclussen, A.C.; Ignatov, Z.; Ignatov, A. GPER functions as a tumor suppressor in triple-negative breast cancer cells. *J. Cancer Res. Clin. Oncol.* **2014**, *140*, 713–723. [CrossRef] [PubMed]

55. Weißenborn, C.; Ignatov, T.; Poehlmann, A.; Wege, A.K.; Costa, S.D.; Zenclussen, A.C.; Ignatov, A. GPER functions as a tumor suppressor in MCF-7 and SKBR-3 breast cancer cells. *J. Cancer Res. Clin. Oncol.* **2014**, *140*, 663–671. [CrossRef] [PubMed]

56. Chan, Q.K.; Lam, H.M.; Ng, C.F.; Lee, A.Y.; Chan, E.S.; Ng, H.K.; Ho, S.M.; Lau, K.M. Activation of GPR30 inhibits the growth of prostate cancer cells through sustained activation of Erk1/2, c-jun/c-fos-dependent

upregulation of p21, and induction of G$_2$ cell-cycle arrest. *Cell Death Differ.* **2010**, *7*, 1511–1523. [CrossRef] [PubMed]

57. Volinia, S.; Galasso, M.; Sana, M.E.; Wise, T.F.; Palatini, J.; Huebner, K.; Croce, C.M. Breast cancer signatures for invasiveness and prognosis defined by deep sequencing of microRNA. *Proc. Natl. Acad. Sci. USA* **2012**, *109*, 3024–3029. [CrossRef] [PubMed]

58. Vidal-Gómez, X.; Pérez-Cremades, D.; Mompeón, A.; Dantas, A.P.; Novella, S.; Hermenegildo, C. MicroRNA as crucial regulators of gene expression in estradiol-treated human endothelial cells. *Cell. Physiol. Biochem.* **2018**, *45*, 1878–1892. [CrossRef] [PubMed]

59. Zhang, H.; Wang, X.; Chen, Z.; Wang, W. MicroRNA-424 suppresses estradiol-induced cell proliferation via targeting GPER in endometrial cancer cells. *Cell. Mol. Biol.* **2015**, *61*, 96–101. [PubMed]

60. Prossnitz, E.R.; Maggiolini, M. Mechanisms of estrogen signaling and gene expression via GPR30. *Mol. Cell. Endocrinol.* **2009**, *308*, 32–38. [CrossRef] [PubMed]

61. Pandey, D.P.; Lappano, R.; Albanito, L.; Madeo, A.; Maggiolini, M.; Picard, D. Estrogenic GPR30 signalling induces proliferation and migration of breast cancer cells through CTGF. *EMBO J.* **2009**, *28*, 523–532. [CrossRef] [PubMed]

62. Durchdewald, M.; Angel, P.; Hess, J. The transcription factor Fos: A Janus-type regulator in health and disease. *Histol. Histopathol.* **2009**, *11*, 1451–1461. [CrossRef]

63. Hess, J.; Angel, P.; Schorpp-Kistner, M. AP-1 subunits: Quarrel and harmony among siblings. *J. Cell Sci.* **2004**, *117*, 5965–5973. [CrossRef] [PubMed]

64. Milde-Langosch, K. The Fos family of transcription factors and their role in tumourigenesis. *Eur. J. Cancer* **2005**, *41*, 2449–2461. [CrossRef] [PubMed]

65. Bancroft, C.C.; Chen, Z.; Yeh, J.; Sunwoo, J.B.; Yeh, N.T.; Jackson, S.; Jackson, C.; Van Waes, C. Effects of pharmacologic antagonists of epidermal growth factor receptor, PI3K and. MEK signal kinases on NF-κB and AP-1 activation and IL-8 and VEGF expression in human head and neck squamous cell carcinoma lines. *Int. J. Cancer* **2002**, *99*, 538–548. [CrossRef] [PubMed]

66. Mishra, A.; Bharti, A.C.; Saluja, D.; Das, B.C. Transactivation and expression patterns of Jun and Fos/AP-1 super-family proteins in human oral cancer. *Int. J. Cancer* **2010**, *126*, 819–829. [CrossRef] [PubMed]

67. Qie, S.; Diehl, J.A. Cyclin D1, cancer progression, and opportunities in cancer treatment. *J. Mol. Med.* **2016**, *94*, 1313–1326. [CrossRef] [PubMed]

68. Han, Y.; Zhang, Y.; Jia, T.; Sun, Y. Molecular mechanism underlying the tumor-promoting functions of carcinoma-associated fibroblasts. *Tumor Biol.* **2015**, *36*, 1385–1394. [CrossRef] [PubMed]

69. Bhowmick, N.A.; Neilson, E.G.; Moses, H.L. Stromal fibroblasts in cancer initiation and progression. *Nature* **2004**, *432*, 332–337. [CrossRef] [PubMed]

70. Cheng, N.; Chytil, A.; Shyr, Y.; Joly, A.; Moses, H.L. Transforming growth factor-beta signaling-deficient fibroblasts enhance hepatocyte growth factor signaling in mammary carcinoma cells to promote scattering and invasion. *Mol. Cancer Res.* **2008**, *6*, 1521–1533. [CrossRef] [PubMed]

71. Zhi, K.; Shen, X.; Zhang, H.; Bi, J. Cancer-associated fibroblasts are positively correlated with metastatic potential of human gastric cancers. *J. Exp. Clin. Cancer Res.* **2010**, *29*, 66. [CrossRef] [PubMed]

72. Di Leva, G.; Cheung, D.G.; Croce, C.M. miRNA clusters as therapeutic targets for hormone-resistant breast cancer. *Expert Rev. Endocrinol. Metab.* **2015**, *10*, 607–617. [CrossRef] [PubMed]

5

Functional Characterization of microRNA171 Family in Tomato

Michael Kravchik, Ran Stav, Eduard Belausov and Tzahi Arazi *

Institute of Plant Sciences, Agricultural Research Organization, Volcani Center, P.O. Box 6, Bet Dagan 50250, Israel; michael.kravchik@mail.huji.ac.il (M.K.); ranstav@volcani.agri.gov.il (R.S.); eddy@volcani.agri.gov.il (E.B.)
* Correspondence: tarazi@agri.gov.il;

Abstract: Deeply conserved plant microRNAs (miRNAs) function as pivotal regulators of development. Nevertheless, in the model crop *Solanum lycopersicum* (tomato) several conserved miRNAs are still poorly annotated and knowledge about their functions is lacking. Here, the tomato miR171 family was functionally analyzed. We found that the tomato genome contains at least 11 *SlMIR171* genes that are differentially expressed along tomato development. Downregulation of sly-miR171 in tomato was successfully achieved by transgenic expression of a short tandem target mimic construct (STTM171). Consequently, sly-miR171-targeted mRNAs were upregulated in the silenced plants. Target upregulation was associated with irregular compound leaf development and an increase in the number of axillary branches. A prominent phenotype of *STTM171* expressing plants was their male sterility due to a production of a low number of malformed and nonviable pollen. We showed that sly-miR171 was expressed in anthers along microsporogenesis and significantly silenced upon *STTM171* expression. Sly-miR171-silenced anthers showed delayed tapetum ontogenesis and reduced callose deposition around the tetrads, both of which together or separately can impair pollen development. Collectively, our results show that sly-miR171 is involved in the regulation of anther development as well as shoot branching and compound leaf morphogenesis.

Keywords: miR171; pollen; STTM; tapetum; callose; tomato

1. Introduction

Plant microRNAs (miRNAs) constitute a major class of endogenous small RNAs and trigger the sequence-specific post-transcriptional repression of one to several target mRNAs with high sequence complementarity. The analysis of miRNAs from various land plants species indicated the presence of at least eight deeply conserved miRNA families in all embryophytes [1]. Studies of these miRNAs suggest that most of them act as master regulators of development by negative regulation of the expression of transcription factors that function in critical developmental processes [2,3].

The miR171 family is deeply conserved and exists in all major land plant groups, including bryophytes, one of the oldest groups of land plants [4]. Known plant genomes contain variable number of *MIR171* genes: from only two in *Citrus sinensis* to staggering 21 in *Glycine max* (miRBase release 22). Members of a miR171 family contain one or more nucleotide changes similar to members from other miRNA families, but unlike other conserved miRNAs they may be offset by three nucleotides relative to each other [5]. This atypical sequence offset may result in different target specificities for certain miR171 members [6]. Hitherto, miR171 members have been demonstrated to guide the cleavage of mRNAs coding for GRAS domain SCARECROW-like transcription factors that belong to the HAIRY MERISTEM (HAM) or NODULATION SIGNALING PATHWAY (NSP) clades [6–8].

In *Nicotiana benthamiana*, spatial characterization of miR171 expression by in-situ hybridization has shown that it is expressed in a wide variety of tissues including the shoot apical meristem

(SAM), leaf primordia, anthers and ovaries, thus hinting on its involvement in their development [9]. In *A. thaliana* (Arabidobsis), ath-miR171b expression was shown to oscillate during the diurnal cycle suggesting a potential role for light in its accumulation [10]. In *Medicago truncatula*, miR171h expression is upregulated in roots during their colonization by arbuscular mycorrhizal fungi or in response to lipochito-oligosaccharides that are released during fungi pre-symbiotic growth, suggesting that miR171h functions to prevent over-colonization of roots by arbuscular mycorrhizal fungi [8]. In rice, it was demonstrated that reduction of osa-miR171b contributes to Rice stripe virus symptoms whereas osa-miR171b overexpression caused opposite effects, suggesting that expression of miR171-targeted mRNAs may facilitate viral infection [11]. In addition, miR171 has been suggested to be involved in the regulation of abiotic stresses based on its upregulation in Arabidopsis seedlings grown under high salinity, cold, and drought conditions [12].

Target mimic is a non-cleavable miRNA complementary sequence embedded within a longer endogenous or artificial RNA. In contrast to overexpression of a miRNA, which will silence its cognate mRNA targets and hence faithfully report on their functions, the target mimic acts as an "miRNA sponge" that titer out complementary miRNAs and hence is suitable for their functional characterization [13]. Furthermore, ectopic expression of miRNA or its cleavage-resistant mRNA target may lead to deceptive identification of miRNA function, due to incorrect spatio-temporal expression. Overexpression of miR171 resulted in various opposite phenotypes such as dwarfed barley with less tillers and taller rice with more tillers, and even silencing of miR171 can lead to different phenotypes between various species [14]. Ath-miR171a downregulation by single target mimic configurations resulted in a range of phenotypes consistent with ath-miR171a involvement in multiple developmental processes. Common phenotypes included closed buds and reduced pollination due to altered sepal development that bent the carpels, and pale green leaves due to reduced chlorophyll accumulation [15,16]. In addition, ath-miR171a-depleted Arabidopsis had larger rosette leaves, a larger root system during the growth in soil and modified leaf angle under limited light conditions [16]. Nevertheless, a single target mimic sequence can only silence complementary miR171 members, but will not effectively bind miR171 members with sequence offset, thus limiting the efficacy of such configuration for functional analysis of miR171. Recently, expression of two target mimic sequences in a single transcript via short tandem target mimic (STTM) configuration was shown to efficiently induce complementary miRNAs degradation in Arabidopsis and tomato [13,17]. This approach, which may be applied to silence two different miRNAs in parallel, is thus highly suitable for the in planta functional characterization of miR171. Indeed, this approach has been successfully applied in rice. Rice STTM171 plants were semidwarf and had semienclosed panicles and drooping flag leaves. These unique phenotypes suggest divergent functions for miR171 in dicots and monocots [14,18].

Tomato is the number one non-starchy vegetable consumed worldwide and also serves as a primary model for fruit development and ripening. Nevertheless, at present, tomato miRNAs are poorly annotated and the functions of most remain elusive. To date, six miR171 members, sly-miR171a-f, were cloned from tomato (miRBase, release 22) [19]. Previously, we have demonstrated that sly-miR171a and sly-miR171b guide the cleavage of *SlHAM* and *SlHAM2* and sly-miR171b also guides the cleavage of the tomato *NSP2* homolog *SlNSP2L*. Sly-miR171a and sly-miR171b overexpression resulted in over-proliferation of meristematic cells in the periphery of meristems and in the organogenic compound leaf rachis, suggesting that sly-miR171-targeted *SlHAMs* function in meristem maintenance and compound leaf morphogenesis [6]. As part of our continuous effort to unravel the identity and roles of tomato miRNAs, in the current study, we re-annotated the tomato miR171 family and utilized the STTM approach to silence abundant members in the family and investigate their functions. Our results revealed the presence of a much more complex miR171 family than previously documented in tomato and its necessity for vegetative, anther, and pollen development.

2. Results and Discussion

2.1. The Tomato miR171 Family of miRNAs

Six sly-miR171 members, sly-miR171a-f were previously cloned from tomato (miRBase, release 22). To identify additional sly-miR171 members, the small RNAs deposited in the Tomato Functional Genomics Database (TFGD; http://ted.bti.cornell.edu/cgi-bin/TFGD/sRNA/sRNA.cgi) and in-house tomato cv. M82 small RNA data [20] were queried with known miR171 sequences. This search detected eleven differentially abundant 21-nucleotide putative sly-miR171 sequences, including the previously cloned sly-miR171a, b, e, f and a 1-nucleotide longer version of sly-miR171d (Figure 1A). Sequence alignment revealed that identified sly-miR171 sequences can be divided into two groups, which are offset by three nucleotides relative to each other, similar to the Arabidopsis miR171 founder sequences ath-miR171a (group A) and ath-miR171c (group B). By mapping the putative sly-miR171 sequences to the tomato genome followed by alignment of the cloned small RNAs to their predicted pre-miRNAs sequences, we identified the miRNA stars (miRNA*) for all, indicating that they are authentic miRNAs (Table S1). In addition, this alignment revealed that one pre-miRNA (SlMIR171a,b) encodes for both sly-miR171a and sly-miR171b and their respective miRNA* strands (Figure 1B; Table S1). Moreover, four additional newly identified miR171 members (iso-sly-miR171a.1, iso-sly-miR171a.2, iso-sly-miR171b, iso-sly-miR171d) were found to be encoded by an identical precursor as that of other miR171 members and overlapped them in sequence (Figure 1B and Table S1), suggesting that they represent iso-miRNAs [21]. Querying the TFGD with these sequences confirmed their expression in tissues other than seedlings supporting their functionality as miRNAs (Figure 1C). It is noteworthy that iso-sly-miR171d miRNA* strand was previously annotated as sly-miR171c (miRBase, release 22). This analysis indicates that the tomato miR171 family is much more complex than previously thought, but it is of medium size compared to other miR171 families such as in *Glycine max* that contains up to 21 members (miRBase, release 22). Nevertheless, such complexity may hint on redundancy and specialization among different sly-miR171 members. Sly-miR171a and sly-miR171b, which represent group A and B sly-miR171 members, respectively, guide the cleavage of *SlHAM* and *SlHAM2*. Sly-miR171b, but not sly-miR171a, also guides the cleavage of the tomato *SlNSP2L* [6]. Prediction of mRNA targets and mining the published tomato degradome data [22] did not reveal strong evidence for additional sly-miR171-guided mRNA cleavage (Supplement 1).

To identify sly-miR171 sites of activity in tomato, analysis of public small RNAseq data at TFGD was performed. This analysis revealed that sly-miR171a, sly-miR171b, iso-sly-miR171d, and sly-miR171e are the most abundant members in the family, but the expression of each is prominent in a distinct tissue or developmental stage: sly-miR171a—leaves, floral buds and anthesis flowers, sly-miR171b—immature green fruit, iso-sly-miR171d—anthesis flowers and sly-miR171e—leaves, mature and ripening fruit (Figure 1C). Hence, the sly-miR171 family functions in vegetative as well as reproductive tissues with possible functional diversification between different members.

2.2. Knockdown of Sly-miR171 Activity Using the STTM Approach

Previously it was demonstrated that STTM configuration, which contains two target mimic sequences separated by a spacer, is very effective in counteracting the activity of several miRNAs in Arabidopsis and tomato [13,17]. Therefore, a similar STTM configuration was chosen to knockdown sly-miR171 family activity and uncover its importance for tomato development. Since group A and group B sly-miR171 members are offset by three nucleotides relative to each other, the STTM171 fragment was designed to comprise two different target mimic sequences, each of which have complementarity suitable to bind most of group A or group B sly-miR171 members, especially the most abundant sly-miR171 members (Figures 2A and S1). The STTM171 fragment was cloned downstream of the CaMV *35S* promoter (*35S:STTM171*) and then transformed into tomato cv. M82. Seventeen independent transgenic *35S:STTM171* plants were regenerated and screened by northern blot for reduced sly-miR171a and sly-miR171b levels. This analysis identified three T0 primary transformants (9, 17, and 19) with significantly reduced sly-miR171 levels compared to control transgenic *35S:GFP* plants (Figure 2B). Plants *35S:STTM171-9* and *35S:STTM171-19*, which accumulated ~27% and ~21% of the total sly-miR171 levels, respectively, produced only few completely seedless fruits. Compared to the control plants, the *35S:STTM171-17* T0 plant, which accumulated ~30% of total sly-miR171 levels, produced smaller fruit in size and number, most of which were seedless and few contained a small number of seeds (Figure S2A,B). The smaller fruit size of *35S:STTM171-17* plants was probably a secondary effect which emerged due to the reduction in seed number [23]. Transformation efforts to produce additional independent fertile transgenic plants with significantly reduced sly-miR171 levels were not successful (data not shown). The sterility of *35S:STTM171-9* and *35S:STTM171-19* plants prevented further analysis of their progeny. Thus, further characterization of STTM171 plants was performed on *35S:STTM171-17* T2 and T3 progeny. Quantitation of sly-miR171-targeted *SlHAM*, *SlHAM2*, and *SlNSP2* transcripts in young leaves of the *35S:STTM171-17* T2 plants revealed significant ~2–2.5 fold upregulation in all (Figure 2C) indicating that both sly-miR171a and sly-miR171b activities were attenuated in these leaves by *35S:STTM171* expression. Indeed, quantitation of sly-miR171a-b in these leaves confirmed their reduced accumulation (Figure 2D).

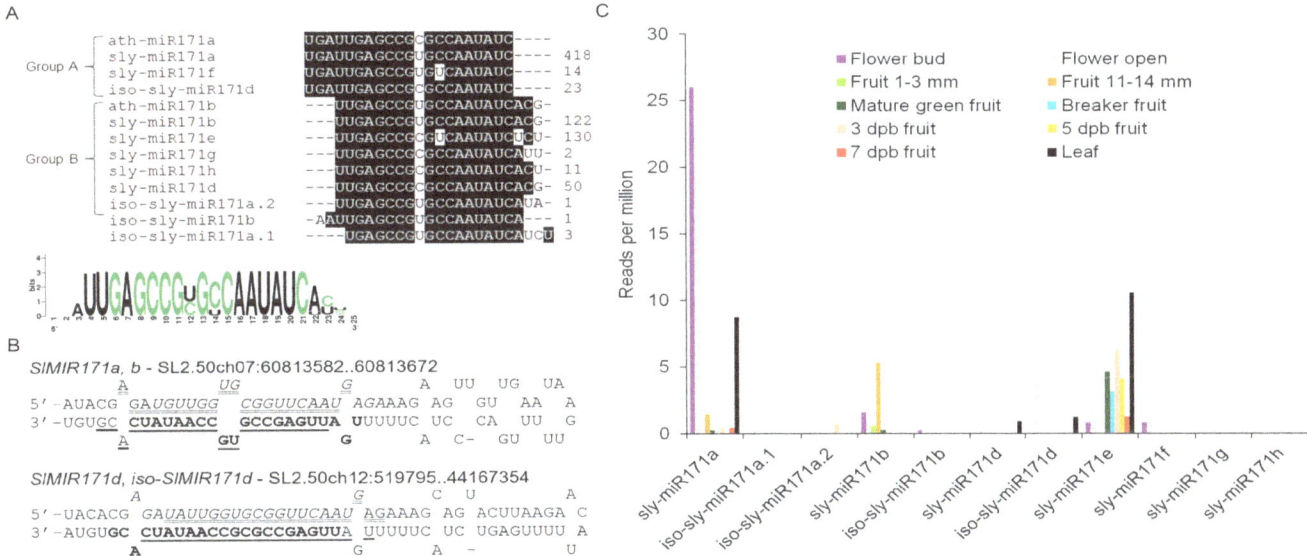

Figure 1. The tomato miR171 family. (**A**) Nucleotide sequence alignment of Arabidopsis (ath-miR171) and tomato (sly-miR171) miR171 members. Relative abundance in seedlings is indicated for each on the right. (**B**) Examples of sly-miR171 precursors that encode two sly-miR171 isoforms. The sequence of *SlMIR171a,b* (SL2.50ch07:60813582..60813672) and *SlMIR171d, iso-SlMIR171d* (SL2.50ch12:519795..519887) stem and loops. The sequences of sly-miR171a/d, sly-miR171a*/d*, sly-miR171b/ sly-miRiso-d and sly-miR171b*/iso-d* are bold-face, italicized, underlined and double-underlined, respectively. (**C**) Accumulation of sly-miR171 members in flower and fruit tissues of tomato cv. Microtome (Flower and fruit) and Heinz (leaf) based on small RNA-seq data deposited in the TFGD database. Dpb—days post breaker.

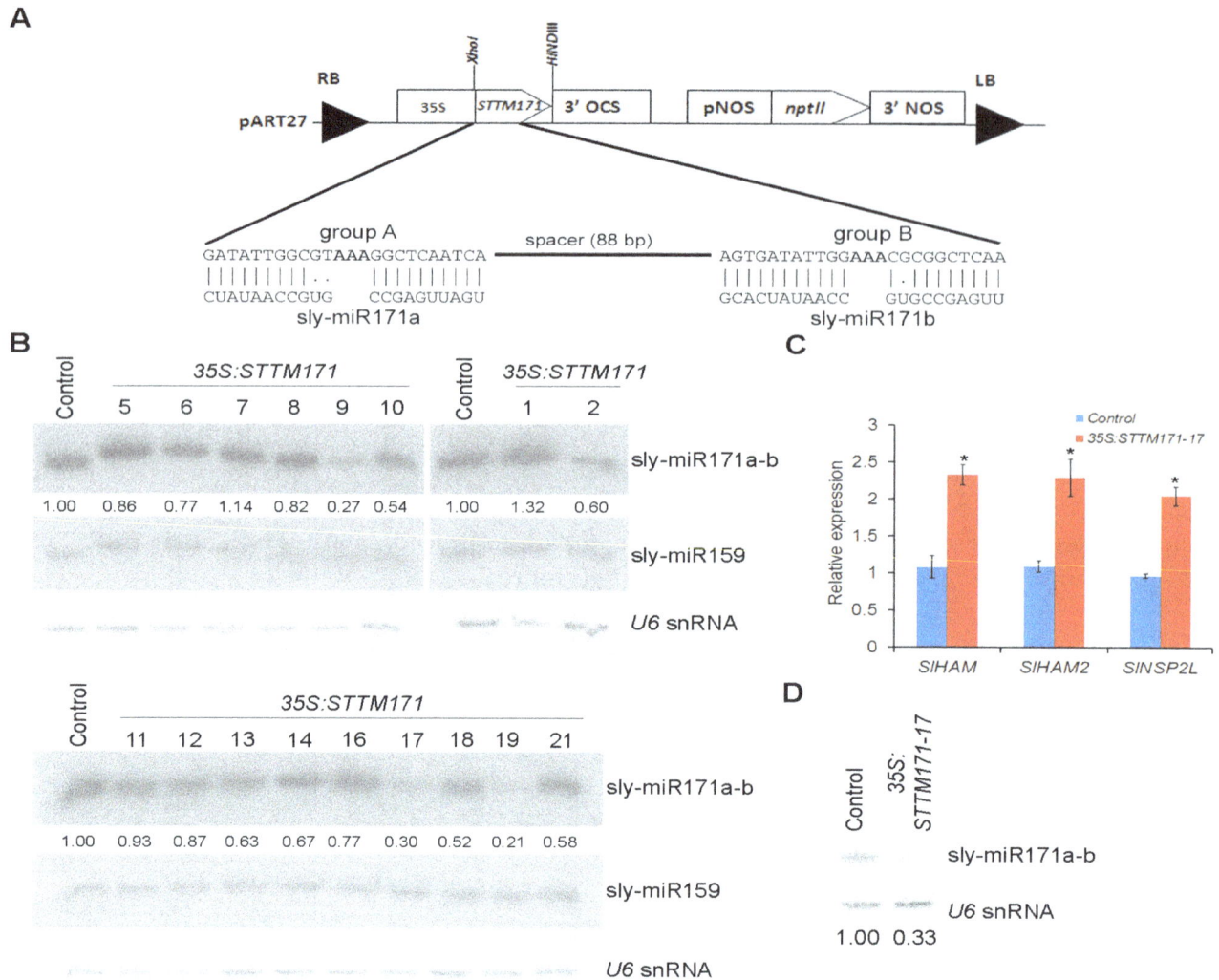

Figure 2. Generation of transgenic STTM171 tomato with reduced sly-miR171 levels. (**A**) A scheme of the Short Tandem Target Mimic construct used for tomato M82 transformation. The Watson–Crick pairings between group A and B target mimic sites and sly-miR171 representative members are shown in the expanded region. (**B**) RNA gel blot analysis of sly-miR171 levels in indicated transgenic T0 plants. Total RNA (5 μg) from leaves was probed by sly-miR171a (sly-miR171a-b), sly-miR159 and *U6* antisense probes. Sly-miR171 expression levels were determined after normalization to sly-miR159 and *U6* snRNA by geometric averaging and are indicated below. (**C**) RT-qPCR analysis of sly-miR171 target transcripts in RNA from young leaves of one-month old T2 35S:STTM171-17 plants. *TIP41* expression values were used for normalization. Error bars indicate ± SD of three biological replicates, each measured in triplicate. Asterisks indicate significant difference relative to *35:GFP* control plants (Tukey–Kramer multiple comparison test; $p < 0.01$). (**D**) RNA gel blot analysis of sly-miR171 in 5 μg total RNA from the samples analyzed in C. The blots were probed with sly-miR171a (sly-miR171a-b) antisense probe. Sly-miR171 expression levels were determined after normalization to *U6* snRNA and are indicated below.

2.3. Sly-miR171 Silencing Affected Compound Leaf Morphogenesis and Increased Branching

During vegetative development the compound leaves of *35S:STTM171-17* plants developed primary leaflets that frequently had a distorted growth angle, were larger and their lobes were deeper than that of the control, implicating sly-miR171 in compound leaf morphogenesis (Figure 3A,B). In addition, compared to control plants, the number of axillary shoots was significantly higher in the *35S:STTM171* plants (Figure 3C). This phenotype is consistent with the increased lateral branch number observed in transgenic tomato plants that ectopically expressed *SlHAM2/SlGRAS24*, which was upregulated in *35S:STTM171* leaves (Figure 2C), and with transgenic Arabidopsis that ectopically expressed the ath-miR171c-ressitant versions of *SCL6-II/HAM1*, *SCL6-III/HAM2*, and *SCL6-IV/HAM4* [24,25]. Both *SlHAM* and *SlHAM2* are abundant in vegetative and reproductive meristems and function in their maintenance [6]. Thus, sly-miR171 may suppress lateral branching by the negative regulation of the expression of *SlHAM2* and apparently also *SlHAM* in axillary meristems.

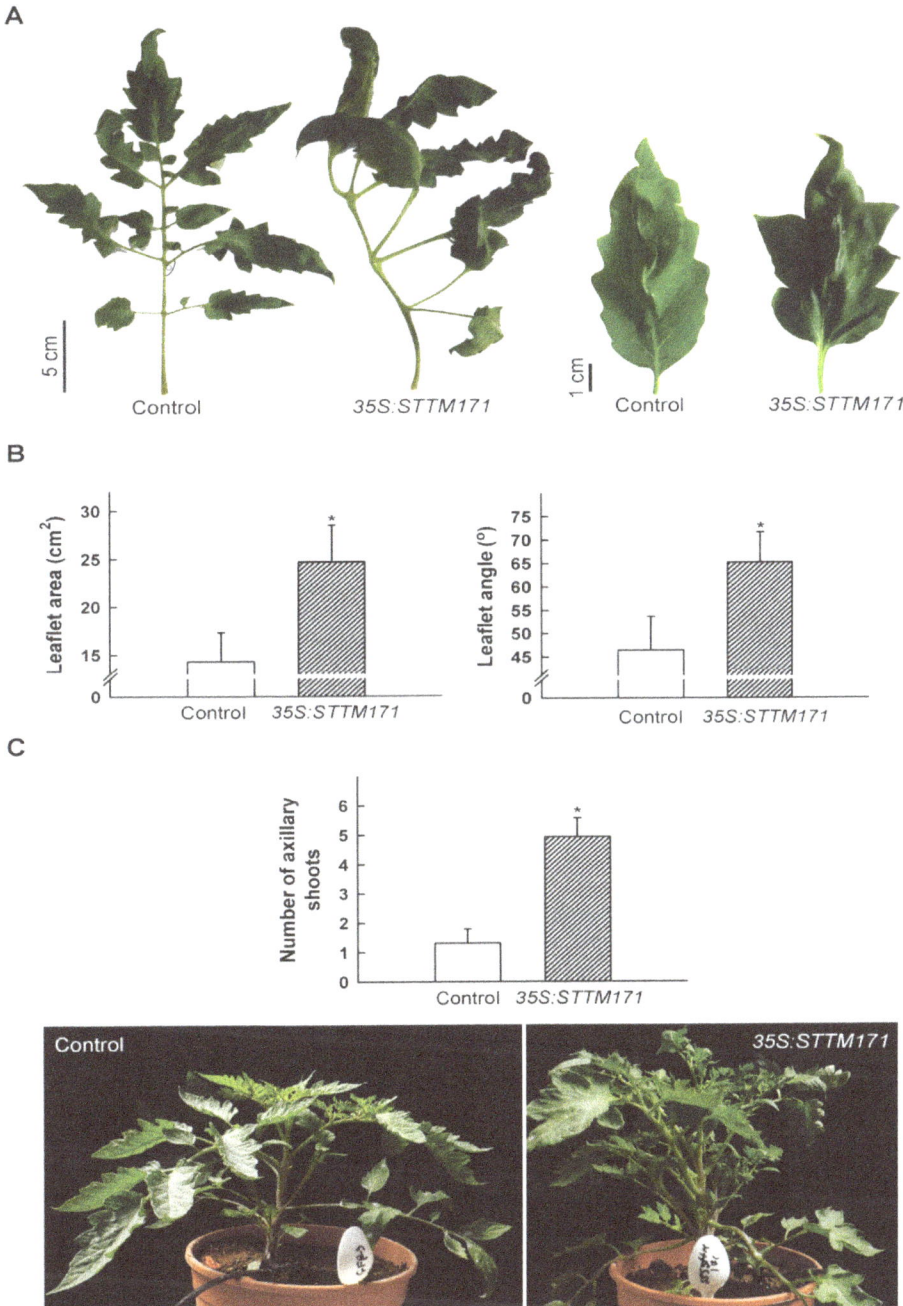

Figure 3. Vegetative phenotypes of *35S:STTM171* plants. (**A**) Photograph of representative fifth leaf and terminal leaflet from 45 DAG plants of indicated genotypes. (**B**) Quantitation of primary leaflet area and petiolule angle (indicated in (A)) in leaves (n = 26) similar to those shown in (A). (**C**) Quantitation of the number of axillary shoots on the main stem (≥0.5 cm) at eight leaf stage plants (n = 13). Error bars indicate ±SD. Asterisks indicate significant difference as determined by Student's *t*-test ($p \leq 0.001$). Representative plant of each genotype is shown below. Pot diameter = 18.8 cm.

2.4. Sly-miR171 Silencing Affected Pollen Morphology and Production

Despite the normal number and morphology of their flowers, *35S:STTM171* plants set only few fruits that were mostly seedless. Whereas manual pollination of *35S:STTM171* flowers with wild-type pollen rarely succeeded, the reciprocal pollination completely failed to induce fruit set, indicating reduced male fertility of the *35S:STTM171* flowers. A similar male sterile phenotype was observed in the transgenic F1 progeny from the cross between wild type and *35S:STTM171* T2 plants suggesting that the *35:STTM171* transgene caused the phenotype. This is also supported by the finding that overexpression of the sly-miR171 target *SlHAM2/SlGRAS24* reduced seed number due to male sterility [25]. To further understand the basis of the male sterility phenotype, we assessed the productivity and quality of pollen grains in anthesis flowers of *35S:STTM171* by differential Alexander staining [26], which distinguishes between aborted and non-aborted pollen, and by testing pollen germination. This analysis indicated that the average total number of *35S:STTM171* pollen grains was reduced by 42% compared to control flowers. Moreover, the majority of the *35S:STTM171* pollen grains were aborted (60.4%), a fraction that is 4-fold higher than that found in control pollen grains. Consistent with that, the number of germinated pollen grains fell by 5.6-fold (Figure 4A). These data suggest that the *35S:STTM171* plants produce relatively small numbers of poor-quality pollen grains compared to control. This is consistent with the apparent sterility of *35S:STTM171* plants and explains why manual fertilization of wild-type tomato flowers with *35S:STTM171* pollen grains was unsuccessful. The observed almost seedless fruit and reduced pollen viability phenotypes in *35S:STTM171* plants are reminiscent to those described in *SlGRAS24/SlHAM2*-overexpressing plants [25]. This similarity suggested that the negative regulation of *SlHAM2* levels by sly-miR171 may be critical for pollen development.

To determine the cause of pollen abortion in *STTM171* expressing plants we initially analyzed pollen morphology by scanning electron microscope (SEM). Wild-type tomato cv. M82 mature pollen grains are psilate and tricolporate [27]. SEM analysis of *35S:STTM171* mature pollen grains revealed that although they remained psilate, they had deformed shapes. These included a collapsed wall, disordered germinal apertures and instead of being tricolporate many were tetracolpate (Figure 4B,C). In contrast to their morphological abnormalities, DAPI staining of *35S:STTM171* developing pollen nuclei did not reveal any nuclear aberrations during the formation of tetrads (Figure 5A), microspores (Figure 5B,C), and bicellular pollen (Figure 5D), suggesting that morphological and not nuclear aberrations underlie the poor quality of the *35S:STTM171* pollen.

2.5. The 35S:STTM171 Anthers Accumulated Reduced Sly-miR171 Levels Associated with Delayed Tapetum Degeneration and Reduced Callose Deposition

Male sterility is frequently associated with deviations in the development of the anthers that contain the sporogenous tissue and its circumjacent tissues, the tapetum and middle layer, which ultimately gives rise to the pollen grains and support pollen development correspondingly [28]. In situ of miR171 in *N. benthamiana* developing flowers has detected high uniform expression in the pollen sacs and surrounding tissues of young anthers [9], suggesting miR171 involvement in pollen development. Northern analysis with sly-miR171a/b/e validated probes (Supplement 2; Figure S3) showed that sly-miR171a/b and sly-miR171e are abundant in anthers throughout their development until maturity (anthesis flower) (Figure S4A). Anther developmental stages were defined according to the study of flower development of tomato by Brukhin et al. [29] and verified by DAPI staining from the tetrad stage (Figure 5). In agreement with the northern analysis, deep sequencing of small RNAs from developing tomato anthers identified sly-miR171a, its group member iso-sly-miR171d and sly-miR171d, a group member of sly-miR171e [30]. Compared to control anthers, the anthers of *35S:STTM171* accumulated significantly reduced levels of sly-miR171a/b in the meiosis (4 mm, stage 9), tetrad stage (5 mm, stages 10–11), free microspores stage (6 mm, stage 12) and mature pollen stage (12 mm, stages 18–19). Moreover, reduced levels were observed for sly-miR171e for which silencing was even more pronounced than for sly-miR171a/b. Depending on the developmental stage, the *35S:STTM171* anthers accumulated only around 5–20% of the control levels of sly-miR171e (Figure S4B).

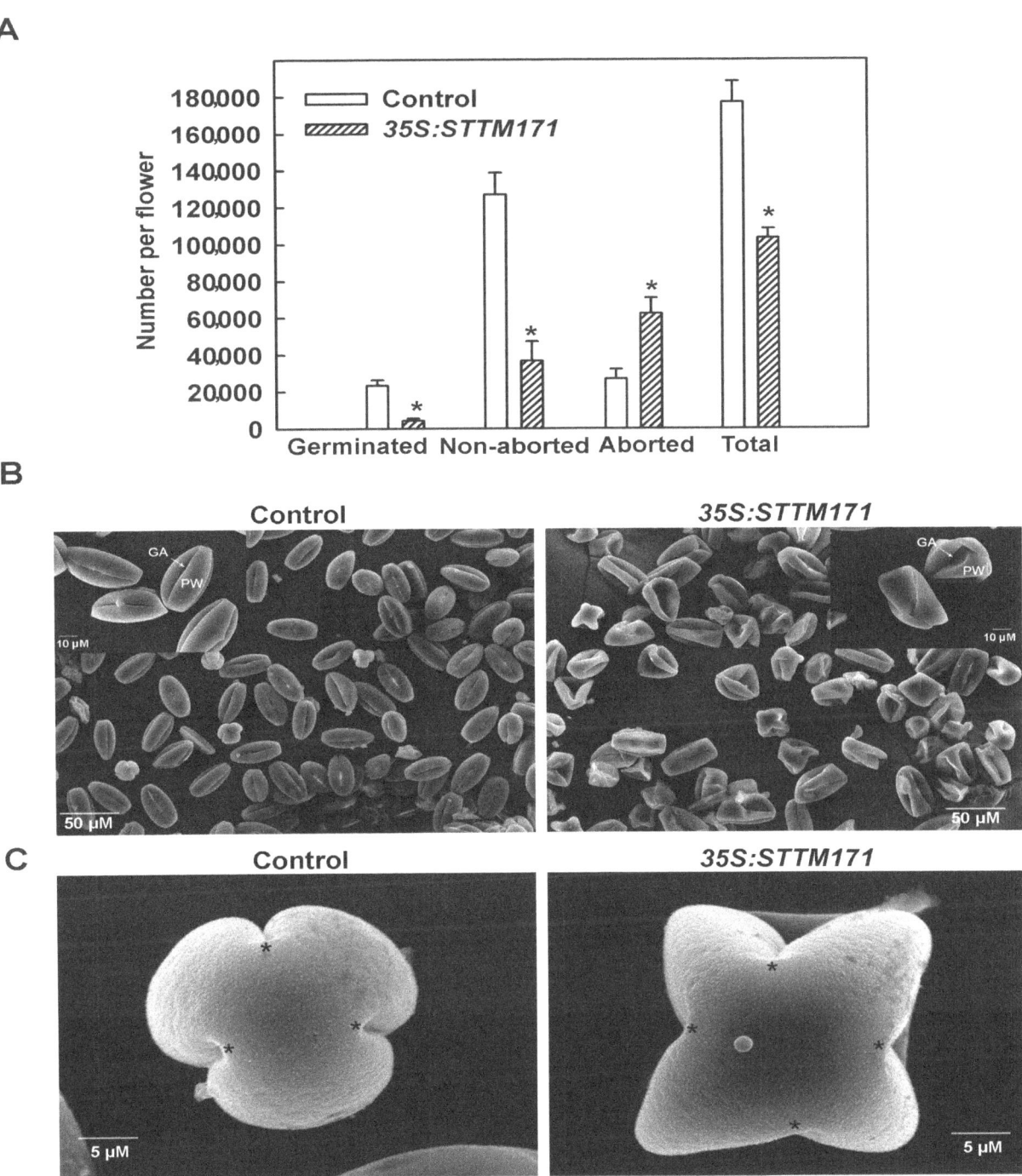

Figure 4. Effect of miR171 family downregulation on pollen quality and quantity. (**A**) Quantitation of total, aborted, non-aborted, and germinated mature pollen grains per anthesis flower of indicated genotype (n = 30). Asterisks indicate significant difference as determined by Student's *t*-test ($p \leq 0.01$). (**B**) Scanning electron micrographs of dehydrated pollen grains from indicated genotypes. Note the high number of collapsed pollen grains in the *35S:STTM171* sample. Inset shows magnified views of few representative pollen grains from each genotype. PW: pollen wall; GA: germinal aperture. (**C**) Scanning electron micrographs of polar view of representative mature pollen grains from indicated genotypes. The locations of the germinal aperture are indicated by asterisks.

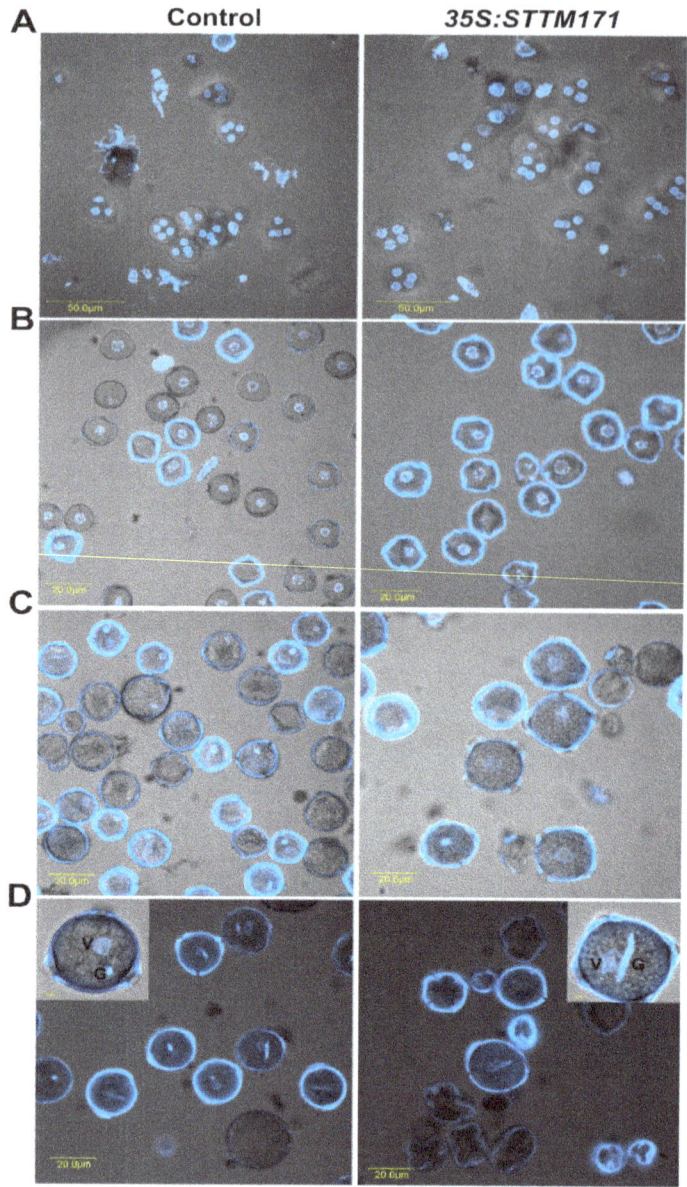

Figure 5. The *35:STTM171* pollen grains contain normal germ unit. DAPI fluorescence micrographs of control and *35:STTM171* tetrads (**A**), microspores (**B**), binucleate microspore (**C**), mature pollen grains (**D**). Inset in (**D**) shows magnified view of a representative pollen grain. The vegetative (V) and generative (G) cells are indicated. Note that the pollen surface fluorescence is due to auto-fluorescence at the same wavelength used for DAPI detection.

Next, we asked whether the reduced accumulation of sly-miR171 in *35S:STTM171* anthers is associated with abnormal development of anther tissues. To answer that, control and *35S:STTM171* anthers at major developmental stages were comparatively examined using transverse section light microscopy. Following examination of anthers under light microscopy showed that at the microsporocyte stage (3 mm buds, stage 8) the control pollen mother cells (PMC) are enclosed by a single layered tapetum (Figure 6A). At this stage, no distinct differences between control and the *35S:STTM171* transgenic anthers were observed (Figure 6B). Morphological differences were initially observed at the meiosis stage (4 mm bud, stage 9, Figure 6C). At that stage the control tapetal layer, which is composed from condensed tapetum cells, as indicated by their shrinkage and deep staining, encloses dividing microsporocytes, whereas in *35S:STTM171* anthers, tapetal cells remain expanded and vacuolated and dividing microsporocytes were not observed (Figure 6D). At the tetrad stage control tapetal cells were completely shrunk (5 mm bud, stage 11, Figure 6E), likely due to the initiation of programmed cell death (PCD) [31], and enclosed separated tetrads. In contrast, in *35S:STTM171* anthers, tapetal cells were not shrunken and instead were enlarged while most tetrads were not separated (Figure 6F). At the microspore stage, degenerated tapetal layer enclosing free microspores was observed in both control (8 mm bud, stages 12–13, Figure 6G) and *35S:STTM171* anthers (Figure 6H), except that in the latter the tapetum was less degenerated. At the bicellular pollen stage the control tapetum was completely degenerated and pollens were mature with characteristic densely stained cytoplasm (10 mm bud, stage 17–18, Figure 6I), whereas in *35S:STTM171* anthers, remnants of the degenerated tapetum were still visible and enclosed aborted pollen (Figure 6J). These observations indicated that the degeneration of the tapetum in the anthers silenced for sly-miR171 was delayed and initiated only after the tetrad stage in comparison to control where tapetum degradation occurred significantly earlier and on time.

Whereas tapetum is still intact at the tetrad stage, its cells export sporopollenin on microspore primexine surface that will provide a basis for future assembling of lipid materials into the pollen coat exine. During this process and earlier meiosis, the PMC and tetrads are surrounded by callose layer that provides protection for developing microspores. Later, the callose layer goes through degradation by tapetum supplied callase and the tapetum goes through PCD aiming to supply additional materials to the microspores [32–34]. The degradation of callose allows microspore release from tetrads and degradation of the tapetum supplies lipidic tapetum-derived materials for pollen exine and nutrients for pollen maturation [33,34]. The timing of tapetal cell death and consistency of the above-described events are critical for pollen development and interference in this process usually results in male sterility [35]. In several studies, delaying tapetum degeneration was shown to affect pollen morphology and results in pollen abortion. The rice mutant *tapetal degeneration retardation (Ostdr)* shows delayed tapetal breakdown resulting in a failure of pollen wall deposition and subsequent microspore degeneration [36]. Mutation in rice *OsACOS12* delays PCD-induced tapetum degradation leading to collapsed aborted pollen [37]. Thus, a likely possibility is that the delayed degeneration of *STTM171* expressing tapetum may be responsible at least in part for the deformed morphology of respective pollen grains.

Callose (β-1,3 glucan) protects PMC and later developing microspores from swelling, rupture, impact of diploid tissues and serves as a mold for future exine layer [38,39]. Often, defective tapetum development plan is accompanied with depletion of callose. To test if the delayed degeneration of *35S:STTM171* tapetum cells affected callose dynamics we performed a lacmoid stain of control and of *35S:STTM171* anthers at the meiosis and tetrad stages. We observed strong staining in control anthers at the meiosis stage (Figure 7A) and much weaker staining at the tetrad stage (Figure 7B) probably as a result of initiated callose degradation by tapetum supplied callase. In contrast, significantly weaker staining was observed in the *35S:STTM171* anthers (Figure 7A,B). Moreover, the weak callose staining was associated with enlarged tapetum cells characteristic of those that have not initiated PCD, rather than condensed cells undergoing PCD, as in the wild-type (Figures 6D,F and 7). In transgenic tobacco plants with delayed tapetum development, male sterility was caused because of premature degradation of callose [40]. In *DISFUNCTIONAL TAPETUM1* mutant enlarged tapetum cells and thin callose layer caused pollen collapse which resulted in male sterility [41]. This suggests that the delayed tapetum development in *35S:STTM171* anthers also delayed the callose deposition. Alternatively, callose depletion might be caused independently of tapetum development as in the *CALLOSE SYNTHASE5* mutants, where failure to produce callose resulted in collapsed pollen [33,42] reminiscent of the *35S:STTM171* pollen phenotype. Taken together these results suggest that the silencing of sly-miR171 in the *35S:STTM171* anthers perturbed tapetum development, callose dynamics and as a result pollen ontogenesis. However, additional studies are required to support this suggestion and determine the mechanism by which these miR171 members regulate anther development.

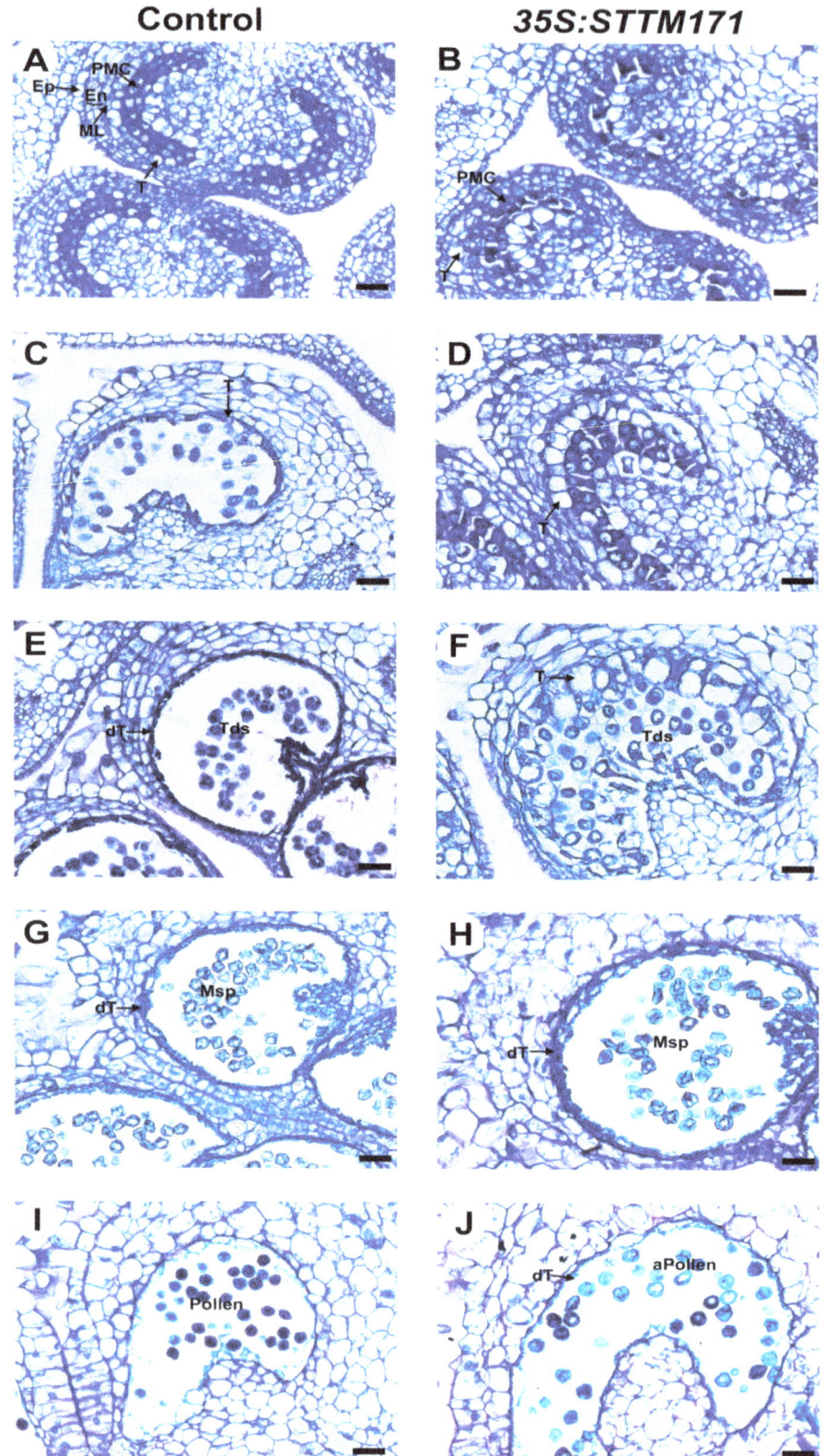

Figure 6. Histological analysis of control and *35S:STTM171* anthers. Pictures of Toluidine-blue stained cross sections of control and *35:STTM171* anthers at subsequent stages of microspore development as follows: (**A,B**) microsporocyte stage, (**C,D**) meiosis stage, (**E,F**) tetrad stage, (**G,H**) microspore stage, (**I,J**) bicellular pollen stage. dT-degenerated tapetum; En-endothecium; Ep-epidermis; ML-middle cell layer; Msp-microspore; MMC-microspore mother cell; T-tapetum; Tds-tetrads; aPollen-aborted pollen. Scale bars = 20 μm.

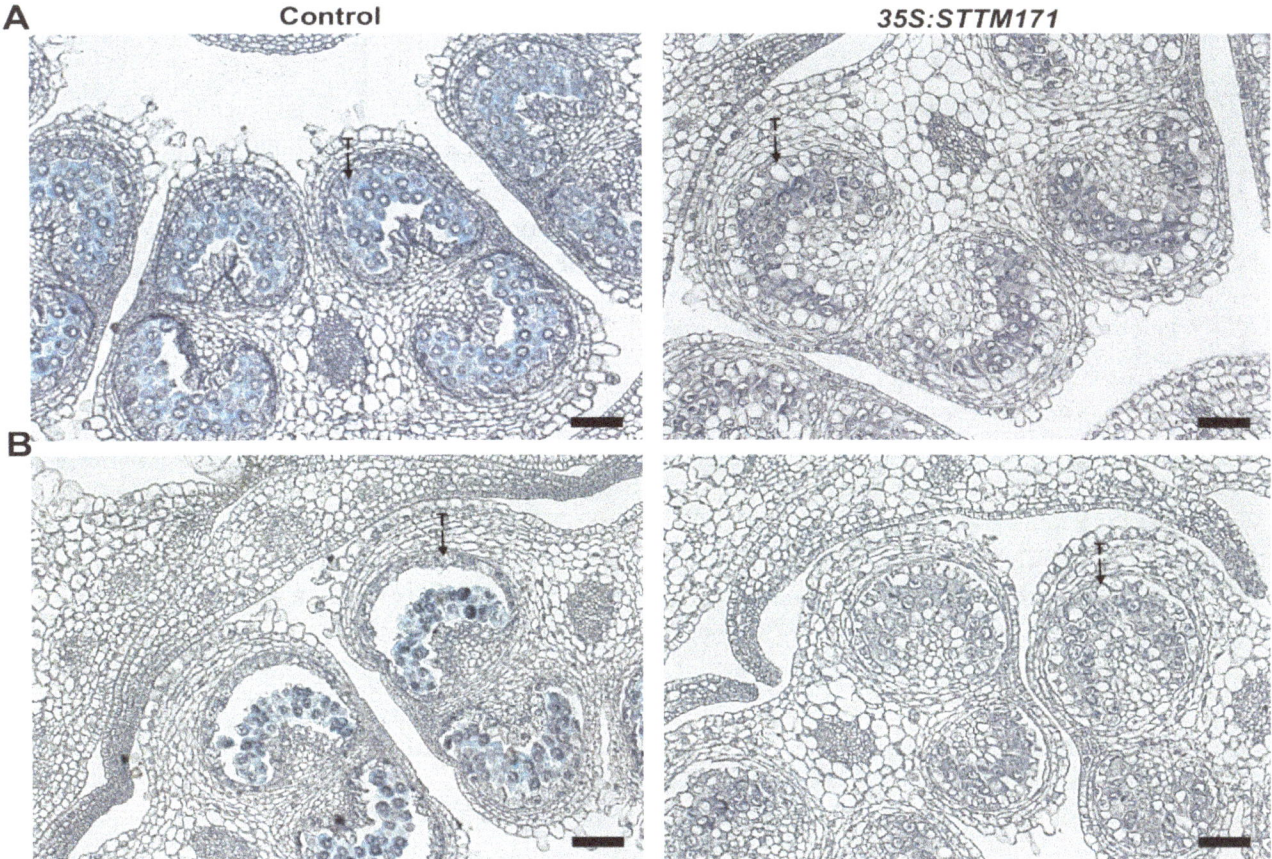

Figure 7. Callose detection in control and *35S:STTM171* anthers. Pictures of Lacmoid stained cross sections of anthers at the meiosis (**A**, 4 mm bud) and tetrad (**B**, 5 mm bud) stages are shown. T-tapetum. Scale bars = 20 μm.

3. Materials and Methods

3.1. Plant Material and Growth Conditions

Tomato cv. M82 plants and seedlings were grown under greenhouse and growth chamber conditions, respectively, as previously described [20].

3.2. Plasmid Construction

The pART27-OP:SlMIR171a and pART27-OP:SlMIR171b responder plasmids were described elsewhere [6]. For the pART27-OP:SlMIR171e responder construct, a 239 bp fragment from *SlMIR171e* including the *pre-miR171e* was amplified with Xho_MIR171e_F and Hind_MIR171e_R which contained *Xho*I and *Hind*III sites at their 5′ ends (for primer sequences, see Supplementary Table S3 online). The amplified fragment was restricted with *Xho*I and *Hind*III and ligated into pART27 binary vector containing the OP array to obtain pART27-OP:SlMIR171e [6]. For the *35S:STTM171* construct, a 136 bp STTM fragment was synthetically synthesized and then PCR amplified with Xho_STTM_171_F and Hind_STTM_171_R primers that contained *Xho*I and *Hind*III sites at their 5′ ends. The amplified fragment was restricted with *Xho*I and *Hind*III and cloned into the appropriate sites of the pART27 binary vector containing the CaMV 35S promoter and *Agrobacterium tumefaciens* octopine synthase terminator (OCS) [17] to obtain pART27-35S:STTM171.

3.3. Transformation of Tomato Plants

The binary vector pART27-35S:STTM171 was transformed into tomato cv. M82 and transgenic plants were selected as described previously [20]. Each kanamycin resistant plant was also subjected to genomic DNA PCR with the primer pair Xho_STTM_171_F and OCS_rev to detect the 35S:STTM171 transgene.

3.4. Total RNA Extraction and RNA Gel-Blot Analysis

Total RNA was extracted from different tomato tissues with Bio-Tri RNA reagent (Bio-Lab, Jerusalem, Israel) according to the manufacturer's protocol. A small-RNA gel-blot analysis of the total RNA was performed as described previously [43] using a complementary radiolabeled oligos as probes (probe sequences are listed in Supplementary Table S3 online).

3.5. cDNA Synthesis and Quantitative RT-PCR Assay

First-strand cDNA was synthesized from 2 µg of total RNA with Maxima first strand cDNA synthesis kit (Thermo Scientific, Waltham, MA, USA) following the manufacturer's instructions. A negative control (-RT) was used to ensure the absence of genomic DNA template in the samples. Three independent biological replicates were used for each sample, and quantification was performed in triplicate. PCR was performed in StepOnePlus Real-Time PCR System (Thermo Fisher Scientific) following the manufacturer's instructions. Primer sequences are listed in Supplementary Table S3 online. Relative expression levels were normalized to *SlTIP41* as a reference gene, and calculated by the standard curve method.

3.6. Analysis of Leaf Morphology and Axillary Shoot Number

Primary leaflet area was calculated by Tomato Analyzer 3.0 software [44]. Leaflet angle, namely the angle between the petiolule and the rachis was measured manually by a protractor. The number of axillary shoots ($\geq$5 mm long) was counted at eight-leaf stage tomato plants.

3.7. Determination of Pollen Quality and Quantity

To determine the pollen quantity and quality, mature pollen was extracted, stained, and counted according to Firon et al. [45]. Briefly, two flowers at the day of anthesis were sampled from control and transgenic plants and three anthers were removed from each flower, sliced in the middle, and immediately placed in a microcentrifuge tube containing germination solution (0.5 mL, 10% sucrose, 2 mM boric acid, 2 mM calcium nitrate, 2 mM magnesium sulfate, and 1 mM potassium nitrate). Then the pollen grains were released by vortex, incubated for 4 h at 25 °C, and stained with Alexander dye, that colors aborted pollen grains in blue-green, and non-aborted pollen grains in magenta-red [26]. The pollen grains were counted under a light microscope in a haemocytometer, eight fields for each sample.

3.8. Scanning Electron Microscopy (SEM) and 4′,6-Diamidino-2-Phenylindole (DAPI) Staining of Pollen Grains

For SEM analysis, pollen grains were collected and placed in FAA (3.7% formaldehyde, 5% acetic acid, 50% EtOH) solution until use. Then the FAA solution was removed and pollen grains were dehydrated in an increasing gradient of ethanol (up to 100%), critical-point-dried, mounted on a copper plate and gold-coated. Samples were viewed in a Jeol 5410 LV microscope (Tokyo, Japan). To stain pollen grains with DAPI, grains at different developmental stages were released by vortex into a DAPI solution (0.1 M sodium phosphate buffer (pH 7), 1 mM EDTA, 0.1% Triton X-100, 0.4 µg/mL DAPI), incubated for 10 min at room temperature and then viewed by Olympus IX81/FV500 laser-scanning confocal microscope (Olympus) at 361 nm maximum absorption, 461 nm maximum emission.

3.9. Histology and Callose Staining

For histological analyses, stamens at different developmental stages were taken and fixed in FAA solution until use, then dehydrated in increasing concentrations of ethanol (70%, 80%, 90%,

95%, and 100%), cleared with histoclear, and embedded in paraffin. Microtome-cut sections (6-µm thick) were spread on microscope slides, and stained with 0.03% Toluidine blue O. For callose staining microtome-cut sections of stamens were stained with 0.2% Lacmoid in 50% ethanol for 48h, then placed in 1% sodium bicarbonate in 50% ethanol for 10 min [46]. Stained slides were examined under bright-field using an Olympus (Olympus, www.olympus-lifescience.com) light microscope equipped with a digital camera.

Supplementary Materials:
Supplement 1. Prediction of target mRNAs for sly-miR171 and corresponding sly-mi171* strands. Supplement 2. Validation of probe specificity of sly-miR171. Figure S1. Alignment between STTM171 and sly-miR171 sequences. Figure S2. Fruit of *35:STTM171* T2 plants. Figure S3. Determination of the specificity of sly-miR171a, b, e RNA gel blot probes. Figure S4. Developing *35:STTM171* anthers accumulate reduced levels of sly-miR171. Table S1: A list of sly-miR171 precursors and corresponding miRNA/miRNA* pairs. Table S2: psRNATarget analysisa of sly-miR171 members and their star strands. Table S3: Primers and probes used in this study.

Author Contributions: M.K. designed, performed experiments, analyzed data, and wrote the manuscript. R.S. did transgenic plants. E.B assisted in confocal microscopy experiments. T.A. supervised the study and wrote the manuscript.

Acknowledgments: We would like to thank Guiliang Tang, Biological Sciences, Michigan Technological University, for his help with the design and cloning of the miR171 STTM construct.

References

1. Axtell, M.J. Classification and comparison of small RNAs from plants. *Annu. Rev. Plant Biol.* **2013**, *64*, 137–159. [CrossRef] [PubMed]

2. Rubio-Somoza, I.; Weigel, D. MicroRNA networks and developmental plasticity in plants. *Trends Plant Sci.* **2011**, *16*, 258–264. [CrossRef] [PubMed]

3. Jones-Rhoades, M.W.; Bartel, D.P. Computational Identification of Plant MicroRNAs and Their Targets, Including a Stress-Induced miRNA. *Mol. Cell* **2004**, *14*, 787–799. [CrossRef] [PubMed]

4. Axtell, M.J.; Bowman, J.L. Evolution of plant microRNAs and their targets. *Trends Plant Sci.* **2008**, *13*, 343–349. [CrossRef] [PubMed]

5. Zhu, X.; Leng, X.; Sun, X.; Mu, Q.; Wang, B.; Li, X.; Wang, C.; Fang, J. Discovery of conservation and diversification of miR171 genes by phylogenetic analysis based on global genomes. *Plant Genome* **2015**, *8*. [CrossRef]

6. Hendelman, A.; Kravchik, M.; Stav, R.; Frank, W.; Arazi, T. Tomato HAIRY MERISTEM genes are involved in meristem maintenance and compound leaf morphogenesis. *J. Exp. Bot.* **2016**, *67*, 6187–6200. [CrossRef]

7. Llave, C.; Kasschau, K.D.; Rector, M.A.; Carrington, J.C. Endogenous and silencing-associated small RNAs in plants. *Plant Cell* **2002**, *14*, 1605–1619. [CrossRef]

8. Lauressergues, D.; Delaux, P.-M.; Formey, D.; Lelandais-Brière, C.; Fort, S.; Cottaz, S.; Bécard, G.; Niebel, A.; Roux, C.; Combier, J.-P. The microRNA miR171h modulates arbuscular mycorrhizal colonization of *Medicago truncatula* by targeting *NSP2*. *Plant J.* **2012**, *72*, 512–522. [CrossRef]

9. Válóczi, A.; Várallyay, É.; Kauppinen, S.; Burgyán, J.; Havelda, Z. Spatio-temporal accumulation of microRNAs is highly coordinated in developing plant tissues. *Plant J.* **2006**, *47*, 140–151. [CrossRef]

10. Siré, C.; Moreno, A.B.; Garcia-Chapa, M.; López-Moya, J.J.; Segundo, B.S. Diurnal oscillation in the accumulation of Arabidopsis microRNAs, miR167, miR168, miR171 and miR398. *FEBS Lett.* **2009**, *583*, 1039–1044. [CrossRef]

11. Tong, A.; Yuan, Q.; Wang, S.; Peng, J.; Lu, Y.; Zheng, H.; Lin, L.; Chen, H.; Gong, Y.; Chen, J.; et al. Altered accumulation of osa-miR171b contributes to rice stripe virus infection by regulating disease symptoms. *J. Exp. Bot.* **2017**, *68*, 4357–4367. [CrossRef]

12. Liu, H.-H.; Tian, X.; Li, Y.-J.; Wu, C.-A.; Zheng, C.-C. Microarray-based analysis of stress-regulated microRNAs in *Arabidopsis thaliana*. *RNA* **2008**, *14*, 836–843. [CrossRef] [PubMed]

13. Yan, J.; Gu, Y.; Jia, X.; Kang, W.; Pan, S.; Tang, X.; Chen, X.; Tang, G. Effective small RNA destruction by the expression of a Short Tandem Target Mimic in Arabidopsis. *Plant Cell* **2012**, *24*, 415–427. [CrossRef]

14. Zhang, H.; Zhang, J.; Yan, J.; Gou, F.; Mao, Y.; Tang, G.; Botella, J.R.; Zhu, J.-K. Short tandem target mimic rice lines uncover functions of miRNAs in regulating important agronomic traits. *Proc. Natl. Acad. Sci. USA* **2017**, *114*, 5277–5282. [CrossRef]

15. Todesco, M.; Rubio-Somoza, I.; Paz-Ares, J.; Weigel, D. A Collection of target mimics for comprehensive analysis of MicroRNA function in *Arabidopsis thaliana*. *PLoS Genet.* **2010**, *6*, e1001031. [CrossRef] [PubMed]

16. Ivashuta, S.; Banks, I.R.; Wiggins, B.E.; Zhang, Y.; Ziegler, T.E.; Roberts, J.K.; Heck, G.R. Regulation of gene expression in plants through miRNA inactivation. *PLoS ONE* **2011**, *6*, e21330. [CrossRef]

17. Damodharan, S.; Zhao, D.; Arazi, T. A common miRNA160-based mechanism regulates ovary patterning, floral organ abscission and lamina outgrowth in tomato. *Plant J.* **2016**, *86*, 458–471. [CrossRef] [PubMed]

18. Peng, T.; Qiao, M.; Liu, H.; Teotia, S.; Zhang, Z.; Zhao, Y.; Wang, B.; Zhao, D.; Shi, L.; Zhang, C.; et al. A resource for inactivation of microRNAs using Short Tandem Target Mimic Technology in model and crop plants. *Mol. Plant* **2018**, *5*, 1400–1417. [CrossRef] [PubMed]

19. Kozomara, A.; Griffiths-Jones, S. miRBase: Annotating high confidence microRNAs using deep sequencing data. *Nucleic Acids Res.* **2014**, *42*, D68–D73. [CrossRef]

20. Kravchik, M.; Sunkar, R.; Damodharan, S.; Stav, R.; Zohar, M.; Isaacson, T.; Arazi, T. Global and local perturbation of the tomato microRNA pathway by a trans-activated DICER-LIKE 1 mutant. *J. Exp. Bot.* **2014**, *65*, 725–739. [CrossRef]

21. Wang, H.; Zhang, X.; Liu, J.; Kiba, T.; Woo, J.; Ojo, T.; Hafner, M.; Tuschl, T.; Chua, N.-H.; Wang, X.-J. Deep sequencing of small RNAs specifically associated with Arabidopsis AGO1 and AGO4 uncovers new AGO functions. *Plant J.* **2011**, *67*, 292–304. [CrossRef]

22. Karlova, R.; van Haarst, J.C.; Maliepaard, C.; van de Geest, H.; Bovy, A.G.; Lammers, M.; Angenent, G.C.; de Maagd, R.A. Identification of microRNA targets in tomato fruit development using high-throughput sequencing and degradome analysis. *J. Exp. Bot.* **2013**, *64*, 1863–1878. [CrossRef]

23. Imanshi, S.; Hiura, I. Relationship between fruit weight and seed content in the tomato. *J. Jpn. Soc. Hortic. Sci.* **1975**, *44*, 33–40. [CrossRef]

24. Wang, L.; Mai, Y.-X.; Zhang, Y.-C.; Luo, Q.; Yang, H.-Q. MicroRNA171c-Targeted SCL6-II, SCL6-III, and SCL6-IV genes regulate shoot branching in Arabidopsis. *Mol. Plant* **2010**, *3*, 794–806. [CrossRef]

25. Huang, W.; Peng, S.; Xian, Z.; Lin, D.; Hu, G.; Yang, L.; Ren, M.; Li, Z. Overexpression of a tomato miR171 target gene SlGRAS24 impacts multiple agronomical traits via regulating gibberellin and auxin homeostasis. *Plant Biotechnol. J.* **2016**, *15*, 472–488. [CrossRef]

26. Alexander, M.P. Differential staining of aborted and nonaborted pollen. *Stain Technol.* **1969**, *44*, 117–122. [CrossRef]

27. Kapp, R. *How to Know Pollen and Spores*; W. C. Brown Co. Publishers: Dubuque, IA, USA, 1969.

28. Gorman, S.W.; McCormick, S.; Rick, D.C. Male sterility in tomato. *Crit. Rev. Plant Sci.* **1997**, *16*, 31–53. [CrossRef]

29. Brukhin, V.; Hernould, M.; Gonzalez, N.; Chevalier, C.; Mouras, A. Flower development schedule in tomato Lycopersicon esculentum cv. sweet cherry. *Sex. Plant Reprod.* **2003**, *15*, 311–320.

30. Omidvar, V.; Mohorianu, I.; Dalmay, T.; Fellner, M. Identification of miRNAs with potential roles in regulation of anther development and male-sterility in 7B-1 male-sterile tomato mutant. *BMC Genom.* **2015**, *16*, 878. [CrossRef]

31. Goldberg, R.B.; Beals, T.P.; Sanders, P.M. Anther development: Basic principles and practical applications. *Plant Cell* **1993**, *5*, 1217–1229. [CrossRef]

32. Owen, H.A.; Makaroff, C.A. Ultrastructure of microsporogenesis and microgametogenesis in Arabidopsis thaliana (L.) Heynh. ecotype Wassilewskija (Brassicaceae). *Protoplasma* **1995**, *185*, 7–21. [CrossRef]

33. Dong, X.; Hong, Z.; Sivaramakrishnan, M.; Mahfouz, M.; Verma, D.P.S. Callose synthase (CalS5) is required for exine formation during microgametogenesis and for pollen viability in Arabidopsis. *Plant J.* **2005**, *42*, 315–328. [CrossRef]

34. Quilichini, T.D.; Douglas, C.J.; Samuels, A.L. New views of tapetum ultrastructure and pollen exine development in *Arabidopsis thaliana*. *Ann. Bot.* **2014**, *114*, 1189–1201. [CrossRef]

35. Kawanabe, T.; Ariizumi, T.; Kawai-Yamada, M.; Uchimiya, H.; Toriyama, K. Abolition of the tapetum suicide program ruins microsporogenesis. *Plant Cell Physiol.* **2006**, *47*, 784–787. [CrossRef]

36. Li, N.; Zhang, D.-S.; Liu, H.-S.; Yin, C.-S.; Li, X.; Liang, W.; Yuan, Z.; Xu, B.; Chu, H.-W.; Wang, J.; et al. The Rice

tapetum degeneration retardation gene is required for tapetum degradation and anther development. *Plant Cell* **2006**, *18*, 2999–3014. [CrossRef]

37. Yang, X.; Liang, W.; Chen, M.; Zhang, D.; Zhao, X.; Shi, J. Rice fatty acyl-CoA synthetase OsACOS12 is required for tapetum programmed cell death and male fertility. *Planta* **2017**, *246*, 105–122. [CrossRef]

38. Zhang, C.; Guinel, F.C.; Moffatt, B.A. A comparative ultrastructural study of pollen development in *Arabidopsis thaliana* ecotype Columbia and male-sterile mutant apt1-3. *Protoplasma* **2002**, *219*, 59–71. [CrossRef]

39. Zhu, J.; Chen, H.; Li, H.; Gao, J.-F.; Jiang, H.; Wang, C.; Guan, Y.-F.; Yang, Z.-N. Defective in Tapetal Development and Function 1 is essential for anther development and tapetal function for microspore maturation in Arabidopsis. *Plant J.* **2008**, *55*, 266–277. [CrossRef]

40. Worrall, D.; Hird, D.L.; Hodge, R.; Paul, W.; Draper, J.; Scott, R. Premature dissolution of the microsporocyte callose wall causes male sterility in transgenic tobacco. *Plant Cell Online* **1992**, *4*, 759–771. [CrossRef]

41. Zhang, W.; Sun, Y.; Timofejeva, L.; Chen, C.; Grossniklaus, U.; Ma, H. Regulation of Arabidopsis tapetum development and function by DYSFUNCTIONAL TAPETUM1 (DYT1) encoding a putative bHLH transcription factor. *Development* **2006**, *133*, 3085–3095. [CrossRef]

42. Nishikawa, S.; Zinkl, G.M.; Swanson, R.J.; Maruyama, D.; Preuss, D. Callose (β-1,3 glucan) is essential for Arabidopsis pollen wall patterning, but not tube growth. *BMC Plant Biol.* **2005**, *5*, 22. [CrossRef]

43. Talmor-Neiman, M.; Stav, R.; Klipcan, L.; Buxdorf, K.; Baulcombe, D.C.; Arazi, T. Identification of trans-acting siRNAs in moss and an RNA-dependent RNA polymerase required for their biogenesis. *Plant J.* **2006**, *48*, 511–521. [CrossRef]

44. Rodríguez, G.R.; Moyseenko, J.B.; Robbins, M.D.; Morejón, N.H.; Francis, D.M.; van der Knaap, E. Tomato Analyzer: A useful software application to collect accurate and detailed morphological and colorimetric data from two-dimensional objects. *J. Vis. Exp. JoVE* **2010**. [CrossRef]

45. Firon, N.; Nepi, M.; Pacini, E. Water status and associated processes mark critical stages in pollen development and functioning. *Ann. Bot.* **2012**, *109*, 1201–1214. [CrossRef]

46. Krishnamurthy, K.V. *Methods in Cell Wall Cytochemistry*; CRC Press: Boca Raton, FL, USA, 1999; pp. 69–70.

6

The Role of Extracellular Vesicles in Cancer: Cargo, Function, and Therapeutic Implications

James Jabalee [1], Rebecca Towle [1] and Cathie Garnis [1,2,]*

[1] Department of Integrative Oncology, British Columbia Cancer Research Center, Vancouver V5Z 1L3, BC, Canada; jjabalee@bccrc.ca (J.J.); rtowle@bccrc.ca (R.T.)

[2] Division of Otolaryngology, Department of Surgery, University of British Columbia, Vancouver V6T 1Z4, BC, Canada

* Correspondence: cgarnis@bccrc.ca;

Abstract: Extracellular vesicles (EVs) are a heterogeneous collection of membrane-bound structures that play key roles in intercellular communication. EVs are potent regulators of tumorigenesis and function largely via the shuttling of cargo molecules (RNA, DNA, protein, etc.) among cancer cells and the cells of the tumor stroma. EV-based crosstalk can promote proliferation, shape the tumor microenvironment, enhance metastasis, and allow tumor cells to evade immune destruction. In many cases these functions have been linked to the presence of specific cargo molecules. Herein we will review various types of EV cargo molecule and their functional impacts in the context of oncology.

Keywords: extracellular vesicles; cancer; therapeutics

1. Introduction

Extracellular vesicles (EVs) are a collection of lipid-bilayer enclosed vesicles secreted by virtually all cell types including cancer cells. EVs can be divided into subtypes based on their biogenesis, size and morphology, and collection method [1]. While other subtypes certainly exist, we use the term "EVs" to refer primarily to exosomes, ectosomes, and apoptotic bodies. Exosomes, the most heavily studied subtype, are ≈50–150 nm EVs formed by invagination of the multivesicular body (MVB) membrane. Once formed, exosomes are released into the extracellular space via fusion of the MVB with the plasma membrane. Ectosomes (sometimes called microvesicles or shedding microvesicles) are more heterogeneous, ranging in size from ≈100–1000 nm, and are formed through an outward blebbing of the plasma membrane. Apoptotic bodies are vesicles secreted by cells undergoing apoptotic cell death and range in size from ≈1000–5000 nm. Much of the work referenced herein is done on populations of EVs isolated via differential ultracentrifugation, which are likely enriched for exosomes compared to other subtypes. However, we use the term EVs to reflect the heterogeneous nature of these vesicles and the imperfect methods used to isolate them (which can lead to a mixture of EV subtypes). We occasionally use the term "exosomes" to refer to EVs pelleted by centrifugation at ≈$100,000 \times g$, "ectosomes" for EVs pelleted at ≈$10,000\ g$, and apoptotic bodies for EVs pelleted at ≈$2000\ g$.

Although initially thought to function exclusively in the removal of unwanted molecules from cells, EVs are now recognized as important mediators of cell–cell communication. EVs play key roles in both normal and disease processes and are important regulators of cancer progression. EVs are known to contain cell-type specific cargo, including RNA, DNA, and protein, which are selectively sorted into EVs [2]. Once released, EVs can interact with cells in the immediate vicinity or at distant locations via transfer through the circulation. EVs interact with recipient cells in a number of ways, including ligand–receptor interaction [3], release of vesicle contents in the extracellular space by bursting [4], direct fusion with the plasma membrane [5], and endocytosis into the cell [6]. The latter mechanisms are of specific interest here as they result in a transfer of molecular cargo from EVs to the recipient cells [2,7]. In cancer, tumor cells both release and receive EVs. This crosstalk between tumor and stromal cells regulates numerous aspects of tumorigenesis, including growth of tumor vasculature [8], recruitment of cancer-associated fibroblasts [9], metastatic potential [8], and evasion of immune destruction [10]. Herein we describe each of the major cargo types associated with EVs, assess the functional impact of EVs on cancer biology, and address the potential clinical uses of EVs relating to their roles as biomarkers and therapeutics.

2. EV Cargo

2.1. EV Isolation and Cargo Profiling

The identification and accurate functional characterization of EV cargo requires appropriate isolation and purification methods. Currently, the most popular method of EV isolation remains differential ultracentrifugation (DU) [11]. DU involves removal of contaminating material through a series of low-speed centrifugations followed by pelleting of EVs at higher speeds. DU is a low-cost, high-throughput method, making it an ideal means of EV isolation in many labs. However, co-purification of non-vesicular proteins and other contaminants is an issue [12]. Combination of DU with other purification methods, such as ultrafiltration or density gradient centrifugation, can improve the purity of the collected vesicles at the expense of particle yield [12]. Furthermore, density gradient centrifugation can be laborious and is unable to separate EVs from contaminants of similar density. The use of alternative techniques, such as size-exclusion chromatography, immunoaffinity capture, microfluidics, and precipitation-based methods are on the rise, each with their own advantages and disadvantages that have been thoroughly examined elsewhere [13]. Briefly, size-exclusion chromatography results in excellent purity, but dilutes samples and therefore requires EVs be re-concentrated following isolation. In contrast, precipitation methods tend to result in high yield but relatively low purity. Immunoaffinity capture and microfluidics can be used to isolate EVs sharing a specific characteristic; for example, all EVs expressing CD63 on their surface [14]. However, markers capable of distinguishing between exosomes, ectosomes, apoptotic bodies, and other EVs have not yet been identified, and markers such as CD63 can be found on various EV subtypes [13]. It is worth noting that because these techniques isolate EVs according to different characteristics (e.g., size, density, presence of specific surface markers, etc.), they may enrich different vesicle subpopulations, potentially generating misleading results [15].

Once EVs have been isolated and purified, consideration must be given to how the cargo of interest (i.e., RNA, protein, DNA, etc.) is to be extracted and profiled. Highlighting the importance of extraction technique, EV RNA yield and size distribution were found to differ greatly depending on the method used, with column-based methods resulting in both the highest yield and broadest size range [16]. In some cases, additional treatment may be required to remove protein and nucleic acids from the outside surface of EVs. Indeed, the International Society of Extracellular Vesicles recommends investigators quantify EV RNA before and after vesicles are treated with proteinase/RNase to determine the contribution of surface-bound RNAs to the total RNA content collected [15]. For additional detail, we refer the reader to excellent recent reviews [13,15]. These studies serve to highlight the need for consistent and detailed reporting of experimental methods for the accurate interpretation of EV studies.

2.2. MicroRNA

Next-generation sequencing (NGS) of EV RNA cargo has revealed the presence of various classes of RNA in EVs derived from normal and cancer cells, including messenger RNA, transfer RNA, ribosomal RNA, microRNA, and more [17–22]. Among these, microRNAs (miRNAs), ≈22-nucleotide non-coding RNA molecules, are perhaps the most intriguing and heavily studied. A single miRNA species can regulate the translation and degradation of numerous mRNA targets, which makes miRNAs powerful regulators of cell phenotype. Although EV-mediated miRNA transfer has been strongly linked to cancer progression, the mechanisms underlying miRNA sorting into EVs are poorly understood.

MiRNA abundance in EVs is highly variable and depends on the cell line under study. MiRNAs were reported to comprise from 5–30% of EV small RNA content in colorectal and breast cancer cell lines [18,19], ≈50% in non-tumorigenic murine hepatocyte cells [23], and <1% in HEK293T cells [22]. The miRNA content of EVs is distinct from that of the parental cell, indicating that specific miRNA are selectively sorted into or excluded from EVs [7]. Although the mechanisms by which sorting occurs remain unclear, recent work has suggested that the recognition of specific miRNA motifs by RNA binding proteins (RBPs) may play a role. Specifically, the RBP hnRNPA2B1 was found to selectively sort miRNAs containing the GGAG motif in the 3′ half of their sequence [24], and the RBP synaptotagmin binding cytoplasmic RNA interacting protein (SYNCRIP) was found to selectively sort miRNAs containing the guanine-guanine-cytosine-uracil (GGCU) motif [23]. Furthermore, hnRNPA2B1 must be attached to small ubiquitin-like modifiers (SUMOylated) in order for sorting to occur, thus adding an additional layer of regulation [24]. How the RBP-miRNA complex is then selected for sorting remains an open question. In some cases, such as for hnRNPA2B1, the protein can be detected within EVs, suggesting the entire complex is sorted [24]. However, this may not be the case for all RBPs. Ubiquitination of the RBP HuR causes the protein to release its bound miRNA; ubiquitinated HuR was found to associate primarily with the MVB, where sorting into exosomes occurs, and had low affinity for its target miRNA compared to the non-ubiquitinated protein [25]. Interestingly, silencing of the RBP hnRNPH1 increased the total RNA in EVs, suggesting that RBPs may also exclude specific miRNAs from EVs [26]. In many cases, RBP-mediated changes in EV miRNA sorting have been linked to tumorigenesis. For example, major vault protein (MVP) regulates sorting of tumor suppressive miR-193a into EVs, effectively removing it from cells and leading to more aggressive disease [27]. The tumor suppressor VPS4A was found to have the opposite effect; in this case, overexpression of VPS4A was found to cause can accumulation of tumor suppressive miRNAs in cells and oncogenic miRNAs in their EVs, thus decreasing the growth, migration, and invasion of the cancer cells [28]. Additional examples of RBPs with known oncogenic or tumor suppressive functions that have been linked to EV miRNA sorting include Annexin A2 (ANXA2) [29], Kirsten rat sarcoma (KRAS) [18], Y-box binding protein 1 (YB-1) [22,30], MEX3C [31], and Argonaute 2 (AGO2) [32–34].

Interestingly, AGO2 is not the only component of the miRNA processing machinery found in EVs. The major components of the RNA-induced silencing complex (RISC)-loading complex machinery, including AGO2, Dicer, and trans-activation responsive RNA-binding protein (TRBP), were found in EVs of breast cancer cells where they actively processed pre-miRNA to mature miRNA, and Dicer sorting was found to be CD43-dependent [33]. In contrast, EVs of cultured monocytes were found to contain single-stranded, mature miRNAs but only low levels of AGO2 [35], suggesting that miRNA sorting occurs independently of RISC. Thus, two independent pathways, one involving the sorting of pre-miRNAs along with the RISC machinery and one involving the sorting of mature miRNAs, may exist [36].

In addition to RBPs, 3′-end nucleotide additions (NTAs) also regulate miRNA sorting. MiRNAs with 3′-end adenylation tend to be overrepresented in cells whereas those with 3′-end uridylation are overrepresented in EVs, particularly exosomes [37]. However, the underlying mechanisms remain unclear. It is unknown if uridylated miRNAs are specifically sorted into EVs and, if so, how sorting occurs. NTAs change the stability and activity of miRNAs, which may in turn affect their availability for sorting. Adenylation stabilizes miRNAs, allowing them to interact with their mRNA targets in the cell, whereas uridylation achieves the opposite [38]. The poor activity of uridylated miRNAs may decrease miRNA-mRNA interaction, allowing for the sorting of the free miRNA. Indeed, altering the expression level of a miRNA or its target mRNA can alter the quantity of that miRNA in EVs [39].

Finally, the biogenesis of EVs and the sorting of miRNA contents is regulated by the membrane lipid ceramide. In the case of exosomes, ceramide is produced via the breakdown of sphingomyelin by neutral sphingomyelinases (nSMases) at the MVB membrane. Inhibition of ceramide generation by the nSMase inhibitor GW4869 greatly reduces the small RNA content of EVs [40] and decreases the quantity of EVs released by cells in vitro [2]. Further, treatment of breast cancer cells with GW4869 decreased EV sorting of pro-angiogenic miR-210, thereby inhibiting tumor angiogenesis [8]. Ceramide is thought to play a role in sorting via the formation of ceramide-rich lipid microdomains. Sorting may occur directly via interaction of miRNA with ceramide-rich microdomains in the MVB membrane or indirectly via microdomain-dependent recruitment of proteins to the site of sorting [41]. Specific miRNA sequences show greater affinity for ceramide than others, thus providing a potential mechanism for sequence-based miRNA sorting [41]. Interestingly, ceramide appears to be required for miRNA sorting by some RBPs. Ceramide has been shown to co-localize with hnRNPA2B1 in the cytoplasm, and GW4869 inhibited HuR-mediated sorting of miR-122 and MEX3C-mediated sorting of miR-451a into exosomes [24,25,31]. Sorting of miR-451a was independent of the endosomal sorting complexes required for transport (ESCRT) pathway [31], discussed below.

EV miRNA content can also be modified by external factors, including hypoxia [42], carcinogens such as asbestos [43] or toluene [44], and infection with oncogenic viruses [45–47]. As with RBPs, these factors can alter EV miRNA content in ways that may promote or inhibit tumorigenesis. Viral infection provides an especially intriguing example of the power of EV miRNAs to promote tumorigenesis by influencing the tumor microenvironment. The genome of Kaposi's sarcoma herpesvirus (KSHV) encodes a set of 12 latency-associated miRNAs, all of which have been found in the EVs of infected host cells [47]. Transfer of the viral miRNAs to non-infected cells via EVs results in a shift toward aerobic glycolysis which supports the growth of cancer cells by providing them with energy-rich metabolites [47]. In this way, virally-encoded EV miRNAs are used to reprogram the tumor microenvironment to enhance growth of KSHV-infected cancer cells.

The sensitivity of EV miRNA content to genetic and environmental stimuli suggests their use as biomarkers. EV miRNA biomarkers show great promise; among their most exciting aspects are stability in the face of non-ideal collection methods, ability to be collected non-invasively, and specificity to certain disease states. With regard to non-invasive collection, potential EV miRNA biomarkers have been identified in numerous body fluids, including blood [43,48–58], urine [59–62], pleural effusion [63,64], and saliva [65]. Intriguingly, EV miRNA biomarkers appear exquisitely sensitive to specific disease states, and have been shown to discriminate among closely-related diseases, such as metastatic versus non-metastatic tumors [50], recurrent versus non-recurrent tumors [60], and high-grade versus low-grade tumors [56,60]. In addition, numerous studies have shown that EV miRNA biomarkers return to normal levels upon surgical resection of the tumor [51,56,61], suggesting their use in monitoring disease progression. While these results are encouraging, large-scale validation studies are required before EV miRNA biomarkers can be approved for clinical use. Along these lines, it is worth noting a few particularly promising biomarkers which have been found numerous times in separate studies. The oncomiR miR-21 has been proposed as a biomarker for various cancer types, including breast [49], prostate [59], bladder [61], brain [56], larynx [66], and liver [67]. MiR-375 [59,60,68] and the miR-200 family [21,51,63] show similar promise.

2.3. mRNA and Other RNA Types

Like miRNA, mRNA appears to make up a minority of EV RNA. In EVs derived from glioma stem-like cultures, mRNA accounted for <10% of the total RNA reads as assayed by NGS, and it is unclear what proportion of mRNA reads can be attributed to full-length mRNAs as opposed to mRNA fragments [18,69]. This may reflect a difference among EV subtypes, since mRNA appears to be more enriched in ectosomes compared to exosomes [69]. Despite uncertainty regarding mRNA abundance, numerous studies have shown that functional mRNAs can be transferred via EVs to recipient cells where they are translated and alter recipient cell phenotype [70–73]. For example, hTERT mRNA, which encodes the catalytic subunit of the telomerase enzyme, is transferred via EVs into nearby fibroblasts, thus increasing their proliferation, extending their life span, postponing senescence, and protecting from DNA damage [72]. Furthermore, hTERT mRNA was found in serum-derived EVs from 67.5% of 133 individuals with various cancers, but none of the 45 healthy controls [74], suggesting its use as a biomarker for detection of multiple cancer types.

Recent studies suggest that the presence of specific sequence motifs play a key role in mRNA sorting into EVs. Interestingly, the YB-1 protein, which has also been linked to miRNA sorting [22,30], has been found to interact specifically with mRNAs whose 3' untranslated region (UTR) contains any of three motifs, while the methyltransferase NSUN interacts with one motif [75,76]. YB-1 is overexpressed in numerous cancer types and drives cell proliferation [77], thus providing a link between cancer progression and EV packaging. Similarly, a 25-nucleotide motif was found to be enriched in the 3' UTR of exosomal mRNAs from glioblastoma cells [78], whereas mRNAs harboring the signal peptide sequence were excluded from EVs [79].

Additional classes of RNA found in EVs include ribosomal RNA, transfer RNA, mitochondrial RNA, long non-coding RNA, piwi-interacting RNA, small nucleolar RNAs, and circular RNA [20,22,80–82]. Recent results have linked EV long non-coding RNAs to chemosensitivity [83] and cancer progression [84]. While outside the scope of the current manuscript, we refer the reader to an excellent recent review on this topic [85].

2.4. DNA

Extracellular vesicles carry DNA, which may be genomic (gDNA) [86,87] or mitochondrial (mtDNA) [88–90] in origin. Depending on the cell line and context, DNA may be single- or double-stranded, and may reside within the lumen or on the surface of EVs [88,91–95]. Surface-bound DNA can alter the ability of EVs to adhere to fibronectin [89], suggesting it may help determine how EVs interact with extracellular matrix molecules, such as those found in the tumor microenvironment or pre-metastatic niche. Luminal DNA, like other cargo types, can be transferred from donor to recipient cells resulting in increased mRNA and protein production, and oncogenes can be distributed among different cell types via this mechanism [92,94]. Intriguingly, EV-mediated spread of oncogenes has been shown to promote disease progression in mice. EVs derived from chronic myeloid leukemia (CML) cells transfer DNA encoding the breakpoint cluster region/Abelson murine leukemia viral oncogene homolog (BCR/ABL) fusion oncogene to the neutrophils of Sprague-Dawley rats and non-obese diabetic/severe combined immunodeficient (NOD/SCID) mice in vivo resulting in increased BCR/ABL mRNA and protein in the recipient murine cells and the eventual onset of CML-like characteristics [96]. Mitochondrial DNA in EVs also appears to be functional in recipient cells, and cancer-associated fibroblast-derived mtDNA has been found to play a role in the resistance of breast cancer cells to hormone therapy [90].

While current work has shed light on the functions of EV DNA, little is known regarding how it is sorted into EVs. Numerous studies have reported that cancer cell-derived EVs contain gDNA from all chromosomes [86,93,95,97], and at least one study has provided evidence for the packaging of the entire mitochondrial genome within EVs [90]. These results suggest that selective sorting of specific DNA sequences may not occur. Interestingly, knockdown of EV release in human fibroblasts and various cancer cell lines results in the accumulation of damaged DNA in the cytoplasm [98]. Such cytoplasmic DNA can be recognized by DNA sensing proteins, the activation of which results in genomic DNA damage, senescence, or apoptosis [98]. Considering this evidence, it is possible that EVs play a role in maintaining cellular homeostasis through non-selective removal of cytoplasmic DNA [98]. In contrast to the idea of non-specific gDNA packaging, apoptotic bodies, ectosomes, and exosomes were found to contain shared and unique DNA sequences [86], suggesting that independent DNA packaging mechanisms for each of the vesicle subtypes may exist. More work is required to clarify the mechanisms underlying EV DNA packaging and its functional relevance to normal and cancer cells.

Intriguingly, cells with specific mutations in their gDNA release EVs containing DNA that harbor identical mutations. Indeed, EV DNA containing mutations identical to the gDNA has been found in cell culture supernatants [86,93,95], the plasma of tumor-bearing mice [95], and the blood (serum and plasma) of human cancer patients [93,97,99–102] (Table 1). Blood-derived EV DNA provides a rich source of clinically relevant information. NGS was used to detect at least 10 potentially clinically actionable mutations in the EV DNA of patients with pancreatic cancer [97], and a similar approach was used to detect specific mutations in three well-known oncogenes in patients with different cancer types with an overall sensitivity of 95% [100]. A major advantage of mutational analysis of EV DNA compared to other techniques, such as standard biopsies, is the ability to collect samples throughout the course of treatment, thus decreasing the need for invasive biopsies. This flexibility allows physicians to monitor genomic changes in the tumor over time.

Table 1. Summary of recent publications using EV DNA for detection of specific mutations in cancer-related genes.

Study	Fluid	Cancer Type	Patients	Technique	Genes Analyzed	Results
Kahlert et al., 2014 [93]	Blood (serum)	Pancreatic ductal adenocarcinoma	Human; 2 cancer (no stage given), 2 healthy	PCR, Sequencing (BigDye terminator kit)	KRAS, TP53	Detected two different KRAS mutations and one TP53 mutation.
Lázaro-Ibáñez et al., 2014 [86]	Blood (plasma)	Prostate	Human; 4 cancer (T stages 1–3), 4 healthy	PCR, sequencing (BigDye terminator kit)	MLH1, PTEN, TP53	Unable to detect specific mutations.
Thakur et al., 2014 [95]	Blood (plasma)	Melanoma	SK-MEL-28 cells xenografted into NOD/SCID mice; EVs collected when tumors reached max allowable size	Allele-specific PCR	BRAF (V600E)	Mutation detected.
San Lucas et al., 2016 [97]	Blood and pleural fluid	Pancreatic ductal adenocarcinoma (PDAC) and ampullary adenocarcinoma	Human; 2 PDAC and 1 ampullary adenocarcinoma	Next-generation sequencing	Whole genome	At least 10 potentially clinically actionable mutations identified in each patient.
Allenson et al., 2017 [99]	Blood (plasma)	Pancreatic ductal adenocarcinoma (PDAC)	Human; 68 PDAC (all stages), 20 PDAC patients whose blood was drawn after resection with curative intent, and 54 healthy controls	Droplet digital PCR	KRAS	Mutations detected in 7.4%, 66.7%, 80%, and 85% of controls, localized, locally advanced, and metastatic PDAC patients.
Möhrmann et al., 2017 [100]	Blood (plasma)	46.5% colorectal, 18.6% melanoma, 14.0% non-small cell lung cancer, 20.9% other	Human; 43 progressing advanced cancers	Next-generation sequencing	$BRAF^{V600}$, $KRAS^{G12/G13}$, $EGFR^{exon19delL858R}$	Mutations in EV DNA which correspond to those in tissue found in 95% of cases. EV DNA did not contain mutations not present in the parental tumor cells.
Yang et al., 2017 [101]	Blood (serum)	Pancreatic ductal adenocarcinoma (PDAC), chronic pancreatitis (CP), intraductal papillary mucinous neoplasm (IPMN)	Human; 48 PDAC, 9 CP, 7 IPMN, 114 healthy controls	Digital PCR	$KRAS^{G12D}$, $TP53^{R273H}$	KRAS mutation detected in 39.6% PDAC, 28.6% IPMN, 55.6% CP, 2.6% healthy controls. TP53 mutation detected in 4.2% PDAC, 14.2% IPMN, 0% CP, 0% healthy controls.
Castellanos-Rizaldos et al., 2018 [102]	Blood (serum)	Non-small cell lung cancer	Human; Training and test cohorts each with 51 mutation positive and 54 mutation negative samples	Allele-specific PCR	$EGFR^{T790M}$	Training: 81% sensitivity, 95% specificity. Test: 92% sensitivity, 89% specificity

Interestingly, recent reports suggest that cell-free circulating tumor DNA (cfDNA), which was previously thought to derive primarily from apoptotic and necrotic tumor cells, is comprised largely of EV DNA [103]. Many companies offer panels for the comprehensive detection of cfDNA mutations in cancer-related genes, and analysis of mutations in cfDNA is currently being used to guide patient management [104]. By simultaneously collecting and analyzing cfDNA, exosomal DNA, and exosomal RNA from the serum of non-small cell lung cancer patients, Castellanos-Rizaldo and colleagues detected the EGFR T790M mutation with 92% specificity and 89% sensitivity when compared to tissue biopsy [102]. These results highlight the exciting clinical applications of EV DNA. A deeper understanding of EV biology-such as how specific molecular targets are selected for packaging and the identification of markers for the separation of tumor and non-tumor-derived EVs-will serve to further refine the utility of EV-based biomarkers in the clinic.

2.5. Protein

In addition to being transferred to recipient cells and influencing cell phenotype, proteins also regulate the sorting of other EV components, determine which cell types are able to receive EVs (i.e., determine tropism), provide markers for the separation of EV subtypes, bind to and activate receptors on recipient cells, and carry out cell-independent reactions inside of EVs after their release. Here we will discuss the sorting of specific proteins followed by a brief discussion of a few of their various other functions in relation to cancer biology.

A key pathway of EV protein sorting involves the endosomal sorting complexes required for transport (ESCRT), a series of four protein complexes (ESCRT-0, I, II, and III), and accessory proteins. ESCRTs recognize and bind ubiquitinated proteins, facilitating their sorting into EVs [105]. In addition to ubiquitination, other post-translation modifications appear to play an important role in EV protein sorting through both ESCRT-dependent and ESCRT-independent pathways, at least for some proteins. Phosphorylation may either promote or inhibit EV sorting, as evidenced by EPHA2 and AGO2, respectively [34,106]. As mentioned above, SUMOylation of the RBP hnRNPA2B1 regulates sorting into exosomes [24]. Other mechanisms of protein sorting into EVs involve dimerization [107], and recruitment via other proteins, such as tetraspanins [108].

The oncogenic activity of EVs is dependent not only upon their intraluminal cargo, but also on the array of proteins that span the EV membrane. One intriguing function of such transmembrane proteins involves determining EV tropism; i.e., which cell types are most likely to take up, and thus be influenced by, EVs. EV tropism has been observed in vitro among different cell types. In vivo, EVs from metastatic cell lines are more likely to be taken up by resident cells at sites to which those lines commonly metastasize [109,110]. Especially important in this process are integrins (ITGs), a class of proteins known to facilitate cell-extracellular matrix interactions. $ITG\alpha_6$, and its partners $ITG\beta_4$ and $ITG\beta_1$, are abundant in EVs that distribute mainly to the lung where they are taken up primarily by S100A4-positive fibroblasts, $ITG\beta_5$ and $ITG\alpha_V$ are abundant in EVs that distribute mainly to the liver where they are taken up primarily by Kupffer cells, and $ITG\beta_3$ is abundant in EVs that distribute mainly to the brain where they are primarily taken up by endothelial cells [110]. The uptake of EVs at specific locations within the body appears to play a key role in determining the location of metasteses, as evidenced by the observation that injection of lung-tropic EVs into mice increased the lung metastatic capacity of breast cancer cells which normally metastasize preferentially to bone [110]. Moreover, EVs can alter gene expression in recipient cells of the pre-metastatic niche, which may include cancer and stromal cells [109–112]. For example, astrocyte-derived EV miRNAs inhibit the tumor suppressor PTEN in cells that metastasize to the brain, thus priming them for metastatic outgrowth [111]. In addition, miR-122-containing breast cancer cell-derived EVs prime the premetastatic niche by decreasing expression of the glucose metabolizing enzyme pyruvate kinase in nearby stromal cells, thus increasing nutrient availability for metastasizing cancer cells [112]. In contrast to integrins, which direct EVs to their preferred targets, additional proteins have been identified which function to modify or block EV uptake. Along these lines, REG3β interferes with the uptake of EVs into target cells by binding to glycoproteins on the EV surface [113]. Similarly, CD47, an anti-phagocytic signal, was found to block uptake of EVs by immune cells, thus prolonging their time in circulation [114]. These results suggest that manipulation of the surface proteins of EVs could be used to alter their tropism and block their pro-tumor effects.

In addition to tropism-determining proteins, the presence of which is dependent upon cell type, other proteins have been shown to be more ubiquitously found in EVs and can provide information on their cellular origin. In an intriguing example, immunoaffinity capture was used to separate A33-positive and EpCAM-positive exosomes secreted from colorectal cancer organoids, and each population was found to be enriched for distinct proteins [115]. These results suggest that even within the EV subpopulation of exosomes, additional subpopulations, which may differ in aspects of their biogenesis and cargo, are likely to exist, a result supported by others [116]. Tetraspanins, a family of membrane-spanning proteins which includes CD63, CD9, and CD81, are among the most commonly cited "exosome markers". Unfortunately, the assumption that such markers are found on all exosomes is an over-simplification; truly specific markers that are found on all exosomes do not exist. CD63 and CD9 were also found, to differing degrees, to be present on other EV subtypes including ectosomes and apoptotic bodies [116], and particles isolated by ultracentrifugation (which are assumed to be enriched for exosomes) can be separated into CD63+, CD9+, CD81+, and non-tetraspanin-bearing subpopulations, although overlap between markers on a single EV is common [116]. The distinction among exosome subtypes may be an important one due to the many key roles tetraspanins play in EV biology, including regulation of biogenesis [117], cargo sorting [117], and tropism [118]. Thus, subpopulations of exosomes that differ in their tetraspanin content may also differ in their biological function. Further detailed investigation into how best to obtain and purify EVs based on protein markers, and whether such distinctions are truly biologically relevant, is warranted.

The separation of EV subtypes based on surface proteins may prove clinically useful. By studying the surface proteins of EVs collected from cancer and non-cancer cell lines, Melo and colleagues found dozens of cancer cell-specific markers [119]. One of these, GPC1, was found to increase in the blood of patients with breast and pancreatic cancer compared to controls, suggesting its use as a disease detection biomarker [119]. Using similar methodology, Castillo and colleagues identified proteins specific to the EVs of pancreatic cancer cells but not normal controls [120]. Unfortunately, GPC1 was not found in pancreatic cancer EVs, and appeared instead to be selectively expressed in non-cancerous tissues [120]. Once refined, separation of cancer and normal EVs will increase the yield of tumor-specific material and decrease unwanted background in downstream analyses.

3. Extracellular Vesicle Function in Cancer

3.1. Impact of EVs on Fibroblasts

Release of EVs by tumor cells is believed to play a major role in intercellular communication, facilitating signaling to surrounding tumor cells and to distant sites via blood or other biological fluid transportation (Figure 1). A primary focus of inquiry has been the impact of EVs on the tumor stroma, including fibroblasts, endothelial cells, and immune cells.

Fibroblasts comprise a major component of the tumor stroma. Under tumorigenic conditions, fibroblasts can undergo morphological changes that confer a phenotype similar to myofibroblasts, which are activated, mobile fibroblasts. Interestingly, tumor-derived EVs are able to induce the transformation of normal stromal fibroblasts into activated cancer-associated fibroblasts (CAFs) [121–125]. For instance, TGFβ-containing prostate cancer-derived EVs are sufficient to induce fibroblast transformation to a CAF-like phenotype [123]. Further, the resultant increases in wound healing and endothelial cell growth were more pronounced in EV-exposed fibroblasts as compared to fibroblasts transformed by soluble TGFβ alone [123]. The ability of EVs to promote CAF activation was found to correlate with the aggressiveness of the tumor cells, with EVs from a more aggressive cell line prompting higher CAF marker expression, proliferation rate, and enzyme release by treated fibroblasts than EVs from a less aggressive cell line [126].

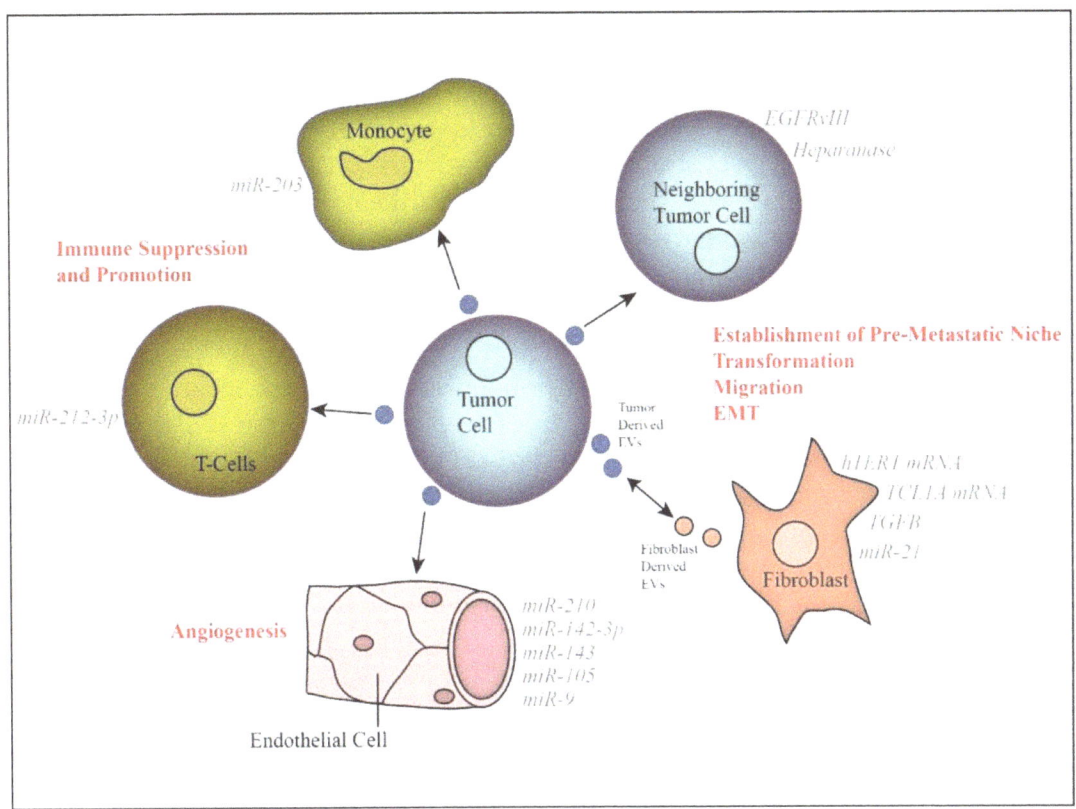

Figure 1. Extracellular vesicle-mediated transfer of specific cargo molecules alters the phenotype of recipient cells, including neighboring tumor cells, fibroblasts, endothelial cells, and immune cells.

Once activated, CAFs secrete EVs that promote tumorigenesis by increasing proliferation, motility, epithelial-mesenchymal transition, migration, and metabolic changes in tumor, endothelial cells, and other fibroblasts [126–129]. Especially intriguing is the role of EVs in mediating resistance to chemotherapy [127,130]. For instance, exposing pancreatic ductal carcinoma cells to conditioned pancreatic fibroblast media was sufficient to confer resistance to gemcitabine, potentially due to up-regulation of snail family transcriptional repressor 1 (SNAIL) and miR-146a in the recipient cells [127]. Gemcitabine treatment also led to an increase in CAF EV secretion, indicating a potential mechanism of drug resistance in pancreatic cancer [127].

3.2. EVs Induce Angiogenesis in Endothelial Cells

The ability to induce angiogenesis is a hallmark of cancer, and recent evidence suggests that EVs are key regulators of tumor vascularization via transfer of pro-angiogenic molecules from tumor to endothelial cells. Indeed, EVs have been shown to increase tube formation, migration, cell–cell adhesion, and proliferation in endothelial cells in a variety of cancer types [8,122,131–138]. For example, activated EGFR found in EVs is sufficient to induce EGFR and VEGFR signaling in recipient endothelial cells, and blocking EV-mediated EGFR transfer decreased tumor growth and angiogenesis [131,139]. Furthermore, EVs produced by hypoxic tumor cells have been shown to have a more pronounced effect on endothelial cells in promoting angiogenesis than those derived from normoxic cells [134,137]. Hypoxia increases the production of tumor and stromal cell-derived EVs and alters their cargo [42,137,140–142]. For example, miR-23a is found in the EVs of hypoxic, but not normoxic, lung cancer cells and promotes angiogenesis through the inhibition of prolyl hydroxylase in recipient endothelial cells [141]. Increased EV production by hypoxic endothelial cells was abrogated by siRNA targeting hypoxia inducible factor 1α, thus providing a clear link between cell response to hypoxia and EV production [142]. Other EV-derived molecules that have been shown to play a role in promoting angiogenesis include miR-9, miR-105, miR-142-3p, miR-210, and H19 lncRNA [132,133,135,138,141,143,144].

Several groups have looked at the impact of sub-populations with cell markers indicative of tumor initiating cells. In renal cell carcinoma cell lines, CD105-positive cells were found to release EVs that increase proliferation, vessel formation, and invasion in HUVEC endothelial cells, whereas CD105-negative cells did not [134]. Similarly, in liver cancer cells, CD90-positive cells were found to secrete EVs that promote tube formation and cell–cell adhesion via transfer of H19 lncRNA [132]. These results highlight the heterogeneity found within tumors and suggest that subsets of tumor cells secrete EVs carrying a unique set of cargo capable of altering stromal cell phenotypes in specific ways.

3.3. Extracellular Vesicles in Immunomodulation

EVs are an important mode of communication among cells of the immune system and are key regulators of the anti-cancer immune response. Initial reports showed that EVs secreted by dendritic cells induce an antitumor immune response, suggesting the use of immune cell-derived EVs as an anti-cancer vaccine as discussed below [145]. This work was strengthened by the observation that EVs contain proteins involved in antigen presentation and immune stimulation, including tumor antigens and major histocompatibility complex (MHC) proteins [146,147]. However, additional work indicated that tumor-derived EVs often promote immune suppression.

Among the most heavily studied immune cell recipients of tumor-derived EVs are dendritic cells and T cells. In many instances, tumor-derived EVs have been found to have an inhibitory effect on dendritic cell function [146,148]. For instance, pancreatic cancer-derived EVs containing miR-203 were found to impair dendritic cell function via reduction of toll-like receptor 4 expression [146]. Furthermore, pancreatic cancer cell-derived EVs were found to alter the transcriptional profile of recipient dendritic cells via transfer of miRNA, specifically miR-212-3p [148]. This led to a decrease in MHCII expression and suppressed immune function. There are conflicting reports on the role of tumor-derived EVs in dendritic cell maturation, with different studies citing either inhibitory or stimulatory effects [149–152].

EVs can also impact T-cell function either directly or via inhibition of other immune cell types, such as dendritic cells [152]. Several studies found that tumor-derived EVs can affect T-cell function, specifically by increasing proliferation, differentiation, and induction of T regulatory cells that function to blunt the immune response [151,153]. Interestingly, direct inhibition of T cell function via EV-associated PD-L1 has also been reported [154]. Furthermore, prostate cancer cell-derived EVs have been found to down-regulate NKG2 in natural T-killer cells and this could contribute to immune suppression [150]. These examples serve to underscore the variety of ways in which tumor-derived EVs can inhibit the immune system.

EVs may also act on cells of the immune system to promote tumor-supportive inflammation. For example, tumor cell-derived EVs can stimulate macrophages to release pro-inflammatory cytokines via activation of the NFκB pathway [155,156]. In breast cancer cells, this pathway activation led to an increase in the secretion of pro-inflammatory cytokines including IL-6, TNFα, GCSF, and CCL2 [156]. Furthermore, tumor EV-associated miR-21 and miR-29a can trigger a pro-inflammatory response in immune cells [157]. Interestingly, these miRs appear to function by acting as EV-associated ligands for toll-like receptors, rather than through their internalization into the cell [157].

3.4. Tumor Promoting Effects of Other Extracellular Vesicles

As noted above, the majority of studies have focused on the functional impact of EV populations collected via ultracentrifugation at $100,000\times g$, which are assumed to be enriched for exosomes. Several papers, however, have also assessed larger EV species, including ectosomes and large oncosomes, which arise from non-apoptotic blebbing of the plasma membrane [124,158,159].

This subset of EVs is obtained by collecting pellets from cell culture media supernatant centrifuged at $10,000 \times g$ and further purified by density gradient centrifugation. Minciacchi and colleagues found that large oncosomes were able to reprogram prostate fibroblasts via alterations in MYC/AKT1 pathways, rather than via TGFβ as was observed by exosomes [123,158]. Thus, different subtypes of EV reprogram fibroblasts using different mechanisms. Interestingly, by comparing the tumorigenic capabilities of the $10,000\ g$ and $100,000\ g$ EV fractions from the same cell line, Lindoso and colleagues found that EVs collected after centrifugation at $10,000 \times g$ were more effective at stimulating angiogenesis, whereas EVs collected after centrifugation at $100,000 \times g$ were more effective at increasing migration of endothelial cells [124]. These results strengthen the conclusion that different EV subtypes perform unique functions within the tumor niche.

4. Therapeutic Implications of Extracellular Vesicles

In addition to EV biomarkers derived from serum or other biological fluids, a topic that has been thoroughly reviewed above and by others [160,161], EVs have significant potential for use in anti-cancer therapy. Strategies include using EVs as potential cancer vaccines or drug delivery systems, developing interventions to sequester tumor-derived EVs in patients, and developing drugs that target factors involved in EV release.

One of the first indications that EVs may have utility as cancer therapeutics was the observation that dendritic cells secrete antigen-presenting vesicles and that tumor peptide-pulsed dendritic cell-derived EVs decrease tumor growth in mice [145]. This finding drove interest in using dendritic cell-derived EVs as tumor vaccines and spurred multiple clinical trials. Three Phase I trials confirmed the safety of use of dendritic cell-derived EVs in anti-cancer treatments; however, the injected EVs exhibited poor potential in stimulating a T-cell response in the patients [162–164]. More recently, a Phase II trial was completed using dendritic cell-derived EVs as a vaccine. This involved EVs derived from IFN-γ-matured dendritic cells rather than immature dendritic cells [165]. Unfortunately, the endpoint goal (4 months of disease-free survival in 50% of patients) was not reached. A major hurdle in the use of EVs as therapeutics involves the standardization of techniques used to collect and analyze EVs and their molecular cargo, as discussed in Section 2.1 [13]. Interestingly, Tkach and colleagues found that EVs derived from immature dendritic cells are functionally heterogeneous, with large $(2000\ g)$ and small $(100,000\ g)$ EVs resulting in different cytokine expression profiles in recipient cells [166]. However, no such heterogeneity was observed for EVs derived from mature dendritic cells [166]. While promising, further work is required to develop a suitable strategy for use of dendritic cell-derived EVs as a form of anti-cancer therapy.

EVs display characteristics that make them ideal options for drug delivery. They are well tolerated in the body, easily taken up by cells, and can be targeted for uptake by specific tissues [167]. A general strategy involves engineering EVs to contain a specific cargo, such as pro-apoptotic proteins, miRNAs, or siRNAs, chemotherapeutic drugs, or molecules targeting specific oncogenes [114,168–173]. Indeed, several papers have described the successful insertion of siRNAs into exosomes [114,169–171]. For example, Alvarez-Erviti and colleagues successfully used self-derived dendritic cell EVs loaded with siRNA to target the brains of mice, finding that this approach did indeed facilitate knockdown of target mRNAs [169]. Further, injection of mice with EVs engineered to contain siRNA targeting mutant KRAS suppressed pancreatic cancer growth and improved overall survival [114].

An additional approach involves counteracting the pro-tumorigenic effects of EVs. One approach is to directly remove EVs from circulation. For example, Marleau and colleagues describe an extracorporeal hemofiltration system which filters blood for components under 200 nm and removes them using affinity agents for target molecules [174]. However, additional studies are required to test the clinical utility of this device. As another example, Nishida-Aoki and colleagues found that treatment of mice with anti-CD9 or anti-CD63 antibodies stimulated EV removal by macrophages, thus greatly decreasing EV concentration in the blood [118]. Although this treatment had no effect on the primary tumor, the authors observed a significant reduction in metastasis [118]. Blocking EV biogenesis in tumor cells by silencing genes encoding EV-related machinery is another potential avenue for inhibiting tumorigenesis. For example, knockdown of *SMPD3* and *RAB27A* resulted in reduced EV secretion and decreased tumorigenesis in mouse models [127,133,143]. However, such strategies may interfere with the normal process of EV-mediated communication; thus, a strategy which serves to minimize off-target effects is required.

5. Summary

In the past few years we have learned a great deal regarding the myriad of cargo molecules contained within EVs and the complex roles EVs play in the tumor microenvironment. The pace of research on this topic has vastly increased in the past couple of years, and we will no doubt make great strides in the years ahead in understanding the complexities underlying the role of EVs in cancer. Though much of this research is still in its infancy, there no doubt there lies many exciting therapeutic and biomarker opportunities ahead.

Author Contributions: Conceptualization, J.J., R.T., and C.G.; Writing-Original Draft Preparation, J.J. and R.T.; Writing-Review & Editing, J.J., R.T., and C.G.

Acknowledgments: We thank Timon Buys for helpful discussion.

References

1. Gould, S.J.; Raposo, G. As we wait: Coping with an imperfect nomenclature for extracellular vesicles. *J. Extracell. Vesicles* **2013**, *2*, 20389. [CrossRef] [PubMed]

2. Trajkovic, K.; Hsu, C.; Chiantia, S.; Rajendran, L.; Wenzel, D.; Wieland, F.; Schwille, P.; Brügger, B.; Simons, M. Ceramide triggers budding of exosome vesicles into multivesicular endosomes. *Science* **2008**, *319*, 1244–1247. [CrossRef] [PubMed]

3. Abusamra, A.J.; Zhong, Z.; Zheng, X.; Li, M.; Ichim, T.E.; Chin, J.L.; Min, W.P. Tumor exosomes expressing fas ligand mediate CD8+ T-cell apoptosis. *Blood Cells Mol. Dis.* **2005**, *35*, 169–173. [CrossRef] [PubMed]

4. Taraboletti, G.; D'Ascenzo, S.; Giusti, I.; Marchetti, D.; Borsotti, P.; Millimaggi, D.; Giavazzi, R.; Pavan, A.; Dolo, V. Bioavailability of VEGF in tumor-shed vesicles depends on vesicle burst induced by acidic pH. *Neoplasia* **2006**, *8*, 96–103. [CrossRef] [PubMed]

5. Montecalvo, A.; Larregina, A.T.; Shufesky, W.J.; Stolz, D.B.; Sullivan, M.L.; Karlsson, J.M.; Baty, C.J.; Gibson, G.A.; Erdos, G.; Wang, Z. Mechanism of transfer of functional microRNAs between mouse dendritic cells via exosomes. *Blood* **2011**. [CrossRef] [PubMed]

6. Tian, T.; Zhu, Y.L.; Zhou, Y.Y.; Liang, G.F.; Wang, Y.Y.; Hu, F.H.; Xiao, Z.D. Exosome uptake through clathrin-mediated endocytosis and macropinocytosis and mediating miR-21 delivery. *J. Biol. Chem.* **2014**, *289*, 22258–22267. [CrossRef] [PubMed]

7. Valadi, H.; Ekström, K.; Bossios, A.; Sjöstrand, M.; Lee, J.J.; Lötvall, J.O. Exosome-mediated transfer of mrnas and microRNAs is a novel mechanism of genetic exchange between cells. *Nat. Cell Biol.* **2007**, *9*, 654–659. [CrossRef] [PubMed]

8. Kosaka, N.; Iguchi, H.; Hagiwara, K.; Yoshioka, Y.; Takeshita, F.; Ochiya, T. Neutral sphingomyelinase 2 (nSMase2)-dependent exosomal transfer of angiogenic microRNAs regulate cancer cell metastasis. *J. Biol. Chem.* **2013**, *288*, 10849–10859. [CrossRef] [PubMed]

9. Webber, J.; Steadman, R.; Mason, M.D.; Tabi, Z.; Clayton, A. Cancer exosomes trigger fibroblast to myofibroblast differentiation. *Cancer Res.* **2010**, *70*, 9621–9630. [CrossRef] [PubMed]

10. Wieckowski, E.U.; Visus, C.; Szajnik, M.; Szczepanski, M.J.; Storkus, W.J.; Whiteside, T.L. Tumor-derived microvesicles promote regulatory T cell expansion and induce apoptosis in tumor-reactive activated CD8+ T lymphocytes. *J. Immunol.* **2009**, *183*, 3720–3730. [CrossRef] [PubMed]

11. Gardiner, C.; Di Vizio, D.; Sahoo, S.; Thery, C.; Witwer, K.W.; Wauben, M.; Hill, A.F. Techniques used for the isolation and characterization of extracellular vesicles: Results of a worldwide survey. *J. Extracell. Vesicles* **2016**, *5*, 32945. [CrossRef] [PubMed]

12. Webber, J.; Clayton, A. How pure are your vesicles? *J. Extracell. Vesicles* **2013**, *2*, 19861. [CrossRef] [PubMed]

13. Ramirez, M.I.; Amorim, M.G.; Gadelha, C.; Milic, I.; Welsh, J.A.; Freitas, V.M.; Nawaz, M.; Akbar, N.; Couch, Y.; Makin, L.; et al. Technical challenges of working with extracellular vesicles. *Nanoscale* **2018**, *10*, 881–906. [CrossRef] [PubMed]

14. Kanwar, S.S.; Dunlay, C.J.; Simeone, D.M.; Nagrath, S. Microfluidic device (exochip) for on-chip isolation, quantification and characterization of circulating exosomes. *Lab Chip* **2014**, *14*, 1891–1900. [CrossRef] [PubMed]

15. Mateescu, B.; Kowal, E.J.; van Balkom, B.W.; Bartel, S.; Bhattacharyya, S.N.; Buzas, E.I.; Buck, A.H.; de Candia, P.; Chow, F.W.; Das, S.; et al. Obstacles and opportunities in the functional analysis of extracellular vesicle RNA—An isev position paper. *J. Extracell. Vesicles* **2017**, *6*, 1286095. [CrossRef] [PubMed]

16. Eldh, M.; Lotvall, J.; Malmhall, C.; Ekstrom, K. Importance of RNA isolation methods for analysis of exosomal RNA: Evaluation of different methods. *Mol. Immunol.* **2012**, *50*, 278–286. [CrossRef] [PubMed]

17. Ji, H.; Chen, M.; Greening, D.W.; He, W.; Rai, A.; Zhang, W.; Simpson, R.J. Deep sequencing of RNA from three different extracellular vesicle (ev) subtypes released from the human lim1863 colon cancer cell line uncovers distinct miRNA-enrichment signatures. *PLoS ONE* **2014**, *9*, e110314. [CrossRef] [PubMed]

18. Cha, D.J.; Franklin, J.L.; Dou, Y.; Liu, Q.; Higginbotham, J.N.; Beckler, M.D.; Weaver, A.M.; Vickers, K.; Prasad, N.; Levy, S. Kras-dependent sorting of miRNA to exosomes. *eLife* **2015**, *4*, e07197. [CrossRef] [PubMed]

19. Fiskaa, T.; Knutsen, E.; Nikolaisen, M.A.; Jørgensen, T.E.; Johansen, S.D.; Perander, M.; Seternes, O.M. Distinct small RNA signatures in extracellular vesicles derived from breast cancer cell lines. *PLoS ONE* **2016**, *11*, e0161824. [CrossRef] [PubMed]

20. Amorim, M.G.; Valieris, R.; Drummond, R.D.; Pizzi, M.P.; Freitas, V.M.; Sinigaglia-Coimbra, R.; Calin, G.A.; Pasqualini, R.; Arap, W.; Silva, I.T. A total transcriptome profiling method for plasma-derived extracellular vesicles: Applications for liquid biopsies. *Sci. Rep.* **2017**, *7*, 14395. [CrossRef] [PubMed]

21. Endzeliņš, E.; Berger, A.; Melne, V.; Bajo-Santos, C.; Soboļevska, K.; Ābols, A.; Rodriguez, M.; Šantare, D.; Rudņickiha, A.; Lietuvietis, V. Detection of circulating miRNAs: Comparative analysis of extracellular vesicle-incorporated miRNAs and cell-free miRNAs in whole plasma of prostate cancer patients. *BMC Cancer* **2017**, *17*, 730. [CrossRef] [PubMed]

22. Shurtleff, M.J.; Yao, J.; Qin, Y.; Nottingham, R.M.; Temoche-Diaz, M.M.; Schekman, R.; Lambowitz, A.M. Broad role for YBX1 in defining the small noncoding RNA composition of exosomes. *Proc. Natl. Acad. Sci. USA* **2017**, *114*, E8987–E8995. [CrossRef] [PubMed]

23. Santangelo, L.; Giurato, G.; Cicchini, C.; Montaldo, C.; Mancone, C.; Tarallo, R.; Battistelli, C.; Alonzi, T.; Weisz, A.; Tripodi, M. The RNA-binding protein SYNCRIP is a component of the hepatocyte exosomal machinery controlling microRNA sorting. *Cell Rep.* **2016**, *17*, 799–808. [CrossRef] [PubMed]

24. Villarroya-Beltri, C.; Gutiérrez-Vázquez, C.; Sánchez-Cabo, F.; Pérez-Hernández, D.; Vázquez, J.; Martin-Cofreces, N.; Martinez-Herrera, D.J.; Pascual-Montano, A.; Mittelbrunn, M.; Sánchez-Madrid, F. Sumoylated HNRNPA2B1 controls the sorting of miRNAs into exosomes through binding to specific motifs. *Nat. Commun.* **2013**, *4*, 2980. [CrossRef] [PubMed]

25. Mukherjee, K.; Ghoshal, B.; Ghosh, S.; Chakrabarty, Y.; Shwetha, S.; Das, S.; Bhattacharyya, S.N. Reversible hur-microRNA binding controls extracellular export of miR-122 and augments stress response. *EMBO Rep.* **2016**, *17*, 1184–1203. [CrossRef] [PubMed]

26. Statello, L.; Maugeri, M.; Garre, E.; Nawaz, M.; Wahlgren, J.; Papadimitriou, A.; Lundqvist, C.; Lindfors, L.; Collen, A.; Sunnerhagen, P.; et al. Identification of RNA-binding proteins in exosomes capable of interacting with different types of RNA: RBP-facilitated transport of RNAs into exosomes. *PLoS ONE* **2018**, *13*, e0195969. [CrossRef] [PubMed]

27. Teng, Y.; Ren, Y.; Hu, X.; Mu, J.; Samykutty, A.; Zhuang, X.; Deng, Z.; Kumar, A.; Zhang, L.; Merchant, M.L. MVP-mediated exosomal sorting of miR-193a promotes colon cancer progression. *Nat. Commun.* **2017**, *8*, 14448. [CrossRef] [PubMed]

28. Wei, J.x.; Lv, L.h.; Wan, Y.l.; Cao, Y.; Li, G.l.; Lin, H.m.; Zhou, R.; Shang, C.z.; Cao, J.; He, H. Vps4A functions as a tumor suppressor by regulating the secretion and uptake of exosomal microRNAs in human hepatoma cells. *Hepatology* **2015**, *61*, 1284–1294. [CrossRef] [PubMed]

29. Hagiwara, K.; Katsuda, T.; Gailhouste, L.; Kosaka, N.; Ochiya, T. Commitment of annexin a2 in recruitment

of microRNAs into extracellular vesicles. *FEBS Lett.* **2015**, *589*, 4071–4078. [CrossRef] [PubMed]

30. Shurtleff, M.J.; Temoche-Diaz, M.M.; Karfilis, K.V.; Ri, S.; Schekman, R. Y-box protein 1 is required to sort microRNAs into exosomes in cells and in a cell-free reaction. *eLife* **2016**, *5*, e19276. [CrossRef] [PubMed]

31. Lu, P.; Li, H.; Li, N.; Singh, R.N.; Bishop, C.E.; Chen, X.; Lu, B. Mex3c interacts with adaptor-related protein complex 2 and involves in miR-451a exosomal sorting. *PLoS ONE* **2017**, *12*, e0185992. [CrossRef] [PubMed]

32. Le, M.T.; Hamar, P.; Guo, C.; Basar, E.; Perdigão-Henriques, R.; Balaj, L.; Lieberman, J. MiR-200–containing extracellular vesicles promote breast cancer cell metastasis. *J. Clin. Investig.* **2014**, *124*, 5109–5128. [CrossRef] [PubMed]

33. Melo, S.A.; Sugimoto, H.; O'Connell, J.T.; Kato, N.; Villanueva, A.; Vidal, A.; Qiu, L.; Vitkin, E.; Perelman, L.T.; Melo, C.A.; et al. Cancer exosomes perform cell-independent microRNA biogenesis and promote tumorigenesis. *Cancer Cell* **2014**, *26*, 707–721. [CrossRef] [PubMed]

34. McKenzie, A.J.; Hoshino, D.; Hong, N.H.; Cha, D.J.; Franklin, J.L.; Coffey, R.J.; Patton, J.G.; Weaver, A.M. KRAS-MEK signaling controls Ago2 sorting into exosomes. *Cell Rep.* **2016**, *15*, 978–987. [CrossRef] [PubMed]

35. Gibbings, D.J.; Ciaudo, C.; Erhardt, M.; Voinnet, O. Multivesicular bodies associate with components of miRNA effector complexes and modulate miRNA activity. *Nat. Cell Biol.* **2009**, *11*, 1143–1149. [CrossRef] [PubMed]

36. Fatima, F.; Nawaz, M. Long distance metabolic regulation through adipose-derived circulating exosomal miRNAs: A trail for RNA-based therapies? *Front. Physiol.* **2017**, *8*, 545. [CrossRef] [PubMed]

37. Koppers-Lalic, D.; Hackenberg, M.; Bijnsdorp, I.V.; van Eijndhoven, M.A.J.; Sadek, P.; Sie, D.; Zini, N.; Middeldorp, J.M.; Ylstra, B.; de Menezes, R.X.; et al. Nontemplated nucleotide additions distinguish the small RNA composition in cells from exosomes. *Cell Rep.* **2014**, *8*, 1649–1658. [CrossRef] [PubMed]

38. Song, J.; Song, J.; Mo, B.; Chen, X. Uridylation and adenylation of RNAs. *Sci. China Life Sci.* **2015**, *58*, 1057–1066. [CrossRef] [PubMed]

39. Squadrito, M.L.; Baer, C.; Burdet, F.; Maderna, C.; Gilfillan, G.D.; Lyle, R.; Ibberson, M.; De Palma, M. Endogenous RNAs modulate microRNA sorting to exosomes and transfer to acceptor cells. *Cell Rep.* **2014**, *8*, 1432–1446. [CrossRef] [PubMed]

40. Kubota, S.; Chiba, M.; Watanabe, M.; Sakamoto, M.; Watanabe, N. Secretion of small/microRNAs including mir-638 into extracellular spaces by sphingomyelin phosphodiesterase 3. *Oncol. Rep.* **2015**, *33*, 67–73. [CrossRef] [PubMed]

41. Janas, T.; Janas, M.M.; Sapon, K.; Janas, T. Mechanisms of RNA loading into exosomes. *FEBS Lett.* **2015**, *589*, 1391–1398. [CrossRef] [PubMed]

42. Zhang, G.; Zhang, Y.; Cheng, S.; Wu, Z.; Liu, F.; Zhang, J. CD133 positive U87 glioblastoma cells-derived exosomal microRNAs in hypoxia-versus normoxia-microenviroment. *J. Neuro-Oncol.* **2017**, *135*, 37–46. [CrossRef] [PubMed]

43. Cavalleri, T.; Angelici, L.; Favero, C.; Dioni, L.; Mensi, C.; Bareggi, C.; Palleschi, A.; Rimessi, A.; Consonni, D.; Bordini, L. Plasmatic extracellular vesicle microRNAs in malignant pleural mesothelioma and asbestos-exposed subjects suggest a 2-miRNA signature as potential biomarker of disease. *PLoS ONE* **2017**, *12*, e0176680. [CrossRef] [PubMed]

44. Lim, J.-H.; Song, M.-K.; Cho, Y.; Kim, W.; Han, S.O.; Ryu, J.-C. Comparative analysis of microRNA and mRNA expression profiles in cells and exosomes under toluene exposure. *Toxicol. In Vitro* **2017**, *41*, 92–101. [CrossRef] [PubMed]

45. Harden, M.E.; Munger, K. Human papillomavirus 16 e6 and e7 oncoprotein expression alters microRNA expression in extracellular vesicles. *Virology* **2017**, *508*, 63–69. [CrossRef] [PubMed]

46. Xiong, L.; Zhen, S.; Yu, Q.; Gong, Z. HCV-E2 inhibits hepatocellular carcinoma metastasis by stimulating mast cells to secrete exosomal shuttle microRNAs. *Oncol. Lett.* **2017**, *14*, 2141–2146. [CrossRef] [PubMed]

47. Yogev, O.; Henderson, S.; Hayes, M.J.; Marelli, S.S.; Ofir-Birin, Y.; Regev-Rudzki, N.; Herrero, J.; Enver, T. Herpesviruses shape tumour microenvironment through exosomal transfer of viral microRNAs. *PLoS Pathog.* **2017**, *13*, e1006524. [CrossRef] [PubMed]

48. Rabinowits, G.; Gerçel-Taylor, C.; Day, J.M.; Taylor, D.D.; Kloecker, G.H. Exosomal microRNA: A diagnostic marker for lung cancer. *Clin. Lung Cancer* **2009**, *10*, 42–46. [CrossRef] [PubMed]

49. Hannafon, B.N.; Trigoso, Y.D.; Calloway, C.L.; Zhao, Y.D.; Lum, D.H.; Welm, A.L.; Zhao, Z.J.; Blick, K.E.; Dooley, W.C.; Ding, W. Plasma exosome microRNAs are indicative of breast cancer. *Breast Cancer Res.* **2016**, *18*, 90. [CrossRef] [PubMed]

50. Li, Z.; Ma, Y.-Y.; Wang, J.; Zeng, X.-F.; Li, R.; Kang, W.; Hao, X.-K. Exosomal microRNA-141 is upregulated in the serum of prostate cancer patients. *OncoTargets Ther.* **2016**, *9*, 139–148.

51. Meng, X.; Müller, V.; Milde-Langosch, K.; Trillsch, F.; Pantel, K.; Schwarzenbach, H. Diagnostic and prognostic relevance of circulating exosomal miR-373, miR-200a, miR-200b and miR-200c in patients with epithelial ovarian cancer. *Oncotarget* **2016**, *7*, 16923–16935. [CrossRef] [PubMed]

52. Ostenfeld, M.S.; Jensen, S.G.; Jeppesen, D.K.; Christensen, L.-L.; Thorsen, S.B.; Stenvang, J.; Hvam, M.L.; Thomsen, A.; Mouritzen, P.; Rasmussen, M.H. MiRNA profiling of circulating epcam+ extracellular vesicles: Promising biomarkers of colorectal cancer. *J. Extracell. Vesicles* **2016**, *5*, 31488. [CrossRef] [PubMed]

53. Caivano, A.; La Rocca, F.; Simeon, V.; Girasole, M.; Dinarelli, S.; Laurenzana, I.; De Stradis, A.; De Luca, L.; Trino, S.; Traficante, A. MicroRNA-155 in serum-derived extracellular vesicles as a potential biomarker for hematologic malignancies-a short report. *Cell. Oncol.* **2017**, *40*, 97–103. [CrossRef] [PubMed]

54. Lan, F.; Qing, Q.; Pan, Q.; Hu, M.; Yu, H.; Yue, X. Serum exosomal miR-301a as a potential diagnostic and prognostic biomarker for human glioma. *Cell. Oncol.* **2018**, *41*, 25–33. [CrossRef] [PubMed]

55. Qu, Z.; Wu, J.; Wu, J.; Ji, A.; Qiang, G.; Jiang, Y.; Jiang, C.; Ding, Y. Exosomal miR-665 as a novel minimally invasive biomarker for hepatocellular carcinoma diagnosis and prognosis. *Oncotarget* **2017**, *8*, 80666–80678. [CrossRef] [PubMed]

56. Santangelo, A.; Imbrucè, P.; Gardenghi, B.; Belli, L.; Agushi, R.; Tamanini, A.; Munari, S.; Bossi, A.M.; Scambi, I.; Benati, D. A microRNA signature from serum exosomes of patients with glioma as complementary diagnostic biomarker. *J. Neuro-Oncol.* **2018**, *136*, 51–62. [CrossRef] [PubMed]

57. Xu, J.-F.; Wang, Y.-P.; Zhang, S.-J.; Chen, Y.; Gu, H.-F.; Dou, X.-F.; Xia, B.; Bi, Q.; Fan, S.-W. Exosomes containing differential expression of microRNA and mRNA in osteosarcoma that can predict response to chemotherapy. *Oncotarget* **2017**, *8*, 75968–75978. [CrossRef] [PubMed]

58. Yan, S.; Han, B.; Gao, S.; Wang, X.; Wang, Z.; Wang, F.; Zhang, J.; Xu, D.; Sun, B. Exosome-encapsulated microRNAs as circulating biomarkers for colorectal cancer. *Oncotarget* **2017**, *8*, 60149–60158. [CrossRef] [PubMed]

59. Koppers-Lalic, D.; Hackenberg, M.; de Menezes, R.; Misovic, B.; Wachalska, M.; Geldof, A.; Zini, N.; de Reijke, T.; Wurdinger, T.; Vis, A. Non-invasive prostate cancer detection by measuring miRNA variants (isomiRs) in urine extracellular vesicles. *Oncotarget* **2016**, *7*, 22566–22578. [CrossRef] [PubMed]

60. Andreu, Z.; Oshiro, R.O.; Redruello, A.; López-Martín, S.; Gutiérrez-Vázquez, C.; Morato, E.; Marina, A.I.; Gómez, C.O.; Yáñez-Mó, M. Extracellular vesicles as a source for non-invasive biomarkers in bladder cancer progression. *Eur. J. Pharm. Sci.* **2017**, *98*, 70–79. [CrossRef] [PubMed]

61. Matsuzaki, K.; Fujita, K.; Jingushi, K.; Kawashima, A.; Ujike, T.; Nagahara, A.; Ueda, Y.; Tanigawa, G.; Yoshioka, I.; Ueda, K. MiR-21-5p in urinary extracellular vesicles is a novel biomarker of urothelial carcinoma. *Oncotarget* **2017**, *8*, 24668–24678. [CrossRef] [PubMed]

62. Xu, Y.; Qin, S.; An, T.; Tang, Y.; Huang, Y.; Zheng, L. MiR-145 detection in urinary extracellular vesicles increase diagnostic efficiency of prostate cancer based on hydrostatic filtration dialysis method. *Prostate* **2017**, *77*, 1167–1175. [CrossRef] [PubMed]

63. Lin, J.; Wang, Y.; Zou, Y.-Q.; Chen, X.; Huang, B.; Liu, J.; Xu, Y.-M.; Li, J.; Zhang, J.; Yang, W.-M. Differential miRNA expression in pleural effusions derived from extracellular vesicles of patients with lung cancer, pulmonary tuberculosis, or pneumonia. *Tumor Biol.* **2016**, *37*, 15835–15845. [CrossRef] [PubMed]

64. Wang, Y.; Xu, Y.-M.; Zou, Y.-Q.; Lin, J.; Huang, B.; Liu, J.; Li, J.; Zhang, J.; Yang, W.-M.; Min, Q.-H. Identification of differential expressed PE exosomal miRNA in lung adenocarcinoma, tuberculosis, and other benign lesions. *Medicine* **2017**, *96*, e8361. [CrossRef] [PubMed]

65. Langevin, S.; Kuhnell, D.; Parry, T.; Biesiada, J.; Huang, S.; Wise-Draper, T.; Casper, K.; Zhang, X.; Medvedovic, M.; Kasper, S. Comprehensive microRNA-sequencing of exosomes derived from head and neck carcinoma cells in vitro reveals common secretion profiles and potential utility as salivary biomarkers. *Oncotarget* **2017**, *8*, 82459–82474. [CrossRef] [PubMed]

66. Wang, J.; Zhou, Y.; Lu, J.; Sun, Y.; Xiao, H.; Liu, M.; Tian, L. Combined detection of serum exosomal miR-21 and HOTAIR as diagnostic and prognostic biomarkers for laryngeal squamous cell carcinoma. *Med. Oncol.* **2014**, *31*, 148. [CrossRef] [PubMed]

67. Wang, H.; Hou, L.; Li, A.; Duan, Y.; Gao, H.; Song, X. Expression of serum exosomal microRNA-21 in human hepatocellular carcinoma. *BioMed Res. Int.* **2014**, *2014*, 864894. [CrossRef] [PubMed]

68. Huang, X.; Yuan, T.; Liang, M.; Du, M.; Xia, S.; Dittmar, R.; Wang, D.; See, W.; Costello, B.A.; Quevedo, F. Exosomal miR-1290 and miR-375 as prognostic markers in castration-resistant prostate cancer. *Eur. Urol.* **2015**, *67*, 33–41. [CrossRef] [PubMed]

69. Wei, Z.; Batagov, A.O.; Schinelli, S.; Wang, J.; Wang, Y.; El Fatimy, R.; Rabinovsky, R.; Balaj, L.; Chen, C.C.; Hochberg, F. Coding and noncoding landscape of extracellular RNA released by human glioma stem cells. *Nat. Commun.* **2017**, *8*, 1145. [CrossRef] [PubMed]

70. Lai, C.P.; Kim, E.Y.; Badr, C.E.; Weissleder, R.; Mempel, T.R.; Tannous, B.A.; Breakefield, X.O. Visualization and tracking of tumour extracellular vesicle delivery and RNA translation using multiplexed reporters. *Nat. Commun.* **2015**, *6*, 7029. [CrossRef] [PubMed]

71. Zomer, A.; Maynard, C.; Verweij, F.J.; Kamermans, A.; Schäfer, R.; Beerling, E.; Schiffelers, R.M.; de Wit, E.; Berenguer, J.; Ellenbroek, S.I.J. In vivo imaging reveals extracellular vesicle-mediated phenocopying of metastatic behavior. *Cell* **2015**, *161*, 1046–1057. [CrossRef] [PubMed]

72. Gutkin, A.; Uziel, O.; Beery, E.; Nordenberg, J.; Pinchasi, M.; Goldvaser, H.; Henick, S.; Goldberg, M.; Lahav, M. Tumor cells derived exosomes contain hTERT mRNA and transform nonmalignant fibroblasts into telomerase positive cells. *Oncotarget* **2016**, *7*, 59173–59188. [CrossRef] [PubMed]

73. Yokoi, A.; Yoshioka, Y.; Yamamoto, Y.; Ishikawa, M.; Ikeda, S.-I.; Kato, T.; Kiyono, T.; Takeshita, F.; Kajiyama, H.; Kikkawa, F. Malignant extracellular vesicles carrying mmp1 mRNA facilitate peritoneal dissemination in ovarian cancer. *Nat. Commun.* **2017**, *8*, 14470. [CrossRef] [PubMed]

74. Goldvaser, H.; Gutkin, A.; Beery, E.; Edel, Y.; Nordenberg, J.; Wolach, O.; Rabizadeh, E.; Uziel, O.; Lahav, M. Characterisation of blood-derived exosomal hTERT mRNA secretion in cancer patients: A potential pan-cancer marker. *Br. J. Cancer* **2017**, *117*, 353. [CrossRef] [PubMed]

75. Kossinova, O.A.; Gopanenko, A.V.; Tamkovich, S.N.; Krasheninina, O.A.; Tupikin, A.E.; Kiseleva, E.; Yanshina, D.D.; Malygin, A.A.; Ven'yaminova, A.G.; Kabilov, M.R. Cytosolic YB-1 and NSUN2 are the only proteins recognizing specific motifs present in mRNAs enriched in exosomes. *Biochim. Biophys. Acta (BBA)-Proteins Proteom.* **2017**, *1865*, 664–673. [CrossRef] [PubMed]

76. Yanshina, D.D.; Kossinova, O.A.; Gopanenko, A.V.; Krasheninina, O.A.; Malygin, A.A.; Venyaminova, A.G.; Karpova, G.G. Structural features of the interaction of the 3′-untranslated region of mRNA containing exosomal RNA-specific motifs with YB-1, a potential mediator of mRNA sorting. *Biochimie* **2018**, *144*, 134–143. [CrossRef] [PubMed]

77. Maurya, P.K.; Mishra, A.; Yadav, B.S.; Singh, S.; Kumar, P.; Chaudhary, A.; Srivastava, S.; Murugesan, S.N.; Mani, A. Role of y box protein-1 in cancer: As potential biomarker and novel therapeutic target. *J. Cancer* **2017**, *8*, 1900–1907. [CrossRef] [PubMed]

78. Bolukbasi, M.F.; Mizrak, A.; Ozdener, G.B.; Madlener, S.; Ströbel, T.; Erkan, E.P.; Fan, J.-B.; Breakefield, X.O.; Saydam, O. MiR-1289 and "zipcode"-like sequence enrich mrnas in microvesicles. *Mol. Ther. Nucleic Acids* **2012**, *1*, e10. [CrossRef] [PubMed]

79. Conley, A.; Minciacchi, V.R.; Lee, D.H.; Knudsen, B.S.; Karlan, B.Y.; Citrigno, L.; Viglietto, G.; Tewari, M.; Freeman, M.R.; Demichelis, F. High-throughput sequencing of two populations of extracellular vesicles provides an mRNA signature that can be detected in the circulation of breast cancer patients. *RNA Biol.* **2017**, *14*, 305–316. [CrossRef] [PubMed]

80. Gezer, U.; Özgür, E.; Cetinkaya, M.; Isin, M.; Dalay, N. Long non-coding RNAs with low expression levels in cells are enriched in secreted exosomes. *Cell Biol. Int.* **2014**, *38*, 1076–1079. [CrossRef] [PubMed]

81. Freedman, J.E.; Gerstein, M.; Mick, E.; Rozowsky, J.; Levy, D.; Kitchen, R.; Das, S.; Shah, R.; Danielson, K.; Beaulieu, L. Diverse human extracellular RNAs are widely detected in human plasma. *Nat. Commun.* **2016**, *7*, 11106. [CrossRef] [PubMed]

82. Li, Y.; Zheng, Q.; Bao, C.; Li, S.; Guo, W.; Zhao, J.; Chen, D.; Gu, J.; He, X.; Huang, S. Circular RNA is enriched and stable in exosomes: A promising biomarker for cancer diagnosis. *Cell Res.* **2015**, *25*, 981–984. [CrossRef] [PubMed]

83. Takahashi, K.; Yan, I.K.; Kogure, T.; Haga, H.; Patel, T. Extracellular vesicle-mediated transfer of long non-coding RNA ror modulates chemosensitivity in human hepatocellular cancer. *FEBS Open Bio* **2014**, *4*, 458–467. [CrossRef] [PubMed]

84. Pan, L.; Liang, W.; Fu, M.; Huang, Z.-H.; Li, X.; Zhang, W.; Zhang, P.; Qian, H.; Jiang, P.-C.; Xu, W.-R. Exosomes-mediated transfer of long noncoding RNA ZFAS1 promotes gastric cancer progression. *J. Cancer*

Res. Clin. Oncol. **2017**, *143*, 991–1004. [CrossRef] [PubMed]

85. Fatima, F.; Nawaz, M. Vesiculated long non-coding RNAs: Offshore packages deciphering trans-regulation between cells, cancer progression and resistance to therapies. *Noncoding RNA* **2017**, *3*, 10. [CrossRef] [PubMed]

86. Lázaro-Ibáñez, E.; Sanz-Garcia, A.; Visakorpi, T.; Escobedo-Lucea, C.; Siljander, P.; Ayuso-Sacido, Á.; Yliperttula, M. Different gdna content in the subpopulations of prostate cancer extracellular vesicles: Apoptotic bodies, microvesicles, and exosomes. *Prostate* **2014**, *74*, 1379–1390. [CrossRef] [PubMed]

87. García-Romero, N.; Carrión-Navarro, J.; Esteban-Rubio, S.; Lázaro-Ibáñez, E.; Peris-Celda, M.; Alonso, M.M.; Guzmán-De-Villoria, J.; Fernández-Carballal, C.; de Mendivil, A.O.; García-Duque, S. DNA sequences within glioma-derived extracellular vesicles can cross the intact blood-brain barrier and be detected in peripheral blood of patients. *Oncotarget* **2017**, *8*, 1416–1428. [CrossRef] [PubMed]

88. Guescini, M.; Genedani, S.; Stocchi, V.; Agnati, L.F. Astrocytes and glioblastoma cells release exosomes carrying mtDNA. *J. Neural Transm.* **2010**, *117*, 1. [CrossRef] [PubMed]

89. Németh, A.; Orgovan, N.; Sódar, B.W.; Osteikoetxea, X.; Pálóczi, K.; Szabó-Taylor, K.É.; Vukman, K.V.; Kittel, Á.; Turiák, L.; Wiener, Z. Antibiotic-induced release of small extracellular vesicles (exosomes) with surface-associated DNA. *Sci. Rep.* **2017**, *7*, 8202. [CrossRef] [PubMed]

90. Sansone, P.; Savini, C.; Kurelac, I.; Chang, Q.; Amato, L.B.; Strillacci, A.; Stepanova, A.; Iommarini, L.; Mastroleo, C.; Daly, L. Packaging and transfer of mitochondrial DNA via exosomes regulate escape from dormancy in hormonal therapy-resistant breast cancer. *Proc. Natl. Acad. Sci. USA* **2017**. [CrossRef] [PubMed]

91. Balaj, L.; Lessard, R.; Dai, L.; Cho, Y.-J.; Pomeroy, S.L.; Breakefield, X.O.; Skog, J. Tumour microvesicles contain retrotransposon elements and amplified oncogene sequences. *Nat. Commun.* **2011**, *2*, 180. [CrossRef] [PubMed]

92. Cai, J.; Han, Y.; Ren, H.; Chen, C.; He, D.; Zhou, L.; Eisner, G.M.; Asico, L.D.; Jose, P.A.; Zeng, C. Extracellular vesicle-mediated transfer of donor genomic DNA to recipient cells is a novel mechanism for genetic influence between cells. *J. Mol. Cell Biol.* **2013**, *5*, 227–238. [CrossRef] [PubMed]

93. Kahlert, C.; Melo, S.A.; Protopopov, A.; Tang, J.; Seth, S.; Koch, M.; Zhang, J.; Weitz, J.; Chin, L.; Futreal, A. Identification of double-stranded genomic DNA spanning all chromosomes with mutated kras and p53 DNA in the serum exosomes of patients with pancreatic cancer. *J. Biol. Chem.* **2014**, *289*, 3869–3875. [CrossRef] [PubMed]

94. Lee, T.H.; Chennakrishnaiah, S.; Audemard, E.; Montermini, L.; Meehan, B.; Rak, J. Oncogenic ras-driven cancer cell vesiculation leads to emission of double-stranded DNA capable of interacting with target cells. *Biochem. Biophys. Res. Commun.* **2014**, *451*, 295–301. [CrossRef] [PubMed]

95. Thakur, B.K.; Zhang, H.; Becker, A.; Matei, I.; Huang, Y.; Costa-Silva, B.; Zheng, Y.; Hoshino, A.; Brazier, H.; Xiang, J. Double-stranded DNA in exosomes: A novel biomarker in cancer detection. *Cell Res.* **2014**, *24*, 766–769. [CrossRef] [PubMed]

96. Cai, J.; Wu, G.; Tan, X.; Han, Y.; Chen, C.; Li, C.; Wang, N.; Zou, X.; Chen, X.; Zhou, F. Transferred bcr/abl DNA from k562 extracellular vesicles causes chronic myeloid leukemia in immunodeficient mice. *PLoS ONE* **2014**, *9*, e105200. [CrossRef] [PubMed]

97. San Lucas, F.; Allenson, K.; Bernard, V.; Castillo, J.; Kim, D.; Ellis, K.; Ehli, E.; Davies, G.; Petersen, J.; Li, D. Minimally invasive genomic and transcriptomic profiling of visceral cancers by next-generation sequencing of circulating exosomes. *Ann. Oncol.* **2015**, *27*, 635–641. [CrossRef] [PubMed]

98. Takahashi, A.; Okada, R.; Nagao, K.; Kawamata, Y.; Hanyu, A.; Yoshimoto, S.; Takasugi, M.; Watanabe, S.; Kanemaki, M.T.; Obuse, C. Exosomes maintain cellular homeostasis by excreting harmful DNA from cells. *Nat. Commun.* **2017**, *8*, 15827. [CrossRef] [PubMed]

99. Allenson, K.; Castillo, J.; San Lucas, F.; Scelo, G.; Kim, D.; Bernard, V.; Davis, G.; Kumar, T.; Katz, M.; Overman, M. High prevalence of mutant KRAS in circulating exosome-derived DNA from early-stage pancreatic cancer patients. *Ann. Oncol.* **2017**, *28*, 741–747. [CrossRef] [PubMed]

100. Möhrmann, L.; Huang, H.; Hong, D.S.; Tsimberidou, A.M.; Fu, S.; Piha-Paul, S.; Subbiah, V.; Karp, D.D.; Naing, A.; Krug, A.K. Liquid biopsies using plasma exosomal nucleic acids and plasma cell-free DNA compared with clinical outcomes of patients with advanced cancers. *Clin. Cancer Res.* **2017**. [CrossRef] [PubMed]

101. Yang, S.; Che, S.P.; Kurywchak, P.; Tavormina, J.L.; Gansmo, L.B.; Correa de Sampaio, P.; Tachezy, M.; Bockhorn, M.; Gebauer, F.; Haltom, A.R. Detection of mutant KRAS and TP53 DNA in circulating exosomes

from healthy individuals and patients with pancreatic cancer. *Cancer Biol. Ther.* **2017**, *18*, 158–165. [CrossRef] [PubMed]

102. Castellanos-Rizaldos, E.; Grimm, D.G.; Tadigotla, V.; Hurley, J.; Healy, J.; Neal, P.L.; Sher, M.; Venkatesan, R.; Karlovich, C.; Raponi, M.; et al. Exosome-based detection of EGFR T790m in plasma from non-small cell lung cancer patients. *Clin. Cancer Res.* **2018**, *24*, 2944–2950. [CrossRef] [PubMed]

103. Fernando, M.R.; Jiang, C.; Krzyzanowski, G.D.; Ryan, W.L. New evidence that a large proportion of human blood plasma cell-free DNA is localized in exosomes. *PLoS ONE* **2017**, *12*, e0183915. [CrossRef] [PubMed]

104. Lanman, R.B.; Mortimer, S.A.; Zill, O.A.; Sebisanovic, D.; Lopez, R.; Blau, S.; Collisson, E.A.; Divers, S.G.; Hoon, D.S.; Kopetz, E.S. Analytical and clinical validation of a digital sequencing panel for quantitative, highly accurate evaluation of cell-free circulating tumor DNA. *PLoS ONE* **2015**, *10*, e0140712. [CrossRef] [PubMed]

105. Williams, R.L.; Urbe, S. The emerging shape of the escrt machinery. *Nat. Rev. Mol. Cell Biol.* **2007**, *8*, 355–368. [CrossRef] [PubMed]

106. Takasugi, M.; Okada, R.; Takahashi, A.; Virya Chen, D.; Watanabe, S.; Hara, E. Small extracellular vesicles secreted from senescent cells promote cancer cell proliferation through EPHA2. *Nat. Commun.* **2017**, *8*, 15729. [CrossRef] [PubMed]

107. Itoh, S.; Mizuno, K.; Aikawa, M.; Aikawa, E. Dimerization of sortilin regulates its trafficking to extracellular vesicles. *J. Biol. Chem.* **2018**, *293*, 4532–4544. [CrossRef] [PubMed]

108. Perez-Hernandez, D.; Gutierrez-Vazquez, C.; Jorge, I.; Lopez-Martin, S.; Ursa, A.; Sanchez-Madrid, F.; Vazquez, J.; Yanez-Mo, M. The intracellular interactome of tetraspanin-enriched microdomains reveals their function as sorting machineries toward exosomes. *J. Biol. Chem.* **2013**, *288*, 11649–11661. [CrossRef] [PubMed]

109. Peinado, H.; Alečković, M.; Lavotshkin, S.; Matei, I.; Costa-Silva, B.; Moreno-Bueno, G.; Hergueta-Redondo, M.; Williams, C.; García-Santos, G.; Ghajar, C.M. Melanoma exosomes educate bone marrow progenitor cells toward a pro-metastatic phenotype through met. *Nat. Med.* **2012**, *18*, 883–891. [CrossRef] [PubMed]

110. Hoshino, A.; Costa-Silva, B.; Shen, T.-L.; Rodrigues, G.; Hashimoto, A.; Mark, M.T.; Molina, H.; Kohsaka, S.; Di Giannatale, A.; Ceder, S. Tumour exosome integrins determine organotropic metastasis. *Nature* **2015**, *527*, 329–335. [CrossRef] [PubMed]

111. Zhang, L.; Zhang, S.; Yao, J.; Lowery, F.J.; Zhang, Q.; Huang, W.C.; Li, P.; Li, M.; Wang, X.; Zhang, C.; et al. Microenvironment-induced pten loss by exosomal microRNA primes brain metastasis outgrowth. *Nature* **2015**, *527*, 100–104. [CrossRef] [PubMed]

112. Fong, M.Y.; Zhou, W.; Liu, L.; Alontaga, A.Y.; Chandra, M.; Ashby, J.; Chow, A.; O'Connor, S.T.; Li, S.; Chin, A.R.; et al. Breast-cancer-secreted miR-122 reprograms glucose metabolism in premetastatic niche to promote metastasis. *Nat. Cell Biol.* **2015**, *17*, 183–194. [CrossRef] [PubMed]

113. Bonjoch, L.; Gironella, M.; Iovanna, J.L.; Closa, D. REG3β modifies cell tumor function by impairing extracellular vesicle uptake. *Sci. Rep.* **2017**, *7*, 3143. [CrossRef] [PubMed]

114. Kamerkar, S.; LeBleu, V.S.; Sugimoto, H.; Yang, S.; Ruivo, C.F.; Melo, S.A.; Lee, J.J.; Kalluri, R. Exosomes facilitate therapeutic targeting of oncogenic KRAS in pancreatic cancer. *Nature* **2017**, *546*, 498–503. [CrossRef] [PubMed]

115. Tauro, B.J.; Greening, D.W.; Mathias, R.A.; Mathivanan, S.; Ji, H.; Simpson, R.J. Two distinct populations of exosomes are released from lim1863 colon carcinoma cell-derived organoids. *Mol. Cell. Proteom.* **2013**, *12*, 587–598. [CrossRef] [PubMed]

116. Kowal, J.; Arras, G.; Colombo, M.; Jouve, M.; Morath, J.P.; Primdal-Bengtson, B.; Dingli, F.; Loew, D.; Tkach, M.; Théry, C. Proteomic comparison defines novel markers to characterize heterogeneous populations of extracellular vesicle subtypes. *Proc. Natl. Acad. Sci. USA* **2016**, *113*, E968–E977. [CrossRef] [PubMed]

117. Hurwitz, S.N.; Nkosi, D.; Conlon, M.M.; York, S.B.; Liu, X.; Tremblay, D.C.; Meckes, D.G. CD63 regulates epstein-barr virus LMP1 exosomal packaging, enhancement of vesicle production, and noncanonical NF-κB signaling. *J. Virol.* **2017**, *91*, e02251-16. [CrossRef] [PubMed]

118. Nishida-Aoki, N.; Tominaga, N.; Takeshita, F.; Sonoda, H.; Yoshioka, Y.; Ochiya, T. Disruption of circulating extracellular vesicles as a novel therapeutic strategy against cancer metastasis. *Mol. Ther.* **2017**, *25*, 181–191. [CrossRef] [PubMed]

119. Melo, S.A.; Luecke, L.B.; Kahlert, C.; Fernandez, A.F.; Gammon, S.T.; Kaye, J.; LeBleu, V.S.; Mittendorf, E.A.; Weitz, J.; Rahbari, N. Glypican-1 identifies cancer exosomes and detects early pancreatic cancer. *Nature* **2015**, *523*, 177–182. [CrossRef] [PubMed]

120. Castillo, J.; Bernard, V.; San Lucas, F.; Allenson, K.; Capello, M.; Kim, D.; Gascoyne, P.; Mulu, F.; Stephens, B.; Huang, J. Surfaceome profiling enables isolation of cancer-specific exosomal cargo in liquid biopsies from pancreatic cancer patients. *Ann. Oncol.* **2017**, *29*, 223–229. [CrossRef] [PubMed]

121. Paggetti, J.; Haderk, F.; Seiffert, M.; Janji, B.; Distler, U.; Ammerlaan, W.; Kim, Y.J.; Adam, J.; Lichter, P.; Solary, E. Exosomes released by chronic lymphocytic leukemia cells induce the transition of stromal cells into cancer-associated fibroblasts. *Blood* **2015**. [CrossRef] [PubMed]

122. Song, Y.H.; Warncke, C.; Choi, S.J.; Choi, S.; Chiou, A.E.; Ling, L.; Liu, H.Y.; Daniel, S.; Antonyak, M.A.; Cerione, R.A.; et al. Breast cancer-derived extracellular vesicles stimulate myofibroblast differentiation and pro-angiogenic behavior of adipose stem cells. *Matrix Biol.* **2017**, *60–61*, 190–205. [CrossRef] [PubMed]

123. Webber, J.P.; Spary, L.K.; Sanders, A.J.; Chowdhury, R.; Jiang, W.G.; Steadman, R.; Wymant, J.; Jones, A.T.; Kynaston, H.; Mason, M.D.; et al. Differentiation of tumour-promoting stromal myofibroblasts by cancer exosomes. *Oncogene* **2015**, *34*, 290–302. [CrossRef] [PubMed]

124. Lindoso, R.S.; Collino, F.; Camussi, G. Extracellular vesicles derived from renal cancer stem cells induce a pro-tumorigenic phenotype in mesenchymal stromal cells. *Oncotarget* **2015**, *6*, 7959–7969. [CrossRef] [PubMed]

125. Baroni, S.; Romero-Cordoba, S.; Plantamura, I.; Dugo, M.; D'Ippolito, E.; Cataldo, A.; Cosentino, G.; Angeloni, V.; Rossini, A.; Daidone, M.G.; et al. Exosome-mediated delivery of miR-9 induces cancer-associated fibroblast-like properties in human breast fibroblasts. *Cell Death Dis.* **2016**, *7*, e2312. [CrossRef] [PubMed]

126. Giusti, I.; Di Francesco, M.; D'Ascenzo, S.; Palmerini, M.G.; Macchiarelli, G.; Carta, G.; Dolo, V. Ovarian cancer-derived extracellular vesicles affect normal human fibroblast behavior. *Cancer Biol. Ther.* **2018**, *19*, 722–734. [CrossRef] [PubMed]

127. Richards, K.E.; Zeleniak, A.E.; Fishel, M.L.; Wu, J.; Littlepage, L.E.; Hill, R. Cancer-associated fibroblast exosomes regulate survival and proliferation of pancreatic cancer cells. *Oncogene* **2017**, *36*, 1770–1778. [CrossRef] [PubMed]

128. Donnarumma, E.; Fiore, D.; Nappa, M.; Roscigno, G.; Adamo, A.; Iaboni, M.; Russo, V.; Affinito, A.; Puoti, I.; Quintavalle, C. Cancer-associated fibroblasts release exosomal microRNAs that dictate an aggressive phenotype in breast cancer. *Oncotarget* **2017**, *8*, 19592–19608. [CrossRef] [PubMed]

129. Zhao, H.; Yang, L.; Baddour, J.; Achreja, A.; Bernard, V.; Moss, T.; Marini, J.C.; Tudawe, T.; Seviour, E.G.; San Lucas, F.A.; et al. Tumor microenvironment derived exosomes pleiotropically modulate cancer cell metabolism. *eLife* **2016**, *5*, e10250. [CrossRef] [PubMed]

130. Au Yeung, C.L.; Co, N.N.; Tsuruga, T.; Yeung, T.L.; Kwan, S.Y.; Leung, C.S.; Li, Y.; Lu, E.S.; Kwan, K.; Wong, K.K.; et al. Exosomal transfer of stroma-derived miR21 confers paclitaxel resistance in ovarian cancer cells through targeting apaf1. *Nat. Commun* **2016**, *7*, 11150. [CrossRef] [PubMed]

131. Al-Nedawi, K.; Meehan, B.; Kerbel, R.S.; Allison, A.C.; Rak, J. Endothelial expression of autocrine VEGF upon the uptake of tumor-derived microvesicles containing oncogenic EGFR. *Proc. Natl. Acad. Sci. USA* **2009**, *106*, 3794–3799. [CrossRef] [PubMed]

132. Conigliaro, A.; Costa, V.; Lo Dico, A.; Saieva, L.; Buccheri, S.; Dieli, F.; Manno, M.; Raccosta, S.; Mancone, C.; Tripodi, M.; et al. CD90+ liver cancer cells modulate endothelial cell phenotype through the release of exosomes containing H19 LncRNA. *Mol. Cancer* **2015**, *14*, 155. [CrossRef] [PubMed]

133. Dickman, C.T.; Lawson, J.; Jabalee, J.; MacLellan, S.A.; LePard, N.E.; Bennewith, K.L.; Garnis, C. Selective extracellular vesicle exclusion of miR-142-3p by oral cancer cells promotes both internal and extracellular malignant phenotypes. *Oncotarget* **2017**, *8*, 15252–15266. [CrossRef] [PubMed]

134. Grange, C.; Tapparo, M.; Collino, F.; Vitillo, L.; Damasco, C.; Deregibus, M.C.; Tetta, C.; Bussolati, B.; Camussi, G. Microvesicles released from human renal cancer stem cells stimulate angiogenesis and formation of lung premetastatic niche. *Cancer Res.* **2011**, *71*, 5346–5356. [CrossRef] [PubMed]

135. Lawson, J.; Dickman, C.; MacLellan, S.; Towle, R.; Jabalee, J.; Lam, S.; Garnis, C. Selective secretion of microRNAs from lung cancer cells via extracellular vesicles promotes CAMK1D-mediated tube formation in endothelial cells. *Oncotarget* **2017**, *8*, 83913–83924. [CrossRef] [PubMed]

136. Schillaci, O.; Fontana, S.; Monteleone, F.; Taverna, S.; Di Bella, M.A.; Di Vizio, D.; Alessandro, R. Exosomes from metastatic cancer cells transfer amoeboid phenotype to non-metastatic cells and increase endothelial permeability: Their emerging role in tumor heterogeneity. *Sci. Rep.* **2017**, *7*, 4711. [CrossRef] [PubMed]

137. Umezu, T.; Tadokoro, H.; Azuma, K.; Yoshizawa, S.; Ohyashiki, K.; Ohyashiki, J.H. Exosomal miR-135b shed from hypoxic multiple myeloma cells enhances angiogenesis by targeting factor-inhibiting HIF-1. *Blood* **2014**, *124*, 3748–3757. [CrossRef] [PubMed]

138. Zhuang, G.; Wu, X.; Jiang, Z.; Kasman, I.; Yao, J.; Guan, Y.; Oeh, J.; Modrusan, Z.; Bais, C.; Sampath, D.; et al. Tumour-secreted miR-9 promotes endothelial cell migration and angiogenesis by activating the Jak-stat pathway. *EMBO J.* **2012**, *31*, 3513–3523. [CrossRef] [PubMed]

139. Al-Nedawi, K.; Meehan, B.; Micallef, J.; Lhotak, V.; May, L.; Guha, A.; Rak, J. Intercellular transfer of the oncogenic receptor EGFRVIII by microvesicles derived from tumour cells. *Nat. Cell Biol.* **2008**, *10*, 619–624. [CrossRef] [PubMed]

140. King, H.W.; Michael, M.Z.; Gleadle, J.M. Hypoxic enhancement of exosome release by breast cancer cells. *BMC Cancer* **2012**, *12*, 421. [CrossRef] [PubMed]

141. Hsu, Y.L.; Hung, J.Y.; Chang, W.A.; Lin, Y.S.; Pan, Y.C.; Tsai, P.H.; Wu, C.Y.; Kuo, P.L. Hypoxic lung cancer-secreted exosomal miR-23a increased angiogenesis and vascular permeability by targeting prolyl hydroxylase and tight junction protein zo-1. *Oncogene* **2017**, *36*, 4929–4942. [CrossRef] [PubMed]

142. Burnley-Hall, N.; Willis, G.; Davis, J.; Rees, D.A.; James, P.E. Nitrite-derived nitric oxide reduces hypoxia-inducible factor 1alpha-mediated extracellular vesicle production by endothelial cells. *Nitric Oxide* **2017**, *63*, 1–12. [CrossRef] [PubMed]

143. Zhou, W.; Fong, M.Y.; Min, Y.; Somlo, G.; Liu, L.; Palomares, M.R.; Yu, Y.; Chow, A.; O'Connor, S.T.; Chin, A.R.; et al. Cancer-secreted mir-105 destroys vascular endothelial barriers to promote metastasis. *Cancer Cell* **2014**, *25*, 501–515. [CrossRef] [PubMed]

144. Zitvogel, L.; Regnault, A.; Lozier, A.; Wolfers, J.; Flament, C.; Tenza, D.; Ricciardi-Castagnoli, P.; Raposo, G.; Amigorena, S. Eradication of established murine tumors using a novel cell-free vaccine: Dendritic cell-derived exosomes. *Nat. Med.* **1998**, *4*, 594–600. [CrossRef] [PubMed]

145. Zhou, M.; Chen, J.; Zhou, L.; Chen, W.; Ding, G.; Cao, L. Pancreatic cancer derived exosomes regulate the expression of TLR4 in dendritic cells via miR-203. *Cell. Immunol.* **2014**, *292*, 65–69. [CrossRef] [PubMed]

146. Wolfers, J.; Lozier, A.; Raposo, G.; Regnault, A.; Thery, C.; Masurier, C.; Flament, C.; Pouzieux, S.; Faure, F.; Tursz, T.; et al. Tumor-derived exosomes are a source of shared tumor rejection antigens for CTL cross-priming. *Nat. Med.* **2001**, *7*, 297–303. [CrossRef] [PubMed]

147. Ding, G.; Zhou, L.; Qian, Y.; Fu, M.; Chen, J.; Chen, J.; Xiang, J.; Wu, Z.; Jiang, G.; Cao, L. Pancreatic cancer-derived exosomes transfer miRNAs to dendritic cells and inhibit rfxap expression via miR-212-3p. *Oncotarget* **2015**, *6*, 29877–29888. [CrossRef] [PubMed]

148. Marton, A.; Vizler, C.; Kusz, E.; Temesfoi, V.; Szathmary, Z.; Nagy, K.; Szegletes, Z.; Varo, G.; Siklos, L.; Katona, R.L.; et al. Melanoma cell-derived exosomes alter macrophage and dendritic cell functions in vitro. *Immunol. Lett.* **2012**, *148*, 34–38. [CrossRef] [PubMed]

149. Lundholm, M.; Schroder, M.; Nagaeva, O.; Baranov, V.; Widmark, A.; Mincheva-Nilsson, L.; Wikstrom, P. Prostate tumor-derived exosomes down-regulate NKG2D expression on natural killer cells and CD8+ T cells: Mechanism of immune evasion. *PLoS ONE* **2014**, *9*, e108925. [CrossRef] [PubMed]

150. Szajnik, M.; Czystowska, M.; Szczepanski, M.J.; Mandapathil, M.; Whiteside, T.L. Tumor-derived microvesicles induce, expand and up-regulate biological activities of human regulatory T cells (TREG). *PLoS ONE* **2010**, *5*, e11469. [CrossRef] [PubMed]

151. Ning, Y.; Shen, K.; Wu, Q.; Sun, X.; Bai, Y.; Xie, Y.; Pan, J.; Qi, C. Tumor exosomes block dendritic cells maturation to decrease the T cell immune response. *Immunol. Lett.* **2018**, *199*, 36–43. [CrossRef] [PubMed]

152. Ye, S.B.; Li, Z.L.; Luo, D.H.; Huang, B.J.; Chen, Y.S.; Zhang, X.S.; Cui, J.; Zeng, Y.X.; Li, J. Tumor-derived exosomes promote tumor progression and T-cell dysfunction through the regulation of enriched exosomal microRNAs in human nasopharyngeal carcinoma. *Oncotarget* **2014**, *5*, 5439–5452. [CrossRef] [PubMed]

153. Yang, Y.; Li, C.W.; Chan, L.C.; Wei, Y.; Hsu, J.M.; Xia, W.; Cha, J.H.; Hou, J.; Hsu, J.L.; Sun, L.; et al. Exosomal PD-L1 harbors active defense function to suppress T cell killing of breast cancer cells and promote tumor growth. *Cell Res.* **2018**. [CrossRef] [PubMed]

154. Wu, L.; Zhang, X.; Zhang, B.; Shi, H.; Yuan, X.; Sun, Y.; Pan, Z.; Qian, H.; Xu, W. Exosomes derived from

gastric cancer cells activate NF-kappaB pathway in macrophages to promote cancer progression. *Tumour Biol.* **2016**, *37*, 12169–12180. [CrossRef] [PubMed]

155. Chow, A.; Zhou, W.; Liu, L.; Fong, M.Y.; Champer, J.; Van Haute, D.; Chin, A.R.; Ren, X.; Gugiu, B.G.; Meng, Z.; et al. Macrophage immunomodulation by breast cancer-derived exosomes requires toll-like receptor 2-mediated activation of NF-kappaB. *Sci. Rep.* **2014**, *4*, 5750. [CrossRef] [PubMed]

156. Fabbri, M.; Paone, A.; Calore, F.; Galli, R.; Gaudio, E.; Santhanam, R.; Lovat, F.; Fadda, P.; Mao, C.; Nuovo, G.J.; et al. MicroRNAs bind to toll-like receptors to induce prometastatic inflammatory response. *Proc. Natl. Acad. Sci. USA* **2012**, *109*, E2110–E2116. [CrossRef] [PubMed]

157. Minciacchi, V.R.; Spinelli, C.; Reis-Sobreiro, M.; Cavallini, L.; You, S.; Zandian, M.; Li, X.; Mishra, R.; Chiarugi, P.; Adam, R.M.; et al. Myc mediates large oncosome-induced fibroblast reprogramming in prostate cancer. *Cancer Res.* **2017**, *77*, 2306–2317. [CrossRef] [PubMed]

158. Minciacchi, V.R.; You, S.; Spinelli, C.; Morley, S.; Zandian, M.; Aspuria, P.-J.; Cavallini, L.; Ciardiello, C.; Sobreiro, M.R.; Morello, M. Large oncosomes contain distinct protein cargo and represent a separate functional class of tumor-derived extracellular vesicles. *Oncotarget* **2015**, *6*, 11327–11341. [CrossRef] [PubMed]

159. Nawaz, M.; Camussi, G.; Valadi, H.; Nazarenko, I.; Ekstrom, K.; Wang, X.; Principe, S.; Shah, N.; Ashraf, N.M.; Fatima, F.; et al. The emerging role of extracellular vesicles as biomarkers for urogenital cancers. *Nat. Rev. Urol.* **2014**, *11*, 688–701. [CrossRef] [PubMed]

160. Nawaz, M.; Fatima, F.; Nazarenko, I.; Ekström, K.; Murtaza, I.; Anees, M.; Sultan, A.; Neder, L.; Camussi, G.; Valadi, H. Extracellular vesicles in ovarian cancer: Applications to tumor biology, immunotherapy and biomarker discovery. *Expert Rev. Proteom.* **2016**, *13*, 395–409. [CrossRef] [PubMed]

161. Escudier, B.; Dorval, T.; Chaput, N.; Andre, F.; Caby, M.P.; Novault, S.; Flament, C.; Leboulaire, C.; Borg, C.; Amigorena, S.; et al. Vaccination of metastatic melanoma patients with autologous dendritic cell (DC) derived-exosomes: Results of the first phase I clinical trial. *J. Transl. Med.* **2005**, *3*, 10. [CrossRef] [PubMed]

162. Morse, M.A.; Garst, J.; Osada, T.; Khan, S.; Hobeika, A.; Clay, T.M.; Valente, N.; Shreeniwas, R.; Sutton, M.A.; Delcayre, A.; et al. A phase I study of dexosome immunotherapy in patients with advanced non-small cell lung cancer. *J. Transl. Med.* **2005**, *3*, 9. [CrossRef] [PubMed]

163. Dai, S.; Wei, D.; Wu, Z.; Zhou, X.; Wei, X.; Huang, H.; Li, G. Phase I clinical trial of autologous ascites-derived exosomes combined with GM-CSF for colorectal cancer. *Mol. Ther.* **2008**, *16*, 782–790. [CrossRef] [PubMed]

164. Besse, B.; Charrier, M.; Lapierre, V.; Dansin, E.; Lantz, O.; Planchard, D.; Le Chevalier, T.; Livartoski, A.; Barlesi, F.; Laplanche, A.; et al. Dendritic cell-derived exosomes as maintenance immunotherapy after first line chemotherapy in NSCLC. *Oncoimmunology* **2016**, *5*, e1071008. [CrossRef] [PubMed]

165. Tkach, M.; Kowal, J.; Zucchetti, A.E.; Enserink, L.; Jouve, M.; Lankar, D.; Saitakis, M.; Martin-Jaular, L.; Thery, C. Qualitative differences in t-cell activation by dendritic cell-derived extracellular vesicle subtypes. *EMBO J.* **2017**, *36*, 3012–3028. [CrossRef] [PubMed]

166. Vader, P.; Mol, E.A.; Pasterkamp, G.; Schiffelers, R.M. Extracellular vesicles for drug delivery. *Adv. Drug Deliv. Rev.* **2016**, *106*, 148–156. [CrossRef] [PubMed]

167. Ohno, S.; Takanashi, M.; Sudo, K.; Ueda, S.; Ishikawa, A.; Matsuyama, N.; Fujita, K.; Mizutani, T.; Ohgi, T.; Ochiya, T.; et al. Systemically injected exosomes targeted to EGFR deliver antitumor microRNA to breast cancer cells. *Mol. Ther.* **2013**, *21*, 185–191. [CrossRef] [PubMed]

168. Alvarez-Erviti, L.; Seow, Y.; Yin, H.; Betts, C.; Lakhal, S.; Wood, M.J. Delivery of siRNA to the mouse brain by systemic injection of targeted exosomes. *Nat. Biotechnol.* **2011**, *29*, 341–345. [CrossRef] [PubMed]

169. Greco, K.A.; Franzen, C.A.; Foreman, K.E.; Flanigan, R.C.; Kuo, P.C.; Gupta, G.N. PLK-1 silencing in bladder cancer by siRNA delivered with exosomes. *Urology* **2016**, *91*, e241–e247. [CrossRef] [PubMed]

170. Wahlgren, J.; De, L.K.T.; Brisslert, M.; Vaziri Sani, F.; Telemo, E.; Sunnerhagen, P.; Valadi, H. Plasma exosomes can deliver exogenous short interfering RNA to monocytes and lymphocytes. *Nucleic Acids Res.* **2012**, *40*, e130. [CrossRef] [PubMed]

171. Mizrak, A.; Bolukbasi, M.F.; Ozdener, G.B.; Brenner, G.J.; Madlener, S.; Erkan, E.P.; Strobel, T.; Breakefield, X.O.; Saydam, O. Genetically engineered microvesicles carrying suicide mRNA/protein inhibit schwannoma tumor growth. *Mol. Ther.* **2013**, *21*, 101–108. [CrossRef] [PubMed]

172. Tian, Y.; Li, S.; Song, J.; Ji, T.; Zhu, M.; Anderson, G.J.; Wei, J.; Nie, G. A doxorubicin delivery platform using engineered natural membrane vesicle exosomes for targeted tumor therapy. *Biomaterials* **2014**, *35*, 2383–2390. [CrossRef] [PubMed]

173. Marleau, A.M.; Chen, C.S.; Joyce, J.A.; Tullis, R.H. Exosome removal as a therapeutic adjuvant in cancer. *J. Transl. Med.* **2012**, *10*, 134. [CrossRef] [PubMed]

174. Bobrie, A.; Krumeich, S.; Reyal, F.; Recchi, C.; Moita, L.F.; Seabra, M.C.; Ostrowski, M.; Thery, C. Rab27a supports exosome-dependent and -independent mechanisms that modify the tumor microenvironment and can promote tumor progression. *Cancer Res.* **2012**, *72*, 4920–4930. [CrossRef] [PubMed]

7

Predicting MicroRNA Mediated Gene Regulation between Human and Viruses

Xin Shu †, Xinyuan Zang †, Xiaoshuang Liu, Jie Yang * and Jin Wang *

The State Key Laboratory of Pharmaceutical Biotechnology and Jiangsu Engineering Research Center for MicroRNA Biology and Biotechnology, NJU Advanced Institute for Life Sciences (NAILS), School of Life Science, Nanjing University, Nanjing 210023, China; cpu_shuxin@126.com (X.S.); hsinring@foxmail.com (X.Z.); xsliunju@foxmail.com (X.L.)

* Correspondence: yangjie@nju.edu.cn (J.Y.); jwang@nju.edu.cn (J.W.)

† These authors contributed equally to this work as first authors.

Abstract: MicroRNAs (miRNAs) mediate various biological processes by actively fine-tuning gene expression at the post-transcriptional level. With the identification of numerous human and viral miRNAs, growing evidence has indicated a common role of miRNAs in mediating the interactions between humans and viruses. However, there is only limited information about Cross-Kingdom miRNA target sites from studies. To facilitate an extensive investigation on the interplay among the gene regulatory networks of humans and viruses, we designed a prediction pipeline, mirTarP, that is suitable for miRNA target screening on the genome scale. By applying mirTarP, we constructed the database mirTar, which is a comprehensive miRNA target repository of bidirectional interspecies regulation between viruses and humans. To provide convenient downloading for users from both the molecular biology field and medical field, mirTar classifies viruses according to "ICTV viral category" and the "medical microbiology classification" on the web page. The mirTar database and mirTarP tool are freely available online.

Keywords: miRNA; virus; host; Cross-Kingdom; target prediction

1. Introduction

MicroRNAs (miRNAs) are a class of small (~24 nt), non-coding RNA molecules that play a critical role in fundamental cellular processes and many types of diseases. They negatively regulate gene expression by binding to the 3′-untranslated regions (3′UTR) of the target mRNAs in cells [1]. Recent studies have found that they are involved in viral infections and play a key role in the host–virus interaction network. Host miRNAs modulate the expression of viral genes by targeting on virus transcripts, while viruses encode miRNAs that protect them from the host's antiviral response by acting on cellular mRNAs [2–5]. Skalsky et al. [5] reported a comprehensive survey of viral and cellular miRNA targetome in Epstein-Barr virus (EBV)-infected lymphoblastoid cell lines using photoactivatable ribonucleoside-enhanced crosslinking and immunoprecipitation (PAR-CLIP) and deep sequencing technique combined with bioinformatics. In this survey, over 500 target sites of EBV miRNAs on cellular transcripts were detected in addition to the cellular miRNA targets on virus. This result may imply that viral miRNAs have a similar mode of multiple targeting as cellular miRNAs. Although the detection of miRNA targets by high throughput techniques remains a big challenge, there has been growing interest in the role of miRNAs in host–virus interactions.

The virus miRNAs can target both the host genes and viral genes in order to contribute to the creation of a propagating environment in the host cell [2]. EBV-encoded miRNA miR-BHRF1-2-5p blocks Interleukin-1 (IL-1) signaling by directly targeting the IL-1 Receptor 1 (IL1R1) [6]. Hancock et al. [7] found that human Cytomegalovirus (HCMV) also uses its own miRNAs, miR-US5-1 and miR-UL112-3p, which bind to IkB kinase (IKK) complex components IKKα and IKKβ, in order to avoid the immune response of the host. Some viral miRNAs show sequence similarity with host miRNAs and thus, may take part in the conserved cellular gene regulation network [8]. In Kaposi's sarcoma-associated herpesvirus (KSHV)-infected human cell line, Manzano et al. [9] identified that KSHV miR-K3+1 and miR-K3 share perfect and offset 5′ homology with cellular miR-23, respectively. KSHV miR-k12-11 is an ortholog of miR-155, which can inhibit the 3′UTR region of BACH-1 [10].

Host miRNAs were found to target the viral RNA transcripts to inhibit viral pathogenesis, which essentially involves being a defense against viral infections [3–5]. It was reported that human miRNA effectively restricts the accumulation of the retrovirus primate foamy virus type 1 (PFV-1) in human cells, which involves hsa-miR-32 inhibiting the proliferation of PFV-1 by targeting PFV-1 F11 sequence. However, PFV-1 also encodes a protein named Tas, which suppresses miRNA-directed functions in mammalian cells and displays Cross-Kingdom anti-silencing activities [4]. This new report focused on an EBV-encoded protein EBNA2, which subverts immune surveillance by downregulating miR-34a that targets an important immune checkpoint PD-L1 in lymphoma B cells [11]. Human liver-specific miRNA hsa-miR-122 can induce hepatitis C virus (HCV) replication by targeting the 5′-non-coding region (NCR) of the viral genome [3]. The human miRNAs let-7b and mir-199a target the 5′UTR of HCV to decrease viral replication [12,13]. Pedersen et al. [14] also found that the overexpression of miR-196 and miR-448 significantly reduced the replication of HCV as they target the NS5A coding region and core of the HCV genome, respectively.

These findings indicate a common role of miRNAs in mediating the diversified interactions between humans and viruses. A total of 2588 mature human miRNAs and 181 mature miRNAs of human-related viruses have been recruited in mirBase so far (release 21) [15]. To facilitate the extensive investigation on the interplay among the gene regulatory network of humans and viruses, computational tools and comprehensive miRNA target repositories pertaining to human–virus interactions is necessary. These resources could provide the researchers with an efficient approach and potential miRNA targets to facilitate the investigation of miRNA function and regulation mechanisms. In particular, in the era of omics when it is possible to obtain a complete set of molecular data of gene expression, prediction tools and database are essential for genome-scale or microbiome-scale data analysis and help to decipher the panorama of the gene regulation network of human–virus interplay. This will ultimately facilitate the discovery of new drug targets for viruses, including HIV [16], HCMV [7] and HCV [3].

MiRNAs suppress interspecies gene expressions by targeting the 3′-UTRs of mRNAs during the infection or antiviral processes. Although many algorithms [17–21] are available for miRNA target prediction, only a few of them can be directly used to predict the interspecies regulation between viruses and hosts [20]. Most of the tools were designed for intra-species application by predicting the miRNA targets on their own genome, such as TargetScan, PicTar, miRanda and DIANA-microT [18,22–24]. In this situation, the databases of Cross-Kingdom miRNA target sites were produced by using the multiple intra-species target prediction tools mentioned above, which may possibly create concerns regarding the methodology and thus, the accuracy.

ViTa [25] provides predicted targets of host miRNAs from humans, mice, rats and chickens (mirBase release 8.2), which are located on 2108 virus species from 23 families. VHoT [26] houses predictions of 271 viral miRNAs on six hosts, which are namely humans, mice, rats, rhesus monkeys and cows. VmiReg [27] contains predicted targets of 169 viral miRNAs (from 10 types of viruses) on humans. VIRmiRNA [28] provides experimentally validated viral miRNAs and their targets on human and other species. All of these databases provide information of interspecies miRNA targeting in one direction only, which include either the target of viral miRNAs on host genes or the target of host miRNAs on viruses. To investigate the complex and dynamic interactions between the gene regulatory network of humans and infectious viruses that are mediated by miRNAs, the database mirTar was constructed that provides a comprehensive miRNA target repository pertaining to 2588 human miRNAs (mirBase release 21) that target 386 genomes of human-related viruses as well as

181 viral miRNAs that target the human genome. The new computational pipeline that was specially designed for human–virus interspecies miRNA target prediction was presented.

2. The prediction Tool and Results

2.1. Data Collection

A total of 2588 mature human miRNAs and 181 mature viral miRNAs were downloaded from mirBase (Release 21). Human genome and virus genomes as well as their classification information and taxonomy annotation were obtained and organized from NCBI [29,30]. Meanwhile, the annotation of gene name and protein name pertaining to the mRNA transcripts were acquired from Ensemble [31]. A total of 386 human-related viral species were collected that are belong to 34 families and fall under the following 7 genome types: (1) Deltavirus; (2) dsDNA viruses, no RNA stage; (3) dsRNA viruses; (4) Retro-transcribing viruses; (5) ssDNA viruses; (6) ssRNA viruses; and (7) unclassified viruses [32]. These viruses are all human infections. Some of them (79/386) are common and medically important viral species as categorized in medical microbiology, which mostly cause diseases of the respiratory tract, gastrointestinal tract and liver.

2.2. The Prediction Tool

Mainstream miRNA target prediction tools were limited to intra-species applications as they were only capable of predicting the miRNA targets on their own genome. Thus, the databases of interspecies miRNA targets were produced by using a combination of these methods as an approach to improve the reliability of the prediction results. For example, ViTa applied miRanda and TargetScan to identify the host miRNA target sites in virus genomes [18,23]. VHot combined five miRNA target prediction tools, which were namely TargetScan, miRanda, RNAhybrid, DIANA-microT and PITA, to form its prediction engine [18,23,24,33,34]. VmiReg predicted targets of viral miRNAs by four established prediction programs, which were namely miRanda, TargetScan, RNAhybrid and PITA [18,23,33,34]. This approach may create problems in inter-species target prediction as the sequence specificity of intra-species miRNA-target interaction are included. In addition, most of these calculations are quite time-consuming and require huge processing resources for a genome-scale prediction. To find miRNA targets across different kingdoms, we designed a prediction pipeline, mirTarP, that directly seeks the potential miRNA target. This can produce results quickly and thus, is very suitable for miRNA target screening using large-scale calculations.

MirTarP was designed by integrating two classical algorithms of sequence analysis, which were Blast [35] and RNAhybrid [34]. They work as the cores of two modules included in mirTarP, which are quick match and duplex assessment. Blast uses heuristics to accelerate searches for similar segments of a sequence. A window of consecutive perfect match can be set when running the algorithm. To improve the calculation efficiency, mirTarP introduced the sequence similarity tool Blast to produce preliminary matches between the miRNA and its target mRNA sequences. The results from the quick match module were subsequently delivered to duplex assessment module, which uses the RNAhybrid program for the calculation of minimum free energy (mfe) of miRNA–mRNA hybridization duplexes based on the principles of thermodynamics. The mfe value stands for the stability of miRNA binding. To assess the influence of local secondary structures on the target accessibility, RNAfold [36] was used to calculate the minimum folding energy around the target sites. The results were listed as the supplementary data of predicted targets. The default parameters set in mirTarP include the 7-consecutive base matches as the seed of targeting and the cutoff of mfe of -25 kcal/mol for local dimer formation. The flow chart of mirTarP is illustrated in Figure 1. The advantage of mirTarP over the current prediction tools is that it operates independent of conservation and thus, can be used to find miRNA targets on virus genomes or obtain other interspecies miRNA target predictions. This tool runs quickly and is easy to use with only 2 parameters to be set. Therefore, it will be helpful to wet-lab researchers dealing with new viruses. A comparison of mirTarP to TargetScan and PITA on a dataset of 221 experimentally validated miRNA-target pairs is included in the website along with the tool mirTarP.

The prediction tool mirTarP is free for downloading in the web page.

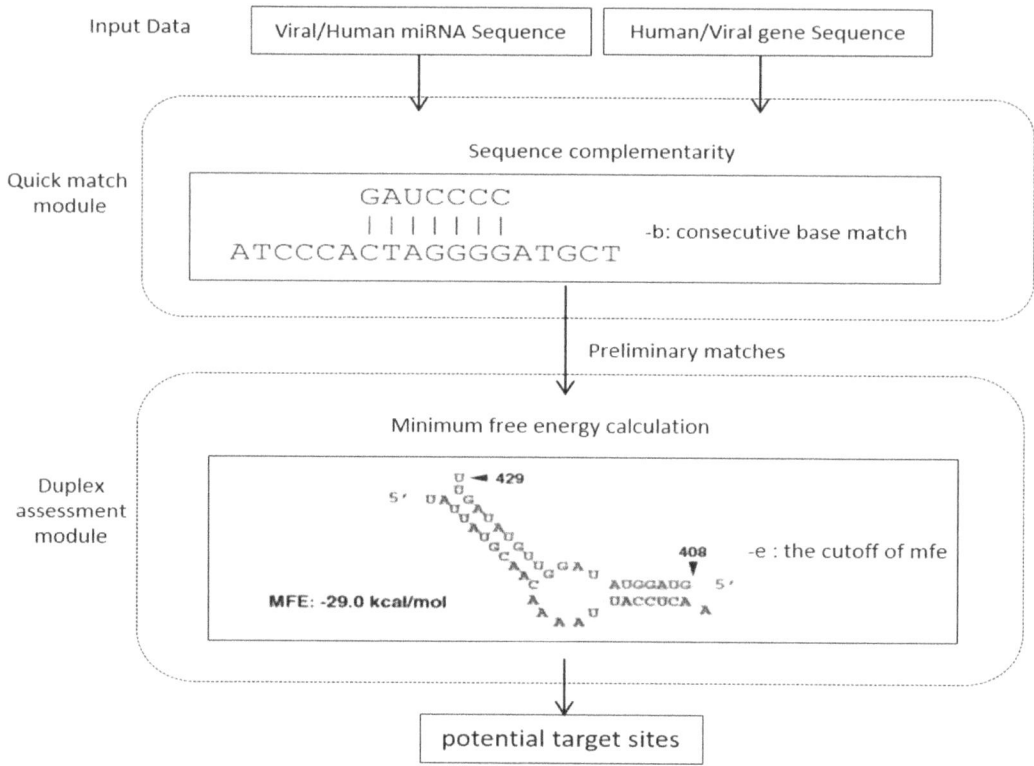

Figure 1. Flow chart of mirTarP for human–virus interspecies miRNA target prediction. The parameter '-b' represents the number of consecutive base matches between miRNA and target sequence, while '-e' represents the cutoff of minimum free energy of miRNA-target binding.

2.3. Prediction of miRNA Targets

By applying mirTarP, 2557 human mature miRNAs were found to have targets in 3133 viral genes, which corresponds to 3376 viral proteins. A total of 181 miRNA records from 13 viral species of 3 families were used for the prediction of targets on human genome. The calculation results showed that these viral miRNAs had potential target sites in 16,439 human genes.

A total of 2,680,194 entries about the miRNAs target sites within human and viral genomes were produced.

3. MirTar Database

3.1. Web Interface Development

MirTar is designed to adapt a wide variety of screen formats and devices (PCs, tablets, smartphones, etc.). All data were organized by MySQL and the website is implemented in PHP, JavaScript and HTML.

3.2. Data Download

The web page provides two ways of data downloads, i.e., customized download and the complete download. The customized download is associated with the items or viruses selected by the user. To provide easy downloading for users from both the molecular biology field and medical field, mirTar database classified the viruses in the following two ways: (1) according to the definition by medical microbiology; and (2) according to ICTV virus category [32]. Currently, the International Committee on the Taxonomy of Viruses (ICTV) provides the most comprehensive, fully annotated compendium of information on virus taxa and taxonomy. Thus, the web page provides a search function for convenient categories when retrieving an input virus. In addition, a python script of the prediction tool mirTarP is available on the web page to facilitate a quick screening of miRNA targets on new viruses. The mirTar database and mirTarP tool are freely available at http://mcube.nju.edu.cn/jwang/mirTar/ docs/mirTar/ or http://118.89.139.70/mirTar/docs/mirTar/. The interface of mirTar is shown in Figure 2.

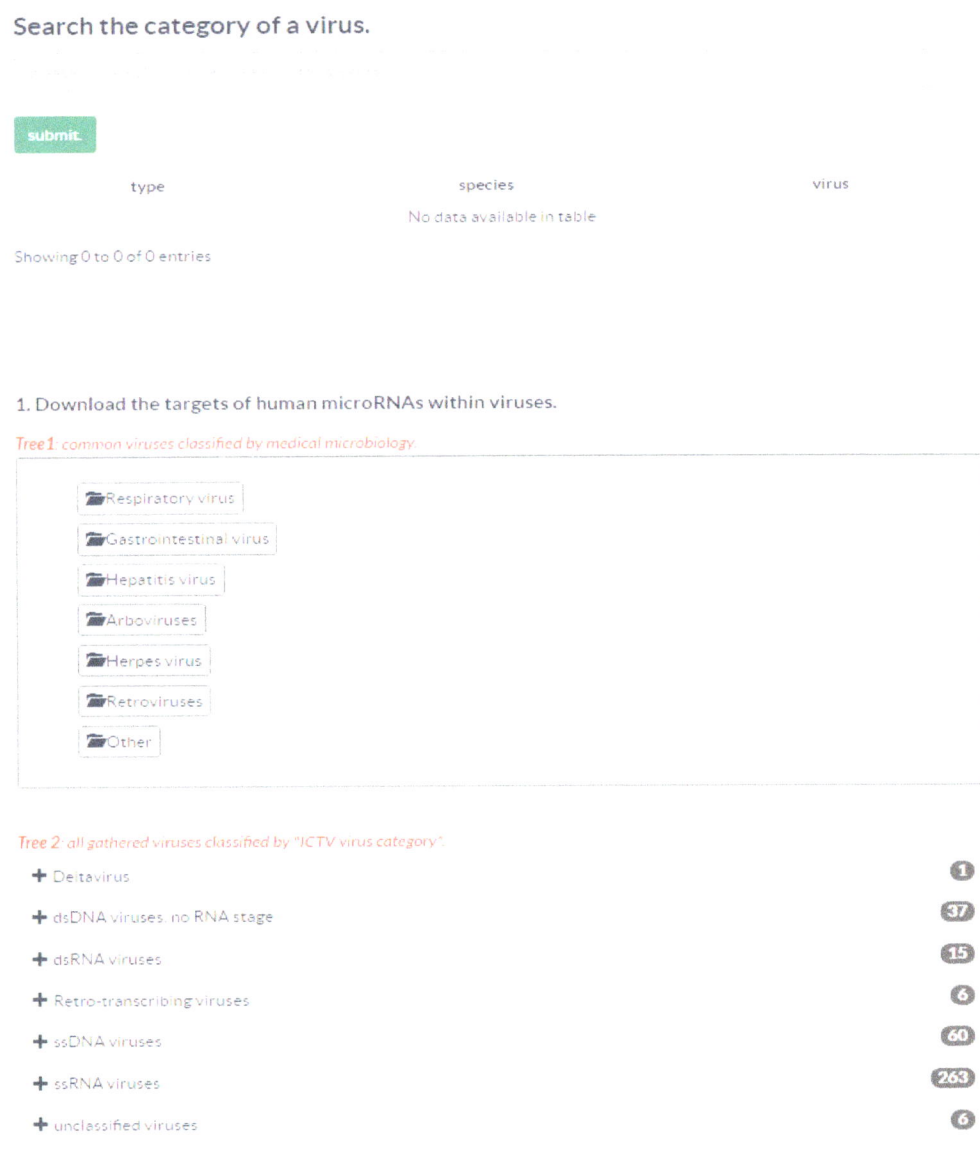

Figure 2. The interface of mirTar database.

4. Conclusions

In this paper, we provide a comprehensive miRNA target database that includes the bidirectional interspecies actions between human and the infectious viruses along with a fast miRNA target prediction program to facilitate a quick screening of miRNA targets on new viruses. The database mirTar contains 2,200,076 candidate target sites on 386 viral genomes for 2577 human mature miRNAs and 480,118 targets of 181 viral mature miRNAs on human genome. The web page of the database was designed for convenient data querying and downloading by classifying the virus species by the two categories of molecular biology and medicine. The database will benefit investigations on the crosstalk between the host and virus gene regulations and the new role of miRNAs in infections and diseases caused by latent viruses, including many cancers.

Author Contributions: Formal analysis, X.Z.; Funding acquisition, J.W.; Methodology, X.S., X.Z. and X.L.; Resources, J.Y., J.W.; Visualization, X.S.; Writing—original draft, X.S.; Writing—review & editing, X.S., X.Z. and J.W.

Acknowledgments: The authors would acknowledge the Center of High Performance Computation of Nanjing University for the support of computational resources.

References

1. Bartel, D.P. Micrornas: Genomics, biogenesis, mechanism, and function. *Cell* **2004**, *116*, 281–297. [CrossRef]
2. Bryan, C.R.; Cullen; Bryan; Vogt, P.K. *Intrinsic Immunity*; Springer: Berlin, Germany, 2013.
3. Jopling, C.L.; Yi, M.; Lancaster, A.M.; Lemon, S.M.; Sarnow, P. Modulation of hepatitis C virus RNA abundance by a liver-specific microRNA. *Science* **2005**, *309*, 1577–1581. [CrossRef] [PubMed]
4. Lecellier, C.-H.; Dunoyer, P.; Arar, K.; Lehmann-Che, J.; Eyquem, S.; Himber, C.; Saïb, A.; Voinnet, O. A cellular microRNA mediates antiviral defense in human cells. *Science* **2005**, *308*, 557–560. [CrossRef] [PubMed]
5. Skalsky, R.L.; Corcoran, D.L.; Gottwein, E.; Frank, C.L.; Kang, D.; Hafner, M.; Nusbaum, J.D.; Feederle, R.; Delecluse, H.-J.; Luftig, M.A. The viral and cellular microRNA targetome in lymphoblastoid cell lines. *PLoS Pathog.* **2012**, *8*, e1002484. [CrossRef] [PubMed]
6. Skinner, C.M.; Ivanov, N.S.; Barr, S.A.; Chen, Y.; Skalsky, R.L. An Epstein-Barr virus microRNA blocks interleukin-1 (IL-1) signaling by targeting IL-1 receptor 1. *J. Virol.* **2017**, *91*. [CrossRef] [PubMed]
7. Hancock, M.H.; Hook, L.M.; Mitchell, J.; Nelson, J.A. Human cytomegalovirus microRNAs miR-US5-1 and miR-UL112-3p block proinflammatory cytokine production in response to NF-κB-activating factors through direct downregulation of IKKα and IKKβ. *mBio* **2017**, *8*, e00109-17. [CrossRef] [PubMed]
8. Ghosh, Z.; Mallick, B.; Chakrabarti, J. Cellular versus viral microRNAs in host-virus interaction. *Nucleic Acids Res.* **2009**, *37*, 1035–1048. [CrossRef] [PubMed]
9. Manzano, M.; Shamulailatpam, P.; Raja, A.N.; Gottwein, E. Kaposi's sarcoma-associated herpesvirus encodes a mimic of cellular miR-23. *J. Virol.* **2013**, *87*, 11821–11830. [CrossRef] [PubMed]
10. Skalsky, R.L.; Samols, M.A.; Plaisance, K.B.; Boss, I.W.; Riva, A.; Lopez, M.C.; Baker, H.V.; Renne, R. Kaposi's sarcoma-associated herpesvirus encodes an ortholog of miR-155. *J. Virol.* **2007**, *81*, 12836–12845. [CrossRef] [PubMed]
11. Anastasiadou, E.; Stroopinsky, D.; Alimperti, S.; Jiao, A.L.; Pyzer, A.R.; Cippitelli, C.; Pepe, G.; Severa, M.; Rosenblatt, J.; Etna, M.P.; et al. Epstein-Barr virus-encoded EBNA2 alters immune checkpoint PD-l1 expression by downregulating miR-34a in B-cell lymphomas. *Leukemia* **2018**. [CrossRef] [PubMed]
12. Cheng, J.C.; Yeh, Y.J.; Tseng, C.P.; Hsu, S.D.; Chang, Y.L.; Sakamoto, N.; Huang, H.D. Let-7b is a novel regulator of hepatitis C virus replication. *Cell. Mol. Life Sci. CMLS* **2012**, *69*, 2621–2633. [CrossRef] [PubMed]
13. Murakami, Y.; Aly, H.H.; Tajima, A.; Inoue, I.; Shimotohno, K. Regulation of the hepatitis C virus genome replication by miR-199a. *J. Hepatol.* **2009**, *50*, 453–460. [CrossRef] [PubMed]
14. Pedersen, I.M.; Cheng, G.; Wieland, S.; Volinia, S.; Croce, C.M.; Chisari, F.V.; David, M. Interferon modulation of cellular microRNAs as an antiviral mechanism. *Nature* **2007**, *449*, 919–922. [CrossRef] [PubMed]
15. Kozomara, A.; Griffiths-Jones, S. miRBase: Annotating high confidence microRNAs using deep sequencing data. *Nucleic Acids Res.* **2013**, *42*, D68–D73. [CrossRef] [PubMed]
16. Balasubramaniam, M.; Pandhare, J.; Dash, C. Are microRNAs important players in HIV-1 infection? An update. *Viruses* **2018**, *10*, 110. [CrossRef] [PubMed]
17. Grün, D.; Wang, Y.-L.; Langenberger, D.; Gunsalus, K.C.; Rajewsky, N. microRNA target predictions across

seven *Drosophila* species and comparison to mammalian targets. *PLoS Comput. Biol.* **2005**, *1*, e13. [CrossRef] [PubMed]

18. Agarwal, V.; Bell, G.W.; Nam, J.-W.; Bartel, D.P. Predicting effective microRNA target sites in mammalian mRNAs. *eLife* **2015**, *4*, e05005. [CrossRef] [PubMed]

19. M. Witkos, T.; Koscianska, E.; J. Krzyzosiak, W. Practical aspects of microRNA target prediction. *Curr. Mol. Med.* **2011**, *11*, 93–109. [CrossRef]

20. Laganà, A.; Forte, S.; Russo, F.; Giugno, R.; Pulvirenti, A.; Ferro, A. Prediction of human targets for viral-encoded microRNAs by thermodynamics and empirical constraints. *J. RNAi Gene Silenc.* **2010**, *6*, 379.

21. Cheng, S.; Guo, M.; Wang, C.; Liu, X.; Liu, Y.; Wu, X. MiRTDL: A deep learning approach for miRNA target prediction. *IEEE/ACM Trans. Comput. Biol. Bioinform.* **2016**, *13*, 1161–1169. [CrossRef] [PubMed]

22. Krek, A.; Grün, D.; Poy, M.N.; Wolf, R.; Rosenberg, L.; Epstein, E.J.; MacMenamin, P.; Da Piedade, I.; Gunsalus, K.C.; Stoffel, M. Combinatorial microRNA target predictions. *Nat. Genet.* **2005**, *37*, 495. [CrossRef] [PubMed]

23. John, B.; Enright, A.J.; Aravin, A.; Tuschl, T.; Sander, C.; Marks, D.S. Human microRNA targets. *PLoS Biol.* **2004**, *2*, e363. [CrossRef] [PubMed]

24. Paraskevopoulou, M.D.; Georgakilas, G.; Kostoulas, N.; Vlachos, I.S.; Vergoulis, T.; Reczko, M.; Filippidis, C.; Dalamagas, T.; Hatzigeorgiou, A.G. Diana-microT web server v5.0: Service integration into miRNA functional analysis workflows. *Nucleic Acids Res.* **2013**, *41*, W169–W173. [CrossRef] [PubMed]

25. Hsu, P.W.-C.; Lin, L.-Z.; Hsu, S.-D.; Hsu, J.B.-K.; Huang, H.-D. Vita: Prediction of host microRNAs targets on viruses. *Nucleic Acids Res.* **2006**, *35*, D381–D385. [CrossRef] [PubMed]

26. Kim, H.; Park, S.; Min, H.; Yoon, S. vHoT: A database for predicting interspecies interactions between viral microRNA and host genomes. *Arch. Virol.* **2012**, *157*, 497–501. [CrossRef] [PubMed]

27. Shao, T.; Zhao, Z.; Wu, A.; Bai, J.; Li, Y.; Chen, H.; Jiang, C.; Wang, Y.; Li, S.; Wang, L. Functional dissection of virus–human crosstalk mediated by miRNAs based on the VmiReg database. *Mol. BioSyst.* **2015**, *11*, 1319–1328. [CrossRef] [PubMed]

28. Qureshi, A.; Thakur, N.; Monga, I.; Thakur, A.; Kumar, M. VIRmiRNA: A comprehensive resource for experimentally validated viral miRNAs and their targets. *Database* **2014**, *2014*. [CrossRef] [PubMed]

29. Brister, J.R.; Ako-Adjei, D.; Bao, Y.; Blinkova, O. NCBI viral genomes resource. *Nucleic Acids Res.* **2014**, *43*, D571–D577. [CrossRef] [PubMed]

30. Federhen, S. The NCBI taxonomy database. *Nucleic Acids Res.* **2011**, *40*, D136–D143. [CrossRef] [PubMed]

31. Aken, B.L.; Ayling, S.; Barrell, D.; Clarke, L.; Curwen, V.; Fairley, S.; Fernandez Banet, J.; Billis, K.; García Girón, C.; Hourlier, T. The Ensembl gene annotation system. *Database* **2016**, *2016*. [CrossRef] [PubMed]

32. Lefkowitz, E.J.; Dempsey, D.M.; Hendrickson, R.C.; Orton, R.J.; Siddell, S.G.; Smith, D.B. Virus taxonomy: The database of the International Committee on Taxonomy of Viruses (ICTV). *Nucleic Acids Res.* **2017**, *46*, D708–D717. [CrossRef] [PubMed]

33. Kertesz, M.; Iovino, N.; Unnerstall, U.; Gaul, U.; Segal, E. The role of site accessibility in microRNA target recognition. *Nat. Genet.* **2007**, *39*, 1278. [CrossRef] [PubMed]

34. Krüger, J.; Rehmsmeier, M. RNAhybrid: MicroRNA target prediction easy, fast and flexible. *Nucleic Acids'Res.* **2006**, *34*, W451–W454. [CrossRef] [PubMed]

35. Altschul, S.F.; Gish, W.; Miller, W.; Myers, E.W.; Lipman, D.J. Basic local alignment search tool. *J. Mol. Boil.* **1990**, *215*, 403–410. [CrossRef]

36. Lorenz, R.; Bernhart, S.H.; Zu Siederdissen, C.H.; Tafer, H.; Flamm, C.; Stadler, P.F.; Hofacker, I.L. ViennaRNA package 2.0. *Algorithms Mol. Biol.* **2011**, *6*, 26. [CrossRef] [PubMed]

8

DRB1, DRB2 and DRB4 are Required for Appropriate Regulation of the microRNA399/ *PHOSPHATE2* Expression Module in *Arabidopsis thaliana*

Joseph L. Pegler, Jackson M. J. Oultram, Christopher P. L. Grof and Andrew L. Eamens *

Centre for Plant Science, School of Environmental and Life Sciences, Faculty of Science, University of Newcastle, Callaghan 2308, New South Wales, Australia; Joseph.Pegler@uon.edu.au (J.L.P.); Jackson.Oultram@uon.edu.au (J.M.J.O.); chris.grof@newcastle.edu.au (C.P.L.G.)
* Correspondence: andy.eamens@newcastle.edu.au;

Abstract: Adequate phosphorous (P) is essential to plant cells to ensure normal plant growth and development. Therefore, plants employ elegant mechanisms to regulate P abundance across their developmentally distinct tissues. One such mechanism is PHOSPHATE2 (PHO2)-directed ubiquitin-mediated degradation of a cohort of phosphate (PO_4) transporters. *PHO2* is itself under tight regulation by the PO_4 responsive microRNA (miRNA), miR399. The DOUBLE-STRANDED RNA BINDING (DRB) proteins, DRB1, DRB2 and DRB4, have each been assigned a specific functional role in the *Arabidopsis thaliana* (*Arabidopsis*) miRNA pathway. Here, we assessed the requirement of DRB1, DRB2 and DRB4 to regulate the miR399/*PHO2* expression module under PO_4 starvations conditions. Via the phenotypic and molecular assessment of the knockout mutant plant lines, *drb1*, *drb2* and *drb4*, we show here that; (1) DRB1 and DRB2 are required to maintain P homeostasis in *Arabidopsis* shoot and root tissues; (2) DRB1 is the primary DRB required for miR399 production; (3) DRB2 and DRB4 play secondary roles in regulating miR399 production, and; (4) miR399 appears to direct expression regulation of the *PHO2* transcript via both an mRNA cleavage and translational repression mode of RNA silencing. Together, the hierarchical contribution of DRB1, DRB2 and DRB4 demonstrated here to be required for the appropriate regulation of the miR399/*PHO2* expression module identifies the extreme importance of P homeostasis maintenance in *Arabidopsis* to ensure that numerous vital cellular processes are maintained across *Arabidopsis* tissues under a changing cellular environment.

Keywords: *Arabidopsis thaliana*; phosphorous (P); phosphate (PO_4) stress; microRNA (miRNA); miR399; *PHOSPHATE2* (*PHO2*); DOUBLE-STRANDED RNA BINDING (DRB) proteins DRB1; DRB2; DRB4; miR399-directed *PHO2* expression regulation; RT-qPCR

1. Introduction

Phosphorous (P) is one of the most limiting factors for plant growth worldwide [1–3], with large quantities of P an essential requirement for numerous processes vital to the plant cell, including energy trafficking, signaling cascades, enzymatic reactions and nucleic acid and phospholipid synthesis [3,4]. Inorganic phosphate (Pi), in the form of PO_4, is the predominant form of P taken up by a plant from the soil, however, soil PO_4 primarily exists in organic or insoluble forms that are largely inaccessible to plant root uptake mechanisms [1]. Therefore, due to limited soil PO_4 availability, combined with the importance of an adequate concentration of P in plant cells to ensure normal growth and development, plants employ elegant mechanisms to spatially regulate P abundance across their developmentally distinct tissues [5,6]. Phosphorous homeostasis is therefore tightly controlled and involves both the remobilization of internal P stores and the increased acquisition of external PO_4 [5,7]. For example,

P limitation triggers the release of organic acids from the plant root system into the soil rhizosphere to chelate with metal ions to promote soluble PO_4 uptake to maintain or increase intracellular P concentration [1,8]. In addition, the P stored in the older leaves of a plant when the plant experiences P stress is remobilized; this allows for (1) continued growth of actively expanding tissues, and (2) the promotion of new growth. Enhanced P trafficking is achieved via promoting the expression of genes encoding PO_4 transporter proteins, and in turn, elevated PO_4 transporter protein abundance generally ensures that the cellular P concentration is maintained irrespective of external PO_4 levels [1,7].

In *Arabidopsis thaliana* (*Arabidopsis*), the first protein identified to be required for the maintenance of P homeostasis under PO_4 limiting conditions was PHOSPHATE1 (PHO1) [9]. The gene encoding PHO1 (*PHO1*; *AT1G14040*) was identified by [9] via their characterization of *pho1* plants, an *Arabidopsis* mutant line demonstrated to over-accumulate P in root tissues due to defective P translocation to the shoot. Although the *Arabidopsis* PHO1 protein, and the PHO1 proteins of other plant species characterized to date, do not closely resemble other PO_4 transporter proteins, PHO1 is indeed central to P movement in plants. The PHO1 protein is essential for PO_4 efflux into the root vascular cylinder; the first step in P transportation to the upper aerial tissues [10,11]. PHOSPHATE2 (PHO2) was the second protein demonstrated essential for the maintenance of P homeostasis with the *pho2* mutant shown to accumulate P to toxic levels in shoot tissues [12,13]. The *PHO2* gene (*AT2G33770*) has since been shown to encode a ubiquitin conjugating enzyme24 (UBC24), with the PHO2 UBC24 proposed to direct ubiquitin-mediated degradation of PO_4 transporters, PHOSPHATE TRANSPORTER1;4 (PHT1;4), PHT1;8 and PHT1;9 [14]. Further, *PHO2* is almost ubiquitously expressed in *Arabidopsis* shoot and root tissues [15], with the loss of PHO2-directed suppression of PHT1;4, PHT1;8 and PHT1;9 abundance in *pho2* plants leading to the enhanced translocation of P from the roots to the shoot tissue [14]. In addition to PHO1 and PHO2, traditional mutagenesis-based approaches have further identified other proteins essential to P homeostasis maintenance, including PHOSPHATE STARVATION RESPONSE1 (PHR1), a MYB domain transcription factor that regulates the expression of numerous P responsive genes [16,17].

More contemporary research, however, has concentrated on documenting the regulatory role directed at the posttranscriptional level by small regulatory RNAs (sRNA), specifically the microRNA (miRNA) class of sRNA, in order to maintain P homeostasis [18,19]. The advent of high throughput sequencing technologies has made sRNA profiling across plant species, and under different growth regimes, including exposure of a plant to abiotic and biotic stress, a routine experimental procedure in modern research [14,20,21]. Such profiling has identified a common suite of conserved miRNAs (miRNAs identified across multiple, evolutionary unrelated plant species) that accumulate differentially when mineral nutrients are lacking, including P, nitrogen (N), copper and sulphur [20,21]. Responsiveness of a single miRNA to multiple mineral nutrient stresses is not surprising considering the considerable overlap in the complex regulation of metal ion transport and/or uptake in plants [14,22,23]. In *Arabidopsis* for example, P and N uptake mechanisms are reciprocally linked to one another, therefore; a miRNA with enhanced accumulation during periods of P stress will usually be reduced in abundance during N starvation [19,24,25].

The miRNA, miR399, has been conclusively linked with the maintenance of P homeostasis and the regulation of PO_4 uptake in *Arabidopsis* [18,19]. In *Arabidopsis*, the miR399 sRNA is processed from six precursor transcripts, namely *PRE-MIR399A* to *PRE-MIR399F*, transcribed from five genomic loci (*MIR399A-MIR399D* and *MIR399E/F*). The miR399 sRNA is unique amongst *Arabidopsis* miRNAs in that it acts as a mobile systemic signal upon PO_4 stress [21,26]. More specifically, when P becomes limited in *Arabidopsis* shoots, *MIR399* gene expression is stimulated by PHR1 [27], and following processing of the now abundant miR399 precursor transcripts by the protein machinery of the *Arabidopsis* miRNA pathway, the mature miR399 sRNA is transported to the roots. Here, miR399 is actively loaded by the miRNA-induced silencing complex (miRISC) to direct miRISC-mediated cleavage of *PHO2*, the target transcript of miR399 [7,21,27]. Reduced PHO2 protein abundance, due to elevated miRISC-mediated cleavage of the *PHO2* transcript, in turn removes the PHO2-mediated suppression of PO_4 transporters, PHT1;4, PHT1;8 and PHT1;9, to ultimately promote root-to-shoot P transport in an attempt to maintain shoot P homeostasis in P limited conditions [28–31]. Additional regulatory complexity to the miR399/*PHO2* expression module is offered by the non-protein-coding RNA, *INDUCED BY PHOSPHATE STARVATION1* (*IPS1*) [32]. Once transcribed, *IPS1* acts as an endogenous target mimic (eTM) of miR399 activity [33]. Specifically, the miR399 target site harbored by *IPS1* contains a three nucleotide mismatch bulge across miR399 nucleotide positions 10 and 11: the position at which the catalytic core of miRISC, ARGONAUTE1 (AGO1), catalyzes the cleavage of miRNA target transcripts [34]. The bulge that forms at this position once miR399-directed AGO1 binds *IPS1*, renders *IPS1* resistant to AGO1-catalyzed cleavage, thereby effectively sequestering away miR399 activity [33].

Three of the five members of the *Arabidopsis* DOUBLE-STRANDED RNA BINDING (DRB) protein family, including DRB1, DRB2 and DRB4, have been assigned functional roles in the *Arabidopsis* miRNA pathway [35–39]. Both DRB1 and DRB4 form functional partnerships with DICER-LIKE (DCL) proteins, RNase III-like endonucleases that cleave molecules of double-stranded RNA (dsRNA). More specifically, the DRB1/DCL1 partnership processes stem-loop structured molecules of imperfectly dsRNA that form post miRNA precursor transcript folding [35–37], and the DRB4/DCL4 partnership is central for the processing of a small subset of miRNA precursor transcripts that fold to form stem-loop structures with high levels of base-pairing due to the almost perfect complementarity of the nucleotide sequences of the stem-loop arms [39]. More recently, DRB2 has also been assigned a functional role in the *Arabidopsis* miRNA pathway due to its demonstrated antagonism and/or synergism with the roles of both DRB1 and DRB4 in sRNA production [37,40]. Here, we therefore assessed the requirement of DRB1, DRB2 and DRB4 in the regulation of the miR399/*PHO2* expression module, both under non-stressed growth conditions and when wild-type *Arabidopsis* plants (ecotype Columbia-0 (Col-0)) and the *drb1*, *drb2* and *drb4* mutant lines are exposed to PO_4 starvation. More specifically, we aimed to determine; (1) the contribution of DRB1, DRB2 and/or DRB4 to miR399 production; (2) the mode of silencing directed by miR399 to regulate *PHO2* expression, and; (3) whether either DRB1, DRB2 or DRB4 are required for P homeostasis maintenance. Phenotypic and molecular assessment of Col-0, *drb1*, *drb2* and *drb4* plants post exposure to a 7-day period of PO_4 starvation, revealed that DRB1 and DRB2 are required for P homeostasis maintenance. Further, DRB1 was established as the primary DRB protein required to regulate miR399 production. However, DRB2 and DRB4 were demonstrated to play a secondary role in miR399 production regulation. Furthermore, miR399 appears to regulate the expression of its targeted transcript, *PHO2*, via both the canonical mechanism of plant miRNA-directed target gene expression repression, target mRNA cleavage, and via the alternative mode of target gene expression regulation, translational repression. Taken together, the hierarchical contribution of DRB1, DRB2 and DRB4 to the regulation of the miR399/*PHO2* expression module in *Arabidopsis* shoots and roots identifies the extreme importance of maintaining P homeostasis to ensure that numerous vital cellular processes are maintained across *Arabidopsis* tissue types and under a changing cellular environment.

2. Results

2.1. The Phenotypic and Physiological Response to PO_4 Stress in the Shoot Tissues of Arabidopsis Plant Lines Defective in DRB Protein Activity

To determine the consequence of loss of DRB activity on P homeostasis maintenance in 15-day old *Arabidopsis* plants post a 7-day period of PO_4 starvation, a series of phenotypic and physiological parameters were assessed in Col-0, *drb1*, *drb2* and *drb4* shoots. The severe developmental phenotype of the *drb1* mutant has been reported previously [36,41,42]. Figure 1A clearly reveals the reduced size of the *drb1* mutant at 15 days of age, compared to Col-0 plants, when both *Arabidopsis* lines are cultivated on standard growth media (P$^+$ media). The retarded development of the *drb1* mutant is further evidenced in Figure 1B where the fresh weight of 8-day old Col-0 and *drb1* seedlings is presented. Specifically, prior to seedling transfer to either P$^+$ or P$^-$ media, the fresh weight of an 8-day old *drb1* seedling (13.5 ± 1.0 mg) is 53.4% less than that a Col-0 seedling (29.0 ± 3.5 mg). Compared to *drb1*, the *drb2* and *drb4* mutants display mild developmental phenotypes [37,42] as evidenced by those displayed by 15-day old *drb2* and *drb4* plants cultivated on P$^+$ growth media (Figure 1A), and by the fresh weights of 8-day old *drb2* (26.8 ± 4.2 mg) and *drb4* (22.9 ± 1.4 mg) seedlings. Although the *drb1* mutant displayed the most severe phenotype, *drb1* development appeared to be the least affected by the 7-day PO$_4$ stress treatment. The fresh weight of P$^-$ *drb1* plants (35.5 ± 1.0 mg) was only reduced by 21.6% compared to P$^+$ *drb1* plants (45.3 mg ± 1.5 mg) (Figure 1C). The development of Col-0, *drb2* and *drb4* plants was negatively impacted to a similar degree by the 7-day PO$_4$ stress treatment, with their fresh weights reduced by 36.6%, 39.1% and 36.3%, respectively (Figure 1C). Determination of rosette area revealed largely similar trends across the *drb* mutant lines analyzed, that is, *drb1* rosette area was reduced by 29.3%, while the rosette development of P$^-$ *drb2* and P$^-$ *drb4* plants was reduced by 48.0% and 38.7%, respectively (Figure 1D). Interestingly, the observed reductions to the rosette area of P$^-$ *drb1*, P$^-$ *drb2* and P$^-$ *drb4* plants was considerably less than the 60.1% reduction to the rosette area of P$^-$ Col-0 plants (11.2 ± 1.7 mm^2) compared to P$^+$ Col-0 plants (28.1 ± 5.5 mm^2) (Figure 1D).

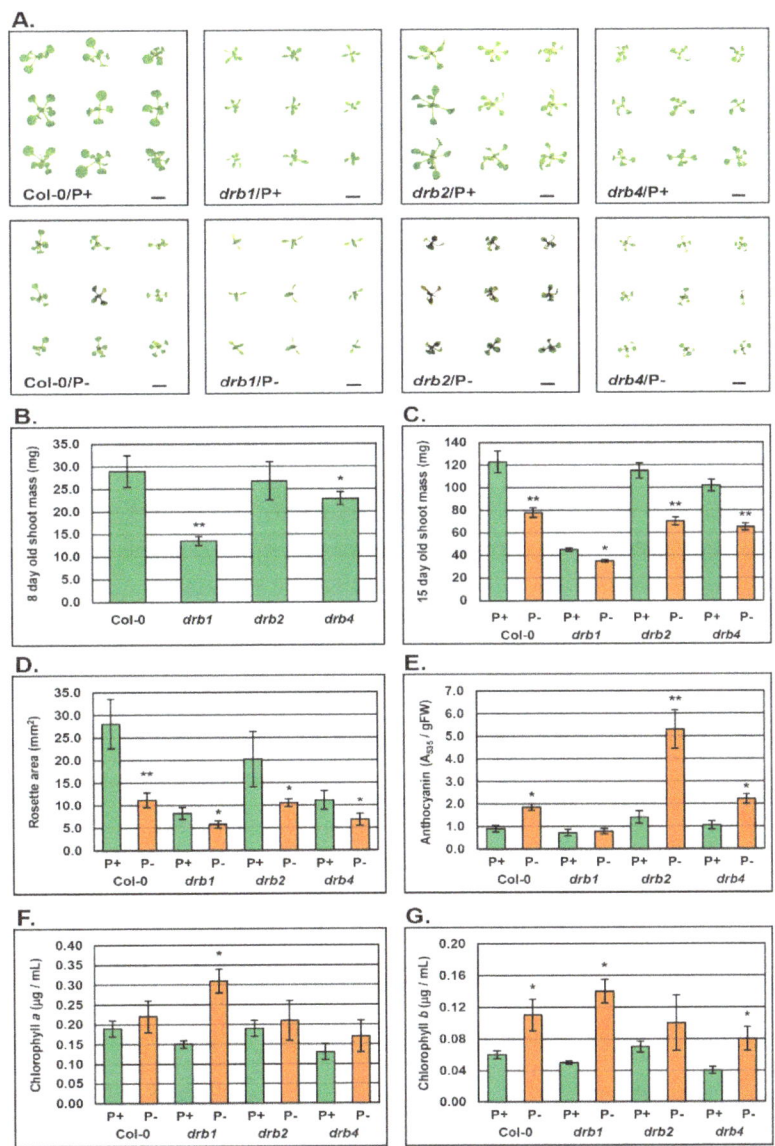

Figure 1. The aerial tissue phenotypes displayed by 15-day old *Arabidopsis* plant lines Col-0, *drb1*, *drb2* and *drb4* post exposure to a 7-day period of PO$_4$ starvation. (**A**) The aerial tissue phenotypes expressed by non-stressed (top row of panels) and PO$_4$-stressed (bottom row of panels) 15-day old Col-0, *drb1*, *drb2* and *drb4* plants. Scale bar = 1cm. (**B**) Quantification of the shoot mass of 8-day old Col-0, *drb1*, *drb2* and *drb4* seedlings germinated and cultivated under standard growth conditions. (**C**) The shoot mass of non-stressed and PO$_4$-stressed 15-day old Col-0, *drb1*, *drb2* and *drb4* plants. (**D**) The rosette area of

non-stressed and PO$_4$-stressed 15-day old *Arabidopsis* lines, Col-0, *drb1*, *drb2* and *drb4*. (**E**) Anthocyanin accumulation in the shoot tissues of 15-day old Col-0, *drb1*, *drb2* and *drb4* plants cultivated under standard growth conditions, or for 7-days under PO$_4$ starvation. (**F** and **G**) Chlorophyll *a* (**F**) and chlorophyll *b* (**G**) abundance in the aerial tissues of non-stressed and PO$_4$-stressed Col-0, *drb1*, *drb2* and *drb4* plants. (**B-G**) Error bars represent the standard deviation of four biological replicates and each biological replicate consisted of a pool of twelve individual plants. The presence of an asterisk above a column represents a statistically significant difference either between non-stressed Col-0 plants and each assessed *drb* mutant post cultivation under either a non-stressed or stressed growth regime (**B**) or between the non-stressed and PO$_4$-stressed sample of each plant line (**C-G**) (*p*-value: * < 0.05; ** < 0.005; *** < 0.001).

Anthocyanin, chlorophyll *a* and chlorophyll *b* content of Col-0, *drb1*, *drb2* and *drb4* shoots was also determined. Phosphate starvation has been previously shown to elevate the levels of PRODUCTION OF ANTHOCYANIN PIGMENT1 (PAP1/MYB75), PAP2 (MYB90) and MYB113, three MYB domain transcription factors that in turn stimulate the expression of a cohort of genes required for anthocyanin production in vegetative tissues [19,43]. These reports, in combination with the readily observable pigmentation that accumulated in the rosette leaves of P$^-$ Col-0, P$^-$ *drb2* and P$^-$ *drb4* plants (Figure 1A), identified anthocyanin as an ideal metric to further assess the response of each *drb* mutant to PO$_4$ starvation. The anthocyanin content of non-stressed Col-0, *drb1*, *drb2* and *drb4* shoots was similar (Figure 1E). However, when PO$_4$ is limited, an approximate 2.0-fold increase in anthocyanin accumulation was detected for P$^-$ Col-0 shoots. Further promotion of anthocyanin accumulation was determined for PO$_4$-stressed *drb2* and *drb4* plants, with anthocyanin content elevated 3.7- and 2.8-fold in P$^-$ *drb2* and P$^-$ *drb4* plants, respectively (Figure 1E). As readily observable in Figure 1A, anthocyanin accumulation was not promoted in the shoot tissue of P$^-$ *drb1* plants. However, spectrophotometry revealed abundance changes for both chlorophyll *a* and chlorophyll *b* in the shoot tissue of P$^-$ *drb1* plants. Specifically, chlorophyll *a* (Figure 1F) and chlorophyll *b* (Figure 1G) abundance was elevated by 2.1- and 2.8-fold in P$^-$ *drb1* shoots, compared to P$^+$ *drb1* shoots. In PO$_4$-stressed Col-0, *drb2* and *drb4* shoots, the chlorophyll *a* level remained largely unchanged compared to the non-stressed counterpart of each plant line (Figure 1F). Chlorophyll *b* accumulation however, was determined to be promoted in Col-0 and *drb4* shoots, by 1.8- and 2.0-fold, by the 7-day PO$_4$ starvation period (Figure 1G).

2.2. Molecular Profiling of the miR399/PHO2 Expression Module in the Shoot Tissues of Arabidopsis Plant Lines Defective in DRB Protein Activity

The results presented in Figure 1 strongly indicated that each *drb* mutant was responding differently to the applied stress and when this finding is considered together with the documented roles of DRB1, DRB2 and DRB4 in the *Arabidopsis* miRNA pathway [35–39], including the demonstrated antagonism between DRB1 and DRB2 [37] and between DRB2 and DRB4 [40] in miRNA production, the miR399/*PHO2* expression module was next profiled via a RT-qPCR-based approach. RT-qPCR profiling was conducted in an attempt to determine if the observed differences in the response of each *drb* mutant line to PO$_4$ stress was a result of dysfunction of the miR399/*PHO2* expression module.

In *Arabidopsis* shoots, *PHR1* promotes *MIR399* gene expression when PO$_4$ supplies become limited, resulting in elevated miR399 abundance [27]. Therefore, RT-qPCR was first used to assess *PHR1* expression in control and PO$_4$-stressed Col-0, *drb1*, *drb2* and *drb4* shoots (Figure 2A). *PHR1* expression was only mildly elevated by 1.5-, 1.6- and 1.7-fold in P$^+$ *drb1*, P$^+$ *drb2* and P$^+$ *drb4* shoots respectively, compared to its levels in non-stressed Col-0 shoots (Figure 2A). RT-qPCR further revealed that PO$_4$ stress only induced mild elevations to *PHR1* expression in P$^-$ Col-0 (1.00 to 1.22 relative expression) and P$^-$ *drb2* shoots (1.62 to 1.74 relative expression) (Figure 2A). This result was not unexpected in view of the previous report of only mild *PHR1* expression induction in PO$_4$-stressed *Arabidopsis* [17]. Interestingly, *PHR1* expression was reduced by 19.6% and 31.2% in P$^-$ *drb1* and P$^-$ *drb4* shoots, respectively (Figure 2A), and not mildly elevated as expected.

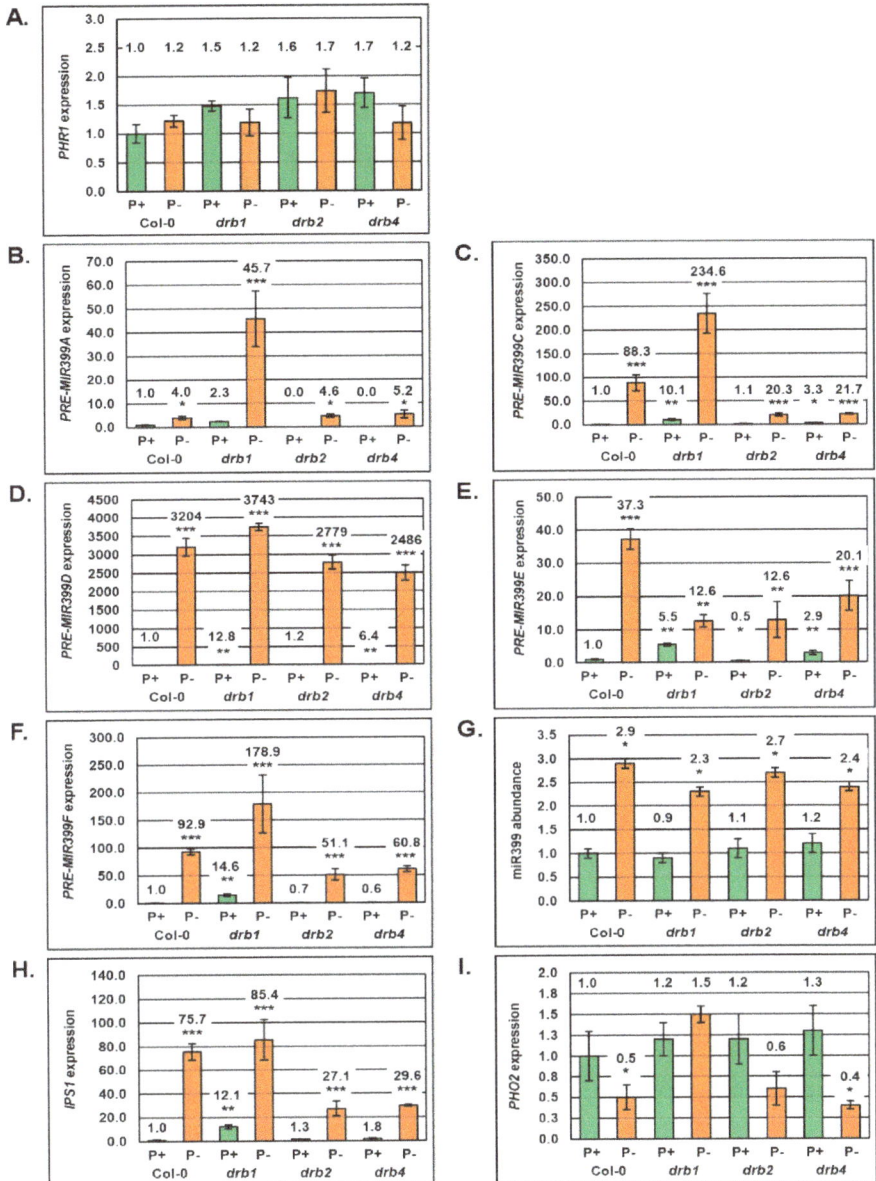

Figure 2. Molecular profiling of the miR399/*PHO2* expression module in the aerial tissues of non-stressed and PO_4-stressed Col-0, *drb1*, *drb2* and *drb4* plants. (**A**) RT-qPCR assessment of the expression of the PO_4 responsive transcription factor PHR1 in the aerial tissues of non-stressed and PO_4-stressed Col-0, *drb1*, *drb2* and *drb4* plants. (**B** to **F**) RT-qPCR profiling of miR399 precursor transcript abundance in the aerial tissues of non-stressed and PO_4-stressed Col-0, *drb1*, *drb2* and *drb4* plants, including precursors *PRE-MIR399A* (**B**), *PRE-MIR399C* (**C**), *PRE-MIR399D* (**D**), *PRE-MIR399E* (**E**) and *PRE-MIR399F* (**F**). (**G**) Quantification of miR399 abundance in the aerial tissues of non-stressed and PO_4-stressed Col-0, *drb1*, *drb2* and *drb4* plants. (**H**) Assessment of the expression of the non-cleavable decoy of miR399 activity, *IPS1*, via RT-qPCR in the aerial tissues of non-stressed and PO_4-stressed *Arabidopsis* lines, Col-0, *drb1*, *drb2* and *drb4*. (**I**) RT-qPCR analysis of the expression of the miR399 target gene, *PHO2*, in the aerial tissues of non-stressed and PO_4-stressed *Arabidopsis* lines, Col-0, *drb1*, *drb2* and *drb4*. (**A–I**) Error bars represent the standard deviation of four biological replicates and each biological replicate consisted of a pool of twelve individual plants. Due to the vastly different levels of each assessed transcript, the relative expression value for each plant line/growth regime is provided above the corresponding column. The presence of an asterisk above a column represents a statistically significant difference between non-stressed Col-0 plants and each of the assessed *drb* mutant lines, post cultivation under either a standard or stressed growth regime (*p*-value: * < 0.05; ** < 0.005; *** < 0.001).

The miR399 sRNA is processed from six structurally distinct precursor transcripts (*PRE-MIR399A* to *PRE-MIR399F*), transcribed from five genomic loci (*MIR399A* to *MIR399D* and *MIR399E/F*) in *Arabidopsis*. RT-qPCR only failed to detect *PRE-MIR399B* expression in Col-0 shoots. RT-qPCR did however clearly reveal that PO_4 stress induced the expression of the five detectable miR399 precursor transcripts by 4.0-, 88.3-, 3204-, 37.3- and 92.9-fold in the shoots of P^- Col-0 plants (Figure 2B–F). Of the three members of the *Arabidopsis* DRB protein family analyzed here, Figure 2B–F clearly show that DRB1 is the primary DRB protein required to regulate miR399 production in *Arabidopsis* shoots with the abundance of *PRE-MIR399A*, *PRE-MIR399C*, *PRE-MIR399D*, *PRE-MIR399E* and *PRE-MIR399F* elevated by 2.3-, 10.1-, 12.8-, 5.5- and 14.6-fold, respectively, in P^+ *drb1* shoots. The primary role of DRB1 in regulating miR399 production in *Arabidopsis* shoots was further highlighted for *PRE-MIR399A*, *PRE-MIR399C*, *PRE-MIR399D* and *PRE-MIR399F* via additional elevations to their respective expression levels, specifically 45.7-, 234.6- 3743- and 178.9-fold increases to transcript abundance in P- *drb1* shoots (Figure 2B–D,F).

Failure to detect the *PRE-MIR399A* precursor by RT-qPCR in P^+ *drb2* shoots, and a similar degree of over-accumulation of this precursor in P^- Col-0 (4.0-fold) and P^- *drb2* shoots (4.6-fold), indicated that DRB2 is not required to regulate miR399 production from this precursor (Figure 2B). Wild-type-like accumulation of *PRE-MIR399C* (1.1-fold) and *PRE-MIR399D* (1.2-fold) in P^+ *drb2* shoots, and a lower degree of over-accumulation of these two precursors in P^- *drb2* shoots, compared to P^- Col-0 shoots, indicated that DRB2 plays a secondary role in regulating miR399 production from these two precursors (Figure 2C,D). A similar level of expression of *PRE-MIE399E* in PO_4-stressed *drb1* and *drb2* shoots suggested that both DRB1 and DRB2 are required for miR399 production from this precursor (Figure 2E). However, lower transcript abundance (0.5 relative expression) in P^+ *drb2* shoots, compared to relative expression levels of 1.0 and 5.5 in P^+ Col-0 and P^+ *drb1* shoots, respectively (Figure 2E), again indicated that under standard growth conditions, DRB2 plays a secondary role in regulating miR399 production from the *PRE-MIR399E* precursor. The abundance of the *PRE-MIR399F* transcript is also reduced in P^+ *drb2* shoots compared to its levels in P^+ Col-0 shoots, and further, the degree of over-accumulation of *PRE-MIR399F* is less in P^- *drb2* shoots compared to its levels in P^- Col-0 shoots (Figure 2F). When these expression trends are considered together with those documented for P^+ and P^- *drb1* shoots, they again indicate a secondary role for DRB2 in regulating miR399 production from this precursor.

As demonstrated for P^+ *drb2* shoots, the *PRE-MIR399A* transcript remained below the detection sensitivity of RT-qPCR in P^+ *drb4* shoots (Figure 2B). RT-qPCR did however, reveal *PRE-MIR399A* expression to be elevated by 5.2-fold in P^- *drb4* shoots, a similar degree of transcript elevation to that observed in P^- Col-0 shoots (4.0-fold increase) (Figure 2B). This indicates that DRB4 is not involved in regulating miR399 production from this precursor. Comparison of the RT-qPCR generated expression trends for *PRE-MIR399C*, *PRE-MIR399D* and *PRE-MIR399E* in P^+ and P^- *drb4* shoots, to those of P^+ Col-0, P^- Col-0, P^+ *drb1* and P^- *drb1* shoots, revealed a secondary role for DRB4 in regulating miR399 production from these three precursor transcripts (Figure 2C–E). DRB4 also appears to play a role in regulating miR399 production from the *PRE-MIR399F* transcript, with *PRE-MIR399F* abundance reduced by 40% in P^+ *drb4* shoots (Figure 2F). RT-qPCR also revealed that the expression of this precursor transcript was elevated to a relative expression level of 60.8 in PO_4-stressed *drb4* shoots; a lower degree of relative expression than observed in either P^- Col-0 (92.9 relative expression) or P^- *drb1* (178.9 relative expression) shoots (Figure 2F). This finding suggests that in the absence of DRB4 activity, miR399 is more efficiently processed from the *PRE-MIR399F* precursor transcript.

RT-qPCR was next applied to quantify miR399 abundance in the shoot material of non-stressed or PO_4-stressed Col-0, *drb1*, *drb2* and *drb4* plants. This analysis revealed that in spite of the considerable variation in precursor transcript abundance in the shoot tissues of P^+ Col-0, P^+ *drb1*, P^+ *drb2* and P^+

drb4 plants, miR399 levels remained largely unchanged (Figure 2G). This was an especially surprising finding for control *drb1* plants, with the *PRE-MIR399A, PRE-MIR399C, PRE-MIR399D, PRE-MIR399E* and *PRE-MIR399F* transcripts demonstrated to over-accumulate by 4.0-, 10.1-, 12.8-, 5.5- and 14.6-fold in P^+ *drb1* shoots, compared to their respective levels in P^+ Col-0 shoots. However, miR399 abundance was only reduced by 10% in P^+ *drb1* shoots. Similarly, although the expression level of the five miR399 precursors varied considerably in P^+ *drb2* and P^+ *drb4* shoots, miR399 abundance was only elevated by 10% and 20%, respectively (Figure 2G). Enhanced miR399 accumulation in P^+ *drb2* and P^+ *drb4* shoots did however further identify that both of these DRB proteins are required to correctly regulate miR399 abundance in *Arabidopsis* shoots. The degree of alteration to miR399 abundance was demonstrated to be higher in the shoot tissues of the four assessed plant lines when these lines were cultivated on PO_4 deplete media. Specifically, RT-qPCR revealed 2.9-, 2.6-, 2.5- and 2.0-fold enhancement to miR399 abundance in PO_4-stressed Col-0, *drb1*, *drb2* and *drb4* shoots, respectively (Figure 2G).

The mild alteration to miR399 abundance quantified by RT-qPCR in non-stressed and PO_4-stressed shoots (Figure 2G) led us to next assess the expression of *IPS1*, the eTM of miR399 [32–34]. Due to *IPS1* being a PO_4 stress-induced gene, it was unsurprising to only observe mild (P^+ *drb2* and P^+ *drb4* shoots) to moderate differences (P^+ *drb1* shoots) in *IPS1* transcript abundance in the shoot tissue of non-stressed Col-0, *drb1*, *drb2* and *drb4* plants (Figure 2H). Further, and as expected, RT-qPCR showed that PO_4 stress induced the expression of *IPS1*, with *IPS1* transcript abundance elevated by 75.7-, 7.1-, 20.8- and 16.4-fold in the shoot tissues of PO_4 stressed Col-0, *drb1*, *drb2* and *drb4* plants, respectively (compared to the non-stressed counterpart of each plant line).

Next, the expression of the target gene of miR399, *PHO2*, was determined by RT-qPCR to largely remain at wild-type levels (P^+ Col-0 shoots) in the shoot tissues of P^+ *drb1*, P^+ *drb2* and P^+ *drb4* plants (Figure 2I). This was an unsurprising result considering that RT-qPCR also revealed only mild changes to miR399 abundance across the three *drb* mutant lines assessed when each plant line was cultivated on standard *Arabidopsis* culture media (Figure 2G). RT-qPCR also revealed that elevated miR399 abundance in P^- Col-0, P^- *drb2* and P^- *drb4* plants, promoted miR399-directed expression repression of *PHO2*, with the abundance of the *PHO2* transcript reduced by 50%, 40% and 60% in the shoot tissues of these three plant lines, respectively (Figure 2I). In P^- *drb1* shoots however, the level of the *PHO2* transcript was increased by 50% (Figure 2I). Elevated *PHO2* expression in P^- *drb1* shoots, a tissue where miR399 abundance was also demonstrated to be elevated, indicated that in the absence of DRB1 activity, miR399-directed mRNA cleavage-mediated regulation of *PHO2* expression is lost.

2.3. The Phenotypic and Physiological Response to PO_4 Stress of the Root System of Arabidopsis Plant Lines Defective in DRB Protein Activity

The unique phenotypic (Figure 1) and molecular (Figure 2) response displayed by *drb1*, *drb2* and *drb4* shoots to PO_4 starvation led us to next repeat these assessments on the root system of each mutant background. As reported for the aerial tissue phenotypes expressed by the *drb1*, *drb2* and *drb4* mutants (Figure 1), Figure 3A again clearly displays the severe developmental phenotype expressed by the *drb1* mutant as well as the comparatively mild phenotypes that result from the loss of either DRB2 or DRB4 activity in *drb2* and *drb4* plants, respectively. The severity of the developmental phenotypes expressed by the three *drb* mutants assessed in this study is further evidenced when the fresh weight of the root system of 8-day old seedlings cultivated on standard growth media was determined. Specifically, the fresh weight of the root system of 8-day old *drb2* and *drb4* seedlings, 7.95 ± 0.20 mg and 8.00 ± 0.15 mg respectively, was equivalent to the fresh weight of the root system of Col-0 plants, 8.25 ± 0.45 mg (Figure 3B). However, the fresh weight of the root system of 8-day old *drb1* plants, 4.25 ± 0.15 mg, was approximately 50% less than that of an 8-day old Col-0 seedling (Figure 3B).

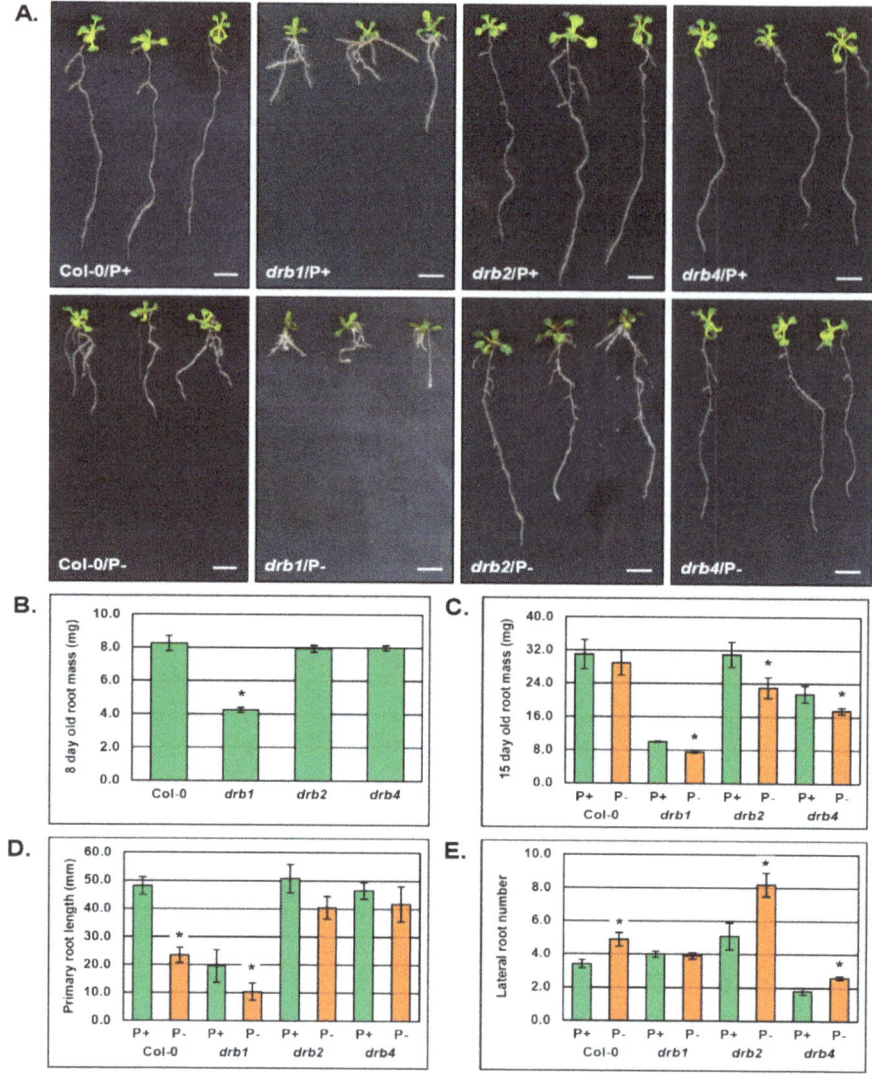

Figure 3. The root system phenotypes displayed by 15-day old *Arabidopsis* plant lines Col-0, *drb1*, *drb2* and *drb4* post exposure to a 7-day period of PO_4 starvation. (**A**) The root system phenotypes expressed by non-stressed (top row of panels) and PO_4-stressed (bottom row of panels) 15-day old Col-0, *drb1*, *drb2* and *drb4* plants. Scale bar = 1cm. (**B**) Quantification of the root mass of 8-day old Col-0, *drb1*, *drb2* and *drb4* seedlings cultivated under standard growth conditions. (**C**) The root mass of non-stressed and PO_4-stressed 15-day old Col-0, *drb1*, *drb2* and *drb4* plants. (**D**) The primary root length of non-stressed and PO_4-stressed 15-day old *Arabidopsis* lines, Col-0, *drb1*, *drb2* and *drb4*. (**E**) The number of lateral roots formed from the primary root of 15-day old Col-0, *drb1*, *drb2* and *drb4* plants cultivated under standard growth conditions, or post the 7-day PO_4 starvation period. (**B–E**) Error bars represent the standard deviation of four biological replicates and each biological replicate consisted of a pool of twelve individual plants. The presence of an asterisk above a column represents a statistically significant difference either between non-stressed Col-0 plants and each assessed *drb* mutant post cultivation under either a non-stressed or stressed growth regime (**B**) or between the non-stressed and PO_4-stressed sample of each plant line (**C-E**) (*p*-value: * < 0.05; ** < 0.005; *** < 0.001).

Figure 3C shows that at the completion of the 7-day PO_4 starvation period, the fresh weight of 15-day old P⁻ Col-0 roots (29.0 ± 3.0 mg) was only reduced by 2.0 mg compared to P⁺ Col-0 roots (31.0 ± 3.5 mg), a mild 6.5% reduction. The fresh weight of the root system of PO_4 stressed *drb1*, *drb2* and *drb4* plants all showed a much greater reduction when compared to their non-stressed counterparts (Figure 3C). That is, the fresh weight of the root system of 15-day old P⁻ *drb1* (7.5 ± 0.15 mg), P⁻ *drb2* (23.0 ± 2.5 mg) and P⁻ *drb4* plants (17.5 ± 0.75 mg) was reduced by 25.0%, 25.8% and 18.6%, respectively (Figure 3C).

Inhibition of primary root length is one of the main phenotypic responses of *Arabidopsis* to PO_4 stress [2,44], and accordingly, Figure 3A,D clearly show that the primary root length of 15-day old P^- Col-0 plants (23.4 ± 2.8 mm) was significantly reduced by 51.2% compared to non-stressed P^+ Col-0 plants (48.1 ± 3.1 mm) (Figure 3D). Although primary root length is already severely inhibited due to detrimental consequences of the loss of DRB1 activity on *Arabidopsis* development, the 7-day stress treatment caused a 46.7% reduction to the primary root length of P^- *drb1* plants (10.4 ± 3.1 mm) compared to P^+ *drb1* plants (19.5 ± 5.9 mm) (Figure 3D). Interestingly, PO_4 stress impacted primary root development to a much lower degree in both the *drb2* and *drb4* mutant backgrounds. Namely, primary root length was reduced by 20.3% and 10.3% in P^- *drb2* (40.5 ± 4.0 mm) and P^- *drb4* (41.8 ± 6.2 mm) plants respectively, compared to the primary root length of P^+ *drb2* (50.8 ± 5.0 mm) and P^+ *drb4* (46.6 ± 2.9 mm) plants (Figure 3D).

In parallel with inhibition to primary root length, promotion of lateral root development is a commonly reported phenotypic response of *Arabidopsis* plants exposed to PO_4 stress [2,44]. It was therefore unsurprising to document a 44% increase in the number of lateral roots that formed on 15-day old P^- Col-0 plants (4.9 ± 0.4) compared to P^+ Col-0 plants (3.4 ± 0.3) (Figure 3E). Interestingly, this phenotypic response to PO_4 stress appeared completely defective in the *drb1* mutant background with both P^+ *drb1* (4.0 ± 0.2) and P^- *drb1* (3.9 ± 0.2) plants forming approximately the same number of lateral roots. Unlike the *drb1* mutant, lateral root development was promoted by ~61% in the *drb2* mutant background with P^- *drb2* plants forming 8.2 ± 0.7 lateral roots compared to P^+ *drb2* plants which formed 5.1 ± 0.8 lateral roots. Lateral root formation was also induced by PO_4 stress in the *drb4* mutant with the number of lateral roots increased by 44% in P^- *drb4* plants (2.6 ± 0.1) compared to their number in P^+ *drb4* plants (1.8 ± 0.2).

2.4. Molecular Profiling of the miR399/PHO2 Expression Module in the Root System of Arabidopsis Plant Lines Defective in DRB Protein Activity

Due to its demonstrated role in inducing *MIR399* gene expression in PO_4 depleted conditions [27], RT-qPCR was initially used to profile *PHR1* expression in PO_4-stressed Col-0, *drb1*, *drb2* and *drb4* roots (Figure 4A). This analysis revealed that compared to the root system of each plant line's non-stressed counterpart, *PHR1* expression remained remarkably constant in P^- Col-0, P^- *drb1*, P^- *drb2* and P^- *drb4* roots (Figure 4A). Although RT-qPCR revealed that *PHR1* expression remained constant in the roots of control and PO_4-stressed plants, RT-qPCR was next applied to profile the expression of the six *MIR399* precursor transcripts in the roots of P^+ and P^- plants. Of the six miR399 precursors, RT-qPCR only allowed for expression quantification of three miR399 precursors, namely *PRE-MIR399A*, *PRE-MIR399C* and *PRE-MIR399D* in *Arabidopsis* roots (Figure 4B–D). In P^- Col-0 roots, RT-qPCR clearly revealed that PO_4 stress induced the expression of the miR399 precursors, *PRE-MIR399A*, *PRE-MIR399C* and *PRE-MIR399D*, by 4.0-, 40.6- and 1546-fold, respectively (Figure 4B–D). When compared to P^+ Col-0 roots, the moderate 2.3- and 3.6-fold elevation in the abundance of *PRE-MIR399A* and *PRE-MIR399C* in P^+ *drb1* roots, identified DRB1 as the primary DRB required for miR399 production regulation from these two precursor transcripts in the roots of wild-type *Arabidopsis* plants (Figure 4B,C). The primary role of DRB1 in *PRE-MIR399A* and *PRE-MIR399C* processing in non-stressed Col-0 roots is further evidenced by the wild-type equivalent accumulation of these two precursors in P^+ *drb2* and P^+ *drb4* roots, and by the highest degree of *PRE-MIR399A* and *PRE-MIR399C* precursor transcript over-accumulation in P^- *drb1* roots (Figure 4B,C). Considering this result, it was therefore of considerable interest to observe the greatest degree of *PRE-MIR399D* over-accumulation, an 8.2-fold increase, in P^+ *drb4* roots and not in P^+ *drb1* roots (4.3-fold increase) (Figure 4D). This finding suggests that in non-stressed wild-type *Arabidopsis* roots, DRB4 is the primary DRB responsible for regulating miR399 production from this precursor transcript. In addition, and under PO_4 stress conditions, the *PRE-MIR399D* transcript increased in its abundance to relative expression values of 829, 849 and 1271 in *drb1*, *drb2* and *drb4* roots, respectively (Figure 4D). Although these determined increases in precursor transcript abundance are all highly significant, they are not as significant as the 1546 relative expression value obtained for the

PRE-MIR399D transcript in P⁻ Col-0 roots. A lower degree of precursor transcript over-accumulation in each assessed *drb* mutant background, compared to the expression induction observed in wild-type roots, indicated that all three DRB proteins potentially play a role in fine-tuning the regulation of miR399 production from the *PRE-MIR399D* precursor in PO₄-stressed *Arabidopsis* roots (Figure 4D).

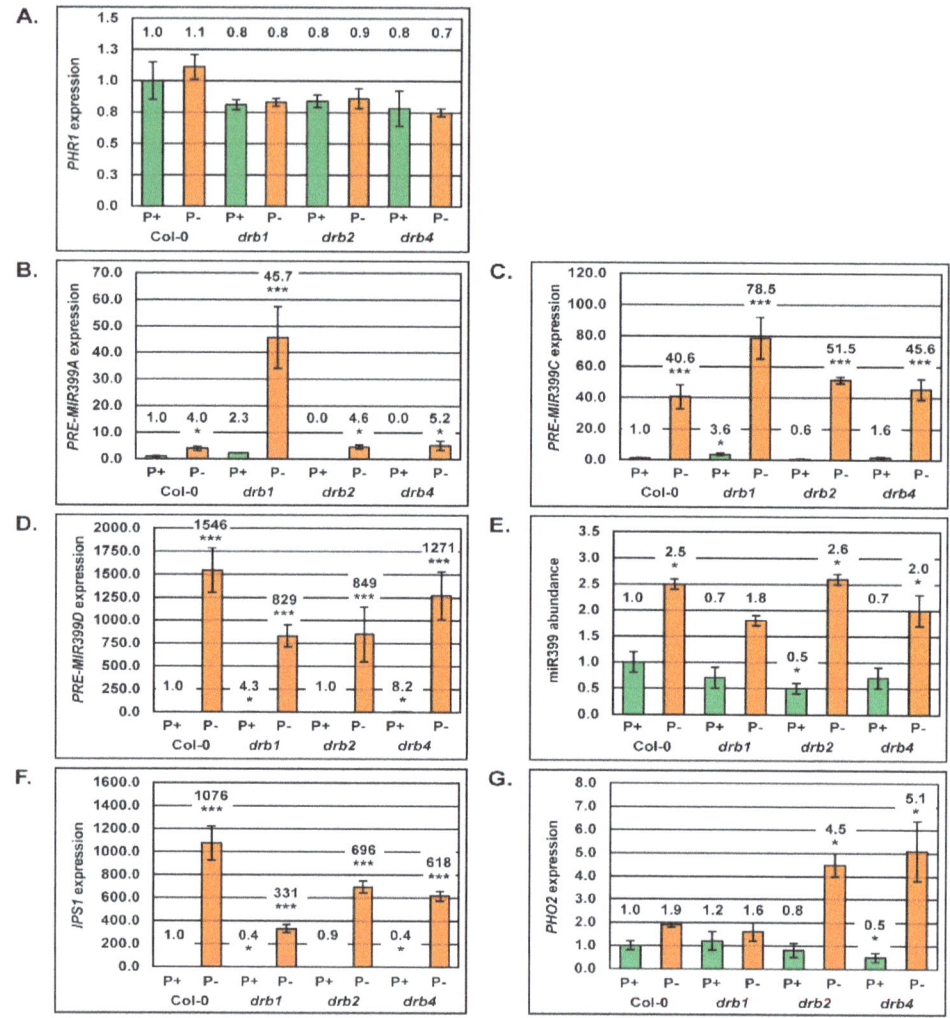

Figure 4. Molecular profiling of the miR399/*PHO2* expression module in the root system of non-stressed and PO₄-stressed Col-0, *drb1*, *drb2* and *drb4* plants. (**A**) RT-qPCR assessment of the expression of the PO₄ responsive transcription factor *PHR1* in the roots of non-stressed and PO₄-stressed Col-0, *drb1*, *drb2* and *drb4* plants. (**B–D**) RT-qPCR profiling of miR399 precursor transcript abundance in the root system of non-stressed and PO₄-stressed Col-0, *drb1*, *drb2* and *drb4* plants, including precursors *PRE-MIR399A* (**B**), *PRE-MIR399C* (**C**) and *PRE-MIR399D* (**D**). (**E**) Quantification of miR399 abundance in the roots of non-stressed and PO₄-stressed Col-0, *drb1*, *drb2* and *drb4* plants. (**F**) Assessment of *IPS1* transcript abundance in the roots of non-stressed and PO₄-stressed *Arabidopsis* lines, Col-0, *drb1*, *drb2* and *drb4*. (**G**) RT-qPCR analysis of *PHO2* expression, the target gene of miR399, in the root system of non-stressed and PO₄-stressed *Arabidopsis* lines, Col-0, *drb1*, *drb2* and *drb4*. (**A–G**) Error bars represent the standard deviation of four biological replicates and each biological replicate consisted of a pool of twelve individual plants. Due to the vastly different level of each assessed transcript, the relative expression value for each plant line/growth regime is provided above the corresponding column. The presence of an asterisk above a column represents a statistically significant difference between non-stressed Col-0 plants and each of the assessed *drb* mutant lines, post cultivation under either a standard or stressed growth regime (*p*-value: * < 0.05; ** < 0.005; *** < 0.001).

Post-establishment of highly variable expression profiles for *PRE-MIR399A*, *PRE-MIR399C* and *PRE-MIR399D* in non-stressed *drb1*, *drb2* and *drb4* roots (Figure 4B–D), miR399 abundance reductions of 30%, 50% and 30% in P$^+$ *drb1*, P$^+$ *drb2* and P$^+$ *drb4* roots, respectively, was expected (Figure 4E). Quantification of miR399 abundance, 2.5-, 1.8-, 2.6- and 2.0-fold elevations, respectively, in the root tissues of PO$_4$-stressed Col-0, *drb1*, *drb2* and *drb4* plants, revealed that the considerable induction to *PRE-MIR399A*, *PRE-MIR399C* and *PRE-MIR399D* expression (Figure 4B–D), did not however, result in an overly altered miR399 accumulation profile (Figure 4E).

Failure to establish a strong correlation between precursor transcript expression and miR399 abundance in either control or PO$_4$-stressed Col-0, *drb1*, *drb2* and *drb4* roots, led us to next assess *IPS1* expression in this tissue (Figure 4F). *IPS1* transcript abundance remained relatively unchanged in the root tissues of non-stressed Col-0 and *drb2* plants (Figure 4F). Interestingly, *IPS1* expression was reduced by 60% in P$^+$ *drb1* and P$^+$ *drb4* roots (Figure 4F). Significant induction of *IPS1* expression was observed in PO$_4$-stressed *drb1*, *drb2* and *drb4* roots, 331-, 696- and 618-fold elevations, respectively. Interestingly, RT-qPCR demonstrated that *IPS1* expression was promoted to its greatest degree, 1076-fold, in PO$_4$-stressed Col-0 roots (Figure 4F).

The expression of the miR399 target gene, *PHO2*, was next quantified by RT-qPCR in non-stressed and PO$_4$-stressed Col-0, *drb1*, *drb2* and *drb4* roots (Figure 4G). In P$^+$ *drb1* and P$^+$ *drb2* roots, RT-qPCR revealed *PHO2* expression to be elevated and reduced by 20%, respectively, and in P$^+$ *drb4* roots, *PHO2* expression was reduced by 30%. Elevated *PHO2* expression in P$^+$ *drb1* roots was expected considering the slight reduction to miR399 abundance observed in this tissue (Figure 4E). However, the reduced *PHO2* transcript levels in P$^+$ *drb2* and P$^+$ *drb4* roots was a surprise finding considering that miR399 abundance was also reduced in these two mutant lines by 50% and 30%, respectively (Figure 4E). *PHO2* expression was demonstrated by RT-qPCR to be elevated by 1.9-, 1.6-, 4.5- and 5.1-fold in PO$_4$-stressed Col-0, *drb1*, *drb2* and *drb4* roots, respectively (Figure 4G). This finding also formed an unexpected result considering that PO$_4$ starvation induced the accumulation of the miR399 sRNA in all four assessed plant lines (Figure 4E).

2.5. Correct Inorganic Phosphate Partitioning Between the Shoot and Root Tissue of Arabidopsis Requires DRB1 and DRB2

The molecular profiling of alterations to the miR399/*PHO2* expression module in the shoot and root tissue of *Arabidopsis* Col-0, *drb1*, *drb2* and *drb4* plants under PO$_4$ stress, in combination with each plant line displaying a unique phenotypic response to this stress, led us to next assess Pi partitioning in the aerial tissue and root system of P$^+$ and P$^-$ Col-0, *drb1*, *drb2* and *drb4* plants. In the shoot tissues of 15-day old plants cultivated in PO$_4$ replete conditions, Pi content was only altered in the *drb2* mutant background, with the Pi content of P$^+$ *drb2* shoots (13.8 μmol/gFW) reduced by 27.4% compared to the Pi content of P$^+$ Col-0 shoots (19.0 μmol/gFW) (Figure 5A). When cultivated in PO$_4$-stress conditions however, only the Pi content of P$^-$ *drb1* shoots (1.15 μmol/gFW) differed to that of P$^-$ Col-0 shoots (1.75 μmol/gFW); a 34.3% reduction (Figure 5A). In non-stressed roots, the Pi content of P$^+$ *drb1* (11.4 μmol/gFW) and P$^+$ *drb2* (9.8 ± 0.8 μmol/gFW) roots was determined to be elevated by 58.3% and 37.5% respectively, compared to P$^+$ Col-0 roots (7.2 μmol/gFW) (Figure 5B). As demonstrated for non-stressed *drb1* and *drb2* roots, the Pi content of P$^-$ *drb1* (1.84 μmol/gFW) and P$^-$ *drb2* (0.65 μmol/gFW) roots also differed to that of PO$_4$-stressed Col-0 roots (1.25 μmol/gFW), elevated and reduced by 47.2% and 48%, respectively (Figure 5B).

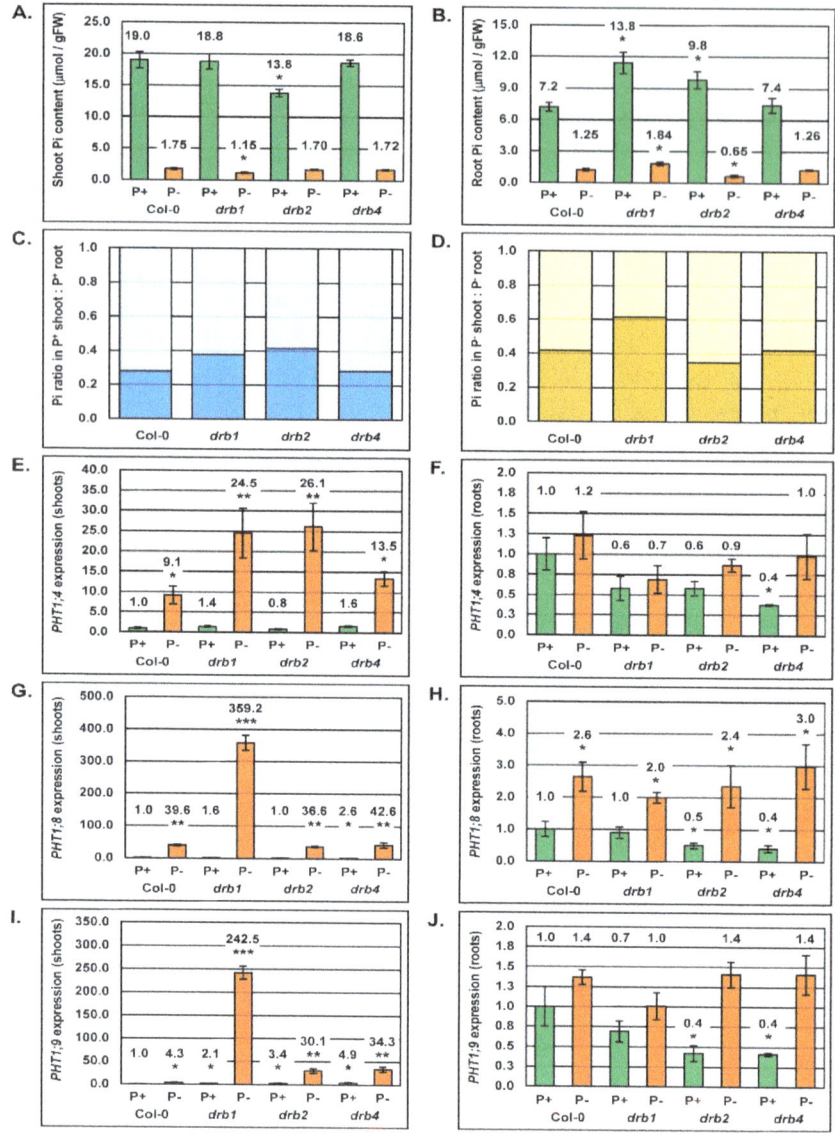

Figure 5. Pi content and PO_4 transporter gene expression in the shoot and root tissue of 15-day old *Arabidopsis* plant lines Col-0, *drb1*, *drb2* and *drb4* cultivated under either a standard growth regime or post-exposure to a 7-day period of PO_4 starvation. (**A,B**) Comparison of the Pi content of the shoots (A) and roots (B) of 15-day old non-stressed and PO_4-stressed Col-0, *drb1*, *drb2* and *drb4* plants. (**C**) Pi content shoot (light blue) to root (dark blue) ratio of 15-day old Col-0, *drb1*, *drb2* and *drb4* plants cultivated under standard growth conditions. (**D**) Pi content shoot (light gold) to root (dark gold) ratio of 15-day old Col-0, *drb1*, *drb2* and *drb4* plants post 7-days of PO_4 starvation. (**E,F**) Quantification of *PHT1;4* expression in the shoot (E) and root (F) tissues of 15-day old Col-0, *drb1*, *drb2* and *drb4* plants cultivated under standard growth conditions or post a 7-day period of PO_4 starvation. (**G,H**) RT-qPCR assessment of *PHT1;8* transcript abundance in the shoots (G) and roots (H) of 15-day old Col-0, *drb1*, *drb2* and *drb4* plants cultivated under either standard or PO_4 stress conditions. (**I,J**) *PHT1;9* expression in the shoot (I) and root (J) material of non-stressed or PO_4-stressed Col-0, *drb1*, *drb2* and *drb4* plants at 15 days of age. (A,B,E–J) Error bars represent the standard deviation of four biological replicates and each biological replicate consisted of a pool of twelve individual plants. Due to the vastly different levels of each assessed transcript, the relative expression value for each plant line/growth regime is provided above the corresponding column. The presence of an asterisk above a column represents a statistically significant difference between the non-stressed and PO_4-stressed sample of each plant line (A,B) or between non-stressed Col-0 plants and each *drb* mutant line, post cultivation under either a standard or stressed growth regime (E–J) (*p*-value: * < 0.05; ** < 0.005; *** < 0.001).

The reduced Pi content of P$^+$ *drb2* shoots (Figure 5A), together with the elevated Pi contents of P$^+$ *drb1* and P$^+$ *drb2* roots (Figure 5B), suggested that Pi partitioning was potentially defective in these two mutant backgrounds. We therefore next determined the Pi content ratio of the shoot and root of non-stressed and PO$_4$-stressed Col-0, *drb1*, *drb2* and *drb4* plants. Figure 5C clearly shows that Pi partitioning between the shoot and root tissue of P$^+$ *drb1* and P$^+$ *drb2* plants is defective, even when these two mutant lines are cultivated on standard *Arabidopsis* growth media. Under PO$_4$ stress conditions, defective Pi partitioning is even more readily evident in the *drb1* mutant background which showed a 0.38:0.62 shoot to root Pi content ratio, compared to the shoot to root Pi content ratio of 0.58:0.42 for P$^-$ Col-0 plants. Although not as striking as determined for P$^+$ *drb2* plants, the altered shoot to root Pi content ratio (0.65:0.35) of PO$_4$-stressed *drb2* plants again indicated that Pi partitioning is defective in this mutant background (Figure 5D).

Altered shoot to root Pi content ratios in *drb1* and *drb2* plants strongly suggested that Pi partitioning is defective in these two mutant backgrounds. Considering that PO$_4$ transporters, PHT1;4, PHT1;8 and PHT1;9, are known targets of PHO2-mediated ubiquitination [7,14], together with our demonstration in Figures 2 and 4 that the miR399/*PHO2* expression module is altered to differing degrees in the shoot and root tissues of each of the three assessed *drb* mutants, RT-qPCR was next applied to profile *PHT1;4*, *PHT1:8* and *PHT1:9* expression in non-stressed and PO$_4$-stressed Col-0, *drb1*, *drb2* and *drb4* plants. RT-qPCR revealed that PO$_4$ starvation promoted *PHT1;4*, *PHT1:8* and *PHT1:9* expression by 9.1-, 39.6- and 4.3-fold in Col-0 shoots (Figure 5E,G,I), and by 1.2-, 2.6- and 1.4-fold in Col-0 roots, respectively (Figure 5F,H,J). In non-stressed *drb1* shoots, the abundance of the *PHT1;4*, *PHT1:8* and *PHT1:9* transcripts were only mildly altered compared to their respective expression levels in P$^+$ Col-0 shoots, returning 1.4-, 1.6- and 2.1-fold changes in expression. A similar mild degree of expression alteration was observed for P$^+$ *drb1* roots. Specifically, compared to P$^+$ Col-0 roots, the *PHT1;4*, *PHT1:8* and *PHT1:9* transcripts returned fold changes in abundance of 0.6, 1.0 and 0.7, respectively. The expression of these three PO$_4$ transporters was significantly induced by the 7-day stress period, returning abundance fold changes of 24.5 (*PHT1;4*), 359.2 (*PHT1:8*) and 242.5 (*PHT1:9*), respectively (Figure 5E,G,I), in P$^-$ *drb1* shoots. In spite of the significant induction of *PHT1* gene expression in P$^-$ *drb1* shoots, *PHT1;4*, *PHT1:8* and *PHT1:9* levels were reduced (0.7-fold), elevated (2.0-fold) and unchanged (1.0-fold), respectively (Figure 5F,H,J) in the root system of PO$_4$-stressed *drb1* roots. As demonstrated for P$^+$ *drb1* shoots, RT-qPCR again revealed that *PHT1;4*, *PHT1:8* and *PHT1:9* expression was mildly altered in P$^+$ *drb2* shoots by 0.8-, 1.0- and 3.4-fold, respectively. In non-stressed *drb2* roots however, the expression of all three PO$_4$ transporters was reduced by 40%, 50% and 60%, respectively, compared to their expression levels in non-stressed Col-0 roots. Furthermore, Figure 5E–J clearly show that the 7-day PO$_4$ starvation period induced the expression of these three PO$_4$ transporter encoding genes in both the P$^-$ *drb2* shoot and root samples, compared to their expression levels in non-stressed *drb2* shoot and roots. Considering that Pi content of non-stressed and PO$_4$-stressed *drb4* shoots and roots was determined to be the same as that of the corresponding tissues in P$^+$ and P$^-$ Col-0 plants, it was unexpected to observe such varied differences in PO$_4$ transporter expression across both assessed tissues/growth conditions. For example, in P$^+$ *drb4* roots, *PHT1;4*, *PHT1;8* and *PHT1;9* levels were each reduced by 60%, compared to P$^+$ Col-0 roots (Figure 5F,H,J), yet the Pi content of non-stressed Col-0 and *drb4* roots was identical (Figure 5B).

3. Discussion

A lack of available P in the soil is a key limitation for plant growth globally [3,45] and as a consequence of P limitation, land plants have evolved highly complex regulatory mechanisms to control both the uptake of external P from the soil, primarily in the form of PO$_4$ (Pi), as well as the remobilization of internal stores of P during periods of low external PO$_4$ availability [46]. These elaborate P responsive mechanisms allow a plant to attempt to (1) maintain growth and development and (2) regulate cellular P content, regardless of external P concentration [1,2,7]. More contemporary research has focused on the regulatory role played by a suite of PO$_4$ responsive miRNA sRNAs that either initiate or maintain PO_4 signaling pathways across the plant kingdom [4,20]. Central to this PO_4 responsive miRNA cohort, is miR399, with the miR399 sRNA required to regulate the abundance of the *PHO2* transcript, to in turn regulate the level of the PHO2 protein, an E2 ubiquitin conjugase that mediates the ubiquitin-directed turnover of a group of PO_4 transporter proteins [7,14,47]. The DRB family members, DRB1, DRB2 and DRB4, have each been ascribed a specific functional role in the *Arabidopsis* miRNA pathway [35–40,48,49]. Therefore, we sought to document the involvement of these three DRBs in the production of the PO_4 responsive miRNA, miR399, and to determine the mode of action directed by the miR399 sRNA during PO_4 starvation to regulate *PHO2* abundance in the *drb1*, *drb2* and *drb4* mutant backgrounds. Specifically, we attempted to determine what effect an altered miR399/*PHO2* expression module profile would have on the response of *drb1*, *drb2* or *drb4* mutant plants to the imposed stress in order to establish the contribution of either DRB1, DRB2 and/or DRB4 to the maintenance of P homeostasis in *Arabidopsis*.

3.1. DRB1 is Required to Maintain Phosphorous Homeostasis in Arabidopsis

Here, it was discovered that the maintenance of P homeostasis is impaired in the *drb1* loss-of-function mutant. The most compelling evidence for this was the documented alteration of the shoot to root Pi content ratio in both non-stressed (Figure 5C) and PO_4-stressed *drb1* plants (Figure 5D), relative to wild-type *Arabidopsis* (P^+ or P^- Col-0 plants). Specifically, the shoot Pi content was reduced to a much greater degree in PO_4-stressed *drb1* plants than the observed reduction to Pi content in P^- Col-0 shoots. Furthermore, Pi was demonstrated to over-accumulate in the roots of both P^+ and P^- *drb1* plants (Figure 5A,B), compared to the Pi content of the corresponding tissue, and growth regime, of Col-0 plants. The maintenance of appropriate P content in plant tissues is essential for the production of macromolecules, energy trafficking and for numerous signaling pathways [1,2,46]. Therefore, alterations to the P content of the shoot and root tissues of *drb1* plants indicated that in the absence of functional DRB1, P partitioning is impaired. Assessment of the expression of PO_4 transporters, *PHT1;4*, *PHT1;8* and *PHT1;9*, revealed that the abundance of each transporter was highly elevated by 24.5- 359.2- and 242.5-fold respectively, in the shoot tissue of P^- *drb1* plants. Phosphate transporter expression was also demonstrated to be altered in both P^+ (*PHT1;4* reduced by 1.7-fold and *PHT1;9* reduced by 1.5-fold) and P^- (*PHT1;4* reduced by 1.5-fold and *PHT1;8* elevated by 2.0-fold) *drb1* roots, expression alterations that when taken together indicated that incorrect Pi partitioning in *drb1* plants potentially results from defective PO_4 transport from the root system to the aerial tissue in this mutant background.

Defective root to shoot PO_4 transport in the *drb1* mutant was further evidenced by the unique phenotypic response displayed by the *drb1* shoot to PO_4 stress. Specifically, the fresh weight of the shoot of 15-day old P^- *drb1* plants was only reduced by 21.6% compared to its non-stressed counterpart (Figure 1C). The rosette area of P^- *drb1* plants was also demonstrated to only be reduced by 29.3% post the 7-day PO_4 stress treatment (Figure 1D). Both responses were comparatively mild compared to the 36.6% and 60.1% reductions to fresh weight and rosette area respectively, documented for Col-0 shoots post the application of PO_4 stress. In addition, anthocyanin failed to change in abundance in the shoot tissues of P^- *drb1* plants compared to the shoots of non-stressed P^+ *drb1* plants (Figure 1E). Anthocyanin production is a general response to a range of abiotic stresses, including PO_4 starvation [19,50]. The impaired ability of *drb1* shoots to produce anthocyanin in response to PO_4 stress may implicate DRB1, and the functional partnership DRB1 forms with DCL1, in the induction of PO_4 responsive gene expression pathways. Considering these mild responses displayed by *drb1* shoots, it was therefore surprising to observe that chlorophyll *a* and *b* overaccumulation was promoted to the greatest extent in the aerial tissues of *drb1* plants starved of PO_4. Altered chlorophyll content in P^+ *drb1* shoots indicated that (1) *drb1* shoots are indeed negatively impacted by the imposed PO_4 stress, and (2) that DRB1 may potentially mediate a PO_4-directed role in regulating photosynthesis in *Arabidopsis* chloroplasts.

Considering the well-established role of the DRB1/DCL1 functional partnership in the production of the majority of miRNAs that accumulate in *Arabidopsis* tissues, it was unsurprising to observe that the miR399 precursors, *PRE-MIR399A*, *PRE-MIR399C*, *PRE-MIR399D*, *PRE-MIR399E* and *PRE-MIR399F*, over-accumulated to the greatest extent in P$^+$ *drb1* shoots (Figure 2A–E). In addition, precursors *PRE-MIR399A*, *PRE-MIR399C*, *PRE-MIR399D* and *PRE-MIR399F* were further demonstrated to be most highly abundant in the shoot tissues of PO$_4$-stressed *drb1* plants. The enhanced abundance of miRNA precursor transcripts in the *drb1* mutant background is most likely the result of inefficient precursor transcript processing by DCL1 in the absence of DRB1 functional assistance, with DRB1 accurately positioning DCL1 on each miRNA precursor to direct accurate processing [48,49]. In spite of the readily observable evidence of inefficient miR399 precursor transcript processing in P$^+$ *drb1* shoots, miR399 levels were only reduced by 10% (Figure 5G). Similarly, although miR399 precursor transcript abundance was elevated to a much greater degree in P$^-$ *drb1* shoots due to a combination of (1) *MIR399* gene expression induction in response to PO$_4$ starvation, and (2) inefficient precursor transcript processing in the absence of DRB1 activity, miR399 abundance was again demonstrated to be only mildly elevated by 2.3-fold in the shoots of PO$_4$-stressed *drb1* plants (Figure 5G). Further, the abundance of the miR399 target transcript, *PHO2*, was only mildly elevated by 1.2-fold in response to the 10% reduction in miR399 levels in P$^+$ *drb1* shoots (Figure 2I). Surprisingly, *PHO2* transcript abundance was elevated by 1.5-fold in response to the 2.3-fold elevation in miR399 accumulation in P$^-$ *drb1* shoots, and not reduced as expected. However, in P$^+$ Col-0 shoots, and as expected, the 2.9-fold enhancement to miR399 abundance led to a 50% reduction in *PHO2* expression (Figure 5G,I). Therefore, elevated *PHO2* abundance in response to enhanced miR399 levels in P$^-$ *drb1* shoots, readily demonstrates that miR399-directed *PHO2* transcript cleavage, to regulate *PHO2* expression, is defective in the absence of DRB1 activity.

Altered PO$_4$ transporter expression in *drb1* roots indicated that the response of the root system of the *drb1* mutant to PO$_4$ stress would differ to that of the root system of wild-type *Arabidopsis*. Accordingly, the fresh weight of PO$_4$-stressed *drb1* roots was reduced by 25.0% compared to the mild 6.5% reduction to the fresh weight of P$^-$ Col-0 roots, a 3.8-fold enhancement to the severity of this phenotypic response (Figure 3C). It was therefore curious to observe a similar degree of reduction to primary root length in P$^-$ *drb1* (46.7%) and P$^-$ Col-0 (51.2%) plants (Figure 3D). A greater degree of reduction to the fresh weight of P$^-$ *drb1* roots, compared to P$^-$ Col-0 roots, could be partially explained by the observation that the induction of lateral root formation by PO$_4$ stress was completely defective in P$^-$ *drb1* roots, compared to a 44.0% increase in lateral root number in P$^-$ Col-0 roots (Figure 3D). Considering that the measurement of fresh weight is largely assessing the moisture content of a plant, the observed reduction to fresh weight of P$^-$ *drb1* roots could potentially be indicating that under PO$_4$ stress conditions, DRB1 is somehow involved in regulating the moisture content of the root system of *Arabidopsis*. However, this was not assessed in this study with the mechanism driving the enhancement of fresh weight reductions requiring further investigation in the future.

Similar to its establishment as the primary DRB protein required to regulate miR399 production from the *PRE-MIR399A*, *PRE-MIR399C*, *PRE-MIR399D*, *PRE-MIR399E* and *PRE-MIR399F* precursors in the aerial tissues of non-stressed *Arabidopsis* plants, DRB1 was again demonstrated to be the primary DRB protein required to regulate miR399 production from the *PRE-MIR399A* and *PRE-MIR399C* precursor transcripts in the *Arabidopsis* root system with both precursors demonstrated to accumulate to the greatest degree in P$^+$ and P$^-$ *drb1* roots (Figure 4B,C). Reduced *PRE-MIR399A* and *PRE-MIR399C* processing efficiency in the absence of DRB1 activity, reduced miR399 abundance by 30% in P$^+$ *drb1* roots (Figure 4E), and in turn, this moderate reduction to miR399 levels led to a mild elevation (1.2-fold) in the expression of the miR399 target gene, *PHO2* (Figure 4G). As documented in P$^-$ *drb1* shoots, the 1.8-fold elevation to miR399 levels in P$^-$ *drb1* roots, resulted in a moderate elevation to *PHO2* transcript abundance (1.6-fold), and not a reduction in target gene expression as would be expected for a miRNA that regulates the expression of its targeted genes solely via a mRNA cleavage mode of RNA silencing. However, considering that a similar miRNA/target gene expression profile of elevated miR399 abundance (2.5-fold), together with enhanced *PHO2* expression (1.9-fold) was also observed in PO$_4$-stressed Col-0 roots, this curious finding indicates that miR399-directed *PHO2* transcript cleavage may not be the predominant mechanism of target gene expression regulation directed by the miR399 sRNA in the *Arabidopsis* root system. Alternatively, elevated *PHO2* expression in P$^+$ Col-0 and P$^+$ *drb1* roots when miR399 abundance is also elevated may result from the enhanced expression of the eTM of miR399 activity, *IPS1*. In P$^-$ Col-0 shoots for example, where elevated miR399 abundance was demonstrated to direct enhanced expression repression of the *PHO2* transcript (Figure 2G,I), *IPS1* abundance was elevated by 75.7-fold, compared to its abundance in P$^+$ Col-0 shoots (Figure 2H). In PO$_4$-stressed roots, however, *IPS1* expression was elevated to a much greater degree, by 1076-fold (Figure 4F). This 14.2-fold greater promotion to *IPS1* expression in P$^-$ Col-0 roots, than that observed in P$^-$ Col-0 shoots, would be expected to sequester a higher amount of miR399, which in turn, could have led to the observed elevation in *PHO2* expression in P$^-$ Col-0 roots in the presence of 2.5-fold greater abundance of the *PHO2* targeting miRNA, miR399.

3.2. DRB2 is Required to Maintain Phosphate Homeostasis in Arabidopsis

As documented for the *drb1* mutant, P homeostasis was determined to be defective in the *drb2* mutant. Specific to *drb2* plants however, was the 27.8% reduction to the Pi content of non-stressed *drb2* shoots (Figure 5A). Of the four *Arabidopsis* plant lines assessed in this study, *drb2* was the only line determined to have a reduced aerial tissue Pi content when cultivated under standard growth conditions. Furthermore, in P$^+$ *drb2* shoots, *PHT1;4* (Figure 5E) and *PHT1;8* (Figure 5G) expression was determined to be reduced and elevated by 1.2- and 3.4-fold respectively, compared to the expression of these two PO$_4$ transporters in P$^+$ Col-0 shoots. In addition, Pi was determined to over-accumulate by 36.1% in P$^+$ *drb2* roots. In P$^+$ *drb2* roots, *PHT1;4*, *PHT1;8* and *PHT1;9* expression was reduced by 1.7-, 2.0- and 2.4-fold respectively, compared to their expression levels in P$^+$ Col-0 roots. Together, (1) reduced Pi content of P$^+$ *drb2* shoots, (2) elevated Pi content in P$^+$ *drb2* roots, and (3) reduced PO$_4$ transporter gene expression in P$^+$ *drb2* roots, indicated that PO$_4$ root to shoot transport is defective in non-stressed *drb2* plants. Based on this finding, it was curious to observe a similar Pi content in P$^-$ *drb2* shoots and P$^-$ Col-0 shoots (Figure 5A), especially considering the document enhancement to *PHT1;4* and *PHT1;9* expression in P$^-$ *drb2* shoots, with the expression of these two PO$_4$ transporters elevated by 2.8- and 7.0-fold respectively, compared to the degree of expression induction observed in P$^-$ Col-0 roots (Figure 5E,I). However, and as demonstrated for P$^+$ *drb2* shoots and roots, the Pi content of the root system of PO$_4$-stressed *drb2* plants was altered, reduced by 48% compared to the Pi content of P$^-$ Col-0 roots. Interestingly, RT-qPCR revealed similar degrees of elevated *PHT1;8* (Figure 5H) and *PHT1;9* (Figure 5J) expression in PO$_4$-stressed Col-0 and *drb2* roots with only the *PHT1;4* transcript returning a slight difference in its expression in P$^-$ Col-0 roots (elevated by 1.2-fold compared to P$^+$ Col-0 roots) and P$^-$ *drb2* roots (reduced by 1.1-fold compared to P$^+$ Col-0 roots). The PO$_4$ transporters, PHT1;1 and PHT1;4, have been demonstrated to be responsible for the import of more than half of the Pi that is taken up from the soil [51]. It therefore seems unlikely that the mild 10% reduction to *PHT1;4* transcript abundance documented in PO$_4$-stressed *drb2* roots, is the sole cause of the considerable reduction to the Pi content of the root system in the *drb2* mutant background.

Considering that the Pi content of PO$_4$-stressed Col-0 and *drb2* shoots was determined to be similar, it was unsurprising to document a similar degree of reduction to fresh weight of the shoot tissues of P$^-$ Col-0 (36.6%) and P$^-$ *drb2* (39.1%) plants (Figure 1C). Rosette area was also decreased by a similar degree in P$^-$ Col-0 (60.1%) and P$^-$ *drb2* (48.0%) plants (Figure 1D). However, compared to PO$_4$-stressed Col-0 shoots, anthocyanin accumulated to considerably higher levels in the aerial tissues of *drb2* plants when exposed to PO$_4$ stress (Figure 1E). The induction of anthocyanin production is a well-characterized response to PO$_4$ starvation [19,50]. Therefore, the considerable enhancement of anthocyanin accumulation in P$^-$ *drb2* shoots, compared to the shoot tissues of PO$_4$-stressed Col-0 plants, suggests that this P-responsive pathway is hyperactivated in the absence of DRB2 activity, as well as potentially implicating DRB2 in mediating a regulatory role in a range of other P-responsive pathways in *Arabidopsis* aerial tissues that were not assessed in this study.

We have previously demonstrated a role for DRB2 in the production stage of the *Arabidopsis* miRNA pathway with the abundance of specific miRNA cohorts altered in the *drb2* mutant background [37]. More specifically, DRB2 can either be antagonistic or synergistic to DRB1 function in the DRB1/DCL1 partnership for the production of specific miRNAs, resulting in miRNA abundance either being enhanced (antagonistic) or reduced (synergistic) in *drb2* plants [37,38]. Reduced precursor transcript abundance in non-stressed *drb2* shoots, indicated that DRB2 plays a secondary role in regulating miR399 production from the *PRE-MIR399A*, *PRE-MIR399E* and *PRE-MIR399F* precursors, potentially via antagonism of DRB1 function (Figure 2B,E,F). The antagonism of DRB2 on the DRB1/DCL1 partnership becomes more readily apparent via the profiling of miR399 precursor transcript expression in P⁻ *drb2* shoots, with lower degrees of expression induction observed for the *PRE-MIR399C*, *PRE-MIR399D*, *PRE-MIR399E* and *PRE-MIR399F* precursors (Figure 2C–F). Reduced precursor transcript abundance in P⁻ *drb2* shoots, compared to the respective abundance of each precursor in either P⁻ Col-0 or P⁻ *drb1* shoots, indicates that in the absence of DRB2 activity, precursor transcript processing efficiency is enhanced due to more precursor transcript being freely available to enter the canonical DRB1/DCL1 production pathway.

As demonstrated in P⁺ *drb1* shoots, significantly altered precursor transcript abundance in P⁺ *drb2* shoots, failed to have a strong influence on the accumulation of miR399, with miR399 levels only mildly elevated by 10% in P⁺ *drb2* shoots, compared to P⁺ Col-0 shoots (Figure 5G). However, DRB2 antagonism was still evidenced by this mild increase to miR399 abundance compared to the 10% reduction in miR399 levels observed in P⁺ *drb1* shoots. The antagonism of DRB2 on miR399 production was further evidenced by the enhanced expression repression of *PHO2* in P⁻ *drb2* shoots (Figure 2I). The abundance of miR399 was elevated by 2.7-fold in P⁻ *drb2* shoots, and therefore, a further degree of reduced *PHO2* expression in P⁻ *drb2* shots, compared to P⁻ *drb1* shoots where miR399 levels were elevated by 2.3-fold and *PHO2* expression was enhanced by 1.5-fold, clearly demonstrated enhanced DRB1-mediated, miR399-directed, *PHO2* transcript cleavage in the absence of DRB2 antagonism. Similarly, it is important to note here that *IPS1* transcript abundance was enhanced to a much lower degree in P⁻ *drb2* shoots (27.1-fold) compared to *IPS1* abundance induction in either PO₄-stressed Col-0 (75.7-fold) or *drb1* (85.4-fold) shoots. This unexpected observation again indicated that in the absence of DRB2 activity, miR399-directed target transcript cleavage was enhanced. Although *IPS1* has been identified as a non-cleavable eTM of miR399 activity, the *IPS1* expression trends presented in Figure 5H suggest that miR399 may well be capable of directing miRISC-catalyzed cleavage of the *IPS1* transcript in addition to solely being sequestered by *IPS1*.

Compared to the mild 6.5% reduction to the fresh weight of P⁻ Col-0 roots, the negative response of the root system of the *drb2* mutant to PO₄ stress was considerably more pronounced at 25.8% (Figure 3C). Considering that the correct regulation of Pi content is dysfunctional in both control and PO₄-stressed *drb2* roots, differing responses to PO₄ stress in *drb2* roots, compared to P⁻ Col-0 roots, was not surprising. Similarly, inhibition of the primary root length of P⁻ *drb2* plants at 20.3% was comparatively mild compared to the severe 51.2% inhibition to the primary root length observed for P⁻ Col-0 plants (Figure 5D). The degree of lateral root induction also differed between PO₄-stressed Col-0 and *drb2* roots (Figure 5E), specifically; lateral root formation was enhanced by ~44% in P⁻ Col-0 plants, and in PO₄-stressed *drb2* plants, lateral root formation was further promoted by 17% with P⁻ *drb2* plants developing ~61% more lateral roots than their non-stressed counterparts. When these phenotypic responses of the root system of PO₄-stressed *drb2* plants are considered together, including a lower degree of primary root length inhibition (2.5-fold less than P⁻ Col-0 plants), and a more pronounced enhancement to lateral root formation (1.4-fold more than P⁻ Col-0 plants), it was highly surprising that the fresh weight of P⁻ *drb2* roots was reduced by a 4.0-fold greater degree than documented for P⁻ Col-0 roots.

Similar levels of expression of *PRE-MIR399A* in both non-stressed and PO$_4$-stressed Col-0 and *drb2* roots revealed that DRB2 does not play a role in regulating miR399 processing from this precursor transcript (Figure 4B). Reduced expression of *PRE-MIR399C* in P$^+$ *drb2* roots (compared to P$^-$ Col-0 and P$^-$ *drb1* roots) and a lower level of precursor over-accumulation in P$^-$ *drb2* roots (compared to P$^-$ *drb1* roots), identified DRB2 as playing a secondary role in regulating miR399 production from this precursor transcript in the *Arabidopsis* root system (Figure 4C) via antagonism of DRB1 function. The expression trend of *PRE-MIR399D* in P$^-$ *drb2* roots additionally identified a secondary role for DRB2 in regulating miR399 production from the third miR399 precursor transcript detected in the root system of the four *Arabidopsis* plants lines assessed in this study. However, for the *PRE-MIR399D* precursor, DRB2 appears to be antagonistic to the DRB4/DCL4 partnership, and not to the canonical DRB1/DCL1 partnership demonstrated to be required for the production of the majority of *Arabidopsis* miRNAs. DRB2 has been demonstrated previously to be antagonistic to DRB4 function in the DRB4/DCL4 partnership for the production of a small subset of newly evolved *Arabidopsis* miRNAs processed from precursor transcripts that fold to form highly complementary stem-loop structures [39,40]. Considering that in P$^+$ *drb2* roots, *PRE-MIR399A* and *PRE-MIR399D* remained at their approximate wild-type levels, and that the *PRE-MIR399C* precursor was reduced in its abundance by 1.7-fold, a finding that initially indicated that this precursor is more efficiently processed by DRB1/DCL1 in the absence of DRB2 activity, the 2.0-fold reduction to miR399 abundance alternatively indicated that *MIR399C* gene expression may in fact be reduced in PO$_4$-stressed *drb2* roots. It was therefore curious to observe *PHO2* expression to be reduced by 1.3-fold in P$^+$ *drb2* roots, and not elevated in response to reduced miR399 abundance as expected. However, this observation is potentially demonstrating that in spite of being reduced in abundance, this lower level of miR399 directs more efficient cleavage of the *PHO2* transcript in the absence of DRB2 activity. In P$^-$ *drb2* roots, miR399 abundance was determined to be elevated by 2.6-fold compared to its abundance in P$^-$ Col-0 roots (Figure 4E). As observed in P$^+$ *drb2* roots, *PHO2* expression scaled in accordance with elevated miR399 abundance, with *PHO2* expression increased by 4.5-fold in PO$_4$-stressed *drb2* roots. It is interesting to note here that *PHO2* expression scaled with miR399 abundance in six out the eight root tissue samples molecularly assessed by RT-qPCR in this study. We have previously demonstrated that DRB2-dependent miRNAs direct a translational repression mode of miRNA-directed target gene expression repression [52], and scaling of miRNA target transcripts together with their targeting miRNA, has been previously reported for miRNA sRNAs that direct a translational repression mode of target gene expression regulation [52–54].

3.3. DRB4 is Required For miR399 Production in Arabidopsis Roots

Profiling of PO$_4$ transporter expression in the shoots and roots of P$^+$ and P$^-$ *drb4* plants revealed considerable alteration to *PHT1;4*, *PHT1;8* and *PHT1;9* transcript abundance across both assessed tissues and growth regimes (Figure 5E–J). However, in spite of these documented differences in PO$_4$ transporter gene expression in *drb4* shoots and roots, the Pi content of non-stressed and PO$_4$-stressed *drb4* tissues remained at levels comparable to P$^+$ and P$^-$ Col-0 shoots and roots (Figure 5A,B). Considering this finding, it was unsurprising that the developmental progression of Col-0 and *drb4* plants was impeded to the same extent when cultivated in the absence of PO$_4$ for a 7-day period. Specifically, the fresh weight of both P$^-$ Col-0 and P$^-$ *drb4* shoots was reduced by ~36% compared to their non-stressed counterparts of the same age (Figure 1C). In addition, anthocyanin, chlorophyll *a* and chlorophyll *b* were all elevated to the same degree in PO$_4$-stressed Col-0 and *drb4* shoots, compared to their respective non-stressed counterparts. It was therefore surprising that the rosette area of P$^-$ *drb4* plants was only reduced by 38.7% compared to the more severe 60.1% reduction observed for P$^-$ Col-0 plants. Although an unexpected finding, this result clearly indicated that some of the responses of the *drb4* mutant to PO$_4$ starvation differ to those of wild-type *Arabidopsis*.

Considering the well-established role of the DRB4/DCL4 partnership in *trans*-acting siRNA (tasiRNA) [55,56] and p4-siRNA [40] production, and for the processing of a small number of newly evolved miRNAs from their highly complementary precursor transcripts [39], it was highly surprising to additionally establish the widespread involvement of DRB4 in regulating the production of the highly conserved miRNA, miR399, in *Arabidopsis* shoots (Figure 2). Specifically, DRB4 was determined to play a secondary role to DRB1 in regulating the efficiency of miR399 production from all five precursors detectable by RT-qPCR in non-stressed *Arabidopsis* shoots. As demonstrated for DRB2, the involvement of DRB4 in miR399 production in *Arabidopsis* shoots is most likely via antagonism of the canonical DRB1/DCL1 partnership. Antagonism of the DRB1/DCL1 partnership by DRB4 was again demonstrated by the accumulation profiles of precursors, *PRE-MIR399C*, *PRE-MIR399D*, *PRE-MIR399E* and *PRE-MIR399F*, in the shoot tissues of PO_4-stressed *drb4* plants (Figure 2C–F). Although precursor transcript abundance was highly variable in *drb4* shoots, miR399 levels were only mildly elevated by 1.2- and 2.4-fold in P^+ *drb4* and P^- *drb4* shoots, respectively (Figure 2G). Surprisingly, in spite of the 20% elevation to miR399 levels in P^+ *drb4* shoots, *PHO2* expression was elevated to a similar degree (30% increase), and not reduced as expected (Figure 5I). In P^- *drb4* shoots, however, the 2.4-fold elevated abundance of the miR399 sRNA was determined, as expected, to reduce the expression of *PHO2* by 2.5-fold. This result clearly indicated that in the absence of DRB4 activity in *Arabidopsis* shoots, the efficiency of DRB1-mediated, miR399-directed cleavage of the *PHO2* transcript is enhanced.

The fresh weight of P^- *drb4* roots was reduced by 18.6% compared to the fresh weight of P^+ *drb4* roots, a 2.9-fold further enhancement of this phenotypic response to PO_4 stress, compared to the mild response of P^- Col-0 roots (6.5% fresh weight reduction compared to P^+ Col-0 roots). The response of the primary root of the *drb4* mutant to PO_4 stress also differed to that of wild-type roots. Namely, the length of P^- *drb4* primary root was only reduced by 10.3% compared to the significant 51.2% reduction to the length of the primary root of P^- Col-0 plants (Figure 3D). Although lateral root development was induced to the same degree (44%) in the root system of PO_4-stressed Col-0 and *drb4* plants, the considerable differences observed for the fresh weight of the *drb4* root system, and the lack of inhibition to primary root length in P^- *drb4* plants, clearly revealed that the *drb4* mutant background is defective in some of its responses to PO_4 starvation, compared to the responses of the Col-0 root system to this stress.

At the molecular level, the wild-type-like expression of the *PRE-MIR399A* precursor in the roots of non-stressed and PO_4-stressed *drb4* plants indicated that DRB4 does not play a role in regulating miR399 production from this precursor in *Arabidopsis* roots. Expression analysis of *PRE-MIR399C* did however identify a secondary role for DRB4 in regulating miR399 production from this precursor, potentially via antagonism of DRB1 function (Figure 4C). Of particular interest stemming from miR399 precursor transcript profiling in non-stressed and PO_4-stressed *Arabidopsis* roots is the unexpected finding that DRB4 appears to be the primary DRB required to regulate miR399 production from the *PRE-MIR399D* precursor (Figure 4D), with the abundance of the *PRE-MIR399D* precursor over-accumulating to its highest levels in both P^+ and P^- *drb4* roots. Curiously, assessment of the stem-loop folding structures of the six precursors from which the miR399 sRNA is liberated does not readily distinguish the *PRE-MIR399D* structure from the folding structures of the other five miR399 precursor transcripts. Therefore, the establishment of a role for DRB4 in regulating miR399 processing efficiency from its precursor transcripts was a highly unexpected finding, a finding that requires additional experimentation in the future to identify the precursor transcript-based sequence and/or structural features that recruits the involvement of DRB4 to the miR399/*PHO2* expression module.

The elevated abundance of the *PRE-MIR399C* and *PRE-MIR399D* precursors in P^+ *drb4* roots indicated reduced precursor transcript processing efficiency in the absence of DRB4. Accordingly, a 30% reduction to miR399 accumulation was observed in P^+ *drb4* roots (Figure 4E). Surprisingly, this 1.4-fold reduction to miR399 levels in P^+ *drb4* roots led to a 2.0-fold reduction to *PHO2* expression (Figure 4G). This result suggested that although miR399 levels were reduced in non-stressed *drb4* roots, the reduced amount of the miR399 sRNA was actually directing enhanced *PHO2* expression repression via unimpeded DRB1-mediated, miR399-directed, *PHO2* cleavage. However, enhanced miR399-directed *PHO2* cleavage appeared to be lost in PO_4-stressed *drb4* roots with both miR399 and *PHO2* levels elevated by 2.0- and 5.1-fold, respectively (Figure 4E,G). Therefore, when taken together, although miR399-directed *PHO2* cleavage appeared to be enhanced in P^+ *drb4* roots, the scaling of *PHO2* expression together with miR399 abundance in PO_4-stressed *drb4* roots, potentially suggests that in a cell type with altered physiology, and where DRB4 function is defective, the miR399 sRNA changed from directing an mRNA cleavage mode of RNA silencing, to directing a translational repression mode of RNA silencing.

3.4. DRB1, DRB2 and DRB4 Are Required to Regulate the miR399/PHO2 Expression Module in Arabidopsis Shoots and Roots

Here we demonstrate that the phenotypic and molecular response to PO_4 starvation were unique to each *drb* mutant background assessed due to the hierarchical contribution of DRB1, DRB2 and DRB4 to the regulation of the miR399/*PHO2* expression module. Specifically, the molecular profiling of miR399 precursor transcript expression identified DRB1 as the primary DRB required for efficient miR399 production from each precursor in non-stressed and PO_4-stressed shoots and roots. Deregulated miR399 precursor transcript processing efficiency in the absence of DRB1 activity was demonstrated to result in defective P homeostasis maintenance, altering the shoot to root ratio of Pi content in the *drb1* mutant background. The maintenance of P homeostasis was also defective in *drb2* plants, with the Pi content shoot to root ratio altered in this mutant background, both under standard growth conditions and in conditions of PO_4 starvation. An altered Pi content in *drb2* tissues appeared to result from defective PO_4 transport between the root system and aerial tissues in the absence of DRB2 function. Further, DRB2 was determined to play a secondary role to DRB1 in regulating miR399 production from the profiled *PRE-MIR399* precursor transcripts. The secondary role of DRB2 in regulating miR399 production from the assessed *PRE-MIR399* precursor transcripts was revealed to most likely be via antagonism of DRB1 function. DRB4 was also determined to play a secondary role in regulating the miR399/*PHO2* expression module in *Arabidopsis* shoots and roots, and as demonstrated for the secondary role of DRB2 in providing additional regulatory complexity to this expression module, DRB4 also appeared to be antagonistic to the primary functional role of DRB1 in regulating miR399 precursor transcript processing efficiency. Furthermore, DRB4 also appeared to be the primary DRB required for miR399 production from the *PRE-MIR399D* precursor in non-stressed and PO_4-stressed *Arabidopsis* roots. When taken together, the hierarchical contribution of DRB1, DRB2 and DRB4 to the regulation of the miR399/*PHO2* expression module documented here, readily demonstrates the crucial importance of maintaining P homeostasis in *Arabidopsis* tissues to ensure the maintenance of a wide range of cellular processes to which P is essential.

4. Materials and Methods

4.1. Plant Material and Phosphate Stress Treatment

The T-DNA insertion knockout mutant lines used in this study, including the *drb1* (*drb1-1*; SALK_064863), *drb2* (*drb2-1*; GABI_348A09) and *drb4* (*drb4-1*; SALK_000736) mutants, have been described previously [42]. The seeds of these three *drb* mutant lines, and of wild-type *Arabidopsis* (ecotype Columbia-0 (Col-0)) plants, were sterilized using chlorine gas and post-sterilization, seeds were plated out onto standard *Arabidopsis* plant growth media (half-strength Murashige and Skoog (MS) salts), and stratified for 48 h at 4 °C in the dark. Post-stratification, the sealed plates were transferred to a temperature-controlled growth cabinet (A1000 Growth Chamber, Conviron® Australia) and cultivated for an 8-day period under a standard growth regime of 16 h light / 8 h dark, and a day/night temperature of 22 °C / 18 °C. Post this initial 8-day cultivation period, equal numbers of Col-0, *drb1*, *drb2* and *drb4* seedlings were transferred under sterile conditions to either fresh standard *Arabidopsis* plant growth media that contained 1.0 mM of PO_4 (P^+ plants; non-stressed controls) or to *Arabidopsis* plant growth media where the PO_4 had been replaced with an equivalent molar amount (1.0 mM) of potassium chloride (KCl) (P^- plants; PO_4 stress treatment). Post seedling transfer, the P^+ and P^- plates for each plant line were returned to the temperature-controlled growth cabinet for an additional 7-day cultivation period. For the tissue-specific analyses performed here, namely the root tissue assessments, additional Col-0, *drb1*, *drb2* and *drb4*, 8-day old seedlings were treated exactly as outlined above, except for the 7-day treatment period, where P$^+$ and P$^-$ plates were orientated for vertical growth. Unless stated otherwise, all the phenotypic and molecular analyses reported here were conducted on 15-day old plants.

4.2. Phenotypic and Physiological Assessments

The fresh weight of 8-day old Col-0, *drb1*, *drb2* and *drb4* whole plants germinated and cultivated on standard *Arabidopsis* plant growth media was initially determined to establish the effect of loss of DRB1, DRB2 or DRB4 activity on *Arabidopsis* development. The fresh weight of 15-day old Col-0, *drb1*, *drb2* and *drb4* plants was also determined to establish the effect of the 7-day PO$_4$ stress treatment on the development of each plant line. The area of the rosette and the length of the primary root of 15-day old Col-0, *drb1*, *drb2* and *drb4* plants was determined via the assessment of photographic images using the ImageJ software. The same photographic images were also used to establish the number of lateral roots formed by P$^+$ and P$^-$ Col-0, *drb1*, *drb2* and *drb4* plants post the 7-day stress treatment period.

A standard methanol:HCl (99:1 v/v) extraction method was applied to extract anthocyanin from P$^+$ and P$^-$ plants, and post extraction, anthocyanin content was determined using a spectrophotometer (Thermo Scientific, Australia) at an absorbance wavelength of 535 nanometers (A$_{535}$). The 99:1 (v/v) methanol:HCl extraction buffer was used as the blanking solution and the A$_{535}$ of each sample was next divided by the fresh weight of the sample to calculate the relative anthocyanin content per gram of fresh weight (A$_{535}$/g FW).

For chlorophyll *a* and *b* content quantification, rosette leaves of 15-day old P$^+$ and P$^-$ Col-0, *drb1*, *drb2* and *drb4* plants were sampled and incubated in 80% acetone for 24 h in the dark. Post incubation, rosette leaf tissue was clarified via centrifugation at $15,000 \times g$ for 7 min at room temperature. The resulting supernatants were immediately transferred to a spectrophotometer and the absorbance of these solutions assessed at wavelengths 646 nm (A$_{646}$) and 663 nm (A$_{663}$) using 80% acetone as the blanking solution. The chlorophyll *a* and *b* content of each sample was then determined using the Lichtenthaler's equations exactly as outlined in [57], and these initially determined values were subsequently converted to micrograms per gram of fresh weight (μg/g FW).

The shoot and root tissue of 15-day old P$^+$ and P$^-$ Col-0, *drb1*, *drb2* and *drb4* plants were carefully separated from each other and then ground into a fine powder under liquid nitrogen (LN$_2$). One milliliter (1.0 mL) of 10% acetic acid (v/v in H$_2$O) was added to the ground powder and the powder thoroughly resuspended via vigorous vortexing. The resulting resuspension was then centrifuged at $15,000 \times g$ for 5 min at room temperature, and post centrifugation, 700 μL of the resulting supernatant was mixed with an equivalent volume of Ames Assay Buffer (6 parts 0.5% ammonium molybdite (v/v in H$_2$O) to 1 part of 2.5% sulphuric acid (v/v in 10% acetic acid)) and incubated at room temperature for 1 h in the dark. The absorbance of each solution was determined using a spectrophotometer at wavelength 820 nm (A$_{820}$) and the Pi content (μmol/gFW) of each sample subsequently determined via the construction of a Pi standard curve.

4.3. Total RNA Extraction for Quantitative Molecular Assessments

For each molecular assessment reported here, total RNA was extracted from four biological replicates (each biological replicate contained tissue sampled from eight individual plants) of 15-day old P$^+$ and P$^-$ Col-0, *drb1*, *drb2* and *drb4* plants using TRIzol^{TM} Reagent according to the manufacturer's (Invitrogen^{TM}) instructions. The quality of the extracted total RNA was visually assessed via a standard electrophoresis approach on a 1.2% (w/v) ethidium bromide stained agarose gel and the quantity of total RNA extracted determined using a NanoDrop spectrophotometer (NanoDrop® ND-1000, Thermo Scientific, Australia).

For the synthesis of a miR399-specific complementary DNA (cDNA), 200 nanograms (ng) of total RNA was treated with 0.2 units (U) of DNase I (New England Biolabs, Australia) according to the manufacturer's instructions. The DNase I-treated total RNA was next used as template for cDNA synthesis with 1.0 U of ProtoScript® II Reverse Transcriptase (New England Biolabs, Australia) and the cycling conditions of 1 cycle of 16 °C for 30 min; 60 cycles of 30 °C for 30 s, 42 °C for 30 s, and 50 °C for 2 s, and; 1 cycle of 85 °C for 5 min.

A global, high molecular weight cDNA library for gene expression quantification was constructed via the initial treatment of 5.0 μg of total RNA with 5.0 U of DNase I according to the manufacturer's protocol (New England Biolabs, Australia). The DNase I-treated total RNA was next purified using an RNeasy Mini Kit (Qiagen, Australia) and 1.0 μg of this preparation used as template for cDNA synthesis along with 1.0 U of ProtoScript® II Reverse Transcriptase (New England Biolabs, Australia) and 2.5 mM of oligo $dT_{(18)}$, according to the manufacturer's instructions.

All generated, single-stranded cDNAs were next diluted to a working concentration of 50 ng/μL in RNase-free H_2O prior to their use as a template for the quantification of the abundance of either the miR399 sRNA or of gene transcripts. In addition, all RT-qPCRs used the same cycling conditions of 1 cycle of 95 °C for 10 min, followed by 45 cycles of 95 °C for 10 s and 60 °C for 15 s, and the GoTaq® qPCR Master Mix (Promega, Australia) was used as the fluorescent reagent for all performed RT-qPCR experiments. miR399 abundance and gene transcript expression was quantified using the $2^{-\Delta\Delta CT}$ method with the small nucleolar RNA, *snoR101*, and *UBIQUITIN10* (*UBI10*; *AT4G05320*) used as the respective internal controls to normalize the relative abundance of each assessed transcript. The sequence of each DNA oligonucleotide used in this study either for the synthesis of a miR399-specific cDNA, or to quantify transcript abundance via RT-qPCR is provided in Supplemental Table S1.

Author Contributions: J.L.P., C.P.G. and A.L.E. conceived and designed the research. J.L.P. and J.M.O. performed the experiments and analyzed the data. J.L.P., C.P.G., J.M.O. and A.L.E. authored the manuscript and, J.L.P., C.P.G., J.M.O. and A.L.E. have read and approved the final version of the manuscript.

Acknowledgments: The authors would like to thank fellow members of the Centre for Plant Science at the University of Newcastle for their guidance with plant growth care and RT-qPCR experiments.

References

1.   Abel, S.; Ticconi, C.A.; Delatorre, C.A. Phosphate sensing in higher plants. *Planta* **2002**, *115*, 1–8. [CrossRef]

2.   Hammond, J.P.; Bennett, M.J.; Bowen, H.C.; Broadley, M.R.; Eastwood, D.C.; May, S.T.; Rahn, C.; Swarup, R.; Woolaway, K.E.; White, P.J. Changes in Gene Expression in *Arabidopsis* Shoot during Phosphate Starvation and the Potential for Developing Smart Plants. *Society* **2003**, *132*, 578–596. [CrossRef] [PubMed]

3.   Raghothama, K.G. Phosphate transport and signaling. *Curr. Opin. Plant Biol.* **2000**, *3*, 182–187. [CrossRef]

4.   Hackenberg, M.; Shi, B.J.; Gustafson, P.; Langridge, P. Characterization of phosphorus-regulated miR399 and miR827 and their isomirs in barley under phosphorus- sufficient and phosphorus-deficient conditions. *BMC Plant Biol.* **2013**, *13*, 214. [CrossRef]

5.   Chiou, T.J.; Lin, S.I. Signaling network in sensing phosphate availability in plants. *Annu. Rev. Plant Biol.* **2011**, *62*, 185–206. [CrossRef] [PubMed]

6.   Wu, P.; Ma, L.; Hou, X.; Wang, M.; Wu, Y.; Liu, F.; Deng, X.W. Phosphate Starvation Triggers Distinct Alterations of Genome Expression in *Arabidopsis* Roots and Leaves. *Plant Physiol.* **2003**, *132*, 1260–1271. [CrossRef] [PubMed]

7.   Lin, S.I.; Chiang, S.F.; Lin, W.Y.; Chen, J.W.; Tseng, C.Y.; Wu, P.C.; Chiou, T.J. Regulatory network of microRNA399 and PHO2 by systemic signaling. *Plant Physiol.* **2008**, *147*, 732–746. [CrossRef] [PubMed]

8.   Jones, D.L. Organic acids in the rhizosphere—A critical review. *Plant Soil* **1998**, *205*, 25–44. [CrossRef]

9.   Poirier, Y.; Thoma, S.; Somerville, C.; Schiefelbein, J. Mutant of *Arabidopsis* deficient in xylem loading of phosphate. *Plant Physiol.* **1991**, *97*, 1087–1093. [CrossRef]

10. Stefanovic, A.; Arpat, A.B.; Bligny, R.; Gout, E.; Vidoudez, C.; Bensimon, M.; Poirier, Y. Over-expression of PHO1 in *Arabidopsis* leaves reveals its role in mediating phosphate efflux. *Plant J.* **2011** *66*, 689–699.

[CrossRef]

11. Wang, Y.; Ribot, C.; Rezzonico, E.; Poirier, Y. Structure and Expression Profile of the *Arabidopsis* PHO1 Gene Family Indicates a Broad Role in Inorganic Phosphate Homeostasis 1. *Plant Physiol.* **2004**, *135*, 400–411. [CrossRef]

12. Delhaize, E.; Randall, P.J. Characterization of a Phosphate-Accumulator Mutant of *Arabidopsis thaliana*. *Plant Physiol.* **1995**, *107*, 207–213. [CrossRef]

13. Dong, B.; Rengel, Z.; Delhaize, E. Uptake and translocation of phosphate by *pho2* mutant and wild-type seedlings of *Arabidopsis thaliana*. *Planta* **1998**, *205*, 251–256. [CrossRef]

14. Park, B.S.; Seo, J.S.; Chua, N.H. Nitrogen Limitation Adaptation recruits PHOSPHATE2 to target the phosphate transporter PT2 for degradation during the regulation of *Arabidopsis* phosphate homeostasis. *Plant Cell* **2014**, *26*, 454–464. [CrossRef]

15. Aung, K. *pho2*, a Phosphate Overaccumulator, Is Caused by a Nonsense Mutation in a MicroRNA399 Target Gene. *Plant Physiol.* **2006**, *141*, 1000–1011. [CrossRef]

16. Nilsson, L.; Müller, R.; Nielsen, T.H. (2007). Increased expression of the MYB-related transcription factor, PHR1, leads to enhanced phosphate uptake in *Arabidopsis thaliana*. *Plant Cell Environ.* **2007**, *30*, 1499–1512. [CrossRef]

17. Rubio, V.; Linhares, F.; Solano, R.; Mart'in, A.C.; Iglesias, J.; Leyva, A.; Paz-Ares, J. A conserved MYB transcription factor involved in phosphate starvation signaling both in vascular plants and in unicellular algae. *Genes Dev.* **2001**, *15*, 2122–2133. [CrossRef]

18. Fujii, H.; Chiou, T.J.; Lin, S.I.; Aung, K.; Zhu, J.K. A miRNA involved in phosphate-starvation response in *Arabidopsis*. *Curr. Biol.* **2005**, *15*, 2038–2043. [CrossRef]

19. Hsieh, L.C.; Lin, S.I.; Shih, A.C.C.; Chen, J.W.; Lin, W.Y.; Tseng, C.Y.; Li, W.H.; Chiou, T.J. Uncovering small RNA-mediated responses to phosphate deficiency in *Arabidopsis* by deep sequencing. *Plant Physiol.* **2009**, *151*, 2120–2132. [CrossRef]

20. Buhtz, A.; Springer, F.; Chappell, L.; Baulcombe, D.C.; Kehr, J. Identification and characterization of small RNAs from the phloem of *Brassica napus*. *Plant J.* **2008**, *53*, 739–749. [CrossRef]

21. Pant, B.D.; Buhtz, A.; Kehr, J.; Scheible, W.R. MicroRNA399 is a long distance signal for the regulation of plant phosphate homeostasis. *Plant J.* **2008**, *53*, 731–738. [CrossRef]

22. Wang, Y.-H.; Garvin, D.F.; Kochian, L.V. Rapid induction of regulatory and transporter genes in response to phosphorus, potassium, and iron deficiencies in tomato roots. Evidence for cross talk and root/rhizosphere-mediated signals. *Plant Physiol.* **2002**, *130*, 1361–1370. [CrossRef]

23. Shin, R.; Berg, R.H.; Schachtman, D.P. Reactive oxygen species and root hairs in *Arabidopsis* root response to nitrogen, phosphorus and potassium deficiency. *Plant Cell Physiol.* **2005**, *46*, 1350–1357. [CrossRef]

24. Kant, S.; Peng, M.; Rothstein, S.J. Genetic regulation by NLA and microRNA827 for maintaining nitrate-dependent phosphate homeostasis in *Arabidopsis*. *PLoS Genet.* **2011**, *7*, e1002021. [CrossRef] [PubMed]

25. Liang, G.; He, H.; Yu, D. Identification of Nitrogen Starvation-Responsive MicroRNAs in *Arabidopsis thaliana*. *PLoS ONE* **2012**, *7*, e48951. [CrossRef] [PubMed]

26. Doerner, P. Phosphate starvation signaling: A threesome controls systemic Pi homeostasis. *Curr. Opin. Plant Biol.* **2008**, *11*, 536–540. [CrossRef]

27. Bari, R.; Pant, B.D.; Stitt, M.; Golm, S.P. PHO2, MicroRNA399, and PHR1 Define a Phosphate-Signaling Pathway in Plants. *Plant Physiol.* **2006**, *141*, 988–999. [CrossRef] [PubMed]

28. Berkowitz, O.; Jost, R.; Kollehn, D.O.; Fenske, R.; Finnegan, P.M.; O'Brien, P.A.; Hardy, G.E.S.J.; Lambers, H. Acclimation responses of *Arabidopsis thaliana* to sustained phosphite treatments. *J. Exp. Bot.* **2013**, *64*, 1731–1743. [CrossRef] [PubMed]

29. Chiou, T.J.; Aung, K.; Lin, S.I.; Wu, C.C.; Chiang, S.F.; Su, C.L. Regulation of phosphate homeostasis by MicroRNA in *Arabidopsis*. *Plant Cell* **2006**, *18*, 412–421. [CrossRef] [PubMed]

30. Huang, T.K.; Han, C.L.; Lin, S.I.; Chen, Y.J.; Tsai, Y.C.; Chen, Y.R.; Chen, J.W.; Lin, W.Y.; Chen, P.M.; Liu, T.Y.; et al. Identification of downstream components of ubiquitin-conjugating enzyme PHOSPHATE2 by quantitative membrane proteomics in *Arabidopsis* roots. *Plant Cell* **2013**, *25*, 4044–4060. [CrossRef]

31. Rouached, H.; Arpat, A.B.; Poirier, Y. Regulation of phosphate starvation responses in plants: Signaling players and cross-talks. *Mol. Plant* **2010**, *3*, 288–299. [CrossRef] [PubMed]

32. Martin, A.C.; Del Pozo, J.C.; Iglesias, J.; Rubio, V.; Solano, R.; De La Peña, A.; Leyva, A.; Paz-Ares, J. Influence of cytokinins on the expression of phosphate starvation responsive genes in *Arabidopsis*. *Plant J.* **2000**, *24*, 559–567. [CrossRef] [PubMed]

33. Franco-Zorrilla, J.M.; Valli, A.; Todesco, M.; Mateos, I.; Puga, M.I.; Rubio-Somoza, I.; Leyva, A.; Weigel, D.; García, J.A.; Paz-Ares, J. Target mimicry provides a new mechanism for regulation of microRNA activity. *Nat. Genet.* **2007**, *39*, 1033–1037. [CrossRef] [PubMed]

34. Mallory, A.C.; Bouché, N. MicroRNA-directed regulation: To cleave or not to cleave. *Trends Plant Sci.* **2008**, *13*, 359–367. [CrossRef] [PubMed]

35. Han, M.H.; Goud, S.; Song, L.; Fedoroff, N. (2004). The *Arabidopsis* double-stranded RNA-binding protein HYL1 plays a role in microRNA-mediated gene regulation. *Proc. Natl. Acad. Sci. USA* **2004**, *101*, 1093–1098. [CrossRef]

36. Eamens, A.L.; Smith, N.A.; Curtin, S.J.; Wang, M.B.; Waterhouse, P.M. The *Arabidopsis thaliana* double-stranded RNA binding protein DRB1 directs guide strand selection from microRNA duplexes. *RNA* **2009**, *15*, 2219–2235. [CrossRef]

37. Eamens, A.L.; Kim, K.W.; Curtin, S.J.; Waterhouse, P.M. DRB2 is required for microRNA biogenesis in *Arabidopsis thaliana*. *PLoS ONE* **2012**, *7*, e35933. [CrossRef]

38. Eamens, A.L.; Kim, K.W.; Waterhouse, P.M. DRB2, DRB3 and DRB5 function in a non-canonical microRNA pathway in *Arabidopsis thaliana*. *Plant Signal. Behav.* **2012**, *7*, 1224–1229. [CrossRef]

39. Rajagopalan, R.; Vaucheret, H.; Trejo, J.; Bartel, D.P. A diverse and evolutionarily fluid set of microRNAs in *Arabidopsis thaliana*. *Genes Dev.* **2006**, *20*, 3407–3425. [CrossRef]

40. Pélissier, T.; Clavel, M.; Chaparro, C.; Pouch-Pélissier, M.N.; Vaucheret, H.; Deragon, J.M. Double-stranded RNA binding proteins DRB2 and DRB4 have an antagonistic impact on polymerase IV-dependent siRNA levels in *Arabidopsis*. *RNA* **2011**, *17*, 1502–1510. [CrossRef]

41. Lu, C.; Fedoroff, N. A mutation in the *Arabidopsis* HYL1 gene encoding a dsRNA binding protein affects responses to abscisic acid, auxin, and cytokinin. *Plant Cell* **2000**, *12*, 2351–2366. [CrossRef]

42. Curtin, S.J.; Watson, J.M.; Smith, N.A.; Eamens, A.L.; Blanchard, C.L.; Waterhouse, P.M. The roles of plant dsRNA-binding proteins in RNAi-like pathways. *FEBS Lett.* **2008**, *582*, 2753–2760. [CrossRef]

43. Luo, Q.J.; Mittal, A.; Jia, F.; Rock, C.D. An autoregulatory feedback loop involving PAP1 and TAS4 in response to sugars in *Arabidopsis*. *Plant Mol. Biol.* **2012**, *80*, 117–129. [CrossRef]

44. Williamson, L.C.; Ribrioux, S.; Fitter, A.H.; Leyser, H.M.O. Phosphate availability regulates root system architecture in *Arabidopsis*. *Plant Physiol.* **2001**, *126*, 875–882. [CrossRef]

45. Schachtman, D.P.; Reid, R.J.; Ayling, S.M. Phosphorus Uptake by Plants: From Soil to Cell. *Plant Physiol.* **1998**, *116*, 447–453. [CrossRef]

46. Yang, X.J.; Finnegan, P.M. Regulation of phosphate starvation responses in higher plants. *Ann. Bot.* **2010**, *105*, 513–526. [CrossRef]

47. Remy, E.; Cabrito, T.R.; Batista, R.A.; Teixeira, M.C.; Sá-Correia, I.; Duque, P. The Pht1;9 and Pht1;8 transporters mediate inorganic phosphate acquisition by the *Arabidopsis thaliana* root during phosphorus starvation. *New Phytol.* **2012**, *195*, 356–371. [CrossRef]

48. Vazquez, F.; Vaucheret, H.; Rajagopalan, R.; Lepers, C.; Gasciolli, V.; Mallory, A.C.; Hilbert, J.L.; Bartel, D.P.; Crété, P. Endogenous *trans*-acting siRNAs regulate the accumulation of *Arabidopsis* mRNAs. *Mol. Cell* **2004**, *16*, 69–79. [CrossRef]

49. Kurihara, Y.; Takashi, Y.; Watanabe, Y. The interaction between DCL1 and HYL1 is important for efficient and precise processing of pri-miRNA in plant microRNA biogenesis. *RNA* **2006**, *12*, 206–212. [CrossRef]

50. Jiang, C.; Gao, X.; Liao, L.; Harberd, N.P.; Fu, X. Phosphate starvation root architecture and anthocyanin accumulation responses are modulated by the gibberellin-DELLA signaling pathway in *Arabidopsis*. *Plant Physiol.* **2007**, *145*, 1460–1470. [CrossRef]

51. Shin, H.; Shin, H.S.; Dewbre, G.R.; Harrison, M.J. Phosphate transport in *Arabidopsis*: Pht1;1 and Pht1;4 play a major role in phosphate acquisition from both low- and high-phosphate environments. *Plant J.* **2004**, *39*, 629–642. [CrossRef] [PubMed]

52. Reis, R.S.; Hart-Smith, G.; Eamens, A.L.; Wilkins, M.R.; Waterhouse, P.M. Gene regulation by translational inhibition is determined by Dicer partnering proteins. *Nat. Plants* **2015**, *1*, 14027. [CrossRef]

53. Yang, L.; Wu, G.; Poethig, R.S. Mutations in the GW-repeat protein SUO reveal a developmental function for microRNA-mediated translational repression in *Arabidopsis*. *Proc. Natl. Acad. Sci. USA* **2012**, *109*, 315–320.

[CrossRef] [PubMed]

54. Xu, M.; Hu, T.; Zhao, J.; Park, M.Y.; Earley, K.W.; Wu, G.; Yang, L.; Poethig, R.S. Developmental Functions of miR156-Regulated *SQUAMOSA PROMOTER BINDING PROTEIN-LIKE (SPL)* Genes in *Arabidopsis thaliana*. *PLoS Genet.* **2016**, *12*, e1006263. [CrossRef]

55. Adenot, X.; Elmayan, T.; Lauressergues, D.; Boutet, S.; Bouché, N.; Gasciolli, V.; Vaucheret, H. DRB4-Dependent *TAS3 trans*-Acting siRNAs Control Leaf Morphology through AGO7. *Curr. Biol.* **2006**, *16*, 927–932. [CrossRef] [PubMed]

56. Nakazawa, Y.; Hiraguri, A.; Moriyama, H.; Fukuhara, T. The dsRNA-binding protein DRB4 interacts with the Dicer-like protein DCL4 in vivo and functions in the *trans*-acting siRNA pathway. *Plant Mol. Biol.* **2007**, *63*, 777–785. [CrossRef] [PubMed]

57. Lichtenthaler, H.K.; Wellburn, A.R. Determination of total carotenoids and chlorophylls *a* and *b* of leaf extracts in different solvents. *Biochem. Soc. Trans.* **1983**, *11*, 591–592. [CrossRef]

9

Gene Regulation Mediated by microRNA-Triggered Secondary Small RNAs in Plants

Felipe Fenselau de Felippes

Science and Engineering Faculty, Queensland University of Technology, Brisbane, Australia;
felipe.felippes@qut.edu.au

Abstract: In plants, proper development and response to abiotic and biotic stimuli requires an orchestrated regulation of gene expression. Small RNAs (sRNAs) are key molecules involved in this process, leading to downregulation of their target genes. Two main classes of sRNAs exist, the small interfering RNAs (siRNAs) and microRNAs (miRNAs). The role of the latter class in plant development and physiology is well known, with many examples of how miRNAs directly impact the expression of genes in cells where they are produced, with dramatic consequences to the life of the plant. However, there is an aspect of miRNA biology that is still poorly understood. In some cases, miRNA targeting can lead to the production of secondary siRNAs from its target. These siRNAs, which display a characteristic phased production pattern, can act in *cis*, reinforcing the initial silencing signal set by the triggering miRNA, or in *trans*, affecting genes that are unrelated to the initial target. In this review, the mechanisms and implications of this process in the gene regulation mediated by miRNAs will be discussed. This work will also explore techniques for gene silencing in plants that are based on this unique pathway.

Keywords: tasiRNA; phasiRNA; miRNA; secondary siRNA

1. Introduction

MicroRNAs (miRNAs) are molecules that play pivotal roles in the control of gene expression, and together with small-interfering RNAs (siRNAs) they form the two major classes of regulatory small RNAs (sRNAs) in plants. Biogenesis of miRNAs relies on the activity of DICER-LIKE 1 (DCL1), an RNAse III enzyme that processes transcripts with imperfectly, self-complementary foldback structures to 21–22 nt long mature miRNA. The miRNA is then loaded into ARGONAUTE (AGO), conferring sequence specificity to the RNA-induced silencing complex (RISC), which promotes cleavage of the target transcript through the slicing activity of AGO. Alternatively, RISC-mediated gene downregulation can also be achieved via translation inhibition, a process still poorly understood in plants [1].

There are many examples of physiological and developmental pathways regulated by the direct action of miRNAs [2], yet there is an aspect of miRNA activity that only now has become more evident, and that is the ability of these molecules to indirectly regulate gene expression through the production of secondary siRNAs. In most cases, the outcome of miRNA-loaded RISC activity is cleavage and subsequent degradation of the target transcript, however, in a few cases, targeting can result in synthesis of a double-stranded RNA (dsRNA), having the target transcript as template for RNA-DEPENDENT RNA POLYMERASE 6 (RDR6) [3–7]. This step also requires the activity of SUPPRESSOR OF GENE SILENCING 3 (SGS3) and SILENCING DEFECTIVE 5 (SDE5) [3–9]. The newly synthetized dsRNA molecule is primarily the substrate for DCL4 to generate a new population of "secondary siRNAs", which have a phased pattern as their main characteristic, with siRNAs being produced in intervals 21 or 24 nt from the miRNA cleavage site [4,5,7,10–12]. These secondary siRNAs can have a dramatic impact on gene regulation mediated by miRNAs. They can act in *cis*, amplifying the silencing effect on their targets, or in *trans*, promoting downregulation of genes that otherwise would not be targeted by the trigger miRNA. Moreover, if the secondary siRNA precursor is a member of a gene family or shares sequences with other transcripts, silencing could spread to other genes, creating a regulatory cascade initially triggered by a single miRNA targeting event [13,14]. In addition to these features, generation of secondary siRNAs by miRNAs can add several other advantages to the regulation of gene expression, which will be discussed in more detail in this review.

The first miRNA-triggered secondary siRNA-producing loci were initially identified and characterized in *Arabidopsis thaliana*, where non-coding RNA molecules were found to give rise to siRNAs suppressing the expression of genes that were unrelated to their precursor molecules; therefore, they were referred as *trans*-acting siRNAs (tasiRNAs) [3,4]. To date, four families of tasiRNA-producing loci (*TAS1-4*) have been described in *A. thaliana*. *TAS1* and *TAS2* are both targeted by miR173, with *TAS3* and *TAS4* tasiRNA biogenesis being dependent on miR390 and miR828, respectively [3–6,15]. Additional *TAS* genes (*TAS5-10*) have been described or predicted in species other than *A. thaliana*, suggesting that many secondary siRNA-producing loci are yet to be discovered [16–19]. Indeed, with the advance of genomic-scale analyses, several phased, secondary siRNA-producing loci were recently identified in various plant species. As for *TAS* transcripts, one or more miRNAs were shown or predicted to target the precursor RNA molecule. However, different from classic tasiRNAs, generation of these newly identified secondary siRNAs can also be associated with protein-coding genes, and their activity in promoting cleavage of their target in *trans* is often not shown. Therefore, these secondary siRNAs are called phased siRNAs (phasiRNAs), and the loci where they come from is referred to as a *PHAS* gene [20,21]. In summary, phased, miRNA-triggered secondary siRNAs are generally referred to as phasiRNAs, while tasiRNAs are a specialized subclass of phasiRNAs for which function has been demonstrated to occur in *trans*. In addition, *TAS* loci are usually considered as noncoding with no function other than being precursor molecules to secondary siRNAs [13,20,21]. This review will focus on the factors leading to miRNA-triggered production of secondary siRNAs as well as the main features and possible advantages of this system for control of gene expression in plants. In addition, different methods to trigger gene downregulation using this silencing mechanism will be discussed.

2. Biogenesis of miRNA-Triggered Secondary siRNAs

The question of how some miRNAs can trigger the production of secondary siRNAs from their targets has been one of the major subjects of phasiRNA research. Two hypotheses have been commonly used to explain this peculiar phenomenon, known simply as "one-hit" and "two-hit" models. However, recent findings have reshaped our understanding of how miRNA-triggered siRNAs are generated (Figure 1). It is worth noting that, despite most of what is known has originated from studies using *TAS* loci as a model, the mechanisms for miRNA-triggered secondary siRNA biogenesis seems to be valid for the majority, if not all, phasiRNAs.

2.1. "Two-Hit" Model

The first attempt to explain biogenesis of phasiRNAs came from the observation that *TAS3* in *Physcomitrella patens* (*PpTAS3*), unlike most of the plant miRNA targets, displayed two miR390 complementary sequences, and the majority of sRNAs produced from this transcript were confined between these two sites [22]. The authors also showed that this pattern in the *PpTAS3* gene could be extended to *A. thaliana* and several other species. Interestingly, the miR390 target site located 5' to the tasiRNAs was not cleavable in *A. thaliana*; however, mutations disrupting this or the cleavable 3' miR390 site resulted in plants showing phenotypes associated with the impairment of *TAS3* function [22]. This and the discovery that other secondary siRNA-producing loci are also flanked by sRNA complementary sites led to a model where dual sRNA hits would act as a trigger for recruitment of RDR6.

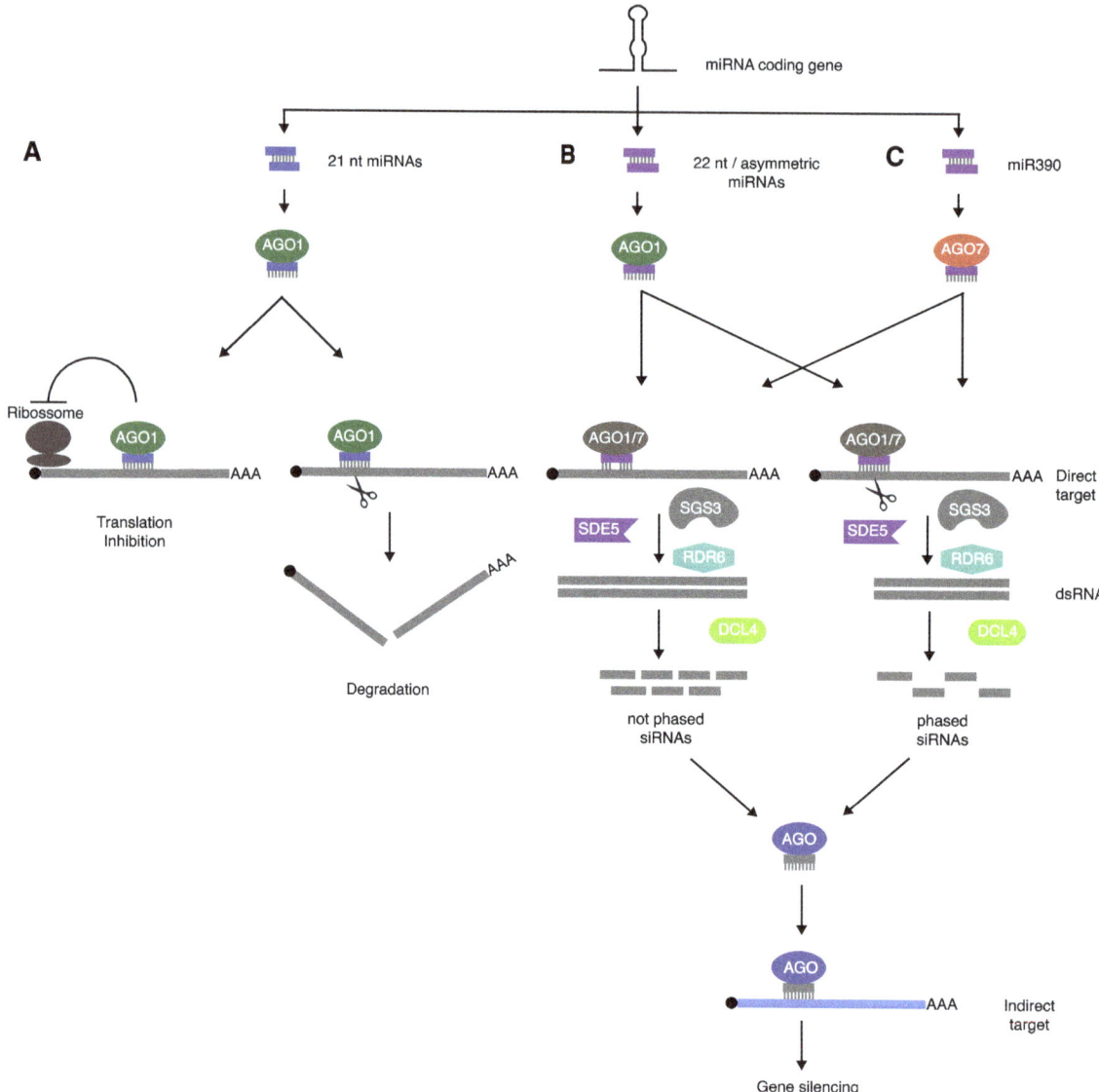

Figure 1. Biogenesis of miRNA-triggered secondary siRNAs. (**A**) Most plant miRNAs are processed into 21 nt long molecules, loaded into AGO1, and promote post-transcriptional gene silencing (PTGS) via translational repression or cleavage, followed by degradation of the target transcript. (**B**) Production of secondary siRNAs occurs when 22 nt long/asymmetric miRNA are bound to AGO1 to target a transcript. (**C**) Alternatively, 21 nt long miRNAs, such as miR390, can also initiate transitivity via interaction with AGO7. In both cases (B and C), cleavage is required for phasing, but not for generation of secondary siRNAs. Biogenesis of secondary siRNAs is dependent on the action of RDR6, SGS3, and SDE5, resulting in the synthesis of a dsRNA, which is mainly processed by DCL4. These siRNAs are loaded into AGOs and can drive gene silencing of their targets.

2.2. "One-Hit" Model

The "two-hit" model, however, was not sufficient to explain biogenesis of secondary siRNAs from other loci, such as *TAS1*, *TAS2* and *TAS4*, in which tasiRNA precursors were all cleaved at a single site upstream of the sRNA production region [3–6,15,22]. One of the initial insights into the mechanism behind tasiRNA generation from these transcripts came from experiments testing the requirements for secondary siRNA biogenesis from *TAS1* in *A. thaliana* [23,24]. These authors have shown that miR173-mediated targeting was not only necessary, but also sufficient to trigger tasiRNA production. This observation suggested that miRNAs, such as miR173, had unique features that differentiated them from the majority of other miRNAs that could not trigger the production of secondary siRNAs from their targets. Indeed, it was later demonstrated that miRNA length and the structure of the miRNA/miRNA* duplex (miRNA* refers to the sequence complementary to the predominant miRNA in the precursor molecule) were key determinants in triggering miRNA-dependent secondary siRNA production [25–27]. In plants, most miRNAs are processed as 21 nt long molecules. Interestingly, genome-wide analysis of sRNAs found that the majority of miRNAs and siRNAs associated with secondary siRNA production were 22 nt in length, indicating that sRNA size could be a crucial aspect for phasiRNA biogenesis. To test this hypothesis, Cuperus et al. [26] and Chen et al. [25] engineered miRNA precursors to produce mature miRNAs, either 21 or 22 nt of length, and tested their ability to initiate secondary siRNA production. For instance, miR173, which initiates tasiRNA production in *TAS1* and *TAS2*, is naturally found as a molecule 22 nt in length; however, its ability to trigger tasiRNA biogenesis was abolished when this molecule was 21 nt in length. Accordingly, turning the 21 nt miR319, which does not trigger secondary siRNAs production, into a 22 nt miRNA conferred this molecule the capacity to start siRNA generation from its target transcript. In the vast majority of cases, the generation of 22 nt miRNAs, instead of the more commonly found 21 nt variety, is caused by the presence of an asymmetric bulge in the pairing between miRNA and miRNA* in the precursor molecule, resulting in maturation, by DCL1, of an miRNA/miRNA* duplex with a 22/21 nt configuration [25,26]. Interestingly, 22 nt miRNAs can also be created by post-processing modification events, as shown in the case of the soybean miR1510. This sRNA, which is able to trigger phasiRNA production, is processed as 21 nt molecule, but accumulates as a 22 nt isoform via monouridylation [28]. Despite this, it is not only miRNA size that seems to account for the ability to initiate transitivity (another name for RDR6-dependent secondary siRNA production). MiRNAs that are 21 nt in length can also trigger transitivity when their miRNA* is found as a 22 nt molecule. It has been proposed that asymmetry in the miRNA/miRNA* duplex, which is also found in miRNAs processed as 22 nt molecules, is sufficient for the initiation of secondary siRNA production in target transcripts. This idea was confirmed by producing an asymmetric miR173/miR173* duplex, where both miRNA and miRNA* were produced as 21 nt entities, demonstrating that this configuration could efficiently trigger transitivity [27].

2.3. A Unified "One-Hit" Model

The models described above have been considered as two independent mechanisms leading to phasiRNA generation in plants. However, a recent study has shown that this might not be the case, and these two processes might be more similar than previously suspected. In *A. thaliana*, miR390 was recently shown to trigger tasiRNA production from the *TAS3* transcript even when only one targeting event occurred, similar to what happened with other *A. thaliana TAS* families [29]. Supporting the idea that "one-hit" is sufficient for tasiRNA production in *TAS3*, many dicots, conifers, and cycads carry a second *TAS3*-related gene, referred to as *TAS3-2*, which, in some species of citrus, chicories, and populous, possess only one miR390 target site [30,31]. In addition, *TAS3* in spruce has been characterized as a large family with 18 members, some of them carrying only one miR390 complementary sequence [32]. Taken together, these observations suggest that "one-hit" might be the basic system behind secondary siRNA production, and the "two-hit" configuration may have evolved as a regulatory mechanism to avoid possible off-targeting incidents by limiting the region from where secondary siRNAs are produced [29]. Another peculiarity of this unified model concerns cleavage of the precursor transcript as a requirement for secondary siRNA generation. Since the discovery of tasiRNAs, the slicing activity of AGO within RISC has been considered essential for the recruitment of RDR6 [13,14,33]. However, secondary siRNAs deriving from *TAS1* and *TAS3* can also be detected, even when the respective transcripts are not cleaved by AGO, indicating that a non-cleavable interaction of RISC with its target is sufficient to trigger efficient phasiRNA production [29,34]. Nonetheless, slicing of the target transcript is still crucial for the proper phase and, therefore, function of tasiRNAs, which is probably due to the lack of a well-defined end of the dsRNA molecule caused by the absence of cleavage.

Despite unification under a same "one-hit" process, tasiRNA biogenesis from *TAS3* and *TAS1/2/4* still differs regarding the initiation mechanism. While *TAS1*, *TAS2*, and *TAS4* give rise to secondary siRNAs after being hit by miRNAs that are 22 nt in length, miR390, which targets *TAS3*, is 21 nt long and does not show asymmetric structures in its precursor [3–6,15,25,26]. Another particularity of the *TAS3*/miR390 system is that miR390 is "assigned" with its own AGO protein. In *A. thaliana* there are 10 AGOs (AGO1-10), with most miRNAs loaded in AGO1, including the bulk of those that can trigger transitivity [33,35,36]. However, miR390 is not only preferentially found associated with AGO7, but it is also the specific ligand of this protein, which seems to select miRNA through recognition of its initial 5′ adenosine residue and the central region of the miR390/miR390* duplex [33,37]. More interestingly, unlike when it is loaded into AGO7, the ability of miR390 to initiate tasiRNA production was abolished when this miRNA was found or forced to interact with AGO1 or AGO2 [33], highlighting the high degree of specialization found among members of this family.

In summary, in this unified model the production of secondary siRNA from miRNA targets is dependent on a targeting event, where the AGO involved is found in a competent status. This condition is achieved when AGO1, for instance, interacts with an miRNA that is 22 nt long and/or asymmetric as a duplex. AGO7, on the other hand, would be a specialized form of this protein that could continually promote the production of secondary siRNAs from its targets, allowing miRNAs neither 22 nt long nor asymmetric to initiate transitivity. How AGO1, loaded with 22 nt /asymmetric miRNAs or the miR390/AGO7 complex, can route its target to the RDR6 pathway is still unknown. One could speculate that once loaded with 22 nt /asymmetric miRNAs, AGO1 would suffer a change in its configuration allowing the onset of transitivity. Such a change in conformation is supported by crystal structure analysis of *Thermus thermophilus* AGO bound to DNA guide strands of different sizes [38]. In the case of AGO7, it is plausible that this protein has evolved to constitutively be found in this competent form. It is clear that further work will be required to test this hypothesis.

2.4. Other Elements Involved in Secondary siRNA Production

The subcellular location where miRNA-triggered secondary siRNAs are produced has also been the focus of investigation. Many components required for phasiRNA biogenesis, such as DCL4, RDR6, SGS3, and AGO7 accumulate in the cytoplasm [39–43]. More specifically, SGS3, RDR6, and AGO7 have been shown to co-localize in cytoplasmic foci, called siRNA bodies, which are distinct from processing-bodies (P-bodies) involved in mRNA turnover [41,42]. In addition, AGO7, miR390, and SGS3 were shown to be present in microsomal fractions and localized in the endoplasmic reticulum (ER), suggesting that phasiRNA production was connected to cytoplasmic membrane structures [40,42]. Indeed, it has been reported that miRNAs, including 22 nt ones, their target transcripts, and AGO1 are found associated to membrane-bound polysomes (MBPs) in the ER. Moreover, miRNA-guided cleavage could also be detected in MBP fractions [44]. Corroborating the view that proper subcellular localization was crucial for miRNA-triggered secondary siRNA production, phasiRNA generation was affected in *ago1-27* plants, most likely from a decrease in association between MBP and AGO1 [44]. Similarly, *TAS3*-tasiRNA biogenesis was impaired when AGO7 was forced to accumulate in the nucleus [42]. It is still unclear how these different components are brought together to the same subcellular compartment, and clearly more research will be necessary.

The role of SGS3 and SDE5 in the production of miRNA-triggered secondary siRNAs is another subject that still remains somewhat enigmatic. SGS3 has been identified from the beginning as an essential component of this system [3–7]. In vitro experiments have shown that SGS3 acts in conjunction with the cleaved transcript, protecting it against degradation and making it available for RDR6 [45]. However, it seems likely that SGS3 has other functions in addition to solely stabilizing the cleaved RNA. As discussed previously, tasiRNA production was shown to be independent of miRNA-mediated cleavage of the precursor transcript, yet SGS3 was still required for the synthesis of secondary siRNAs under non-slicing conditions [29,34]. This scenario is corroborated by the association of SGS3 with a slicing-defective RISC that binds uncut target RNAs [45]. A possible additional function of SGS3 in the production of secondary siRNAs could be in the proper placement of factors involved in transitivity in the same subcellular location, as suggested by the interaction of SGS3 with RDR6 and colocalization with

AGO7 in specialized cytoplasmic siRNA bodies [41,42]. As with *sgs3*, the accumulation of secondary siRNAs is also abolished in *sde5* mutants, suggesting a key, although, to date, unclear role for this protein in transitivity [8,9]. SDE5 encodes for a putative RNA export protein, and its role has been suggested to involve the traffic of mRNAs between the nucleus and cytoplasm and/or to route RNA to RDR6 [8,9]. Genetic experiments have placed SDE5 function downstream of SGS3, but upstream of RDR6 activity [46]. Nonetheless, the mode-of-action of these proteins still needs to be investigated in more detail.

In addition to the core components of the pathway, there are other elements that are not essential for the production of miRNA-dependent secondary siRNAs but still have an influence on the biogenesis of these molecules. Components of the THO/TREX complex, which is involved in the intercellular trafficking of mRNAs, have been shown to affect tasiRNA synthesis. In mutant plants where this complex had been disrupted, some tasiRNAs accumulated at lower levels when compared to the wild type [9,47]. It has been suggested that the THO/TREX complex is involved in the transport of *TAS* precursors from their production site to subcellular locations where secondary siRNA biogenesis takes place. Although tasiRNA precursors are considered to be non-coding transcripts, some *TAS* genes have short open reading frames (ORFs) located just upstream of the tasiRNA-producing region, which could potentially give rise to small peptides. Indeed, it has been described that some of these ORFs interact with ribosomes and are actually translated [44,46,48,49]. Interestingly, this process seems to be important for the proper accumulation of tasiRNAs from the transcripts involved. In *TAS2* and *TAS3*, mutations affecting these ORFs result in reduction of tasiRNA accumulation, most likely because of decreased stability of the *TAS* precursor caused by lower levels of association with ribosomes [46,49]. Alternatively, ribosome occupancy has been suggested as a factor that defines the regions of a transcript giving origin to secondary siRNAs [44]. The importance of translation in the production of secondary siRNAs is corroborated by the observation that production of synthetic tasiRNAs (syn-tasiRNAs) is improved with the introduction of a stop codon immediately before the miR173 target site [50].

3. Features and Advantages of miRNA-Triggered Secondary siRNA Gene Regulation

In *A. thaliana*, miRNA targeting events leading to transitivity are uncommon, with only a few cases described, and are better exemplified by the *TAS1–4* families. Nonetheless, gene regulation promoted by miRNA-triggered secondary siRNAs have an important impact on plant development and physiology. For instance, miR390 targeting of *TAS3* results in the production of tasiRNAs that can regulate the expression of different auxin response factor genes (*ARFs*), affecting important functions such as leaf morphology, the transition from juvenile to adult phase, and flower and root formation to mention a few [3,5,6,11,51–55]. In recent years, with the popularization and expansion of genomic-based studies, several loci that spawn phased, miRNA-triggered secondary siRNAs were identified in numerous other species. In many cases, these sRNA populations seem to have a role in a variety of pathways related to development, response to stresses, and disease resistance (for more details on these pathways and the miRNAs/phasiRNAs involved, please see this recent review [21]). But what would be the benefits of such an indirect role of miRNAs in the control of gene expression? In the second part of this review, some features and putative added values of indirect gene regulation by miRNAs via secondary siRNAs will be discussed.

The obvious consequence resulting from the production of tasi- and phasiRNAs is the amplification and potential enhancement of the silencing signal (Figure 2A). From a single miRNA targeting event, a population of secondary siRNAs is produced, all with the potential to silence in *cis*, multiplying the number of molecules that could cause downregulation of the precursor loci and, therefore, increase silencing pressure on the target. Indeed, evidence of secondary siRNA targeting in *cis* are quite common. For instance, in watermelon, *Medicago*, and citrus, several phasiRNAs were reported to target their precursor transcripts [20,56,57]. Despite their function mainly being associated with the silencing of unrelated genes in *trans*, tasiRNAs are also known to promote cleavage of the transcript of origin in *cis*. TasiRNA-5D2(-), one of the tasiRNAs emerging from *TAS3*, has been shown to cut its precursor transcript in different species [5,20,22,29,58,59]. In this case however, because it involves a non-coding transcript, it is possible that the re-attack of tasiRNA-5D2(-) acts more as a feedback regulatory mechanism, fine-tuning tasiRNA levels.

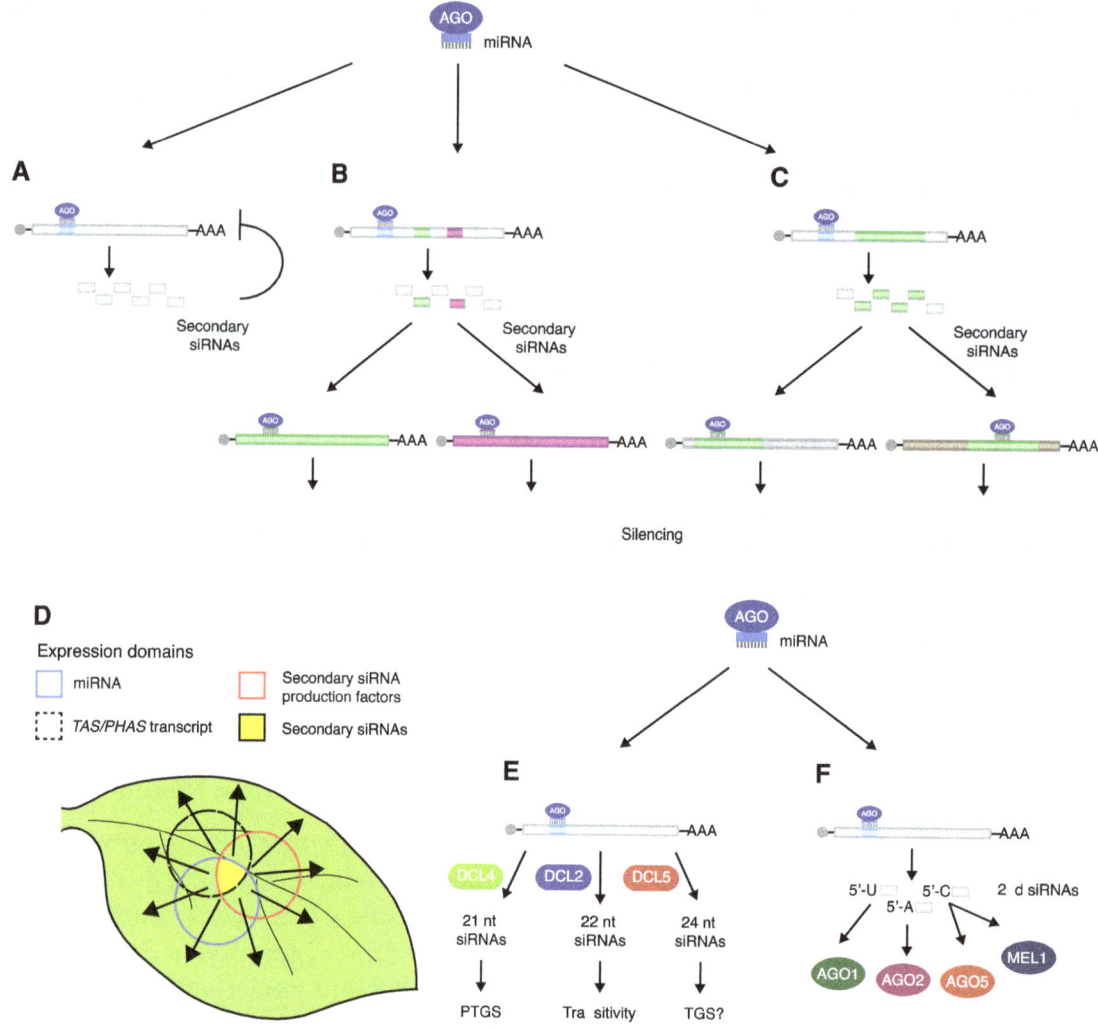

Figure 2. Features and advantages of gene regulation via miRNA-triggered secondary siRNAs. (**A**) By producing secondary siRNAs, miRNAs can increase the silencing pressure on their targets. (**B**) Secondary siRNAs targeting distinct transcripts can be produced from the same precursor, increasing the silencing range of the trigger miRNA. (**C**) The number of genes indirectly regulated by an miRNA can be increased if secondary siRNAs are produced from regions containing a conserved sequence shared by different loci. (**D**) Production of secondary siRNAs is restricted to regions where all elements participating in their biogenesis are present; however, they could later spread to neighboring cells to function non-cell autonomously (as indicated by the arrows). (**E**) The dsRNA synthetized by RDR6 is mainly processed by DCL4, generating 21 nt long siRNAs that are involved in PTGS. Alternatively, the dsRNA can also be the substrate for other DCLs, such as DCL2 and DCL5, resulting in the biogenesis of secondary siRNAs with different characteristics and functions. (**F**) Compared to most miRNAs, which have a 5′ terminal uridine and are loaded into AGO1, secondary siRNAs show an increased diversity on their 5′ extremity, allowing for sorting into different AGO proteins, with possible consequences to their activities.

Another interesting aspect of gene regulation mediated by secondary siRNAs is the possibility that one miRNA could affect the expression of several genes that otherwise would not have been targeted (Figure 2B). This is well illustrated in cotton, where cleavage of *MYB2* by miR828 results in the production of tasiRNAs that have been predicted to target several unrelated genes, such as sucrose synthase, histone acetyltransferase, and glutamate receptor, none of which are targeted by the triggering miRNA [60]. The *P. patens* miR390/*TAS3* system is an additional case where one miRNA can promote downregulation of several genes that are unrelated, sequence- and function-wise. In addition to the well-characterized, interspecies-conserved, tasiRNA targeting *ARF*-like mRNAs, *P. patens* *TAS3* transcripts also give origin to secondary siRNAs that can promote downregulation of three AP2 domain-containing transcripts [61]. Alternatively, phasiRNAs can also increase the number of genes regulated by a single miRNA if these secondary siRNAs are produced from conserved regions (Figure 2C). In this scenario, the newly generated siRNAs could function not only in *cis*, but also in *trans*, with the potential to affect any transcript that shares this same conserved region. The best example of such a regulatory network has been described in *Medicago* and involves the generation of phasiRNAs from nucleotide-binding leucine-rich repeat (*NB-LRR*) disease-resistance genes. In this species, three miRNA families (miR1507, miR2109, and miR2118) were described to target different NB-LRR-conserved motifs in 74 transcripts, leading to the biogenesis of phasiRNAs with the potential to regulate 60% of the estimated 540 *NB-LRR* genes [20]. Another mechanism resulting in the expansion of the miRNA activity range is through the production of secondary siRNAs that are 22 nt in size (Figure 2E). In peaches, two miR7122-triggered tasiRNAs, which are predominantly 22 nt in length, have been described to initiate phasiRNA generation from their targets [62], similar to what has been reported for the miR173/*TAS2* pathway in *Arabidopsis* [25].

As discussed previously, mechanisms leading to the formation of phasiRNAs require elements of the miRNA pathway as well as new components, such as RDR6, SGS3, SDE5, and AGO7. This increase in complexity brings new possibilities of regulation with interesting consequences to the indirect function of miRNAs in controlling gene expression. In *Arabidopsis*, tasiRNA production from *TAS3* is dependent on the activity of miR390 and AGO7 [5,33]. Interestingly, the *TAS* transcript, the AGO protein, and miRNA have distinct expression patterns and as a consequence; synthesis of secondary siRNAs from *TAS3* is restricted to cells where all these elements are present (Figure 2D) [33,54,63,64]. This spatio-temporal coordination has been shown to be important for the proper development of leaves. Abaxial/adaxial fate specification is a result of asymmetric expression of *ARFs* in the leaf, caused by the polarized accumulation pattern of *TAS3*-tasiRNAs [63–65]. This localized accumulation of tasiRNAs is only possible because of a delimited presence of AGO7 and *TAS3* to the adaxial side, restricting the biogenesis of secondary siRNAs to this region, despite the broader miR390 expression domain [63,64].

With few exceptions, most miRNA precursors are processed by DCL1 into mature molecules 21/22 nt in length, loaded into AGO1, and promote post-transcription gene silencing [1]. However, a whole new level of plasticity can be added to the control of gene expression when miRNA-triggered secondary siRNAs are employed (Figure 2E). The dsRNA molecule synthetized by RDR6 from *TAS* and *PHAS* transcripts are primarily processed by DCL4 into 21-nt-long siRNAs, which like miRNAs, act post-transcriptionally [7,10,11]. Nevertheless, grasses possess an additional DCL enzyme, DCL5 (formerly known as DCL3b), which is responsible for the production of phasiRNAs 24 nt in length from transcripts targeted by miR2275 [66,67]. Interestingly, this is the same size of siRNAs that interact with AGO4, the main effector of the transcriptional gene silence (TGS) pathway, which results in DNA methylation and subsequent silencing [1]. The implications of this discovery are still elusive. These 24 nt phasiRNAs were first described in rice and maize reproductive tissues. They accumulated in meiotic-stage anthers and, therefore, were believed to be involved in reproduction [12,68]. More importantly, given the size of these siRNAs, it is tempting to speculate that 24 nt long, DCL5-dependent phasiRNAs can be associated with AGO4 to promote DNA methylation, adding a new layer to gene regulation mediated by miRNAs. Supporting this view, Xia and colleagues [69] found that the miR2275/24 nt phasiRNAs pathway is not only present in monocots but also in eudicots plants. However, differently from the former group, miR2275-dependent, 24 nt long phasiRNA production in eudicots does not rely on the activity of a specific protein, such as DCL5, but instead it most likely requires the action of DCL3. This is the same enzyme responsible for producing the 24 nt siRNA associated with AGO4 and involved in TGS [1]. Reflecting the high level of specialization and conservation of the pathway, the vast majority of mature miRNAs have a 5′ terminal uridine (U), which has been shown to be a key determinant for the sorting of sRNAs into AGO1 [33,35,36]. In contrast, AGO2 and AGO4 prefer sRNAs that contain an adenosine (A) at the 5′ end, while AGO5 is more often associated to molecules that have a cytosine (C). Many of the secondary siRNAs with conserved functions, such as tasiRNAs that target *ARF* genes, are similar to miRNAs, having an uridine at the 5′ extremity of the mature molecule and, thus, are loaded into AGO1 [5]. Nonetheless, many phasiRNAs do not follow this trend, with many of them found associated to other AGOs such as AGO2 [33,35,36]. In rice, MEL1 is a specialized protein ortholog of AGO5, and it has been described to preferentially bind phasiRNAs that begin with a cytosine [70]. The function of these secondary siRNAs is still poorly understood, but nevertheless, MEL1 has been shown to mediate sporophytic germ-cell development and meiosis, suggesting that these sRNAs might play a direct role in these processes [71]. In summary, the variability of features found among the different AGOs can also be explored when miRNAs that trigger transitivity are involved in gene regulation (Figure 2F).

In addition to the features discussed above, there may be other unknown or poorly understood characteristics of phasiRNAs that could add extra value to miRNA-regulated pathways. For instance, compared to other classes of sRNAs, tasiRNA have been described to display extended cell-to-cell mobility, suggesting that, by initiating transitivity, miRNAs could increase their range of activity (Figure 2D) [72]. Indeed, an artificial miRNA (amiRNA) designed to be produced as a molecule 22 nt in length was reported to start secondary siRNA biogenesis from its target, resulting in silencing in tissues that otherwise would not be affected by amiRNAs of regular size or siRNAs produced from a hairpin construct [73].

4. Utilizing miRNA-Triggered Secondary siRNAs to Promote Directed Gene Silencing

Silencing promoted by sRNAs is not only an important mechanism to control gene expression, but it has also been used as a powerful tool to downregulate transcripts in both academic and applied purposes [74]. Despite not being as popular as founding techniques, such as artificial miRNAs (amiRNAs) and hairpin RNA interference (hpRNAi), systems based on the ability of miRNAs to start transitivity also exist and are undoubtedly a valuable addition to the collection of methods aiming to control gene activity (Figure 3).

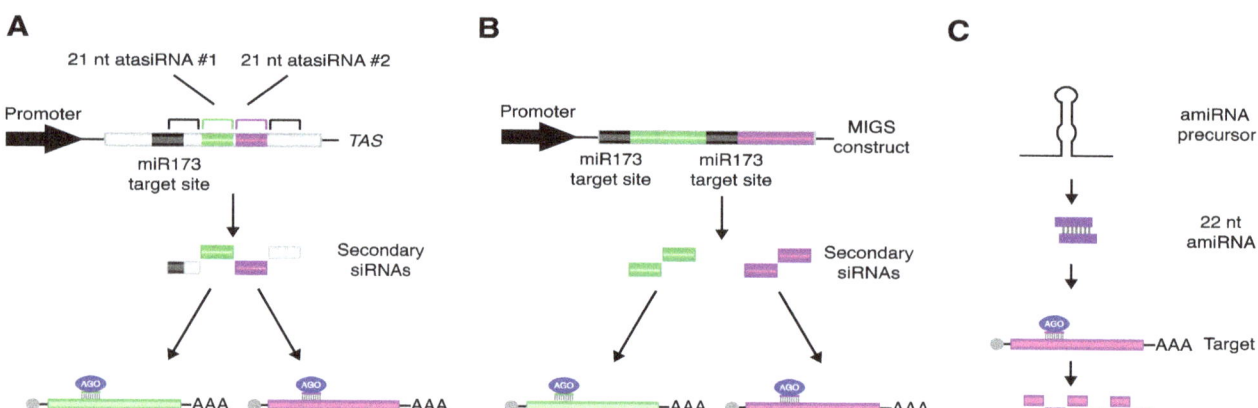

Figure 3. Methods for gene silencing based on miRNA-triggered secondary siRNAs. (**A**) Silencing using atasiRNAs consists of replacing one or more of the tasiRNAs in the *TAS* gene for a sequence designed to target the gene of interest. (**B**) miRNA-induced gene silencing (MIGS) constructs can be generated by placing the sequence recognized by an miRNA that can start transitivity in front of a fragment of the target gene (e.g., miR173). Downregulation of more than one gene using this technique can be easily accomplished by repeating the same pattern with different gene fragments. (**C**) By using specific precursors (such as *MIR173*), 22 nt long amiRNAs can be produced. These molecules can then initiate secondary siRNA synthesis from their targets, adding new features to the original method.

Artificial tasiRNA (atasiRNA), also known as synthetic tasiRNA (syn-tasiRNA), was the first method developed based on miRNA-triggered biogenesis of secondary siRNAs. It has been used to successfully reduce gene expression of endogenous sequences and to interfere with viroid infections [23,24,33,75–77]. This approach consists of replacing one or more tasiRNAs in a *TAS* transcript for sequences devised to target the gene of interest, in a process similar to the design of amiRNAs (Figure 3A) [75,78]. One of the most useful features of this technique is the possibility of having one precursor producing several atasiRNAs, each targeting different sequences, which could be located in the same or distinct transcripts [76]. AtasiRNAs share many of the advantages and limitations of amiRNAs. One distinct advantage is that high levels of specificity can be achieved, decreasing the chance of off-targeting. However, to efficiently design atasiRNA molecules that specifically downregulate one or just a few genes, with minimal chances of silencing unwanted transcripts, it is important that information about the entire genome is made available. Moreover, compared to approaches that make use of whole gene fragments for generation of the silencing construct, such as hpRNAi, this method is more susceptible to the effects of target accessibility that could reduce the effectiveness of the sRNA [79].

An additional system exploring the transitivity initiated by miRNAs is referred to as miRNA-induced gene silencing (MIGS), which has as its main characteristic the easiness of design [80,81]. With a single PCR step, the target site of an miRNA triggering phasiRNA production can be fused upstream of the fragment of a gene of interest. Upon transcript cleavage by RISC, the newly synthetized secondary siRNAs can subsequently promote silencing of related target sequences. MIGS is also a powerful tool to downregulate multiple genes using a single vector (Figure 3B). This is achieved simply by linking fragments of different targets, each with their own miRNA target site [80]. Since this method does not require genome-wide data, it is an interesting alternative to be used in species where this information is still lacking. Despite the aforementioned advantages, the risk of off-targeting needs to be considered when employing this system. Differently to atasiRNAs, MIGS constructs give rise to a population of siRNAs, all with the capacity to silence related sequences. Therefore, depending on the degree of conservation present in the fragment used, genes (other than the intended target) that share similar sequences could also become silenced. To date, MIGS has been shown to be an effective tool to silence genes in several species, including *Arabidopsis*, *Nicotiana benthamiana*, *Medicago*, soybean, rice, and petunia [80,82–86].

A common theme between atasiRNAs and MIGS is the requirement of a trigger miRNA. Therefore, it is important to take into consideration that the spatio-temporal expression pattern of the miRNA could influence the way in which atasiRNAs and MIGS-derived siRNAs are produced. In addition, some species might not code for the miRNA initiating the production of secondary siRNAs from the silencing constructs. To overcome this issue, a collection of plasmids has been created to allow the co-expression of the triggering miRNA and the MIGS/atasiRNA construct from a single vector [80,87]. Alternatively, designing an amiRNA to be produced as a 22 nt long molecule could also be a way to silence genes via secondary siRNAs, without the limitation of a two-component system (Figure 3C). McHale and colleagues [73] have demonstrated that a 22 nt long amiRNA targeting *CHALCONE SYNTHASE* (*CHS*) was able to cause widespread silencing due to the production of secondary siRNAs by RDR6.

5. Conclusion and Final Remarks

This review has discussed some of the crucial aspects related to the production of secondary siRNA triggered by miRNA and how this process can add valuable features to the control of gene expression mediated by sRNAs. Moreover, different methods to promote gene silencing in plants that are based on this unique ability of certain miRNAs were discussed, showing that they can be important alternatives to well-established systems, such as amiRNA and hpRNAi.

In recent years, our understanding of the mechanisms leading to miRNA-triggered secondary siRNA generation, and the importance of these molecules in plant physiology and development, has increased rapidly, yet this pathway is still one of the least understood among different processes involving sRNAs. It is still unknown, for example, how 22 nt long miRNA loaded into AGO1 or the miR390/AGO7 complex can lead to the recruitment of RDR6 to the target transcript. Also, the role and molecular mechanisms behind the activity of many phasiRNAs recently described in different plant species remain elusive. The elucidation of these and other aspects related to miRNA-triggered secondary siRNAs will greatly improve our understanding of how sRNAs impact the proper development of plants and the response to abiotic and biotic stresses. In addition, this new knowledge could be useful for the development of new technologies for biotechnological applications.

Acknowledgments: The author would like to thank Samanta Bolzan de Campos and Dr. Christopher Andrew Brosnan for critical reading and suggestions on the preparation of this review.

References

1. Bologna, N.G.; Voinnet, O. The diversity, biogenesis, and activities of endogenous silencing small rnas in arabidopsis. *Annu. Rev. Plant Biol.* **2014**, *65*, 473–503. [CrossRef] [PubMed]

2. Li, S.; Castillo-González, C.; Yu, B.; Zhang, X. The functions of plant small rnas in development and in stress responses. *Plant J.* **2017**, *90*, 654–670. [PubMed]

3. Peragine, A.; Yoshikawa, M.; Wu, G.; Albrecht, H.L.; Poethig, R.S. Sgs3 and sgs2/sde1/rdr6 are required for juvenile development and the production of trans-acting sirnas in arabidopsis. *Genes Dev.* **2004**, *18*, 2368–2379. [CrossRef]

4. Vazquez, F.; Vaucheret, H.; Rajagopalan, R.; Lepers, C.; Gasciolli, V.; Mallory, A.C.; Hilbert, J.-L.; Bartel, D.P.; Crété, P. Endogenous trans-acting sirnas regulate the accumulation of arabidopsis mrnas. *Mol. Cell* **2004**, *16*, 69–79. [CrossRef]

5. Allen, E.; Xie, Z.; Gustafson, A.M.; Carrington, J.C. Microrna-directed phasing during trans-acting sirna biogenesis in plants. *Cell* **2005**, *121*, 207–221. [CrossRef] [PubMed]

6. Williams, L.; Carles, C.C.; Osmont, K.S.; Fletcher, J.C. A database analysis method identifies an endogenous trans-acting short-interfering rna that targets the arabidopsis arf2, arf3, and arf4 genes. *Proc. Natl. Acad. Sci. USA* **2005**, *102*, 9703–9708. [CrossRef]

7. Yoshikawa, M.; Peragine, A.; Park, M.Y.; Poethig, R.S. A pathway for the biogenesis of trans-acting sirnas in arabidopsis. *Genes Dev.* **2005**, *19*, 2164–2175. [CrossRef]

8. Hernandez-Pinzon, I.; Yelina, N.E.; Schwach, F.; Studholme, D.J.; Baulcombe, D.; Dalmay, T. Sde5, the putative homologue of a human mrna export factor, is required for transgene silencing and accumulation of trans-acting endogenous sirna. *Plant J.* **2007**, *50*, 140–148. [CrossRef] [PubMed]

9. Jauvion, V.; Elmayan, T.; Vaucheret, H. The conserved rna trafficking proteins hpr1 and tex1 are involved in the production of endogenous and exogenous small interfering rna in arabidopsis. *Plant Cell* **2010**, *22*, 2697–2709. [CrossRef] [PubMed]

10. Gasciolli, V.; Mallory, A.C.; Bartel, D.P.; Vaucheret, H. Partially redundant functions of arabidopsis dicer-like enzymes and a role for dcl4 in producing trans-acting sirnas. *Curr. Biol.* **2005**, *15*, 1494–1500. [CrossRef] [PubMed]

11. Xie, Z.; Allen, E.; Wilken, A.; Carrington, J.C. Dicer-like 4 functions in trans-acting small interfering rna biogenesis and vegetative phase change in arabidopsis thaliana. *Proc. Natl. Acad. Sci. USA* **2005**, *102*, 12984–12989. [CrossRef] [PubMed]

12. Johnson, C.; Kasprzewska, A.; Tennessen, K.; Fernandes, J.; Nan, G.-L.; Walbot, V.; Sundaresan, V.; Vance, V.; Bowman, L.H. Clusters and superclusters of phased small rnas in the developing inflorescence of rice. *Genome Res.* **2009**, *19*, 1429–1440. [CrossRef] [PubMed]

13. Fei, Q.; Xia, R.; Meyers, B.C. Phased, secondary, small interfering rnas in posttranscriptional regulatory networks. *Plant Cell* **2013**, *25*, 2400–2415. [CrossRef]

14. Vazquez, F.; Hohn, T. Biogenesis and biological activity of secondary sirnas in plants. *Scientifica* **2013**, *2013*, 783253. [CrossRef] [PubMed]

15. Rajagopalan, R.; Vaucheret, H.; Trejo, J.; Bartel, D.P. A diverse and evolutionarily fluid set of micrornas in arabidopsis thaliana. *Genes Dev.* **2006**, *20*, 3407–3425. [CrossRef]

16. Arif, M.A.; Fattash, I.; Ma, Z.; Cho, S.H.; Beike, A.K.; Reski, R.; Axtell, M.J.; Frank, W. Dicer-like3 activity in physcomitrella patens dicer-like4 mutants causes severe developmental dysfunction and sterility. *Mol. Plant* **2012** *5*, 1281–1294. [CrossRef]

17. Li, F.; Orban, R.; Baker, B. Somart: A web server for plant mirna, tasirna and target gene analysis. *Plant J.* **2012**, *70*, 891–901. [CrossRef]

18. Zhang, C.; Li, G.; Wang, J.; Fang, J. Identification of trans-acting sirnas and their regulatory cascades in grapevine. *Bioinformatics* **2012**, *28*, 2561–2568. [CrossRef]

19. Zuo, J.; Wang, Q.; Han, C.; Ju, Z.; Cao, D.; Zhu, B.; Luo, Y.; Gao, L. Srnaome and degradome sequencing analysis reveals specific regulation of srna in response to chilling injury in tomato fruit. *Physiol. Plant* **2017**, *160*, 142–154. [CrossRef]

20. Zhai, J.; Jeong, D.-H.; De Paoli, E.; Park, S.; Rosen, B.D.; Li, Y.; González, A.J.; Yan, Z.; Kitto, S.L.; Grusak, M.A.; et al. Micrornas as master regulators of the plant nb-lrr defense gene family via the production of phased, trans-acting sirnas. *Genes Dev.* **2011**, *25*, 2540–2553. [CrossRef]

21. Deng, P.; Muhammad, S.; Cao, M.; Wu, L. Biogenesis and regulatory hierarchy of phased small interfering rnas in plants. *Plant Biotechnol. J.* **2018**, *16*, 965–975. [CrossRef]

22. Axtell, M.J.; Jan, C.; Rajagopalan, R.; Bartel, D.P. A two-hit trigger for sirna biogenesis in plants. *Cell* **2006**, *127*, 565–577. [CrossRef]

23. Montgomery, T.A.; Yoo, S.J.; Fahlgren, N.; Gilbert, S.D.; Howell, M.D.; Sullivan, C.M.; Alexander, A.; Nguyen, G.; Allen, E.; Ahn, J.H.; et al. Ago1-mir173 complex initiates phased sirna formation in plants. *Proc. Natl. Acad. Sci. USA* **2008**, *105*, 20055–20062. [CrossRef]

24. Felippes, F.F.; Weigel, D. Triggering the formation of tasirnas in arabidopsis thaliana: The role of microrna mir173. *EMBO Rep.* **2009**, *10*, 264–270. [CrossRef]

25. Chen, H.-M.; Chen, L.-T.; Patel, K.; Li, Y.-H.; Baulcombe, D.C.; Wu, S.-H. 22-nucleotide rnas trigger secondary sirna biogenesis in plants. *Proc. Natl. Acad. Sci. USA* **2010**, *107*, 15269–15274. [CrossRef]

26. Cuperus, J.T.; Carbonell, A.; Fahlgren, N.; Garcia-Ruiz, H.; Burke, R.T.; Takeda, A.; Sullivan, C.M.; Gilbert, S.D.; Montgomery, T.A.; Carrington, J.C. Unique functionality of 22-nt mirnas in triggering rdr6-dependent sirna biogenesis from target transcripts in arabidopsis. *Nat. Struct. Mol. Biol.* **2010**, *17*, 997–1003. [CrossRef]

27. Manavella, P.A.; Koenig, D.; Weigel, D. Plant secondary sirna production determined by microrna-duplex structure. *Proc. Natl. Acad. Sci. USA* **2012**, *109*, 2461–2466. [CrossRef] [PubMed]

28. Fei, Q.; Yu, Y.; Liu, L.; Zhang, Y.; Baldrich, P.; Dai, Q.; Chen, X.; Meyers, B.C. Biogenesis of a 22-nt microrna in phaseoleae species by precursor-programmed uridylation. *Proc. Natl. Acad. Sci. USA* **2018**, *115*, 8037–8042. [CrossRef]

29. de Felippes, F.F.; Marchais, A.; Sarazin, A.; Oberlin, S.; Voinnet, O. A single mir390 targeting event is sufficient for triggering tas3-tasirna biogenesis in arabidopsis. *Nucleic Acids Res.* **2017**, *45*, 5539–5554. [CrossRef] [PubMed]

30. Krasnikova, M.S.; Milyutina, I.A.; Bobrova, V.K.; Ozerova, L.V.; Troitsky, A.V.; Solovyev, A.G.; Morozov, S.Y. Novel mir390-dependent transacting sirna precursors in plants revealed by a pcr-based experimental approach and database analysis. *J. Biomed. Biotechnol.* **2009**, *2009*, 952304–952309. [CrossRef]

31. Xia, R.; Zhu, H.; An, Y.-Q.; Beers, E.P.; Liu, Z. Apple mirnas and tasirnas with novel regulatory networks. *Genome Biol.* **2012**, *13*, R47. [CrossRef]

32. Xia, R.; Xu, J.; Arikit, S.; Meyers, B.C. Extensive families of mirnas and phas loci in norway spruce demonstrate the origins of complex phasirna networks in seed plants. *Mol. Biol. Evol.* **2015**, *32*, 2905–2918. [CrossRef]

33. Montgomery, T.A.; Howell, M.D.; Cuperus, J.T.; Li, D.; Hansen, J.E.; Alexander, A.L.; Chapman, E.J.; Fahlgren, N.; Allen, E.; Carrington, J.C. Specificity of argonaute7-mir390 interaction and dual functionality in tas3 trans-acting sirna formation. *Cell* **2008**, *133*, 128–141. [CrossRef] [PubMed]

34. Arribas-Hernández, L.; Marchais, A.; Poulsen, C.; Haase, B.; Hauptmann, J.; Benes, V.; Meister, G.; Brodersen, P. The slicer activity of argonaute1 is required specifically for the phasing, not production, of trans-acting short interfering rnas in arabidopsis. *Plant Cell* **2016**, *28*, 1563–1580. [CrossRef]

35. Mi, S.; Cai, T.; Hu, Y.; Chen, Y.; Hodges, E.; Ni, F.; Wu, L.; Li, S.; Zhou, H.; Long, C.; et al. Sorting of small rnas into arabidopsis argonaute complexes is directed by the 5′ terminal nucleotide. *Cell* **2008**, *133*, 116–127. [CrossRef]

36. Takeda, A.; Iwasaki, S.; Watanabe, T.; Utsumi, M.; Watanabe, Y. The mechanism selecting the guide strand from small rna duplexes is different among argonaute proteins. *Plant Cell Physiol.* **2008**, *49*, 493–500. [CrossRef]

37. Endo, Y.; Iwakawa, H.O.; Tomari, Y. Arabidopsis argonaute7 selects mir390 through multiple checkpoints

during risc assembly. *EMBO Rep* **2013**, *14*, 652–658. [CrossRef]

38. Wang, Y.; Sheng, G.; Juranek, S.; Tuschl, T.; Patel, D.J. Structure of the guide-strand-containing argonaute silencing complex. *Nature* **2008**, *456*, 209–213. [CrossRef]

39. Glick, E.; Zrachya, A.; Levy, Y.; Mett, A.; Gidoni, D.; Belausov, E.; Citovsky, V.; Gafni, Y. Interaction with host sgs3 is required for suppression of rna silencing by tomato yellow leaf curl virus v2 protein. *Proc. Natl. Acad. Sci. USA* **2008**, *105*, 157–161. [CrossRef]

40. Elmayan, T.; Adenot, X.; Gissot, L.; Lauressergues, D.; Gy, I.; Vaucheret, H. A neomorphic sgs3 allele stabilizing mirna cleavage products reveals that sgs3 acts as a homodimer. *FEBS J.* **2009**, *276*, 835–844. [CrossRef]

41. Kumakura, N.; Takeda, A.; Fujioka, Y.; Motose, H.; Takano, R.; Watanabe, Y. Sgs3 and rdr6 interact and colocalize in cytoplasmic sgs3/rdr6-bodies. *FEBS Lett.* **2009**, *583*, 1261–1266. [CrossRef] [PubMed]

42. Jouannet, V.; Moreno, A.B.; Elmayan, T.; Vaucheret, H.; Crespi, M.D.; Maizel, A. Cytoplasmic arabidopsis ago7 accumulates in membrane-associated sirna bodies and is required for ta-sirna biogenesis. *EMBO J.* **2012**, *31*, 1704–1713. [CrossRef] [PubMed]

43. Pumplin, N.; Sarazin, A.; Jullien, P.E.; Bologna, N.G.; Oberlin, S.; Voinnet, O. DNA methylation influences the expression of dicer-like4 isoforms, which encode proteins of alternative localization and function. *Plant Cell* **2016**, *28*, 2786–2804. [CrossRef] [PubMed]

44. Li, S.; Le, B.; Ma, X.; Li, S.; You, C.; Yu, Y.; Zhang, B.; Liu, L.; Gao, L.; Shi, T.; et al. Biogenesis of phased sirnas on membrane-bound polysomes in arabidopsis. *elife* **2016**, *5*, e22750. [CrossRef] [PubMed]

45. Yoshikawa, M.; Iki, T.; Tsutsui, Y.; Miyashita, K.; Poethig, R.S.; Habu, Y.; Ishikawa, M. 3′ fragment of mir173-programmed risc-cleaved rna is protected from degradation in a complex with risc and sgs3. *Proc. Natl. Acad. Sci. USA* **2013**, *110*, 4117–4122. [CrossRef] [PubMed]

46. Yoshikawa, M.; Iki, T.; Numa, H.; Miyashita, K.; Meshi, T.; Ishikawa, M. A short open reading frame encompassing the microrna173 target site plays a role in trans-acting small interfering rna biogenesis. *Plant Physiol.* **2016**, *171*, 359–368. [CrossRef] [PubMed]

47. Yelina, N.E.; Smith, L.M.; Jones, A.M.; Patel, K.; Kelly, K.A.; Baulcombe, D.C. Putative arabidopsis tho/trex mrna export complex is involved in transgene and endogenous sirna biosynthesis. *Proc. Natl. Acad. Sci. USA* **2010**, *107*, 13948–13953. [CrossRef] [PubMed]

48. Hou, C.Y.; Lee, W.C.; Chou, H.C.; Chen, A.P.; Chou, S.J.; Chen, H.M. Global analysis of truncated rna ends reveals new insights into ribosome stalling in plants. *Plant Cell* **2016**, *28*, 2398–2416. [CrossRef]

49. Bazin, J.; Baerenfaller, K.; Gosai, S.J.; Gregory, B.D.; Crespi, M.; Bailey-Serres, J. Global analysis of ribosome-associated noncoding rnas unveils new modes of translational regulation. *Proc. Natl. Acad. Sci. USA* **2017**, *114*, E10018–E10027. [CrossRef] [PubMed]

50. Zhang, C.; Ng, D.W.; Lu, J.; Chen, Z.J. Roles of target site location and sequence complementarity in trans-acting sirna formation in arabidopsis. *Plant J.* **2012**, *69*, 217–226. [CrossRef]

51. Adenot, X.; Elmayan, T.; Lauressergues, D.; Boutet, S.; Bouché, N.; Gasciolli, V.; Vaucheret, H. Drb4-dependent tas3 trans-acting sirnas control leaf morphology through ago7. *Curr. Biol.* **2006**, *16*, 927–932. [CrossRef] [PubMed]

52. Fahlgren, N.; Montgomery, T.A.; Howell, M.D.; Allen, E.; Dvorak, S.K.; Alexander, A.L.; Carrington, J.C. Regulation of auxin response factor3 by tas3 ta-sirna affects developmental timing and patterning in arabidopsis. *Curr. Biol.* **2006**, *16*, 939–944. [CrossRef]

53. Yoon, E.K.; Yang, J.H.; Lim, J.; Kim, S.H.; Kim, S.-K.; Lee, W.S. Auxin regulation of the microrna390-dependent transacting small interfering rna pathway in arabidopsis lateral root development. *Nucleic Acids Res.* **2009**, *38*, 1382–1391. [CrossRef] [PubMed]

54. Marin, E.; Jouannet, V.; Herz, A.; Lokerse, A.S.; Weijers, D.; Vaucheret, H.; Nussaume, L.; Crespi, M.D.; Maizel, A. Mir390, arabidopsis tas3 tasirnas, and their auxin response factor targets define an autoregulatory network quantitatively regulating lateral root growth. *Plant Cell* **2010**, *22*, 1104–1117. [CrossRef] [PubMed]

55. Matsui, A.; Mizunashi, K.; Tanaka, M.; Kaminuma, E.; Nguyen, A.H.; Nakajima, M.; Kim, J.-M.; Nguyen, D.V.; Toyoda, T.; Seki, M. Tasirna-arf pathway moderates floral architecture in arabidopsis plants subjected to drought stress. *BioMed Res. Int.* **2014**, *2014*, 303451. [CrossRef] [PubMed]

56. Liu, Y.; Ke, L.; Wu, G.; Xu, Y.; Wu, X.; Xia, R.; Deng, X.; Xu, Q. Mir3954 is a trigger of phasirnas that affects flowering time in citrus. *Plant J.* **2017**, *92*, 263–275. [CrossRef] [PubMed]

57. Liu, L.; Ren, S.; Guo, J.; Wang, Q.; Zhang, X.; Liao, P.; Li, S.; Sunkar, R.; Zheng, Y. Genome-wide identification and comprehensive analysis of micrornas and phased small interfering rnas in watermelon. *BMC Genomics* **2018**, *19*, 111. [CrossRef]

58. Jagadeeswaran, G.; Zheng, Y.; Li, Y.-F.; Shukla, L.I.; Matts, J.; Hoyt, P.; Macmil, S.L.; Wiley, G.B.; Roe, B.A.; Zhang, W.; et al. Cloning and characterization of small rnas from medicago truncatula reveals four novel legume-specific microrna families. *New Phytol.* **2009**, *184*, 85–98. [CrossRef] [PubMed]

59. Rajeswaran, R.; Aregger, M.; Zvereva, A.S.; Borah, B.K.; Gubaeva, E.G.; Pooggin, M.M. Sequencing of rdr6-dependent double-stranded rnas reveals novel features of plant sirna biogenesis. *Nucleic Acids Res.* **2012**, *40*, 6241–6254. [CrossRef]

60. Guan, X.; Pang, M.; Nah, G.; Shi, X.; Ye, W.; Stelly, D.M.; Chen, Z.J. Mir828 and mir858 regulate homoeologous myb2 gene functions in arabidopsis trichome and cotton fibre development. *Nat. Commun.* **2014**, *5*, 3050. [CrossRef]

61. Axtell, M.J.; Snyder, J.A.; Bartel, D.P. Common functions for diverse small rnas of land plants. *Plant Cell* **2007**, *19*, 1750–1769. [CrossRef]

62. Xia, R.; Meyers, B.C.; Liu, Z.; Beers, E.P.; Ye, S.; Liu, Z.; Liu, Z. Microrna superfamilies descended from mir390 and their roles in secondary small interfering rna biogenesis in eudicots. *Plant Cell* **2013**, *25*, 1555–1572. [CrossRef] [PubMed]

63. Chitwood, D.H.; Nogueira, F.T.S.; Howell, M.D.; Montgomery, T.A.; Carrington, J.C.; Timmermans, M.C.P. Pattern formation via small rna mobility. *Genes Dev.* **2009**, *23*, 549–554. [CrossRef]

64. Schwab, R.; Maizel, A.; Ruiz-Ferrer, V.; Garcia, D.; Bayer, M.; Crespi, M.; Voinnet, O.; Martienssen, R.A. Endogenous tasirnas mediate non-cell autonomous effects on gene regulation in arabidopsis thaliana. *PLoS ONE* **2009**, *4*, e5980. [CrossRef]

65. Pekker, I.; Alvarez, J.P.; Eshed, Y. Auxin response factors mediate arabidopsis organ asymmetry via modulation of kanadi activity. *Plant Cell* **2005**, *17*, 2899–2910. [CrossRef] [PubMed]

66. Margis, R.; Fusaro, A.F.; Smith, N.A.; Curtin, S.J.; Watson, J.M.; Finnegan, E.J.; Waterhouse, P.M. The evolution and diversification of dicers in plants. *FEBS Lett.* **2006**, *580*, 2442–2450. [CrossRef] [PubMed]

67. Song, X.; Li, P.; Zhai, J.; Zhou, M.; Ma, L.; Liu, B.; Jeong, D.-H.; Nakano, M.; Cao, S.; Liu, C.; et al. Roles of dcl4 and dcl3b in rice phased small rna biogenesis. *Plant J.* **2011**, *69*, 462–474. [CrossRef] [PubMed]

68. Zhai, J.; Zhang, H.; Arikit, S.; Huang, K.; Nan, G.L.; Walbot, V.; Meyers, B.C. Spatiotemporally dynamic, cell-type-dependent premeiotic and meiotic phasirnas in maize anthers. *Proc. Natl. Acad. Sci. USA* **2015**, *112*, 3146–3151. [CrossRef] [PubMed]

69. Xia, R.; Chen, C.; Pokhrel, S.; Ma, W.; Huang, K.; Patel, P.; Wang, F.; Xu, J.; Liu, Z.; Li, J.; et al. 24-nt reproductive phasirnas are broadly present in angiosperms. *Nat Commun* **2019**, *10*, 627. [CrossRef]

70. Komiya, R.; Ohyanagi, H.; Niihama, M.; Watanabe, T.; Nakano, M.; Kurata, N.; Nonomura, K.-I. Rice germline-specific argonaute mel1 protein binds to phasirnas generated from more than 700 lincrnas. *Plant J.* **2014**, *78*, 385–397. [CrossRef]

71. Nonomura, K.I.; Morohoshi, A.; Nakano, M.; Eiguchi, M.; Miyao, A.; Hirochika, H.; Kurata, N. A germ cell specific gene of the argonaute family is essential for the progression of premeiotic mitosis and meiosis during sporogenesis in rice. *Plant Cell* **2007**, *19*, 2583–2594. [CrossRef] [PubMed]

72. de Felippes, F.F.; Ott, F.; Weigel, D. Comparative analysis of non-autonomous effects of tasirnas and mirnas in arabidopsis thaliana. *Nucleic Acids Res.* **2011**, *39*, 2880–2889. [CrossRef] [PubMed]

73. McHale, M.; Eamens, A.L.; Finnegan, E.J.; Waterhouse, P.M. A 22-nt artificial microrna mediates widespread rna silencing in arabidopsis. *Plant J.* **2013**, *76*, 519–529. [CrossRef] [PubMed]

74. Pandey, P.; Senthil-Kumar, M.; Mysore, K.S. Advances in plant gene silencing methods. In *Plant Gene Silencing*; Mysore, K.S., Senthil-Kumar, M., Eds.; Springer: New York, NY, USA, 2015; Volume 1287, pp. 3–23.

75. de la Luz Gutiérrez-Nava, M.; Aukerman, M.J.; Sakai, H.; Tingey, S.V.; Williams, R.W. Artificial trans-acting sirnas confer consistent and effective gene silencing. *Plant Physiol.* **2008**, *147*, 543–551. [CrossRef] [PubMed]

76. Carbonell, A.; Takeda, A.; Fahlgren, N.; Johnson, S.C.; Cuperus, J.T.; Carrington, J.C. New generation of artificial microrna and synthetic trans-acting small interfering rna vectors for efficient gene silencing in arabidopsis. *Plant Physiol.* **2014**, *165*, 15–29. [CrossRef] [PubMed]

77. Carbonell, A.; Darós, J.-A. Artificial micrornas and synthetic trans-acting small interfering rnas interfere with viroid infection. *Mol. Plant Pathol.* **2017**, *18*, 746–753. [CrossRef] [PubMed]

78. Schwab, R.; Ossowski, S.; Riester, M.; Warthmann, N.; Weigel, D. Highly specific gene silencing by artificial micrornas in arabidopsis. *Plant Cell* **2006**, *18*, 1121–1133. [CrossRef]

79. Ossowski, S.; Schwab, R.; Weigel, D. Gene silencing in plants using artificial micrornas and other small rnas. *Plant J.* **2008**, *53*, 674–690.

80. de Felippes, F.F.; Wang, J.; Weigel, D. Migs: Mirna-induced gene silencing. *Plant J.* **2012**, 541–547. [CrossRef]

81. de Felippes, F.F. Downregulation of plant genes with mirna-induced gene silencing. In *Sirna Design*; Taxman, D., Ed.; Humana Press: Totowa, NJ, USA, 2013; Volume 942, pp. 379–387.

82. Benstein, R.M.; Ludewig, K.; Wulfert, S.; Wittek, S.; Gigolashvili, T.; Frerigmann, H.; Gierth, M.; Flügge, U.-I.; Krueger, S. Arabidopsis phosphoglycerate dehydrogenase1 of the phosphoserine pathway is essential for development and required for ammonium assimilation and tryptophan biosynthesis. *Plant Cell* **2013**, *25*, 5011–5029. [CrossRef] [PubMed]

83. Imin, N.; Mohd-Radzman, N.A.; Ogilvie, H.A.; Djordjevic, M.A. The peptide-encoding cep1 gene modulates lateral root and nodule numbers in medicago truncatula. *J. Exp. Bot.* **2013**, *64*, 5395–5409. [CrossRef] [PubMed]

84. Han, Y.; Zhang, B.; Qin, X.; Li, M.; Guo, Y. Investigation of a mirna-induced gene silencing technique in petunia reveals alterations in mir173 precursor processing and the accumulation of secondary sirnas from endogenous genes. *PLoS ONE* **2015**, *10*, e0144909–e0144916. [CrossRef] [PubMed]

85. Jacobs, T.B.; Lawler, N.J.; LaFayette, P.R.; Vodkin, L.O.; Parrott, W.A. Simple gene silencing using the trans-acting sirna pathway. *Plant Biotechnol. J.* **2015**, *14*, 117–127. [CrossRef]

86. Zheng, X.; Yang, L.; Li, Q.; Ji, L.; Tang, A.; Zang, L.; Deng, K.; Zhou, J.; Zhang, Y. Migs as a simple and efficient method for gene silencing in rice. *Front. Plant Sci.* **2018**, *9*, 662. [CrossRef]

87. Baykal, U.; Liu, H.; Chen, X.; Nguyen, H.T.; Zhang, Z.J. Novel constructs for efficient cloning of srna-encoding DNA and uniform silencing of plant genes employing artificial trans- acting small interfering rna. *Plant Cell Rep.* **2016**, *35*, 2137–2150. [CrossRef] [PubMed]

10

Inferring Novel Autophagy Regulators Based on Transcription Factors and Non-Coding RNAs Coordinated Regulatory Network

Shuyuan Wang [1,†], **Wencan Wang** [1,†], **Qianqian Meng** [1], **Shunheng Zhou** [2], **Haizhou Liu** [2], **Xueyan Ma** [1], **Xu Zhou** [1], **Hui Liu** [1], **Xiaowen Chen** [1,*] and **Wei Jiang** [2,*]

[1]   College of Bioinformatics Science and Technology, Harbin Medical University, Harbin 150081, China; bioccwsy@163.com (S.W.); wangwencan1314@163.com (W.W.); mqq1992hmu@163.com (Q.M.); 18345550297@163.com (X.M.); biomathzx@163.com (X.Z.); liuhui870320@163.com (H.L.)

[2]   College of Automation Engineering, Nanjing University of Aeronautics and Astronautics, Nanjing 211106, China; zhoushunheng@163.com (S.Z.); liuhaizhou2015@126.com (H.L.)

*   Correspondence: hrbmucxw@163.com (X.C.); weijiang@nuaa.edu.cn (W.J.)

†   These authors contributed equally to this work.

Abstract: Autophagy is a complex cellular digestion process involving multiple regulators. Compared to post-translational autophagy regulators, limited information is now available about transcriptional and post-transcriptional regulators such as transcription factors (TFs) and non-coding RNAs (ncRNAs). In this study, we proposed a computational method to infer novel autophagy-associated TFs, micro RNAs (miRNAs) and long non-coding RNAs (lncRNAs) based on TFs and ncRNAs coordinated regulatory (TNCR) network. First, we constructed a comprehensive TNCR network, including 155 TFs, 681 miRNAs and 1332 lncRNAs. Next, we gathered the known autophagy-associated factors, including TFs, miRNAs and lncRNAs, from public data resources. Then, the random walk with restart (RWR) algorithm was conducted on the TNCR network by using the known autophagy-associated factors as seeds and novel autophagy regulators were finally prioritized. Leave-one-out cross-validation (LOOCV) produced an area under the curve (AUC) of 0.889. In addition, functional analysis of the top 100 ranked regulators, including 55 TFs, 26 miRNAs and 19 lncRNAs, demonstrated that these regulators were significantly enriched in cell death related functions and had significant semantic similarity with autophagy-related Gene Ontology (GO) terms. Finally, extensive literature surveys demonstrated the credibility of the predicted autophagy regulators. In total, we presented a computational method to infer credible autophagy regulators of transcriptional factors and non-coding RNAs, which would improve the understanding of processes of autophagy and cell death and provide potential pharmacological targets to autophagy-related diseases.

Keywords: autophagy regulator; transcriptional factor; non-coding RNA; regulatory network; RWR algorithm

1. Introduction

Autophagy is a process of cytoplasmic degradation that is essential in homeostasis and stress-response, as well as in protein degradation and organelles turnover [1]. The regulation of autophagy is critical in human health and disease. Both its insufficient and overdriven activity can disturb the body functions, including causing cancers. For example, autophagy deficiency causes oxidative stress and genome instability which is a known cause of cancer initiation and progression [2] and up-regulation of autophagy in RAS-transformed cancer cells promotes their growth, survival, tumorigenesis invasion, and metastasis [3]. The process of autophagy involves multiple kinds of regulators, including autophagy-related (*ATG*) genes, ATG proteins and non-coding RNAs (ncRNAs). For instance, the autophagy database archived a list of 582 experimentally demonstrated ATG proteins [4] and Wu et al. provided a comprehensive bioinformatics resource to dissect ncRNA-mediated autophagy interactions [5]. In addition, regulation of autophagy by targeting autophagy regulators is a promising strategy for cancer therapy [6]. For example, temsirolimus could significantly prolong progression-free survival of mantle cell lymphoma (MCL) patients by inhibiting the mechanistic target of rapamycin (mTOR) protein, a post-translational autophagy regulator [7].

Currently, post-translational autophagy regulators, such as ATG proteins, are well known while limited information is available about transcriptional and post-transcriptional regulators, such as transcription factors (TFs) and ncRNAs [8]. Inferring that novel transcriptional and post-transcriptional autophagy regulators will help to dissect the autophagy regulation mechanisms and provide possible pharmacological targets to regulate autophagy. The TFs and ncRNAs coordinated regulatory (TNCR) network has demonstrated its power as a tool to study biological issues such as regulatory pathways in human diseases, classifiers for drug resistance and so on [9–11]. For example, Liang et al. performed deconvolution on the transcriptional network and demonstrated that BACH1 was the master regulator of breast cancer bone metastasis [12]. Wang et al. identified disease-related regulatory cascades by dissecting the TF and miRNA regulatory network, which helped understand the pathogenesis [13]. Recently, lncRNAs were found to be targeted by miRNAs and functioned as miRNA sponges to attenuate the inhibition ability of miRNAs to mRNAs. Furthermore, lncRNAs were also shown to play crucial roles in the regulation of gene expression at transcriptional and post-transcriptional levels [14,15]. Thus, lncRNAs introduce an extra layer of complexity to the TNCR network, enhancing the analytical ability of the regulatory network.

In this study, we proposed a computational method to predict novel autophagy-associated TFs, miRNAs and lncRNAs based on the TNCR network. First, experimentally verified transcriptional and post-transcriptional regulatory relationships among TFs, miRNAs and lncRNAs were collected and a comprehensive regulatory network was constructed. Next, the known autophagy-associated TFs, miRNAs and lncRNAs were gathered from public data resources. The random walk with restart (RWR) algorithm was implemented on the regulatory network to prioritize autophagy regulators. Leave-one-out cross-validation (LOOCV) achieved an area under the curve (AUC) of 0.889. Functional enrichment analyses and extensive literature surveys demonstrated the credibility of predicted regulators. Altogether, we presented a computational method of inferring credible autophagy regulators and we believed that this would help improve the understanding of the autophagy regulation mechanisms.

2. Materials and Methods

2.1. Construction of a Comprehensive TNCR Network

We integrated five types of experimentally verified transcriptional and post-transcriptional regulatory relationships among TFs, miRNAs and lncRNAs, including TF-miRNA, TF-lncRNA, miRNA-lncRNA, miRNA-TF, lncRNA-TF. The TFs regulations of miRNAs were downloaded from the database TransmiR, which manually surveyed literature and recorded experimentally supported TF-miRNA regulation [16]. The TFs regulations of lncRNAs were obtained from the database ChIPBase, which decoded the transcriptional regulation of lncRNAs from ChIP-seq data in diverse tissues and cell lines [17]. Here, only TF-lncRNA regulations that were identified in more than 20 datasets were retained. In order to improve the credibility of the regulations, we also used the TRANSFAC Match program to assure transcription factor binding sites (TFBS) in lncRNA sequences [18] using minimum false-positive profiles of vertebrate high quality matrices. The final TF-lncRNA regulations were obtained by intersecting the ChIPBase data source with the TRANSFAC results. The miRNAs regulations of TFs were integrated from two databases, miRecords [19] and miRTarBase [20]. Both of these two databases collected experimentally validated miRNA-target interactions, and we retained the union set of the relationships presented in these two databases. The miRNAs regulations of lncRNAs were derived from LncBase v2 which provided experimentally supported and in silico predicted miRNA recognition elements (MREs) on lncRNAs [21]. We retained the interactions presented in the experimental module and the prediction scores should have been equal to or greater than 0.95. The lncRNAs regulations of TFs were downloaded from LncReg [22] and LncRNA2Target [23]. The database LncReg collected validated lncRNA-associated regulatory entries while LncRNA2Target curated differentially expressed genes after the lncRNA knockdown or overexpression. We kept the union set of the lncRNAs regulations of TFs which were provided by these two databases. Integrating all of the above regulations, we constructed a comprehensive TNCR network.

2.2. Collection of Known Autophagy Regulators

The known autophagy-associated TFs, miRNAs and lncRNAs were collected from public data resources. We first obtained human genes in autophagy related Gene Ontology (GO) terms from the AmiGO-2 database. Next, we downloaded the human autophagy-associated genes from the autophagy database [4], a multifaceted online resource providing information on genes and proteins related to autophagy across several eukaryotic species. The union set of these two gene sets were regarded as known autophagy-associated genes. As for autophagy-associated miRNAs and lncRNAs, we resorted to the database ncRDeathDB, a comprehensive bioinformatics resource archiving ncRNA-associated cell death interactions and picked up the autophagy-associated miRNAs and lncRNAs [5]. All the autophagy-associated genes, miRNAs and lncRNAs we obtained were mapped onto the TNCR network, and the intersections were regarded as seeds for RWR algorithm.

2.3. Prioritization of Novel Autophagy Regulators with the RWR Method

We performed the RWR method on the constructed TNCR network to prioritize novel autophagy regulators. The RWR method simulates a random walker that starts on given seed nodes and transits randomly from the current node to neighboring nodes in the network with the restart probability to teleport to the start nodes. Here, the known autophagy regulators were used as seed nodes. We denoted P_0 as the initial probability vector and P_t as a vector in which the i-th element held the probability of finding the random walker at node i in step t. Let α be the restart probability of the random walk in each step at the source nodes. W denotes the probability transition matrix and is derived from the adjacency matrix of the TNCR network. The formula is defined as:

$$w(i,j) = \begin{cases} A(i,j)/\sum_j A(i,j), & if \ \sum_j A(i,j) \neq 0 \\ 0, & otherwise \end{cases} \tag{1}$$

where $w(i,j)$ represents the element in the probability transition matrix, and $A(i,j)$ represents the element in the adjacency matrix. The probability vector in step $t+1$ can be described as follows:

$$p_{t+1} = (1-\alpha)wp_{t+1} + \alpha p_0 \tag{2}$$

Based upon the previous work, the restart probability (α) was set as 0.5, and the initial probability (P_0) of each seed node was set as $1/n$ (where n is the number of seed autophagy regulators) while the initial probability of all non-seed nodes was set as zero [24,25]. With the iteration steps going on, the probability of the RWR algorithm will become stable. We defined the stable probability as P_∞ when the difference between P_t and P_{t+1} was less than 10^{-10}. The stable probability of P_∞ can be used as a measure of proximity to the seed regulators. If $P_\infty(node_i) > P_\infty(node_j)$, then $node_i$ will be in closer proximity to the seed regulators in the regulatory network than $node_j$. As a result, all candidate nodes in the regulatory network can be ranked according to P_∞ and the top ranked elements can be expected to have a high probability of being associated with autophagy.

2.4. Functional Analysis for Predicted Autophagy Regulators

To demonstrate the credibility of the proposed prediction method, we performed functional analysis for the predicted autophagy regulators. We first retrieved the top 100 ranked regulator candidates (excluding seeds), including TFs, miRNAs and lncRNAs, and performed separately the functional enrichment analyses. For the obtained TFs, we used DAVID to perform GO and Kyoto Encyclopedia of Genes and Genomes (KEGG) pathway enrichment analysis [26]. For the obtained miRNAs, we collected the experimentally verified miRNA targets from the miRecords [19] and miRTarBase [20]; we then used the union set of the miRNA targets to perform GO and KEGG pathway enrichment analysis with DAVID. For the obtained lncRNAs, we utilized the recently developed function annotation tool of non-coding RNA (FARNA), a knowledgebase of inferred functions of human ncRNA transcripts, to implement function annotation analysis. We searched the FARNA database by using each obtained lncRNA, and retrieved promoter-associated transcription factors and transcription co-factors for the lncRNA. Then, all the obtained transcription factors and transcription co-factors were inputted into DAVID to perform GO and KEGG pathway enrichment analysis. In addition, we also performed GO enrichment analysis for the known autophagy-associated TFs, miRNAs and lncRNAs separately, as described above. The union set of the significant GO categories were considered as the autophagy related GO terms. All these DAVID analyses adopted the same criteria that the biological process (BP) category was used for GO analysis, and the significance of enrichment was set at p-value < 0.05. Finally, we calculated the functional similarity scores between the GO terms enriched in the predicted autophagy regulators and the autophagy related GO terms. The computational procedure was implemented using R package GOSemSim [27] and the rcmax method was chosen as a combined method for aggregating multiple GO terms. We also performed 1000 random tests to evaluate the significance of obtained functional similarity scores. In each random test, we randomly chose the same number of GO terms as in the real situation and calculated the functional similarity scores as above. The statistical p-value was calculated as the ratio of random functional similarity scores higher than the real functional similarity score.

3. Results

3.1. Characteristics of the TNCR Network

In this study, we integrated five types of experimentally verified transcriptional and post-transcriptional regulatory relationships from public data resources and constructed a comprehensive TNCR network (see Materials and Methods for details). The TNCR network comprised of 4529 edges, including 155 TFs, 681 miRNAs and 1332 lncRNAs (Figure 1A, Supplementary Table S1). To get an overview of the TNCR network, we examined the degree distribution of the network. As shown in Figure 1B, most nodes (50.4%) had degree one and few nodes had a high degree. In addition, the power-law distribution of the forms $y = 327.4 \times 10^{-1.31}$ $(R^2 = 0.823)$, $y = 157.4 \times 10^{-1.19}$ $(R^2 = 0.773)$ and $y = 224.4 \times 10^{-1.36}$ $(R^2 = 0.774)$ were fitted for degree, out-degree and in-degree respectively. These results indicated that the TNCR network satisfied approximate scale-free topology which is the common feature of most biological networks [28]. Next, we further investigated the in-degree and out-degree distributions for TFs, miRNAs and lncRNAs, respectively (Figure 1C). In general, few nodes had very high degrees and many had low degrees, regardless of TFs, miRNAs or lncRNAs in-degree and out-degree. Furthermore, TFs had a higher median in-degree and out-degree than miRNAs and lncRNAs, which meant that TFs more likely acted as hubs in the TNCR network.

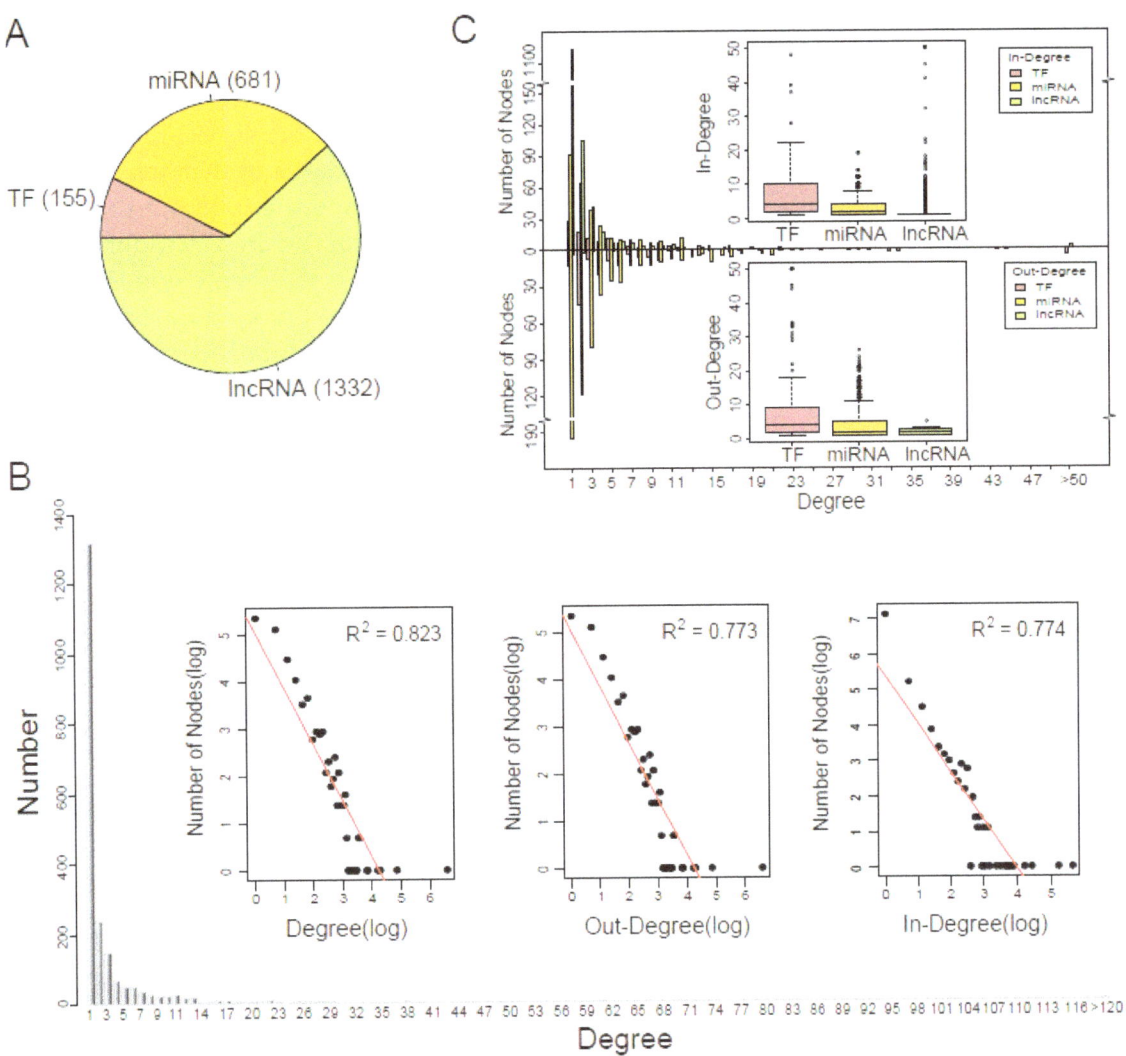

Figure 1. Characteristics of the TFs and ncRNAs coordinated regulatory (TNCR) network. (**A**) Proportion of transcription factor (TF), microRNA (miRNA) and long non-coding RNA (lncRNA) in the TNCR network. (**B**) Degree distribution of all nodes in the TNCR network and the log-log plots for the degree, out-degree and in-degree distributions of all nodes. (**C**) In-degree and out-degree distributions of TFs, miRNAs and lncRNAs in the TNCR network.

3.2. Performance Evaluation of the Proposed Method

By integrating data from AmiGO-2, the autophagy database and the ncRDeathDB, we obtained 1222 known autophagy regulators in total (Supplementary Table S2). After mapping these regulators onto the TNCR network, we finally got 178 autophagy regulators as seeds, including 25 TFs, 152 miRNAs and 1 lncRNAs (Supplementary Table S3). By performing the RWR method on the TNCR network with the seeds, we finally prioritized novel autophagy regulators.

In order to evaluate the performance of our method for inferring autophagy regulators, we performed LOOCV analysis. Each known autophagy regulator was left out in turn as the test case and the other known autophagy regulators were taken as seeds. All the other nodes in the TNCR network were regarded as candidate autophagy regulators. Sensitivity and specificity were calculated for each threshold. Finally, a receiver operating characteristic (ROC) curve was plotted by varying the threshold and then the value of the AUC was calculated. Our method, tested on already known autophagy regulators, achieved an AUC of 0.889 (Figure 2), exhibiting excellent performance. Here, the TNCR network incorporated three kinds of regulators (TFs, miRNAs and lncRNAs) and five kinds of regulations (TF-miRNA, TF-lncRNA, miRNA-lncRNA, miRNA-TF and lncRNA-TF). To demonstrate the effectivity and reliability of the TNCR network, we compared the performance of partial TNCR networks. The AUCs were calculated for a TNCR-ML network (miRNAs and lncRNAs only) and a TNCR-TM network (TFs and miRNAs only) separately by performing LOOCV (the TNCR-TL network (TFs and lncRNAs only) was not analyzed because of missing seed regulators). The AUCs were 0.697 and 0.544 respectively, which were lower than those using the TNCR network (Figure 2). To further determine whether the results of the cross validation might have been generated by chance, we performed randomization tests. The seeds were generated randomly from candidate nodes in all three networks and the AUC values were calculated by performing LOOCV, as above. The AUC values under randomized tests were much lower than those in real situations (0.530, 0.549 and 0.519, respectively, for these three conditions), confirming the valid and reliable performance of autophagy regulator seeds in our method (Figure 2). We also performed RWR on 1000 degree-preserving randomized TNCR networks and the average value of the AUCs was calculated. As shown in Figure 2, the result based on the real TNCR network and the real seed nodes performed best.

The prioritization of all candidate autophagy regulators is provided in Supplementary Table S4. The top 100 ranked candidate regulators, including 55 TFs, 19 miRNAs and 26 lncRNAs, were further validated by literature mining, in which 52 regulators had been verified to be associated with autophagy in published papers (Supplementary Table S5). For example, the fifth ranked regulator MYC was recently proved to mitigate its oncogenic activity by chaperone-mediated autophagy (CMA) regulation [29] and the ninth ranked regulator XIST was determined to increase autophagy activity in non-small-cell lung cancer by regulation of ATG7 [30]. The extensive literature surveys demonstrated the feasibility of our method to predict autophagy regulators.

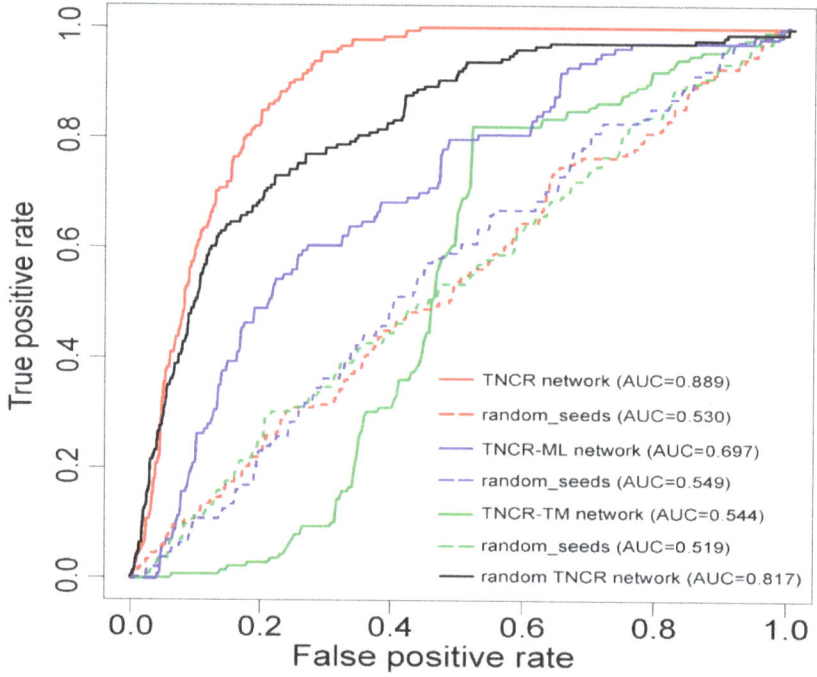

Figure 2. Receiver operating characteristic (ROC) curves and area under the curve (AUC) values for the random walk with restart (RWR) method on the whole, partial and random TNCR networks with real seeds and random seeds. The ROC curves were plotted and AUC values were calculated separately by leave-one-out cross-validation (LOOCV) for the TNCR network, TNCR-ML (miRNAs and lncRNAs only) network, TNCR-TM (TFs and miRNAs only) network and the random TNCR network with real and random seeds.

3.3. Functional Characteristics of Predicted Autophagy Regulators

The top 100 ranked candidate autophagy regulators were retrieved, including 55 TFs, 19 miRNAs and 26 lncRNAs (Supplementary Table S4), then the functional analyses were performed separately for these predicted autophagy regulators (see Materials and Methods for details). The top 20 significantly enriched GO terms and KEGG pathways for TFs are shown in Figure 3. We observed that some cell death related GO terms, such as cell cycle arrest and negative regulation of cell proliferation, were enriched by these top ranked TFs. Several significantly enriched KEGG pathways were also related to cell death, for instance, cell cycle and adherens junction. In addition, some cancer related pathways, such as colorectal cancer, prostate cancer and thyroid cancer, were also enriched, indicating that the autophagy regulators played important roles in cancer. This was consistent with previous studies [11,31,32]. The top 20 significantly enriched GO terms and KEGG pathways by the top ranked miRNAs and lncRNAs are shown in Figure S1 and Figure S2. Similar to the top ranked TFs, the cell death related GO terms and KEGG pathways, such as apoptotic process and cell proliferation, were also enriched by top ranked miRNAs and lncRNAs. Cancer related pathways, such as pancreatic cancer and small cell lung cancer, were enriched by top ranked miRNAs and lncRNAs. We observed that there were obvious overlaps among GO terms and KEGG pathways enriched by top ranked TFs, miRNAs and lncRNAs (Figures 3 and 4A). All of the significantly enriched GO terms and KEGG pathways (*p*-value < 0.05) for top ranked TFs, miRNAs and lncRNAs were shown in Supplemental Table S6.

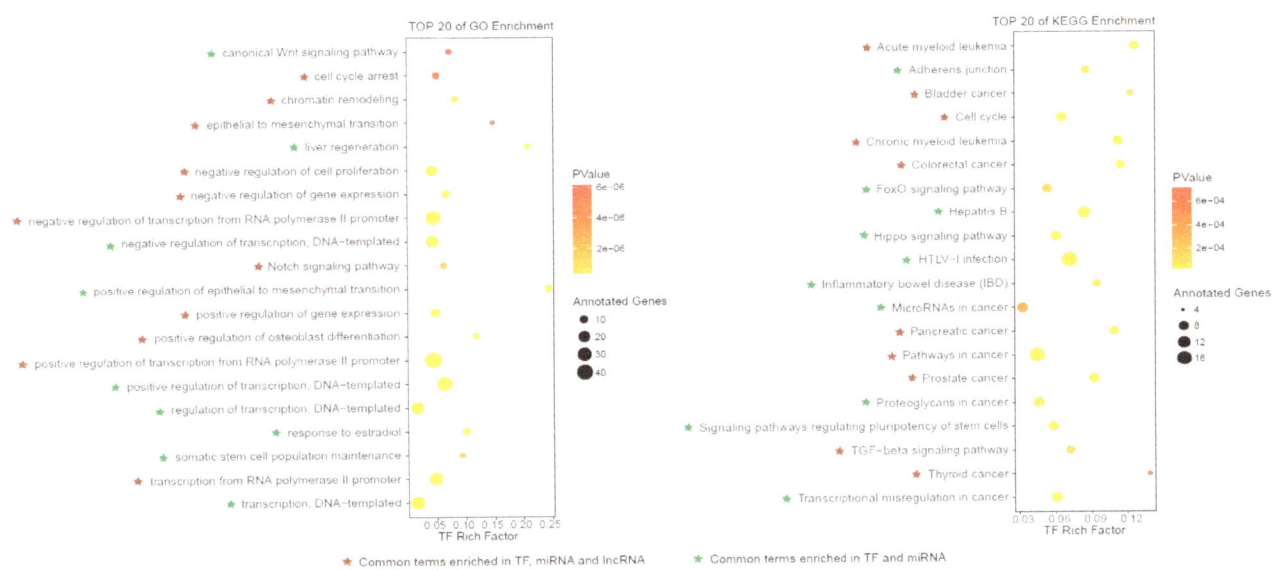

Figure 3. The top 20 Gene Ontology (GO) enrichment and Kyoto Encyclopedia of Genes and Genomes (KEGG) enrichment results for top ranked TFs. The common enriched GO terms and KEGG pathways among top ranked TFs, miRNAs and lncRNAs are marked.

To further evaluate the top ranked regulators associated with autophagy, we compared the GO terms enriched by the top 100 ranked regulators with those enriched by known autophagy-associated factors (including protein-coding genes, miRNAs and lncRNAs). As shown in Figure 4A, the numbers of overlapping enriched GO terms among top-ranked TFs, miRNAs, lncRNAs and known autophagy-associated factors were high (the significantly enriched GO terms for known autophagy-associated factors were shown in Supplemental Table S7). We calculated the functional similarity scores between the GO terms enriched by the top 100 ranked regulators and the autophagy related GO terms. The functional similarity scores between the autophagy related GO terms and those enriched by top ranked TFs, miRNAs, lncRNAs were 0.970, 0.978 and 0.949, respectively. The random functional similarity scores for each kind of regulators, which were calculated by randomly choosing the same number of GO terms as in the real situation, were significantly lower than the real scores (Figure 4B, Figure S3). All these p-values were less than 2.2×10^{-16} (see Materials and Methods for details). This meant that the top ranked regulators were significantly associated with autophagy. The functional characteristics of the top ranked regulators indicated that our method was capable of identifying novel autophagy regulators.

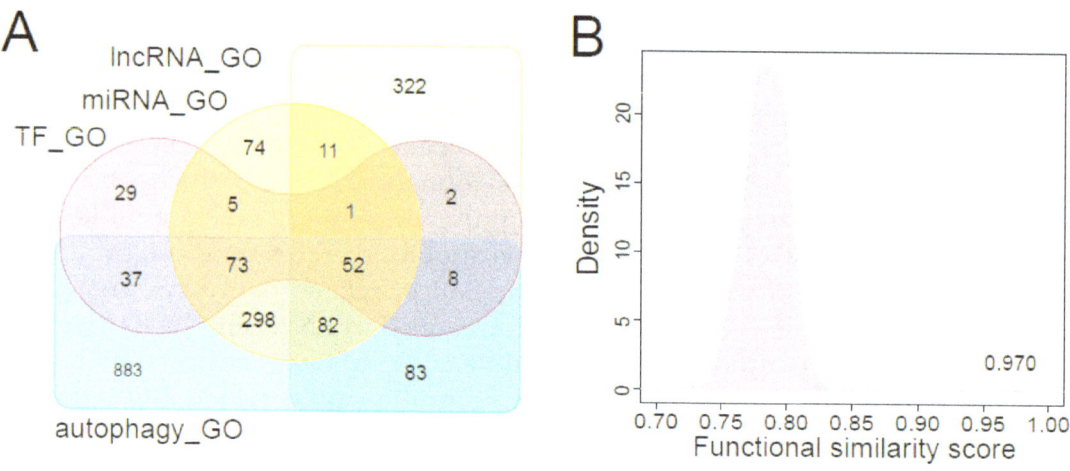

Figure 4. Evaluation of the top ranked regulators associated with autophagy. (**A**) Venn plot for the GO functional annotation comparison among the top ranked TFs, miRNAs, lncRNAs and the known autophagy-associated factors. (**B**) Distribution of random functional similarity scores for the top ranked TFs and the autophagy-associated factors. The triangle indicates the true functional similarity score for top ranked TFs and the known autophagy-associated factors.

4. Discussion

Autophagy is an intracellular catabolic process for maintaining homeostasis and involved systematic regulation at post-translational, transcriptional, and post-transcriptional levels [33]. Both its insufficient and overdriven functions can disturb intracellular homeostasis [34]. Thus, the regulation of autophagy is critical for body cells normal function. Although the knowledge of autophagy regulation is making certain progress, the landscape of autophagy regulators is far from completeness. In addition, autophagy demonstrates a promising therapeutic target in several pathologies [35]. Thus, identification of novel autophagy regulators is beneficial to targeted therapy of complex human diseases. Regulatory networks provide global views of the transmission of genetic information, and are proved to be powerful tools for studying biological issues. In this study, we conducted a computational method to infer novel autophagy regulators based on the regulatory network. We first constructed a comprehensive regulatory TNCR network that incorporated transcriptional and post-transcriptional regulators, including TFs, miRNAs and lncRNAs. Network topological analysis revealed that the degree distribution of the TNCR network approximately followed the power-law distribution. Then, the candidate autophagy regulators were ranked by implementing the RWR method on the TNCR network using the known autophagy regulators as seed nodes. The AUC values determined by LOOCV achieved 0.889, demonstrating the high credibility of our method for recovering known autophagy regulators. Furthermore, functional enrichment analyses revealed that the predicted autophagy regulators were associated with cell death related functional categories such as negative regulation of cell proliferation, cell death and cell cycle arrest. Significantly high functional semantic similarity scores were obtained between the obtained GO terms and the autophagy related GO terms. In addition, extensive literature surveys demonstrated that the top ranked regulators were verified to have associations with autophagy. All these results indicate that our approach is effective in inferring transcriptional and post-transcriptional autophagy regulators and that it would help to improve the understanding of the autophagy regulation mechanisms.

In the past several years, the landscape of TNCR networks has been described elaborately [12,13]. Several experimentally verified transcriptional and post-transcriptional regulatory databases have been developed, such as TransmiR [16], ChIPBase [17], miRTarBase [20] and so on. However, the exhaustive transcriptional and post-transcriptional regulatory relationships still need further elucidation. For example, the characterization of lncRNAs regulation of TFs is still at a primary level [36]. Furthermore, the competing endogenous RNA (ceRNA) relationships involved in TFs, miRNAs and lncRNAs provide further complex regulations among transcriptional and post-transcriptional factors which should be considered in the future analysis of TNCR network [37]. Our approach in this study was based on the general regulatory network TNCR; however, the autophagy plays tissue-specific and double-edged roles in the cellular homeostasis and survival. We believe that the performance of our approach would be improved if we use the data of a specific cancer. In addition, the comprehensiveness of seeds is critical for the performance of the RWR algorithm [38]. Currently, protein-coding regulators of the autophagic machinery are relatively well known, while few studies have been conducted on the non-coding RNA regulators, especially lncRNAs. With the abundance of research of autophagy related regulators, we will obtain comprehensive seed autophagy regulators, and provide more credible, verifiable autophagy regulators.

Author Contributions: Conceptualization, X.C. and W.J.; data curation, Q.M. and X.Z.; formal analysis, S.W. and W.W.; funding acquisition, S.W. and W.J.; investigation, S.W.; methodology, S.Z.; supervision, W.J.; validation, Q.M., H.L. and H.L.; visualization, W.W. and X.M.; writing—original draft, S.W.; writing—review and editing W.J.

References

1. Mizushima:, N. Autophagy: Process and function. *Genes Dev.* **2007**, *21*, 2861–2873. [CrossRef] [PubMed]
2. Mathew, R.; Kongara, S.; Beaudoin, B.; Karp, C.M.; Bray, K.; Degenhardt, K.; Chen, G.; Jin, S.; White, E. Autophagy suppresses tumor progression by limiting chromosomal instability. *Genes Dev.* **2007**, *21*, 1367–1381. [CrossRef] [PubMed]
3. Lock, R.; Kenific, C.M.; Leidal, A.M.; Salas, E.; Debnath, J. Autophagy-dependent production of secreted factors facilitates oncogenic ras-driven invasion. *Cancer Discov.* **2014**, *4*, 466–479. [CrossRef] [PubMed]
4. Homma, K.; Suzuki, K.; Sugawara, H. The autophagy database: An all-inclusive information resource on autophagy that provides nourishment for research. *Nucleic Acids Res.* **2011**, *39*, D986–D990. [CrossRef] [PubMed]
5. Wu, D.; Huang, Y.; Kang, J.; Li, K.; Bi, X.; Zhang, T.; Jin, N.; Hu, Y.; Tan, P.; Zhang, L.; et al. Ncrdeathdb: A comprehensive bioinformatics resource for deciphering network organization of the ncrna-mediated cell death system. *Autophagy* **2015**, *11*, 1917–1926. [CrossRef] [PubMed]
6. Janku, F.; McConkey, D.J.; Hong, D.S.; Kurzrock, R. Autophagy as a target for anticancer therapy. *Nat. Rev. Clin. Oncol.* **2011**, *8*, 528–539. [CrossRef] [PubMed]
7. Galimberti, S.; Petrini, M. Temsirolimus in the treatment of relapsed and/or refractory mantle cell lymphoma. *Cancer Manag. Res.* **2010**, *2*, 181–189. [CrossRef] [PubMed]
8. Turei, D.; Foldvari-Nagy, L.; Fazekas, D.; Modos, D.; Kubisch, J.; Kadlecsik, T.; Demeter, A.; Lenti, K.; Csermely, P.; Vellai, T.; et al. Autophagy regulatory network-a systems-level bioinformatics resource for studying the mechanism and regulation of autophagy. *Autophagy* **2015**, *11*, 155–165. [CrossRef] [PubMed]
9. Jiang, W.; Zhang, Y.; Meng, F.; Lian, B.; Chen, X.; Yu, X.; Dai, E.; Wang, S.; Liu, X.; Li, X.; et al. Identification of active transcription factor and mirna regulatory pathways in Alzheimer's disease. *Bioinformatics* **2013**, *29*, 2596–2602. [CrossRef] [PubMed]
10. Jiang, W.; Mitra, R.; Lin, C.C.; Wang, Q.; Cheng, F.; Zhao, Z. Systematic dissection of dysregulated transcription factor-mirna feed-forward loops across tumor types. *Brief. Bioinform.* **2016**, *17*, 996–1008. [CrossRef] [PubMed]
11. Dai, E.; Wang, J.; Yang, F.; Zhou, X.; Song, Q.; Wang, S.; Yu, X.; Liu, D.; Yang, Q.; Dai, H.; et al. Accurate prediction and elucidation of drug resistance based on the robust and reproducible chemoresponse communities. *Int. J. Cancer* **2018**, *142*, 1427–1439. [CrossRef] [PubMed]

12. Liang, Y.; Wu, H.; Lei, R.; Chong, R.A.; Wei, Y.; Lu, X.; Tagkopoulos, I.; Kung, S.Y.; Yang, Q.; Hu, G.; et al. Transcriptional network analysis identifies bach1 as a master regulator of breast cancer bone metastasis. *J. Biol. Chem.* **2012**, *287*, 33533–33544. [CrossRef] [PubMed]

13. Wang, S.; Li, W.; Lian, B.; Liu, X.; Zhang, Y.; Dai, E.; Yu, X.; Meng, F.; Jiang, W.; Li, X. Tmrec: A database of transcription factor and mirna regulatory cascades in human diseases. *PloS. ONE* **2015**, *10*, e0125222. [CrossRef] [PubMed]

14. Martens, J.A.; Laprade, L.; Winston, F. Intergenic transcription is required to repress the saccharomyces cerevisiae *ser3* gene. *Nature* **2004**, *429*, 571–574. [CrossRef] [PubMed]

15. Carrieri, C.; Cimatti, L.; Biagioli, M.; Beugnet, A.; Zucchelli, S.; Fedele, S.; Pesce, E.; Ferrer, I.; Collavin, L.; Santoro, C.; et al. Long non-coding antisense rna controls uchl1 translation through an embedded sineb2 repeat. *Nature* **2012**, *491*, 454–457. [CrossRef] [PubMed]

16. Wang, J.; Lu, M.; Qiu, C.; Cui, Q. Transmir: A transcription factor-microrna regulation database. *Nucleic Acids Res.* **2010**, *38*, D119–122. [CrossRef] [PubMed]

17. Yang, J.H.; Li, J.H.; Jiang, S.; Zhou, H.; Qu, L.H. Chipbase: A database for decoding the transcriptional regulation of long non-coding rna and microrna genes from chip-seq data. *Nucleic Acids Res.* **2013**, *41*, D177–D187. [CrossRef] [PubMed]

18. Matys, V.; Kel-Margoulis, O.V.; Fricke, E.; Liebich, I.; Land, S.; Barre-Dirrie, A.; Reuter, I.; Chekmenev, D.; Krull, M.; Hornischer, K.; et al. Transfac and its module transcompel: Transcriptional gene regulation in eukaryotes. *Nucleic Acids Res.* **2006**, *34*, D108–D110. [CrossRef] [PubMed]

19. Xiao, F.; Zuo, Z.; Cai, G.; Kang, S.; Gao, X.; Li, T. Mirecords: An integrated resource for microrna-target interactions. *Nucleic Acids Res.* **2009**, *37*, D105–D110. [CrossRef] [PubMed]

20. Chou, C.H.; Shrestha, S.; Yang, C.D.; Chang, N.W.; Lin, Y.L.; Liao, K.W.; Huang, W.C.; Sun, T.H.; Tu, S.J.; Lee, W.H.; et al. Mirtarbase update 2018: A resource for experimentally validated microrna-target interactions. *Nucleic Acids Res.* **2018**, *46*, D296–D302. [CrossRef] [PubMed]

21. Paraskevopoulou, M.D.; Vlachos, I.S.; Karagkouni, D.; Georgakilas, G.; Kanellos, I.; Vergoulis, T.; Zagganas, K.; Tsanakas, P.; Floros, E.; Dalamagas, T.; et al. Diana-lncbase v2: Indexing microrna targets on non-coding transcripts. *Nucleic Acids Res.* **2016**, *44*, D231–D238. [CrossRef] [PubMed]

22. Zhou, Z.; Shen, Y.; Khan, M.R.; Li, A. Lncreg: A reference resource for lncrna-associated regulatory networks. *Database (Oxford)* **2015**, *2015*. [CrossRef] [PubMed]

23. Jiang, Q.; Wang, J.; Wu, X.; Ma, R.; Zhang, T.; Jin, S.; Han, Z.; Tan, R.; Peng, J.; Liu, G.; et al. Lncrna2target: A database for differentially expressed genes after lncrna knockdown or overexpression. *Nucleic Acids Res.* **2015**, *43*, D193–D196. [CrossRef] [PubMed]

24. Li, Y.; Patra, J.C. Genome-wide inferring gene-phenotype relationship by walking on the heterogeneous network. *Bioinformatics* **2010**, *26*, 1219–1224. [CrossRef] [PubMed]

25. Chen, X.; Shi, H.; Yang, F.; Yang, L.; Lv, Y.; Wang, S.; Dai, E.; Sun, D.; Jiang, W. Large-scale identification of adverse drug reaction-related proteins through a random walk model. *Sci. Rep.* **2016**, *6*, 36325. [CrossRef] [PubMed]

26. Huang da, W.; Sherman, B.T.; Lempicki, R.A. Systematic and integrative analysis of large gene lists using david bioinformatics resources. *Nat. Protoc.* **2009**, *4*, 44–57. [CrossRef] [PubMed]

27. Yu, G.; Li, F.; Qin, Y.; Bo, X.; Wu, Y.; Wang, S. Gosemsim: An r package for measuring semantic similarity among go terms and gene products. *Bioinformatics* **2010**, *26*, 976–978. [CrossRef] [PubMed]

28. Barabasi, A.L.; Oltvai, Z.N. Network biology: Understanding the cell's functional organization. *Nat. Rev. Genet.* **2004**, *5*, 101–113. [CrossRef] [PubMed]

29. Gomes, L.R.; Menck, C.F.M.; Cuervo, A.M. Chaperone-mediated autophagy prevents cellular transformation by regulating myc proteasomal degradation. *Autophagy* **2017**, *13*, 928–940. [CrossRef] [PubMed]

30. Sun, W.; Zu, Y.; Fu, X.; Deng, Y. Knockdown of lncrna-xist enhances the chemosensitivity of nsclc cells via suppression of autophagy. *Oncol. Rep.* **2017**, *38*, 3347–3354. [CrossRef] [PubMed]

31. Yu, T.; Guo, F.; Yu, Y.; Sun, T.; Ma, D.; Han, J.; Qian, Y.; Kryczek, I.; Sun, D.; Nagarsheth, N.; et al. Fusobacterium nucleatum promotes chemoresistance to colorectal cancer by modulating autophagy. *Cell* **2017**, *170*, 548–563. [CrossRef] [PubMed]

32. Plantinga, T.S.; Tesselaar, M.H.; Morreau, H.; Corssmit, E.P.; Willemsen, B.K.; Kusters, B.; van Engen-van Grunsven, A.C.; Smit, J.W.; Netea-Maier, R.T. Autophagy activity is associated with membranous sodium iodide symporter expression and clinical response to radioiodine therapy in non-medullary thyroid cancer.

Autophagy **2016**, *12*, 1195–1205. [CrossRef] [PubMed]

33. Das, G.; Shravage, B.V.; Baehrecke, E.H. Regulation and function of autophagy during cell survival and cell death. *Cold Spring Harb. Perspect. Biol.* **2012**, *4*, a008813. [CrossRef] [PubMed]

34. Thorburn, A. Autophagy and its effects: Making sense of double-edged swords. *PLoS biology* **2014**, *12*, e1001967. [CrossRef] [PubMed]

35. Rubinsztein, D.C.; Codogno, P.; Levine, B. Autophagy modulation as a potential therapeutic target for diverse diseases. *Nat. Rev. Drug Discov.* **2012**, *11*, 709–730. [CrossRef] [PubMed]

36. Kopp, F.; Mendell, J.T. Functional classification and experimental dissection of long noncoding rnas. *Cell* **2018**, *172*, 393–407. [CrossRef] [PubMed]

37. Tay, Y.; Rinn, J.; Pandolfi, P.P. The multilayered complexity of cerna crosstalk and competition. *Nature* **2014**, *505*, 344–352. [CrossRef] [PubMed]

38. Zhang, S.W.; Shao, D.D.; Zhang, S.Y.; Wang, Y.B. Prioritization of candidate disease genes by enlarging the seed set and fusing information of the network topology and gene expression. *Mol. Biosyst.* **2014**, *10*, 1400–1408. [CrossRef] [PubMed]

11

MicroRNAs at the Interface between Osteogenesis and Angiogenesis as Targets for Bone Regeneration

Leopold F. Fröhlich

Department of Cranio-Maxillofacial Surgery, University of Münster, Albert-Schweitzer-Campus 1, 48149 Münster, Germany; leopold.froehlich@ukmuenster.de;

Abstract: Bone formation and regeneration is a multistep complex process crucially determined by the formation of blood vessels in the growth plate region. This is preceded by the expression of growth factors, notably the vascular endothelial growth factor (VEGF), secreted by osteogenic cells, as well as the corresponding response of endothelial cells, although the exact mechanisms remain to be clarified. Thereby, coordinated coupling between osteogenesis and angiogenesis is initiated and sustained. The precise interplay of these two fundamental processes is crucial during times of rapid bone growth or fracture repair in adults. Deviations in this balance might lead to pathologic conditions such as osteoarthritis and ectopic bone formation. Besides VEGF, the recently discovered important regulatory and modifying functions of microRNAs also support this key mechanism. These comprise two principal categories of microRNAs that were identified with specific functions in bone formation (osteomiRs) and/or angiogenesis (angiomiRs). However, as hypoxia is a major driving force behind bone angiogenesis, a third group involved in this process is represented by hypoxia-inducible microRNAs (hypoxamiRs). This review was focused on the identification of microRNAs that were found to have an active role in osteogenesis as well as angiogenesis to date that were termed "CouplingmiRs (CPLGmiRs)". Outlined representatives therefore represent microRNAs that already have been associated with an active role in osteogenic-angiogenic coupling or are presumed to have its potential. Elucidation of the molecular mechanisms governing bone angiogenesis are of great relevance for improving therapeutic options in bone regeneration, tissue-engineering, and the treatment of bone-related diseases.

Keywords: bone angiogenesis; osteogenesis; angiogenic-osteogenic coupling; microRNAs; bone regeneration; bone formation; bone tissue-engineering; angiomiRs; osteomiRs; hypoxamiRs

1. Introduction

The replacement of large bone defects and the availability of adequate tissue-engineered bone remains a major clinical challenge. This tremendous demand results from the high incidence of large segmental bone defects due to trauma, congenital malformations, ageing, or bone-related diseases such as osteoporosis, inflammation or tumors [1]. However, the development of gene therapy approaches in recent years demonstrated that tissue-engineered bone offers new therapeutic strategies to repair tissue defects. Thereby, one of the major disadvantages in the clinical use of engineered bone constructs so far, i.e., the inability to provide sufficient blood supply in the initial phase after implantation which leads to insufficient cell integration and cell death, could be overcome [2]. As an explanation for this shortcoming, the function of angiogenesis—the process of forming new blood vessels from pre-existing vasculature—in bone regeneration is still poorly defined, and the molecular mechanisms that regulate angiogenesis in bone are only just starting to be unraveled. Angiogenesis, a term that was coined in 1935 to describe the formation of blood vessels in the placenta, occurs during normal vertebrate embryogenesis but also as a response to pathophysiological circumstances during the processes of tumor formation, wound healing and tissue regeneration [3,4]. Elucidating the wide orchestrating variety of signal pathways and stimuli linking angiogenesis and bone formation on the molecular level is therefore of great interest. In this context, the gained information will yield improved bone replacement in fracture healing, or prevention of bone loss in osteoporosis and therapeutically-induced reparative angiogenesis.

Bone, with its main function in supporting and withstanding mechanical forces, is a mineralized mesenchymal tissue that also possesses an important role in maintaining mineral homeostasis and the energy metabolism of the organism [5]. The formation of bone, which is either developed by the endochondral or by the intramembranous ossification program, is a complex process which depends on physiological interaction with blood vessels that are simultaneously formed [6–9]. Endochondral ossification designates the process of long bone formation which results from the intermediate generation of a primordial cartilage skeleton composed of mesenchymal stem cells (MSCs) that differentiate into chondrocytes and only at a later stage gradually transform into mature bone recruited by different types of bone-forming cells [6]. Intramembranous ossification, in contrast, denominates the process of establishing flat bones where condensed MSCs directly differentiate into osteoblasts, forming an ossification center. This pathway leads to the formation of craniofacial, calvarial, and clavicle bones [10]. In either type of bone development, angiogenesis—the invasion of small blood vessels derived from preexisting blood vessels—is required [11–13]. Bone angiogenesis is induced by growth factors expressed by osteogenic cells such as hypertrophic chondrocytes and osteoblasts at an early stage during osteogenesis [14]. Thereby, the transport of oxygen and nutrients, as well as the further recruitment of osteoprogenitor cells (MSCs) and osteoblasts, are facilitated. In the later phase of bone formation, angiogenesis is essential for trabecular (spongy, cancellous) bone formation and for maturation of the newly-formed bone by close coordination of mineralization and vascularization in either type of bone. In addition, endochondral angiogenesis is particularly important for the replacement of cartilaginous structures at the primary ossification center which generates the bone marrow cavity, and, at a later stage, in establishing the secondary ossification center at the epiphyses (the distal end of long bones) [15,16]. Recent in vitro experiments in transgenic mice demonstrated that this task is accomplished by a specialized type of blood vessel, i.e., the so-called H-type that is present in long bones [17]. This underlines the fact that, comparable to other organs, the acquisition and maintenance of specialized properties by endothelial cells (ECs) is very important for the functional homeostasis in bone.

Bone is a tissue that has to undergo permanent remodeling and requires a counterbalanced process between the anabolic activities of bone formation (osteogenic) cells and the catabolic activities of bone resorption (osteoclast) cells. This activity enforces continuous self-renewal of bone, thereby maintaining an appropriate bone mass and calcium equilibrium. Interference with bone homeostasis could prevent tissue formation, leading to immature or abnormal bone formation [18]. Impaired blood vessel formation, as found in age-related or disease-induced bone loss, therefore, could also result in imbalanced or defective bone formation [19]. In support of this, a reduced density of blood vessels altering the microcirculation was found to be present in osteoporotic bone in mice and humans, that may lead to local abnormal bone metabolism and provoke an increased risk of fracturing [20,21]. Furthermore, the genetic program of bone angiogenesis needs reactivation during callus formation in fracture healing, which represents a complex process that has not yet been fully elucidated [22].

2. Molecular Regulation of Bone Angiogenesis

Bone angiogenesis is mainly governed by a spectrum of transcription factors and growth factors which have been mostly elaborated for endochondral ossification so far [23,24]. It involves interactive signaling between cells of the skeletal system, namely chondrocytes and osteoblasts, and cells derived from the bone vascular system, primarily ECs. Initially, the formation of blood vessels is promoted by osteogenic cells producing pro-angiogenic factors which, in turn, later support the settlement of osteoprogenitor cells [25]. One of the major driving forces behind angiogenic-osteogenic coupling in bone, where oxygen concentrations below 1% are encountered, is the necessity of supplying the tissue with oxygen [15,26,27]. Via tightly hypoxia-regulated induction of the transcriptional activator hypoxia inducible factor (HIF), a cascade of target genes which are involved in a wide variety of biological processes including energy metabolism, erythropoiesis, cell survival, apoptosis and angiogenesis including the major angiogenic regulator vascular endothelial growth factor (VEGF) are expressed in osteogenic cells [28]. The direct mediator of oxygen-dependent modifications of the HIF factors is the von Hippel-Lindau tumor suppressor protein (pVHL), an E3 ubiquitin ligase that targets HIF-1a for proteasomal degradation. During endochondral ossification, VEGF-A is produced by both chondrocytes, particularly in a later stage in which they undergo hypertrophy, and by osteoblasts [29]. As proof of principle, it has been demonstrated that in mice which produce only altered expression levels of a soluble form of VEGF (VEGF120), delayed blood vessel invasion into the primary ossification center and altered osteoblast differentiation in vitro occurs [11]. In addition to HIFs also the fibroblast growth factor (FGF) family of signaling molecules with the members FGF2 and FGF9, and their receptors FGFR1 and 2, were found to be involved in the transcriptional regulation of VEGFA and VEGFR2 expression during formation of blood vessels in bone [30]. Recently, Notch signaling has been implicated as a response to VEGF signaling in ECs in osteogenic-angiogenic coupling [31]. While in other tissues NOTCH expression generally negatively regulates angiogenesis, it seems to have the opposite role in enchondral bone formation by promoting EC proliferation and vessel growth. In a reciprocal fashion, ECs respond by an angiocrine release of NOGGIN, an antagonist of the bone morphogenetic protein (BMP) pathway that stimulates the maturation of hypertrophic chondrocytes expressing VEGF in the growth plate. This could be nicely demonstrated in mice lacking Notch specifically in ECs, which demonstrated reduced bone angiogenesis due to a loss of type-H blood vessels, as well as mutant bone formation and VEGF expression. Other major players are represented by members of matrix metalloproteases (MMPs) due to their proteolytic activity originating from osteoclasts and vascular cells [32]. MMPs also seem to mediate intracellular signaling involving extracellular matrix-integrin interactions necessary during bone angiogenesis and bone remodeling. Besides VEGFA, the placental growth factor (PlGF/PGF), a member of the VEGF family which binds to VEGFR1, has been found to play an exclusively important role in callus remodeling during fracture healing [33,34]. Other known modulating factors of angiogenesis during bone repair include transforming growth factor beta (TGF-β), BMPs, and growth differentiation factor (GDF) [35].

3. The Role of MicroRNAs

MicroRNAs (miRs/miRNAs) are a newly discovered expanding class of endogenous small, non-coding RNAs that positively or negatively regulate gene expression and cellular processes via the RNA interference pathway [5,36–44]. By targeting messenger RNA transcripts post-translationally, they provoke either translational repression or degradation, depending on the degree of sequence homology. Upon precursor transcription from intronic or polycistronic genomic loci by RNA polymerase II, biogenesis of the primary miRNA (pri-miRNA) transcript takes place by a two-step processing mechanism involving the RNAses Drosha and DICER (DGCR8 RNase III complex) [45–47]. Thereby, single or multiple miRNAs that form hairpin-like structures are exported to the cytoplasm by an exportin 5- and RAN-GTP-dependent process, and cleaved. However, in an alternative non-canonical pathway, miRNAs can be configured by direct transcription or refolded spliced introns as endo-shRNAs (endogenous short hairpin RNAs) or as mirtrons, respectively [48]. Subsequently, the targeting strand of the double-stranded mature miRNAs that are 18 to 22 nucleotides in length is integrated into the RNA-induced silencing complex (RISC) with the help of Argonaute proteins. RISC can finally bind specific target (or so-called "seed") sequences of mRNAs represented by 2 to 8 nucleotides located mostly in their 3′-untranslated regions (UTRs) [49,50]. Up- or down- regulation of the miRNA itself by stage- and tissue-specific expression patterns during development can lead to modified expression of its target genes. Thus, miRNAs function as decisive regulatory molecules in many different cellular activities such as development, proliferation, and differentiation, metabolism, or apoptotic cell death, or even cell fate determination and maintenance (e.g., pluripotency control of embryonic stem cells) [51–60]. Moreover, relevant to bone biology, miRNAs have also been identified to acquire endocrine or paracrine functions by their secretion into the blood stream where they subsequently circulate [61]. Consistent with this finding, it was determined that a single miRNA can be involved in coordinating genetic networks by simultaneously regulating the endogenous expressions of multiple target genes. While miRNAs seem to function rather as auxiliary factors during normal physiological processes, their task seems to become more important under stress or disease-related conditions. Accordingly, disturbed miRNA expression is increasingly identified in a number of pathological conditions such as tumorigenesis or viral infection [53,62].

4. MicroRNAs in Bone Angiogenesis: OsteomiRs, AngiomiRs, and HypoxamiRs

It is now well established that miRNAs are physiologically relevant to all steps of bone as well as blood vessel formation during embryonic development and in maintenance during adulthood [63]. OsteomiRs have been identified to regulate chondrocyte, osteoblast, and osteoclast differentiation by positively targeting the principal osteogenic transcription factors and signaling molecules of osteogenesis [5,38,64–69]. In addition to regulating MSC commitment—i.e., the differentiation of precursor cells into chondrocytic and osteoblastic lineages—several studies showed that miRNAs also contribute to the maturation and function of these cells, suggesting also important roles in bone regeneration. Nevertheless, the exact mechanisms of skeletal miRNAs governing the complex interactions and signaling pathways of different bone-forming cells are only beginning to be elucidated. Deregulated miRNAs expression or even genetic variation by mutations or single nucleotide polymorphisms in miRNAs or their binding sites have been identified furthermore in bone disorders such as osteoporosis, osteosarcoma, osteopetrosis, osteogenesis imperfecta, osteoarthritis, and furthermore in bone fracture [63,70–74].

Increasing evidence further indicates that miRNAs act as pro- and anti-angiogenic regulators of adaptive blood vessel growth in normal cardiovascular development and in tumor angiogenesis [75–80]. The role of these so-called angiomiRs or vascular microRNAs in angiogenic development was initially discovered by the detection of severely disrupted blood vessel formation and delayed angiogenic capabilities of ECs in mid-gestational lethal mouse mutants for the miRNA precursor-processing enzyme Dicer [81]. These mutants die around embryonic day 12 to 14 of development due to vascular defects in the embryo and the yolk sac. Furthermore, a smooth muscle-specific *Dicer* deletion in the mouse exerted late embryonic lethality associated with extensive internal hemorrhage which could be explained by a significant loss of vascular contractile function, smooth muscle cell (SMC) differentiation, and vascular remodeling [82]. Knockdown experiments of *Dicer* in zebrafish moreover provoked a phenotype of pericardial edema and inadequate circulation. But also, loss-of-function of the EC-specific miR-126 in homozygous deficient mice caused defects in vascular integrity and angiogenesis [83]. These findings suggested that angiomiRs modulate crucial target genes in cells derived from angioblastic precursor cells and SMC, which are indispensable during embryonic angiogenesis. By investigating the function of Dicer in adult mice and human cells, considerable dysregulated angiogenesis related to growth factor release, ischemia, and wound healing could be revealed, reflecting important postnatal angiogenic functions [80,84,85]. To date, miRNA have been implicated in a long list of cardiovascular diseases comprising myocardial infarction, heart failure, stroke, peripheral and coronary artery disease and several more [86,87]. Nevertheless, the pathological implications of angiomiRs surfaced also with the help of endothelium-specific Dicer-deficient mice, as the ablation led to reduced tumor progression due to diminished angiogenesis, which is a prerequisite for tumor development [88]. For example, two miRNAs induced by VEGF expression (miRs-296, miRs-132) have been identified as candidates supporting the angiogenic switch during tumor formation i.e., the transition from a pre-vascular to a vascularized tumor phenotype [89,90]. In conclusion, the combination of Dicer-deficient angiogenic phenotypes suggests crucial roles for miRNAs in regulating structure and function of embryonic and postnatal blood vessel development.

In the context of angiogenesis, an additional, very important category is a specialized subset of hypoxia-inducible miRNAs, whose increasing number of representatives was also termed hypoxamiRs [91–96]. Thus, reduced oxygen supply in ossification centers of bone stimulate the expression of VEGF and other angiogenic factors that lead to the development of blood vessel structures [97]. Additionally, hypoxia-regulated pathways have been attributed to regulatory functions such as smooth muscle cell proliferation and contractility, cardiac remodeling, cardiac metabolism and ischemic cardiovascular diseases [94]. Together with a variety of other target genes which are important for physiological low oxygen adaption, their expression is initiated by upregulation of the transcription factor hypoxia-inducible factor alpha (HIF) [98]. One group of hypoxamiRs are therefore upregulated following HIF expression (HIF-dependent hypoxamiRs), with the master hypoxamiR-210 being the most prominent example [99,100]. Hypoxia-dependently expressed miRNAs that affect HIF expression itself also belong to hypoxamiRs. Thus, for the adaptation to low oxygen conditions and induction of angiogenesis, HIF displays a unique role by controlling further upregulation of hypoxamiR-424 in ECs, which promotes its own protein stabilization [101]. A last group of hypoxamiRs, moreover, influences HIF expression in the absence of hypoxia. As an example, miR-31 decreases HIF-1α expression via the "factor-inhibiting HIF (FIH)" while the miR17-92 cluster suppresses HIF-1α upon c-MYC induction [102,103].

5. Specific MicroRNAs Implicated in Angiogenic-Osteogenic Coupling

Taken together, the functions of osteomiRs, angiomiRs, and hypoxamiRs suggest the possibility that miRNAs will also have crucial roles in bone angiogenesis. Subsequently, miRNAs will be outlined that were found to have a significant function in osteogenesis as well as angiogenesis, and therefore represent miRNAs that have already been identified to have an active role in angiogenic-osteogenic coupling or are presumed to have its potential (Figure 1, Table 1). Collectively, these may also be referred to as "CouplingmiRs/CPLGmiRs". MiRNAs with a confirmed function in this process could be employed as therapeutic targets in bone regeneration. Consequently, they could improve the coordination and enhancement of the endogenous osteogenesis and angiogenesis process. Elucidation of the molecular mechanisms governing osteoblast differentiation and angiogenesis are furthermore of great importance for improving the treatment of bone-related diseases.

5.1. MiR-9

miR-9 is a highly conserved microRNA, but exhibits a divergent expression pattern, and seems to modulate different targets in a cellular context- and developmental stage-specific manner [104]. The studies of Han et al. have demonstrated a regulatory role for miR-9 in the development and differentiation of human bone marrow derived MSCs (hBM-MSCs) and neural progenitor cells on proliferation, migration and differentiation [105]. In this regard, cell-autonomous effects have been described in vertebrates by effecting Notch, Wnt, and BAF53a expression. Furthermore, miR-9 has also been related to tissue repair processes involving human MSCs (hMSCs). Qu et al. provided evidence that increased miR-9 expression levels closely correlate with enhanced differentiation of MC3T3-E1 osteoblasts [106]. The effects of miR-9 on angiogenesis were studied by the same authors by using human umbilical vein endothelial cells (HUVECs). Here, miR-9 mimics transfection effectively increased VEGF, VE-cadherin, and FGF concentrations in the culture medium, leading to increased EC migration and capillary tube formation in vitro. The underlying molecular mechanism for both the regulation of osteoblast differentiation and angiogenesis was found in the activation of AMP-activated (AMPK) signaling. Conclusively, the results of Qu et al. imply a potential important role of miR-9 in regulating the process of bone injury repair, and therefore, a potential therapeutic target for the treatment of bone injury-related diseases.

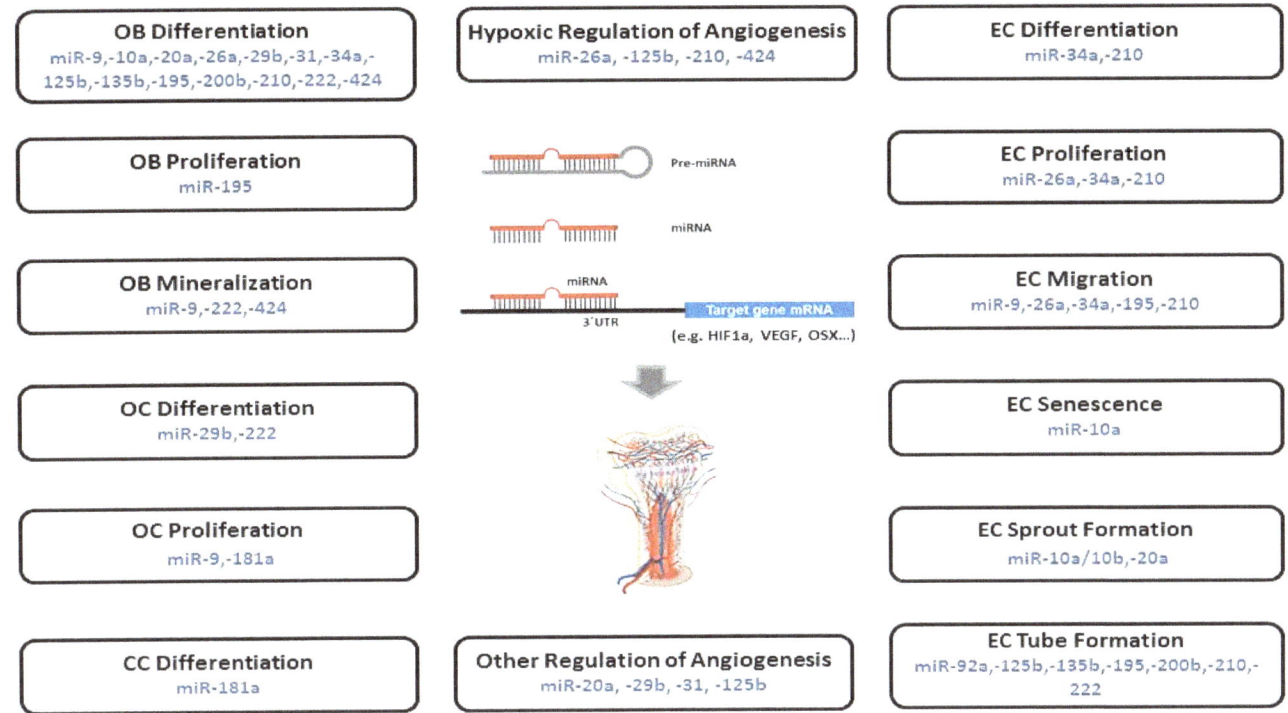

Figure 1. MicroRNAs (miRs/miRNAs) involved in the regulation and coupling of bone angiogenesis ("CouplingmiRs/CPLGmiRs"). Reported miRNAs contributing to the formation of blood vessels during the processes of formation, repair and regeneration of bone were allocated with the individual functions of their target genes during osteogenesis, angiogenesis, or hypoxic regulation of bone angiogenesis. OB, osteoblast; OC, osteoclast; CC, chondrocyte; EC, endothelial cell.

A different study that analyzed the function of miR-9 in C2C12 mesenchymal cells further supported the role of miR-9 in osteoblast differentiation. By significantly decreasing the expression of DKK1 protein, but not of its mRNA, miR-9 stimulated alkaline phosphatase (ALP) activity and osteoblast mineralization, as well as the expression of several osteoblast marker genes, such as COL I (collagen I), OCN (osteocalcin), and BSP (bone sialoprotein) [107]. MiR-9 furthermore was detected as a tumor-secreted pro-angiogenic miRNA that promoted EC migration and tumor angiogenesis in vitro. Mechanistically, this was explained by interference with the expression levels of SOCS5, and thereby, the activation of the JAK-STAT pathway [108].

Moreover, altered expression of miR-9 was found during screening osteoarthritis cartilage involved in the control of tumor necrosis factor α (TNFα) expression. Here, miR-9 has been implicated as a key regulator in the process of endochondral ossification, since its expression varied significantly between the early and late stage of chondrocyte development [73,109]. Functional experiments in a mouse tibial plateau fracture model implicated that miR-9 and miR-181a significantly downregulated Bim concentration. Thereby, osteoclast survival was stimulated and the migration ability of osteoclasts was effected [110].

5.2. MiR-10a

The function of miR-10a during osteoblast differentiation and angiogenesis in vitro was analyzed in MC3T3-E1 and MUVEC (mouse umbilical vein endothelial cells) by the research group of Li et al., respectively [111]. Upon BMP2-induced osteoblast differentiation of MC3T3-E1 cells, miR-10a was downregulated. In contrast, when miR-10a was overexpressed, a suppressive effect on β-catenin and LEF1 expression could be demonstrated. Overexpression inhibited osteogenic differentiation, as demonstrated by reduced expression of the osteoblast-differentiation markers ALP, runt-related transcription factor 2 (RUNX2), Osterix (OSX), and distal-less homeobox 5 gene (DLX5). Moreover, it led to a decrease in MUVEC proliferation, migration and tube formation in combination with reduced concentrations of the angiogenesis-related genes VEGF, VE-cadherin, cyclin D1, and MMP2. As the canonical WNT/β-catenin signaling pathway was found to play an important role in osteogenic cell proliferation, differentiation, and bone regeneration, miR-10a offers a potential therapeutic target for the treatment of bone regeneration and bone-related diseases [112].

A further report referred to the role of miR-10a in regulating endothelial progenitor cell (EPC) senescence in the mouse [113]. Zhu et al. provided evidence that upon upregulation of miR-10A and miR-21, Hmga2 (High-mobility group AT-hook 2) expression is gradually decreased during aging in bone-marrow cells that were enriched for EPCs. Suppression of miR-10A* and miR-21 in aged EPCs, on the other hand, increased Hmga2 expression and improved EPC angiogenesis in vitro and in vivo. This could be demonstrated by rejuvenated EPCs, which resulted in decreased senescence-associated β–galactosidase expression, increased self-renewal potential and decreased p16Ink4a/p19Arf expression. In conclusion, the study demonstrates that the miR-10A*/miR-21–Hmga2–P16Ink4A/P19Arf axis controls EPC senescence and angiogenesis and may represent a potential therapeutic intervention target for improving EPC-mediated angiogenesis and vascular repair.

5.3. MiR-10a/10b

Hassel et al. investigated the function of miR-10 in zebrafish blood vessel formation. During embryogenesis, the knockdown of miR-10a/10b impaired blood vessel outgrowth due to an altered tip cell differentiation behavior, and led to defects of intersegmental vessel growth by modulating fms-related tyrosine kinase 1 (flt1) levels post-transcriptionally [114]. However, as the knockdown of flt1 did not fully rescue the angiogenic phenotypes in miR-10 mutant zebrafish, as well as in miR-10-deficient HUVECs, the authors concluded that in ECs, flt1 could not represent a direct exclusive target of miR-10. They provided evidence that miR-10a/10b regulated angiogenesis in a Notch-dependent manner by directly targeting mib1 (mindbomb E3 ubiquitin protein ligase 1) in zebrafish ECs. Inhibition of mib1 and Notch signaling partially rescued the angiogenic defects in miR-10 morphants, suggesting that the observed angiogenic defects in miR-10a/10b morphants are caused by up-regulation of Notch signaling [115].

5.4. MiR-20a

As a member of the extensively studied miR-17-92 cluster with prominent roles in tissue and organ development, the role of miR-20a was elucidated during osteogenesis by Zhang et al. [116]. They disclosed that, together with several osteoblast markers (BMP2, BMP4, RUNX2, OSX, OCN and OPN), miR-20a was upregulated during osteogenic differentiation of hBM-MSCs derived from bone marrow from differently aged persons. Adipocyte markers, however, such as PPARγ and the osteoblast antagonists, BAMBI and CRIM1, were down-modulated. By introducing miR-20a mimics and lentiviral-miR20a-expression vectors into hBM-MSCs, they verified that miR-20a enhances osteogenic differentiation. Simultaneous direct interaction with all the aforementioned positive and negative effectors of BMP/RUNX2 signaling could be confirmed.

In a report by Doebele et al., miR-20a was investigated with respect to its cell-intrinsic angiogenic activity in EC, as different members of the miR-17-92 cluster that are highly expressed in tumor cells were found to be expressed at increased levels during ischemic conditions [117]. They determined that in vitro overexpression of miR-20a (together with miR-17, -18a, -19a) rigorously inhibited EC sprout formation, whereas their inhibition using antagomiR treatment led to an increase of spheroid sprouting (irrespective of miR-19a). Interestingly, in vivo matrigel plug assays employing antagomiRs for miRs-17-20a as inhibitors demonstrated enhanced angiogenic sprouting, but in contrast to in vitro results, they were unaltered in tumor angiogenesis, indicating context-sensitive regulation. In particular, the pro-angiogenic target gene Janus kinase 1, but also the cell cycle inhibitor p21 and the S1P receptor EDG, were shown to be downregulated by miR-17/20a.

Deng et al. provide evidence that miR-20a and miR-31 serve as stimulators of angiogenesis [118]. As an underlying mechanism, they found that the expression of both miRNA molecules is upregulated via AKT and ERK signals that are themselves activated by the angiogenic factor VEGF. As target genes, miR-20a and miR-31 were found to directly associate with the 3′-UTR of the tumor necrosis factor superfamily-15 (TNFSF15) gene, thereby clarifying its by then unknown mechanistic role in vascular homeostasis. TNFSF15 is expressed in ECs of mature vasculature, and is a known inhibitor of angiogenesis. Interestingly, VEGF-stimulated downregulation of TNFSF15 could be attenuated by treatment of HUVECs with AKT inhibitor LY294002, leading to reduced miR-20a and miR-31 levels, while ERK inhibitor U0126 prevented VEGF-induced expression of miR-20a only. In contrast, inactivation of either ERK or AKT signals restored TNFSF15 gene expression and elevated miR-20a or miR-31 levels which led to enhanced capillary-like tube formation in an in vitro angiogenesis assay.

5.5. MiR-26a/b

Luzi et al. investigated and confirmed an important function of miR-26a during human adipose-derived MSCs (hADSCs) differentiation towards the osteogenic lineage induced by treatment with dexamethasone, ascorbic acid, and beta-glycerol phosphate [119]. Upon inhibition of miR-26a by antisense RNA, upregulation of the transcription factor SMAD1—which was predicted in silico—and its regulated osteogenic differentiation marker genes could be observed in treated osteoblasts. In a follow-up study of Luzi et al., these results were extended to include the interaction between menin and miR-26a as regulators of osteogenic differentiation in hADSCs [120]. Menin is a presumable transcriptional regulator that modulates mesenchymal cell commitment to the myogenic or osteogenic lineages. It is encoded by the *MEN1* oncosuppressor gene which causes the multiple endocrine neoplasia type-1 syndrome. The results demonstrated orchestrated down-regulation of *MEN1* mRNA and miR-26a, with a consequent up-regulation of SMAD1 protein in hADSCs.

Su et al., however, reported an opposing role of miR-26 in hBM-MSC during osteogenic differentiation, suggesting distinct post-transcriptional regulation of tissue-specific hMSC differentiation [121]. Using bioinformatics and functional assays, they confirmed that miR-26a directly regulates *SMAD1*, but added GSK3β as a target to regulate BMP and WNT signaling pathways. The distinct activation pattern and comparative analysis revealed that miR-26a significantly inhibited *SMAD1* to suppress BMP signaling for interfering with the osteogenic differentiation of hADSCs, whereas it targeted on GSK3β to activate WNT signaling for promoting osteogenic differentiation of hBM-SCs. Overall, they concluded that the BMP pathway was more essential for promoting osteogenic differentiation of hADSCs, whereas WNT signaling was enhanced more potently and played a more important role than BMP signaling in osteogenic differentiation of hBM-SCs. In conclusion, although miR-26a enhances osteogenic differentiation in both cell types, different signaling pathways were employed in hBM-MSCs and hADSCs.

In to addition unrestricted somatic stem cells (USSC), the studies of Trompeter et al. investigated a rare population in human cord blood with respect to osteogenic differentiation. Gene expression profiling of two different USSC cell lines (SA5/73 and SA8/25) identified, among other candidates, miRs-26a/b and miR-29b to be consistently upregulated during osteogenic differentiation [122]. As osteo-inhibitory targets of these miRNAs, CDK6 and HDAC4 were evaluated that were downregulated during osteogenic differentiation of USSC, whereas SMAD1 was found as an osteo-promoting target. During osteogenic differentiation of USSC or following ectopic expression of miR-26a/b, SMAD1 exhibited an unchanged expression level, however.

Entangling of miR-26a in pathological and physiological angiogenesis was investigated by Icli et al. in ECs [123]. They studied the effects of modifying the expression of BMP/SMAD1 signaling. Upregulation of miR-26a led to EC cycle arrest, inhibited EC migration, sprouting angiogenesis, and network tube formation in matrigel. Upon inhibiting miR-26a expression, a contrasting phenotype could be detected. At the molecular level, Icli et al. demonstrated direct binding of miR-26 to the 3′-UTR of *SMAD1* thereby reducing its mRNA levels, which subsequently suppressed *ID1* expression and increased *p21WAF/CIP* and *p27* protein expression.

5.6. MiR-29b

MiRNA profiling of MC3T3 preosteoblastic cells derived from fetal mouse calvaria and differentiated to osteoblasts led to the identification of miR-29b, among other members of the miR-29, miR-let-7, and miR-26 families by Li et al. [124]. Versatile effects of miR-29b were found to promote osteoblastogenesis at multiple stages as a key regulator. One mechanism pursues the silencing of negative regulators of osteogenic differentiation, such as TGF-β3, HDAC4, ACTVR2A, CTNNBIP1, and DUSP2 that involve particularly the osteogenic function of RUNX2, as well as the SMAD, ERK, p38 MAPK, and WNT signaling pathways. A second path seeks the suppression of extracellular matrix protein synthesis relevant to bone development (such as COL1A1, COL5A3, and COL4A2) to preserve the differentiated phenotype during mineralization of mature osteoblasts. This alternative mechanism seems to enhance mineral deposition and to prevent fibrosis.

Rossi et al. also disclosed a link of miR-29b to osteoclastogenesis and proposed it for the treatment of multiple myeloma-related bone disease as its expression declined increasingly during human osteoclast differentiation and affected proper bone resorption [125]. Several findings indicated that miR-29b is a negative regulator of human OCL differentiation and activity. Thus, lentiviral transduction of miR-29b into OCLs was associated with diminished tartrate acid phosphatase expression, lacunae generation, and collagen degradation. Attenuated resorptive osteoclast capabilities, due to miR-29b inhibition of proteolytic enzymes, were documented by reduced cathepsin K, MMP-9, and MMP-2 expression. Overall, downstream phenotypic effects along the M-CSF and RANK-L axes that led to impaired action of the master transcription factor NAFTc-1 were explained by miR-29b targeting of c-FOS.

Zhang et al. found inhibitory activity of miR-29b on VEGF secretion via the anti-angiogenic cytokine TNFSF15 (VEGI; TL1A) in the mouse EC line bEnd.3, which defines a new angiogenesis-related signaling pathway [126]. In contrast, down-modulation of TNFSF15 activity by a specific siRNA against its receptor DR3/TNFRSF25, or a neutralizing antibody against TNFSF15, reinstated VEGF generation but suppressed miR-29b expression. TNFSF15-enhanced activation of the JNK-GATA3 signaling pathway was furthermore able to stimulate miR-29b expression but silenced VEGF production, as demonstrated by a specific JNK inhibitor or siRNA.

Li et al. described an anti-angiogenic and anti-tumorigenic role for miR-29b by the regulation of AKT3 expression [127]. AKT is known to induce tumor vascularization via VEGF, and cancer cell activity via c-MYC arrest, in breast tumor. In vitro and in vivo ectopic expression of miR-29b therefore blocked angiogenesis, as well as tumor cell formation, evidencing it as a potential useful anti-cancer therapeutic agent.

5.7. MiR-31

Granchi et al. detected miR-31 by profiling miRNA expression during osteogenic differentiation and mineralization of hBM-MSCs that were derived from three individual donors [128]. As an identified direct target gene, miR-31 differentially modulated the expression of the bone-specific transcription factor OSX during osteogenic differentiation [129]. This could be demonstrated by an inverse miRNA-target expression ratio in osteosarcoma cell lines and an increase in OSX expression upon specific miR-31 inhibition.

Deng et al. provided evidence that the expression of miR-31 increasingly declined during the osteogenic differentiation of hBM-MSC cells [130]. This regulation coincided with increased ALP activity, mineralization of hBM-MSC cultures and expression of the osteogenic transcription factors OPN, BSP, OSX, and OCN, with the exception of RUNX2. Mechanistically, they uncovered a RUNX2, SATB2, and miR-31 regulatory feedback loop that determined hBM-MSC differentiation using inhibitors and mimics of miR-31. RUNX2 directly regulates miR-31 expression levels which itself controls the translation of SATB2 protein.

In a study of Suarez et al. that investigated TNF-mediated induction of endothelial adhesion molecules, the expression of miR-155, -31, -17, and -191 were found to be increased without a change of miR-20a, -222, and -126 levels in HUVECs [84]. By miRNA target prediction algorithms, the 3′-UTR of E-selectin was identified as target molecule, which was verified experimentally by reporter assays using miR-31 mimics. miR-31-mediated regulation of E-selectin expression not only regulates binding of neutrophil granulocytes to HUVECs, but is also involved in inhibition of angiostatin-induced angiogenesis by affecting cell migration [131].

An angiogenesis-related function of miR-31 reported by Deng et al. has already been described above in the subsection of miR-20a [118].

5.8. MiR-34a

Opposing roles for miR-34a in differentiation of hMSC towards osteoblast have been reported. Chen et al. identified miR-34a in a microarray screening, and stated that it inhibits osteoblast differentiation in hMSC and in vivo bone formation in a preclinical model of heterotopic bone formation in mice [132]. Thus, when miR-34a was overexpressed in hMSC, it inhibited early and late OB commitment, as well as differentiation and hMSC proliferation, while anti-miR oligonucleotide treatment reversed these effects. In addition to several cell cycle regulator and cell proliferation proteins (including CDK4, CDK6, and Cyclin D1), JAGGED1, a NOTCH1 receptor ligand, was elicited as a target gene of miR-34a that was regulated at both the transcriptional and translational levels, as determined by RNA interference. JAGGED1 has previously been implicated in human bone biology, as its deficiency causes skeletal abnormalities in the Alagille syndrome.

Kang et al., however, claimed that miR-34a-5p-induced activation of the Notch signaling pathway is a positive regulator of glucocorticoid-mediated osteogenic differentiation of hMSCs [133]. They demonstrated dexamethasone-inhibited osteoblastic differentiation of murine BM-MSC via miR-34a-5p-mediated gene silencing of coincidently identical target genes, as published by Chen et al. Differences in both reports were discussed to be due to different roles of the Notch signaling pathway in osteogenic differentiation of hMSCs that were derived from different species.

A more recent publication by Fan et al. found miR-34a upregulation during osteogenic differentiation of hADSCs [134]. Elevated levels of miR-34a in hADSCs promoted mineralization, ALP activity, and the expression of the key regulatory osteogenic transcription factor *RUNX2* by targeting the retinoblastoma binding protein 2 (RBP2); furthermore, heterotopic bone formation was enhanced in vivo. Expression of *NOTCH1* and *Cyclin D1* genes, that were also involved in this coregulatory network, were found to be downregulated on the other hand, which facilitates cell cycle exit [135,136]; this is the consequence of suppressed proliferation but enforced terminal maturation of osteoblasts by RUNX2.

When investigating the role of miR-34a in glucocorticoid-induced osteonecrosis of the femoral head (GIOFH), Zha et al. verified the findings of Kang et al., but extended this knowledge [137]. By investigating dexamethasone-treated rat subjected to miR-34a-overexpressing lentiviruses, decreased blood vessel development was observed, indicating that VEGF presents a regulatory target of miR-34a. In vitro, miR-34a overexpression enhanced the inhibitory effects of dexamethasone on the viability and activity of ECs and downregulated VEGF protein expression levels.

A study by Zhao et al. investigated the angiogenic role of miR-34a in EPCs derived from adult male Spraque-Dawley rats [138]. The rationale behind this project was the previous implication of miR-34a in targeting silent information regulator 1 (*SIRT1*), provoking cell cycle arrest or apoptosis. The results confirmed the inhibitory effects of raised miR-34a expression levels on *SIRT1* and, thus, on EPC-mediated angiogenesis by inducing senescence. Mechanistic causes could be found in increased acetylated levels of the FOXO1 transcription factor, regulated by *SIRT1*, which could be demonstrated by knockdown of *SIRT1*. The angiogenesis-promoting role of miR-34a and Sirt1 in this context seems to lie in its previously discovered function in vascular endothelial homeostasis by preventing stress-induced senescence in health and disease [139].

Table 1. Summary of microRNAs with a presumed role in osteogenic-angiogenic coupling paving the way for bone angiogenesis ("CouplingmiRs/CPLGmiRs").

MicroRNAs	Targets [1]	Regulatory Role	Effects	Study Models	Ref.
MiR-9	VEGF, VE-CAD (CD144)	AMPK signaling pathway	Enhanced osteogenic diff. & mineral.; increased angiogenesis	MC3T3-E1	[106]
	DKK1	COL1, OCN, BSP; ALP activity	OB diff. & mineralization	C2C12 cells	[107]
	SOCS5	JAK-STAT signaling pathway	Promotion of EC migration & angiogenesis	Primary microvascular ECs, HUVECs	[108]
	Cbl	Bim ubiquitination, apoptosis	Promotion of OC survival	OC, OC precursor cells (RAW264.7)	[110]
MiR-10a	β-catenin, LEF1, VEGF, VE-CAD (CD144), cyclin D1, MMP2	Wnt signaling; angiogenesis-related gene expression	Inhibition of osteogenic diff. & blood vessel formation	MC3T3-E1 MUVECS	[111]
	HMGA2	β-galactosidase expr; p16Ink4a/p19Arf expression	EPC senescence & angiogenesis; self-renewal potential	lin−BM-MSCs	[113]
MiR-10a/10b	MIB1	Notch signaling	Regulating blood vessel outgrowth/tip cell behavior	HUVECs	[115]
MiR-20a	BMP2, BMP4, RUNX2	Effects BMP/RUNX2 signaling positively; blocks OB inhibitors & PPARγ	Enhances osteogenic differentiation; suppresses adipogenesis	hBM-MSC	[116]
	JAK1; p21, S1P receptor EDG	Downregulation of proangiogenic JAK 1 & cell cycle inhibitors	Inhibits EC sprout formation	HUVECs	[117]
	TNFSF15	VEGF-AKT/ERK –miR20a/31 signaling	Stimulation of angiogenesis	HUVECs	[118]
MiR-26a	VEGF, ANG1, RUNX2, BMP2 OCN, ALP; GSK3β	WNT signaling activation	Enhanced angiogenesis & bone regeneration	Primary hBM-MSC, MC3T3-E1	[76,98]
	VEGF	PIK3C2α/AKT/HIF-α/VEGFA pathway	Inhibition of angiogenesis;	HUVECs	[141]
	SMAD1	BMP signaling inhibition	OB differentiation	hADSCs	[119]
	SMAD1	BMP signaling	Inhibits EC growth, proliferation, migration; regulates early angiogenesis	HUVECs	[123]
MiR-29b	TGF-β3, HDAC4, ACTVR2A, CTNNBIP1, DUSP2; COL1A1, 5A3, 4A2	Silences neg. osteogenic regulators suppresses ECM protein synthesis	Promotes osteoblastogenesis at multiple stages	MC3T3 pre-OB	[124]
	c-FOS	Reduced TRAP expr, lacunae generation, collagen degradation	Neg. regulator of human OC differentiation and activity	OC (CD14 +)	[125]

Table 1. *Cont.*

MicroRNAs	Targets [1]	Regulatory Role	Effects	Study Models	Ref.
	TNFSF15	TNFSF15-enhanced JNK-GATA3 signal. & VEGF inhibition	Suppression of VEGF secretion	Mouse EC line bEnd.3	[126]
	AKT3	Inhibition of tumor vascularization via VEGF & cancer cell activity via c-MYC	Anti-angiogenic and anti-tumorigenic role	HUVECs, Breast cancer cells	[127]
	OSX	Downregulation of OSX	Influences osteogenic differentiation	hMSC; Osteosarcoma cell	[129]
MiR-31	Satb2 protein	Inhibition by RUNX2; Upregulation of Satb2 protein & osteogenic TF	Induces BM-MSC osteogenic differentiation	hBM-MSC	[130]
	E-selectin	Regulation of E-selectin expression	Inhibition of angiostatin-induced angiogenesis; TNF-mediated induction of endothelial adhesion	HUVECs	[84]
	TNFSF15	VEGF-AKT/ERK –miR20a/31 signaling	Stimulation of angiogenesis	HUVECs	[118]
	Jagged1	Regulation of cell cycle regulator & proliferation proteins & Jagged1	Inhibition of osteoblast differentiation	hMSC; mouse heterotopic bone formation model	[132]
	JAGGED1	Activation of Notch signaling	Induction of glucocorticoid-mediated osteogenic differentiation	hMSC	[133]
MiR-34a	RBP2	Promotes mineral, ALP activity & RUNX2 expression; downreg. NOTCH1 & Cyclin D1 expr.	Promotion of osteogenic differentiation; enhanced heterotopic bone formation	hADSCs; mouse heterotopic bone formation model	[134]
	VEGF	Inhibitory effects of dexamethasone on EC viability & VEGF	Decreased blood vessel development	Rat Glucocorticoid-induced osteonecrosis	[137]
	SIRT1	Increased SIRT1 expr. & FOXO1 acetylation regulating vascular EC homeostasis	Inhibition of EPC-mediated angiogenesis	Rat EPC	[138]
	E2F3a, survivin	Interference with VEGF secretion, EC proliferation & migration	Dysregulated tumor angiogenesis	HNSCC tumors & cells	[140]
	?	?	Enhanced fracture healing & inhib. of neovascularization	Mice with femoral fracture	[142]
MiR-92a	HGF, ANGPT1	ITGA5, MEK4	Inhibition of tube formation by HUVECs	hADSCs	[143]
	?	integrin a5, sirtuin1, eNOS	Attenuates neointimal lesion by accelerating re-endothelialization	MiR-92a knockout mice	[144]

Table 1. *Cont.*

MicroRNAs	Targets [1]	Regulatory Role	Effects	Study Models	Ref.
	OSX	RUNX2, a-SMC, ALP, matrix mineralization	Calcification of vascular smooth muscle cells	HCASMCs	[145]
	ErbB2	?	Inhibits OB diff by downreg. of cell proliferation	ST2 cells (mMSCs)	[146]
MiR-125b	VEGF, ERBB2		Regulation of angiogenesis during wound healing	HUVECS	[147]
	Cbf-beta	ALP, OCN, OPN	Inhibition of osteogenic differentiation	C3H10T1/2	[148]
	SMAD4	ALP, RUNX2	Downregulation of osteogenic differentiation	hMSCs	[149]
	VE-Cadherin		Inhibition of blood vessel (tube) formation	HUVECs	[150]
	?	?	OB differentiation	hBM-SCs	[151]
MiR-135b	HIF-1	?	Enhanced endothelial tube formation	Human MM cells; HUVECs	[152]
	SMAD5	?	Impaired osteogenic differentiation	hMSCs	[153]
	?	CCN1, aggrecan	Maintaining homeostasis of chondrocytes	Human HCS-2/8 cells	[154]
	COL10A1		Chondrocyte differentiation	hMSC	[155]
MiR-181a	RGS16	CXCR4 signaling; VEGE, MMP1	Angiogenesis & metastasis in chondrosarcoma	Xenograft mice; JJ chondrosarc. cells	[156]
	?	VEGF expression	Chondrosarcoma-associated angiogenesis	JJ chondrosarc. cell line	[157]
	Cbl	Bim ubiquitination, apoptosis	promote OC survival	OC, OC precursor cells (RAW264.7)	[110]
MiR-195	?	VEGF	Osteogenic diff. & proliferation; control of angiogenesis	hMSC(MC3T3) chick chorio-allantoic membrane	[158]
	?	VEGF, VAV2 CDC42	HCC-associated angiogenesis & metastasis; migration & capillary tube form. of ECs	QGY-7703, MHCC-97H HCC cells; HUVECs	[159]

Table 1. *Cont.*

MicroRNAs	Targets [1]	Regulatory Role	Effects	Study Models	Ref.
MiR-200b	ZEB1	ZEB1-TF target genes	Inhibits proliferation, migration & invasion of osteosarcoma cells	OsteosarcomaU2OS, Saos2, HOS, MG63	[160]
	VEGF-A; ZEB2, ETS1, KDR,GATA2	Decreases VEGF-A expression & TF-target genes	Inhibition of VEGF-A induced osteogenesis; Inhibition of TF-activated angiogenesis	Rat BM-MSC & HUVEC coculture	[161]
	VEGF, FLT-1, and KDR	VEGF-induced phosph. of ERK1/2	Inhibition of angiogenesis; red. capillary formation	A549 cells, HUVECs	[162]
	AcvR1b	Inhib. of TGFb/activin signaling	Promotes OB differentiation	ST2 stromal cells	[163]
MiR-210	VEGF	PPARgamma, ALP, OSX	Promoteion of OB diff., inhibition of adipocyte diff.	hBM-SCs, 17β-estradiol (E2)treated OB	[164]
	EFNA3	VEGF-expression mediated angiogenesis	EC survival, diff., migration; stim. of tubulogen. & chemotaxis	HUVECs	[165]
	SMAD 1, 5, 8 protein & phosphoryl.	Decreased SMAD5-RUNX2 signaling & OSX, ALP, and OC levels & mineral.	Neg. regulator of osteogenic differentiation	hBM-SC	[166]
MiR-222	c-Src, Dcstamp	RANKL-induced expression of TRAP & cathepsin K	Inhibitory regulator of c-Src-mediated osteoclastogenesis	RAW264.7 pre-OC cells	[167]
	c-KIT	Suppression of tube formation, wound healing, cell migration via SCF	Inhibitory regulation of in vitro angiogenesis	HUVEC	[168]
	RUNX, CBFβ, BMP	Osteogenic diff. of hMSCs	Bone formation	hMSCs	[169]
	MAPK, WNT & insulin signal.	OB differentiation of hMSCs	Bone formation	hMSCs	[170]
MiR-424	FGF-2; via FOXO1	Decrease of ALP, mineralization & osteog. markers	Enhances proliferation & osteogenic differentiation of hMSCs	Pigs, cellular oxidative stress model	[171]
	CUL2; via RUNX-1→ C/EBPα→ PU.1	Stabilization of HIF-1α	Regulation of Angiogenesis	ECs, ischemic tissues	[101]

[1] Identified target genes or downstream effectors; HUVECs, human umbilical vein endothelial cells; ?, unknown molecular target(s) and/or regulatory role; MUVECs, mouse umbilical vein endothelial cells; EC, endothelial cells; EPC, endothelial progenitor cells; hMSCs, human mesenchymal stem cells; hBM-SC, bone-marrow derived stem cells; hADSCs, human adipose-derived mesenchymal stem cells; TNFSF15; cytokine tumor necrosis factor superfamily 15; OB, osteoblasts; OC, osteoclasts; TF, transcription factor.

A negative regulatory role for miR-34a could also be documented by Kumar et al. in tumor angiogenesis in head and neck squamous cell carcinoma (HNSCC) [140]. MiR-34a expression was markedly decreased in HNSCC tumors and cell lines, and ectopic expression of the miR-34a therefore affected cell proliferation and migration in HNSCC cell lines and in a SCID mouse xenograft model. These effects seemed to be mediated by the regulation of survivin expression via the transcription factor E2F3a that is critical for cell cycle progression. Furthermore, tumor angiogenesis was found to be dysregulated by interference of miR-34a with VEGF secretion in tumor cells, as well as EC proliferation, migration, and tube formation, by downregulating a number of key proteins including E2F3, SIRT1, survivin, and CDK4.

5.9. MiR-92a

The autosomal dominant Feingold syndrome that is characterized by microcephaly, short stature, and digital anomalies was identified in individuals carrying hemizygous deletions of the miR-17-92 cluster [172]. To dissect this complex phenotype that could be partially mirrored in mice deficient of the miR-17-92 cluster, and due to postnatal lethality, a single miR-92a targeted mouse line was established by Penzkofer et al. [173]. Surviving mice exhibited reduced body weight and skeletal defects represented by reductions in body length, skull, and tibia length, as well as metacarpal bone size; these defects were presumably also the result of osteoblast proliferation and differentiation phenotypes, as found in hemizygous miR-17-92 mouse mutants [174]. In contrast, however, single miR-92a deletion does not cause low bone mineral density attributed to reduced type-I collagen mRNA, ALP activity, and mineralization ability. Phenotypic differences could be explained by the design of the genomic deletion of miR-92a that partially attenuated expression levels of miR-18a in skeletal tissue. The direct molecular targets of miR-92a responsible for regulation of osteogenesis have not yet been elicited.

A recent study of Mao et al., however, identified aggrecanase-1 and aggrecanase-2 (ADAMTS4/5) as direct miR-92a targets in chondrogenic hMSCs and human chondrocytes by reporter assays [175]. These two members of the ADAMTS family represent MMPs that are important for normal chondrocyte differentiation, but which also promote the progression of osteoarthritis by cartilage degeneration. In comparison to normal cartilage, real-time-PCR analyses also revealed higher miR-92a-3p expression levels in chondrogenic hMSCs, whereas markedly reduced miRNA expression could be detected in ostearthritis cartilage. Moreover, miR-92a-3p-modified expression of ADAMTS-4/5 could be downregulated by IL-1β transfected primary human chondrocytes. The study was preceded by previous findings about miR-92a-3p upregulation in hADSC-derived chondrocytes and chondrogenic hMSCs, as well as osteoarthritis cartilage, where histone deacetylase 2 (HDAC2) was identified as a target gene of miR-92a-3p [176,177].

MiR-92a was first reported as a part of the miR-17-92 cluster by Bonauer et al., and was found to be abundant in human ECs [178]. MiR-92 was established as a suppressor of angiogenesis which targets the expression of several proangiogenic proteins, including the integrin subunit alpha 5 (ITGA5) that is well known for severe vascular defects in gene targeted mice. Systemic administration of miR-92a antagomiR therefore improved the growth of blood vessels and functional recovery of damaged tissue in mouse models of limb ischemia and myocardial infarction, possibly by indirectly inhibiting apoptosis [58]. Thus, miR-92a may serve as a valuable therapeutic target in the setting of ischemic disease.

Daniel et al. further elaborated the function of miR-92a upon re-endothelialization and neointimal formation after wire-induced injury of the femoral artery in mice [144]. By using specific LNA (locked nucleic acids)-based antimiRs as well as miR-92a-deficient ECs in the mouse, they found enhanced re-endothelialization and inhibited neointimal formation induced by de-repression of the miR-92a targets integrin a5 and sirt1 [179]. Thus, an important role can be attributed to miR-92a in blood vessel regeneration in ischemic tissues and after vascular injury.

In a subsequent study, Zhang et al. investigated the ability of miR-92a to influence apoptosis and angiogenesis in ECs in the presence of oxidative stress [180]. It was previously determined that senescent ECs that undergo apoptosis, and which are frequent in ageing and atherosclerosis, have diminished miR-92a levels [181]. The results provided evidence that pre-miR-92a treatment of HUVECs prevented oxidative stress-induced apoptosis in EC, whereas capillary tube formation i.e., angiogenesis, was maintained. Mechanistically, it was determined that pre-miR-92 directly repressed PTEN, and thereby, activated the AKT signaling pathway that regulates EC apoptosis and angiogenesis.

In a study of Kalinina et al., paracrine effects of miR-92a on hMSCs were studied [143]. Conditioned medium of hMSCs transfected with pre-miR-92a prevented tube formation by HUVECs, which could be attributed to significantly lower secretion of hepatocyte growth factor (HGF) and angiopoetin-1 independent of VEGF secretion. HGF is a factor required for stimulation of proliferation of ECs, whereas angiopoietin-1 regulates vessel stabilization and maturation. As neither gene was predicted as direct miR-92a targets, these still have to be identified. Restoration of tube formation was achieved by replenishment of HGF, but not with anti-miR-92a treatment of hMSCs. This led to the conclusion that miR-92a suppresses hMSC-induced angiogenesis by downregulating the secretion of HGF and, therefore, is involved in the control of anti-angiogenic activities in hMSCs.

5.10. MiR-125b

MiR-125b is an early discovered miRNA that plays a key role in cellular functions. Experiments by Mizuno et al. using BMP-4-induced or exogenous miRNA-transfected murine MSC ST2 cells found miR-125b to inhibit osteoblastic (and adipogenic) differentiation through modulating cell proliferation [146]. Further experiments with transfection of siRNA identified ErbB2 as a target gene. miR125b-gene targeted mice, generated by Lee et al., identified further supported the important function of miR-125b in regulating stem cell directional differentiation, but did not reveal its molecular role [182].

Decreasing levels of miR-125b were also identified by Huang et al. during the differentiation of C3H10T1/2 cells, and reporter gene assays led to the identification of a putative target binding site in the 3′-UTR of the Cbfβ gene, a master regulatory gene of osteogenesis [148]. Therefore, silencing of miR-125b increased the mRNA levels of Cbfβ and of osteoblastic marker genes ALP, OCN, and OPN. Conclusively, RUNX2 is considered as an indirect target of miR-125b as well.

Transient expression of miR-125b expression in hypoxic VEGF-or or bFGF stimulated ECs was moreover shown by Muramatsu et al. to directly bind and block the translation of vascular endothelial (VE)-cadherin mRNA and therefore inhibit in vitro tube formation by ECs [150]. Thus, miR-125b may be tested in tumor therapy for inducing disruption of blood vessel formation.

5.11. MiR-135b

During the osteogenic differentiation of several USSC, miR-135b was identified as the most consistently down-regulated candidate by Schaap-Oziemlak et al. [151]. Retroviral overexpression resulting in decreased mineralization confirmed a function of miR-135b in osteogenesis of USSC. Furthermore, quantitative RT-PCR analysis of USSC that overexpressed miR-135b showed decreased expression of the bone mineralization markers IBSP and OSX.

In a profiling study by Xu et al. that investigated the exosomal content of hBM-MSCs during osteogenic differentiation, miR-135b was found to be significantly increased [183]. Bioinformatic analysis led to the conclusion that several important pathways related to osteoblastic differentiation were engaged, and that exosomal miRNA is a regulator thereof.

By establishing hypoxia-resistant multiple myeloma (MM) cells through exposure to chronic hypoxia, Umezu et al. presumably identified a new mechanism of hypoxia-induced angiogenesis targeting the FIH-1/HIF-1 signaling pathway via exosome-contained miRNAs [152,184]. In contrast to the transiently hypoxia-upregulated miR-210 transcriptional levels which decline gradually under normoxic conditions, miR-135b levels were maintained high, even under normoxic conditions. By delivering exosomes to HUVECs, miR-135b is enabled to target the factor-inhibiting hypoxia-inducible factor-1 gene (FIH-1) that encodes an asparaginyl hydroxylase enzyme which inhibits HIF-1a. The positive correlation between miR-135b, HIF-1a, and microvessel density was initially identified by Zhang et al. in a HNSCC model [185]. As a result, endothelial tube formation could be promoted under hypoxic as well as normoxic environments.

5.12. MiR-181a

Among other functions, the miR-181 family has been implicated in genetic regulation of early hematopoiesis and lymphangiogenesis [60,186,187]. Microarray analysis of human HCS-2/8 cells led to the characterization of miR-181a. Sumiyoshi et al. accredited miR-181a with an important function in the maintenance of cartilaginous metabolism by a negative feedback system employing repression of the CCN family member 1 (CCN1) and aggrecan (ACAN) genes, which are both known to be involved in chondrocyte differentiation [154].

In another study, the examination of synovial fluid cells in a mouse model of tibial plateau fractures led to the detection of two downregulated miRNAs, miR-9 and miR-181a [110]. Cbl, an important E3 ubiquitin ligase for bone resorption that was tested as a putative target gene of miR-9 and miR-181a, elicited increased amounts of ubiquitinated Bim, a pro-apoptotic gene in mouse primary osteoclast cells. Therefore, Wang et al. concluded that upregulated Cbl might regulate the survival rate of primary mouse osteoclast cells, as previously Cbl-dependent apoptosis via ubiquitinated Bim was reported [188].

Sun et al. implicated miR-181a as a potential oncomiR (cancer-associated miRNA) in the angiogenesis and metastasis of chondrosarcoma in a xenograft mouse model [156]. They found that miR-181 increases VEGF and MMP1 expression, as well as CXCR4 signaling, via negatively regulating RGS16 (regulator of G-protein signaling 16) under hypoxic cell culture conditions. RGS16 is an inhibitor of CXCR4 which regulates, upon amplified signaling, angiogenesis, invasion, and metastasis in chondrosarcoma. The therapeutic usefulness of this mechanism was proven by miR-181a antagomiR treatment, which decreased proangiogenic gene expression as well as tumor growth and metastasis in a xenograft mouse model.

5.13. MiR-195

Together with miR-497, miR-195 was downregulated in a microarray screen of primary hMSCs performed by Almeida et al. [158]. hMSCs underwent induced osteogenic differentiation with the aim of identifying candidates that are capable of contributing to bone fracture repair. Osteogenic markers were therefore found to be diminished upon overexpression or increased upon inhibition of miR-195 in hMSCs. Using the chicken CAM assay, studying the paracrine effects of hMSCs, the authors furthermore demonstrated decreased blood vessel formation in vivo. VEGF was identified as the target gene that mediates this phenotype, at least in part. MiR-195 interacted with the VEGF 3´-UTR in bone cancer cells and also regulated mRNA and protein expression levels.

Reduced expression of miR-195 furthermore resulted in increased angiogenesis and metastasis in HCC tissues, whereas either loss-of-function or gain-of-function abrogated the ability of HCC cells to migrate and induce capillary tube formation of ECs, as reported by Wang et al. [159]. In xenograft tumors, upregulated miR-195 expression provoked reduced microvessel densities and the formation of metastases. Detailed molecular investigations disclosed VEGF and the prometastatic factors VAV2 and CDC42 as direct targets of miR-195, which could be proven by mirroring the phenotype either by knockdown or overexpression of these target genes. The group also demonstrated that higher VEGF levels due to miR-195 down-regulation promoted EC-mediated tumor angiogenesis by the involvement of VEGF receptor 2 signaling.

5.14. MiR-200b

miR-200b was identified by microarray screening as a miRNA that exhibits downregulated expression upon exposure of osteoblasts to collagen and to silicate-based periodontal grafting material (PerioGlas, P-15) that is used to promote bone formation [189]. Rønbjerg et al. demonstrated miR-200b as a potent regulator of the target gene ZFHX1B via direct interaction [190]. ZFHX1B is a transcriptional repressor involved in the regulation of the TGFβ signaling pathway and epithelial-mesenchymal transitions by E-cadherin, a mediator of cell-cell adhesion, in mesenchymal cells.

In addition to the known communication of interacting cells by the exosomal release of miRNAs (e.g., mir-135b [153,191]), Fan et al. discovered angiogenic-osteogenic cell coupling and reciprocal interactions via gap junctions [161,192]. They provided evidence that in a direct co-culture, miR-200b was transferred in a TGF-β-stimulated process from rat BM-MSCs to HUVECs through gap junctions formed of connexin 43. By this transfer and decrease of miR-200b, VEGF-A-induced expression enhanced osteogenic differentiation in BM-MSCs. In HUVECs, increasing miR-200b levels down-modulated ZEB2, ETS1, KDR and GATA2 transcription factors, which led to a decline of the angiogenic potential of HUVECs, in contrast. In vitro angiogenesis in this co-culture could therefore be partially rescued by employing the TGF-β inhibitor SB431542 or TGF-β-neutralizing antibody. These findings could provide a new strategy for cell-based bone regeneration.

In lung epithelial carcinoma cells (A549 cells), direct negative regulation of VEGF, VEGFR1 (Flt-1) and VEGFR2 (KDR) by binding of miR-200b to the corresponding 3′-UTR could be demonstrated by Liu et al. [193]. This interaction could be furthermore confirmed in an in vitro angiogenesis assay by transfection of HUVECs, which resulted in reduced capillary tube formation and significantly reduced VEGF-induced phosphorylation of ERK1/2. In addition, miR-200b targets the Ets-1 transcription factor, that might concomitantly downregulate VEGFR2, as found in human mammalian epithelial cells by Chan et al. [194]. Thus, miR-200b may be used as a therapeutic angiogenesis inhibitor.

5.15. MiR-210

Upregulated expression of miR-210 was detected in BMP-4-induced osteoblastic differentiation of murine stromal BM-MSC ST2 cells by Mizuno et al. [163]. Transfection experiments of sense and antisense miR-210 therefore promoted or repressed osteogenesis, respectively. As a target gene mediating the positive regulation, the activin A receptor type-1B (AcvR1b) gene was elicited that effects the TGF-β/activin signaling pathway negatively.

Studies of Fasanaro et al. proved that the expression of miR-210 progressively increases upon exposure to hypoxia. The overexpression of miR-210 in HUVECs using miRNA mimics stimulated the formation of capillary-like structures and enhanced *VEGF*-induced cell chemotaxis [165,195,196]. Thus, miR-210 up-regulation is a crucial hypoxia response element of ECs, affecting cell survival, migration, and differentiation and might participate in the modulation of the angiogenic response to ischemia [99,197,198]. As a consistently reported target gene mediating these effects, Ephrin-A3 receptor tyrosine kinase ligand (*EFNA3*) was validated. *EFNA3* was previously shown to play a role in the regulation of angiogenesis and VEGF signaling during vascular development and remodeling via EFNA1/EphA2 interaction [199,200]. Overexpression of an Ephrin-A3 allele that is not targeted by miR-210 therefore prevented miR-210-mediated stimulation of tube formation and EC migration.

In its function of promoting osteoblast differentiation by increased VEGF, ALP and OSX expression in rat MSCs and suppression of adipocyte differentiation, due to decreased PPAR-γ in vitro, miR-210 was also implicated in the regulation of postmenopausal osteoporosis [164]. Although the exact mechanism still needs to be elaborated, Liu et al. also found that HIF-1α and VEGF expression was increased in 17β-estradiol (E2)-treated osteoblasts.

5.16. MiR-222

Yan et al. identified miR-222-3p as a negative regulator involved in osteogenic differentiation of hBM-MSCs [166]. Enhancement of miR-222-3p function in hBM-MSCs was blocking protein levels of SMAD5 and RUNX2. miR-222-3p-specific inhibition via lentivirus infection, in contrast, led to their enhanced expression and increased phosphorylation of Smad proteins (1, 5, 8) that are responsible for expression of osteogenic genes. Thus, in addition to the Smad5-RUNX2 signaling pathway elevated levels of the osteoblast markers OSX, ALP, and OC, as well as increased matrix mineralization, could be detected.

The role of miR-222-3p in c-Src-mediated regulation of osteoclastogenesis was proven by Takigawa et al. [167]. Depending on the use of miR-222-3p inhibitor or mimics in RAW264.7 pre-osteoclastic cells, either upregulation, or downregulation of the mRNA expression levels of osteoclast marker genes NFATc1 or TRAP, were observed, respectively. Inhibition of of c-Src activity and activation of osteoclastogenesis via miR-222-3p was implemented by increased amounts of RANKL-induced expression of TRAP and cathepsin K protein levels. Thereby, the number of multi-nucleated osteoclasts and their pit formation was reduced.

Poliseno et al. investigated the role of miR-221 and miR-222 during in vitro angiogenesis [168]. Both miRNAs were identified among 15 upregulated miRNAs that allegedly target receptors of angiogenic factors in a large-scale screen of HUVECs. Mechanistically, they were found to post-translationally modify the angiogenic effects of stem cell factor (SCF) by targeting its receptor, c-KIT. Together with other angiogenic growth factors, such as VEGF, bFGF, or HGF, SCF is able to stimulate proliferation and migration of ECs and induce capillary-like tube formation when performing in vitro angiogenesis assays [201]. Consistently, tube formation, wound healing, or cell migration was suppressed in miR-221/miR-222– transfected and SCF-treated HUVECs.

5.17. MiR-424

miR-424 was identified by differential screening in a miRNA microarray of isolated primary hMSCs from four individuals that were, or were not, osteo-differentiated using osteogenic differentiation medium. In combination with miR-31, miR-106a, and miR-148a, Gao et al. found that miR-424 was suppressed, and predicted it to target RUNX2, CBFB (core-binding factor, beta subunit), and BMPs [169]. In a similar experimental setting of Vimalraj et al., miR-424, together with miR-106a, miR-148a, let-7i and miR-99a, were detected as hMSC-specific miRNAs that were found to be expressed only in undifferentiated hMSCs [170]. Here, bioinformatics analysis mostly predicted the MAPK, WNT, and insulin signaling pathways as targets.

A recent study of Li et al. elucidated further specific functional mechanisms of miR-424 during bone formation and oxidative stress [171]. It was reported that miR-424 was downregulated by the transcription factor FOXO1 using consensus binding sites in the promoter. Subsequently, via miR-424 mediated upregulation of FGF2 under oxidative stress, proliferation and osteogenic differentiation of hMSCs was accomplished. Uncovering the miR-424/FGF2 pathway revealed a new mechanism how FOXO1 promotes bone formation and could potentially enhance bone repair.

miR-424 was furthermore recognized by Gosh et al. as an important hypoxamiR in human ECs by revealing a novel pathway for HIF regulation [101]. Under hypoxic conditions levels miR-424 or its rodent homolog, mu-miR-322, were found to be elevated in ECs and in ischemic tissues undergoing vascular remodeling in an experimental myocardial infarction rat model. It was elicited that the target of miR-424 is represented by cullin 2 (CUL2), which serves the purpose of stabilizing the hypoxic transcription factor HIF-1α. CUL2 is an essential component for assembling the ubiquitin ligase system, normally leading to its continuous degradation. MiR-424 expression was experimentally evaluated to be regulated by the transcription factor PU.1 that itself was found to be regulated by RUNX-1 and C/EBPα transcription factors. Ectopic expression of miR-424 in retrovirally-transduced HUVECs therefore stimulated angiogenesis in vivo in athymic mice that were subcutaneously implanted.

6. Outlook and Future Directions: MiRNAs in Therapeutic Applications

Bone is a highly vascularized tissue, and is thus reliant on the coordinated interaction between osteogenesis and angiogenesis which occurs between osteoblastic and ECs during development, remodeling and regeneration of the skeletal system [22,27,202]. Deciphering the molecular nature of the mechanisms that couple bone formation to blood vessel formation should therefore be of great interest to enhance bone formation capabilities in vitro and in vivo [203]. These specific underlying mechanisms are gaining growing attention under physiologic and pathologic conditions such as fracture healing, prevention of osteoporosis, tissue engineering or bone regeneration. Thus far, mostly a combination of different growth factors controlling osteogenesis and angiogenesis (e.g., such VEGF, angiopoietins, BMPS, RUNX2) have succeeded in achieving bone formation to a certain degree [33,204–206]. Nevertheless, each of these factors has individual roles at certain stages of development, and may disturb the subtle orchestration of required regulation steps.

MicroRNAs provide potent modifiers which coordinate a broad spectrum of biological processes. In contrast to transcription factors, single miRNAs may only modestly affect individual mRNA target expression. However, on the one hand a specific target mRNA can experience increased repression if it has multiple binding sites in the 3'-UTR. On the other hand, the same 3'-UTR of a gene can be targeted by many different miRNAs simultaneously that intervene with its regulation of expression [207]. Bioinformatic analysis of target prediction databases, such as TargetScan, miRanda and Pictar, suggest that a single average miRNA is capable of modulating up to several hundred target genes by perfect or imperfect base-pairing in combination with tissue-specific expression [208–211]. Additionally, by feedforward or feedback loops that form a regulatory network, miRNA effects can be amplified [212]. And it is estimated that approximately one-third of genome-encoded proteins are effected by miRNA regulation [213]. Therefore, these regulatory RNAs which control multiple endogenous signaling processes simultaneously possess a unique capacity to interfere with all cellular processes and have become an important tool in biological and medical research. As the application of miRNA-based methods for the treatment and monitoring of different pathological conditions is constantly becoming more prevalent, their therapeutic engagement could establish a refined method to stimulate inartificial bone development.

There is increasing evidence that miRNAs play important roles in controlling osteogenesis and angiogenesis. One study identified mutations in miR-2861 in two related adolescents that likely contributed to primary osteoporosis [214]. miR-2861 regulates osteoblast differentiation by targeting HDAC5, which enhances RUNX2 degradation. Moreover, a recent study showed accelerated osteoblast differentiation of hMSCs in a three-dimensional scaffold in vitro through manipulation of miR-148b and miR-489 expression [215]. Several studies also explored the use of miRNA mimicking or inhibitory agents for bone regeneration in animal studies. However, only a few studies have found that miRNAs are positive regulators of these processes. As outlined, miR-29b has been reported to promote osteoblast differentiation by targeting several well characterized inhibitors of bone formation in vitro [124]. Single site-specific delivery of miR-29b into a two-week post fracture callus by Lee et al. significantly improved mouse femoral fracture healing [216]. This was documented radiographically by a decrease of callus width and area; histomorphometrical and micro-computed tomographical analyses demonstrated increased bone volume fraction and bone mineral density of the callus. A single report further delineated the use of miRNAs as a therapeutic strategy to modulate angiogenesis-osteogenesis coupling during bone regeneration or repair in a subcutaneous assay in the mouse. The studies of Li et al. preferred miR-26a over miR-21 and miR-29b, which were all identified as the most potent candidates to mediate both angiogenesis and osteogenesis by microarray profiling of primary osteoblasts [217]. miR-26a was found to be a factor that is upregulated in newly-formed bone, and itself stimulates the expression of osteoblast-specific makers RUNX2, ALP, and OCN, mineralization of osteoblasts as well as VEGF secretion in murine primary BM-MSCs and in MC3T3-E1 cells in vitro. The enhancement of miR-26a expression in vivo by transfection of BM-MSCs with miR-26a mimicking agents led to complete repair of a critical-size calvarial bone defect, mainly due to simultaneously regulating endogenous angiogenesis-osteogenesis coupling. To date, its ability to integrate multiple signaling pathways thus makes miR-26a an ideal candidate for regenerating bone by miRNA-based therapy. Murata et al. detected significantly decreased levels of miR-92a in patients with trochanteric fractures, and investigated its significance in a mouse femoral fracture model [142]. Systemic as well as local administration of antimir-92a via LNA-stabilized oligonucleotide increased callus volume and enhanced fracture healing in the early phase by promoting neovascularization in the mouse femur. Nevertheless, as outlined previously, the direct osteogenesis-related molecular targets for this mechanism, which offers therapeutic potential for repairing bone, still remain to be clarified. In addition to the basic osteogenic functions of miR-31, Deng et al. investigated the therapeutic potential of hAD-MSC combined with beta-calcium triphosphate scaffolds in repairing a rat critical-sized calvarial defect [130,217]. The group reported that a knockdown of miR-31 significantly enhanced the repair of the defect, as could be noticed by an increased bone volume and mineral density in combination with a decreased scaffold. As the molecular basis of the osteogenic differentiation and bone regeneration program, in vitro results using lentiviral expression constructs revealed a BMP-2-inducible regulatory loop between Runx2, miR-31, as well as the miR-31 target gene, AT-rich sequence-binding protein 2 (Satb2). As a negative modulator of angiogenesis, Yoshizuka et al. locally administered miR-222 inhibitor mixed with atelocollagen to a rat femoral transverse fracture with the purpose of enhancing bone healing by stimulating osteogenesis, chondrogenesis, as well as angiogenesis [218]. Bone union at the fracture site with increased capillary density could be confirmed by radiographic, μCT and histological evaluation at 8 weeks after administration. Inhibition of miR-222 promoted osteogenic (RUNX2, COL1A1, OCN) or chondrogenic (COL2A1, aggrican, SOX9) differentiation in hMSCs, as determined by expression of osteogenic or chondrogenic markers, respectively.

In this review, miRNAs or "CouplingmiRs/CPLGmiRs" were identified that are known so far to regulate, either positively or negatively, angiogenesis as well as osteogenesis function—as potent molecular managers—that may simultaneously regulate multiple endogenous signaling cascades. This has been summarized in Figure 2, where the described miRNAs have also been depicted with regard to their allocation during the developmental fate of osteogenic or angiogenic cells. As for many miRNAs in general, also for coupling miRs/CPLGmiRs, most mechanisms of action are negative/inhibitory in nature for the regulation of osteogenesis as well as angiogenesis. Also noteworthy, but not surprising, is the finding that most miRNAs potentially involved in angiogenic-osteogenic coupling are encountered in the osteoblastic lineage rather than in chondroblasts or osteoclasts. And miRNAs are particularly present during the differentiation of osteoblasts, while they exert their modulatory effects during angiogenesis mostly in mature endothelial cells. Delivery of miRNAs may provide a way to maximally mimic the native bone development environment, and thus possess the therapeutic potential to enhance bone regeneration and repair. In contrast, miR inhibitors, i.e., antisense oligonucleotides directed against miRNAs, can be applied in the form of antagomiRs or LNA-antimiRs [58,179,219]. RNA/DNA hybrids modified by this locked nucleic acid-technology represent single stranded modified antisense oligonucleotides with increased stability and affinity and also facilitate cell penetration. Furthermore, they are biocompatible, as they cause no cellular immune response, are highly soluble, and have already been tested by various in vitro and in vivo studies as well as clinical trials (e.g., miR-122 LNA anti-miRNA oligonucleotides for Hepatitis C virus or miR-34 for solid cancers) [220,221]. Additionally, any new findings in this field of research might additionally yield considerable anti-angiogenic targets for tumor therapies, as tumor angiogenesis—i.e., the formation of tumor-associated angiogenic vessels—is a key requirement in tumor growth, progression, and metastasis. As an example, it would be very favorable to identify angiomiRs that negatively regulate VEGF expression in combination with current anti-VEGF therapies. Inversely, the identification of novel oncomiRs could also pave the way and lead to the description of unknown miRNAs with a function in bone angiogenesis [222–224]. Currently, microRNA-targeting therapy is still in development due to new challenges compared to conventional drugs, and results of experimental studies in animal models need to be transferred to clinical applications. Nevertheless, in addition to their role as potential angiogenic or bone regenerative therapeutic targets, miRNAs are emerging as distinguished disease biomarkers relevant to specific physiologic or pathologic conditions which could serve for immediate use in cardiovascular and bone diseases.

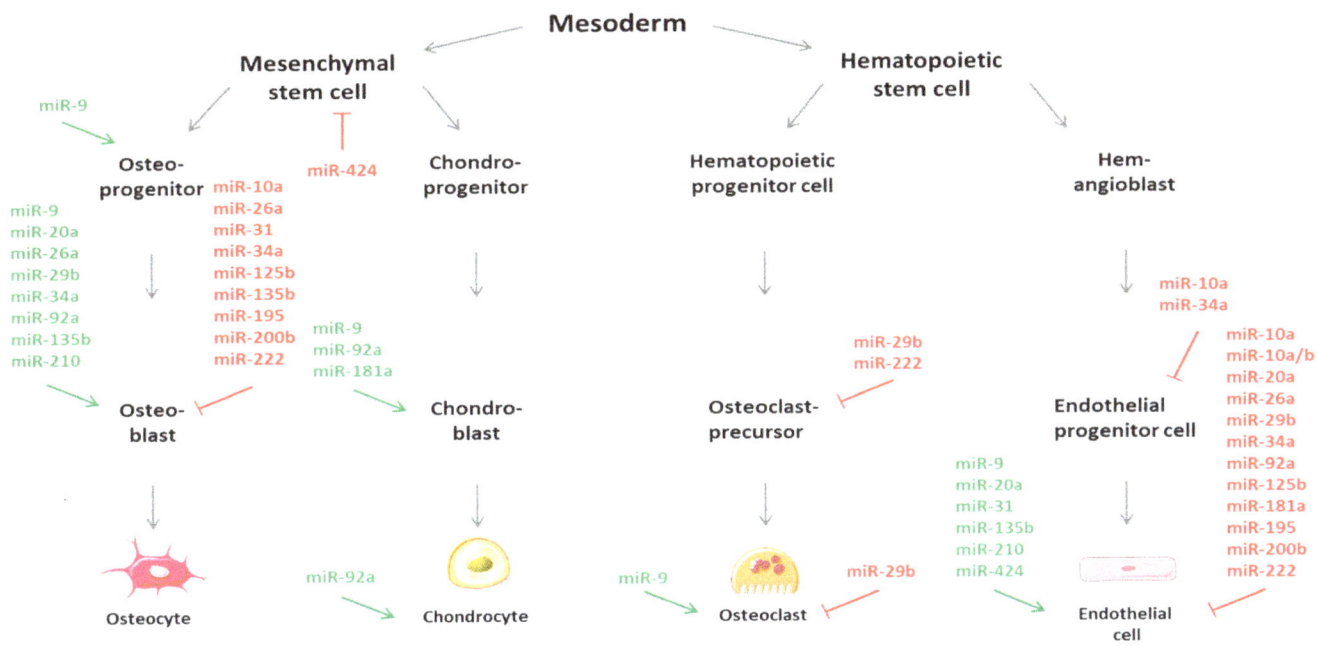

Figure 2. Regulating effects of microRNAs (miRs/miRNAs) involved in the regulation and coupling of bone angiogenesis ("CouplingmiRs") during cell fate determination. MiRNAs were assigned with their individual positive/stimulatory (green colored) or negative/inhibitory (red colored) function and occurrence during the specific differentiation steps of osteogenic and angiogenic cells. Cell images from Servier Medical Art by Servier licensed under a Creative Commons Attribution 3.0.

Author Contributions: L.F.F. was responsible for collecting literature, drafting, writing, reviewing, and illustration of the manuscript.

References

1. Crane, G.; Ishaug, S.; Mikos, A. Bone tissue engineering. *Nat. Med.* **1995**, *1*, 1322–1324. [CrossRef] [PubMed]
2. Rouwkema, J.; Rivron, N.C.; Blitterswijk, C.A. Vascularization in tissue engineering. *Trends Biotechnol.* **2008**, *26*, 434–441. [CrossRef] [PubMed]
3. Hertig, A. Angiogenesis in the early human chorion and the primary placenta of the macaque monkey. *Contrib. Embryol.* **1935**, *25*, 37–81.
4. Chung, A.S.; Ferrara, N. Developmental and pathological angiogenesis. *Annu. Rev. Cell Dev. Biol.* **2011**, *27*, 563–584. [CrossRef]
5. Lian, J.B.; Stein, G.S.; van Wijnen, A.J.; Stein, J.L.; Hassan, M.Q.; Gaur, T.; Zhang, Y. MicroRNA control of bone formation and homeostasis. *Nat. Rev. Endocrinol.* **2013**, *8*, 212–227. [CrossRef] [PubMed]
6. Kronenberg HM Developmental regulation of the growth plate. *Nature* **2003**, *423*, 332–336. [CrossRef] [PubMed]
7. Berendesen, A.; Olsen, B. Bone development. *Bone* **2015**, *80*, 14–18. [CrossRef]
8. Olsen, B.; Reginato, A.; Wang, W. Bone development. *Annu. Rev. Cell Dev. Biol.* **2000**, *16*, 191–220. [CrossRef]
9. Karsenty, G. The complexities of skeletal biology. *Nature* **2003**, *423*, 316–318. [CrossRef]
10. Helms, J.; Schneider, R. Cranial skeletal biology. *Nature* **2003**, *423*, 326–331. [CrossRef]

11. Maes, C.; Carmeliet, P.; Moermans, K.; Stockmans, I.; Smets, N. Impaired angiogenesis and endochondral bone formation in mice lacking the vascular endothelial growth factor isoforms VEGF 164 and VEGF 188. *Mech Dev.* **2002**, *111*, 61–73. [CrossRef]

12. Gerber, H.; Ferrara, N. Angiogenesis and Bone Growth. *TCM* **2000**, *10*, 223–228. [CrossRef]

13. Brandi, M.; Collin-Osdoby, P. Vascular biology and the skeleton. *J. Bone Miner. Res.* **2006**, *21*, 183–192. [CrossRef] [PubMed]

14. Sivaraj, K.K.; Adams, R.H. Blood vessel formation and function in bone. *Development* **2016**, *143*, 2706–2715. [CrossRef] [PubMed]

15. Schipani, E.; Maes, C.; Carmeliet, G.; Semenza, G.L. Regulation of osteogenesis-angiogenesis coupling by HIFs and VEGF. *J. Bone Miner. Res.* **2009**, *24*, 1347–1353. [CrossRef] [PubMed]

16. Schipani, E.; Wu, C.; Rankin, E.B.; Giaccia, A.J. Regulation of Bone Marrow Angiogenesis by Osteoblasts during Bone Development and Homeostasis. *Front. Endocrinol.* **2013**, *4*, 85. [CrossRef]

17. Kusumbe, A.P.; Ramasamy, S.K.; Adams, R.H. Coupling of angiogenesis and osteogenesis by a specific vessel subtype in bone. *Nature* **2014**, *507*, 323–328. [CrossRef]

18. Carmeliet, P.; Ferreira, V.; Breier, G.; Pollefeyt, S.; Kieckens, L.; Gertsenstein, M.; Fahrig, M.; Vandenhoeck, A.; Harpal, K.; Eerhardt, C.; et al. Abnormal blood vessel development and lethality in embryos lacking a single VEGF allele. *Nature* **1996**, *380*, 435–439. [CrossRef]

19. Carulli, C.; Innocenti, M.; Brandi, M.L. Bone vascularization in normal and disease conditions. *Front. Endocrinol.* **2013**, *4*, 1–10. [CrossRef]

20. Ding, W.-G.; Yan, W.; Wei, Z.-X.; Liu, J.-B. Difference in intraosseous blood vessel volume and number in osteoporotic model mice induced by spinal cord injury and sciatic nerve resection. *J. Bone Miner. Metab.* **2012**, *30*, 400–407. [CrossRef]

21. Wang, L.; Zhou, F.; Zhang, P.; Wang, H.; Qu, Z.; Jia, P.; Yao, Z.; Shen, G.; Li, G.; Zhao, G.; et al. Human type H vessels are a sensitive biomarker of bone mass. *Cell Death Dis.* **2017**, *8*, e2760. [CrossRef] [PubMed]

22. Stegen, S.; Van Gastel, N.; Carmeliet, G. Bringing new life to damaged bone: The importance of angiogenesis in bone repair and regeneration. *Bone* **2014**, *70*, 19–27. [CrossRef] [PubMed]

23. Hu, K.; Olsen, B.R. Osteoblast-derived VEGF regulates osteoblast differentiation and bone formation during bone repair. *J. Clin. Invest.* **2016**, *126*, 509–526. [CrossRef]

24. Hu, K.; Olsen, B.R. The roles of vascular endothelial growth factor in bone repair and regeneration. *Bone* **2016**, *91*, 30–38. [CrossRef] [PubMed]

25. Maes, C.; Kobayashi, T.; Selig, M.; Torrekens, S.; Roth, S.; Mackem, S.; Carmeliet, G.; Kronenberg, H. Osteoblast precursors, but not mature osteoblasts, move into developing and fractured bones along with invading blood vessels. *Dev. Cell* **2010**, *19*, 329–344. [CrossRef] [PubMed]

26. Maes, C.; Carmeliet, G.; Schipani, E. Hypoxia-driven pathways in bone development, regeneration and disease. *Nat. Rev. Rheumatol.* **2012**, *8*, 358–366. [CrossRef] [PubMed]

27. Wang, Y.; Wan, C.; Deng, L.; Liu, X.; Cao, X.; Gilbert, S.R.; Bouxsein, M.L.; Faugere, M.; Guldberg, R.E.; Gerstenfeld, L.C.; et al. The hypoxia-inducible factor α pathway couples angiogenesis to osteogenesis during skeletal development. *J. Clin. Invest.* **2007**, *117*, 1616–1626. [CrossRef]

28. Pugh, C.W.; Ratcliffe, P.J. Regulation of angiogenesis by hypoxia: Role of the HIF system. *Nat. Med.* **2003**, *9*, 677–684. [CrossRef]

29. Ferrara, N.; Gerber, H.; Lecouter, J. The biology of VEGF and its receptors. *Nat. Med.* **2003**, *9*, 669–676. [CrossRef]

30. Ornitz, D.; Marie, P. Fibroblast growth factor signaling in skeletal development and disease. *Genes Dev.* **2015**, *29*, 1463–1468. [CrossRef]

31. Ramasamy, S.K.; Kusumbe, A.P.; Wang, L.; Adams, R.H. Endothelial Notch activity promotes angiogenesis and osteogenesis in bone. *Nature* **2014**, *507*, 376–380. [CrossRef] [PubMed]

32. Ortega, N.; Behonick, D.; Werb, Z. Matrix remodeling during endochondral ossification. *Trends Cell Biol.* **2004**, *14*, 86–93. [CrossRef] [PubMed]

33. Kleinheinz, J.; Stratmann, U.; Joos, U.; Wiesmann, H.-P. VEGF-Activated Angiogenesis During Bone Regeneration. *J. Oral Maxillofac. Surg.* **2005**, *63*, 1310–1316. [CrossRef] [PubMed]

34. Maes, C.; Coenegrachts, L.; Stockmans, I.; Daci, E.; Luttun, A.; Petryk, A.; Gopalakrishnan, R.; Moermans, K.; Smets, N.; Verfaillie, C.M.; et al. Placental growth factor mediates mesenchymal cell development, cartilage turnover, and bone remodeling during fracture repair. *J. Clin. Invest.* **2006**, *116*, 16–18. [CrossRef]

35. Kingsley, D. What do BMPs do in mammals? Clues from the mouse short-ear mutation. *Trends Genet.* **1994**, *10*, 16–21. [CrossRef]

36. Hassan, M.Q.; Tye, C.E.; Stein, G.S.; Lian, J.B. Non-coding RNAs: Epigenetic regulators of bone development and homeostasis. *Bone* **2015**, *81*, 746–756. [CrossRef] [PubMed]

37. Papaioannou, G.; Mirzamohammadi, F.; Kobayashi, T. MicroRNAs involved in bone formation. *Cell Mol. Life Sci.* **2014**, *71*, 4747–4761. [CrossRef] [PubMed]

38. Papaioannou, G. miRNAs in Bone Development. *Curr. Genom.* **2015**, *16*, 427–434. [CrossRef] [PubMed]

39. Ambros, V. MicroRNAs: Tiny regulators with great potential. *Cell* **2001**, *107*, 823–826. [CrossRef]

40. Lau, N.C.; Lim, L.P.; Weinstein, E.G.; Bartel, D.P. An abundant class of tiny RNAs with probable regulatory roles in Caenorhabditis elegans. *Science* **2001**, *294*, 858–862. [CrossRef]

41. Bartel, D.P.; Lee, R.; Feinbaum, R. MicroRNAs: Genomics, Biogenesis, Mechanism, and Function Genomics: The miRNA Genes. *Cell* **2004**, *116*, 281–297. [CrossRef]

42. Bartel, D.P. MicroRNAs: Target Recognition and Regulatory Functions. *Cell* **2009**, *136*, 215–233. [CrossRef] [PubMed]

43. Lee, R.C.; Ambros, V. An extensive class of small RNAs in Caenorhabditis elegans. *Science* **2001**, *294*, 858–862. [CrossRef] [PubMed]

44. Macfarlane, L.A.; Murphy, P.R. MicroRNA: Biogenesis, Function and Role in Cancer. *Curr. Genom.* **2010**, *11*, 537–561. [CrossRef] [PubMed]

45. Hutvagner, G.; McLachlan, J.; Pasquinelli, A.E.; Balint, E.; Tuschl, T.; Zamore, P.D. A cellular function for the RNA-interference enzyme Dicer in the maturation of the let-7 small temporal RNA. *Science* **2001**, *293*, 834–838. [CrossRef] [PubMed]

46. Grishok, A.; Pasquinelli, A.E.; Conte, D.; Li, N.; Parrish, S.; Ha, I.; Baillie, D.L.; Fire, A.; Ruvkun, G.; Mello, C. Genes and mechanisms related to RNA interference regulate expression of the small temporal RNAs that control C. elegans developmental timing. *Cell* **2001**, *106*, 23–34. [CrossRef]

47. Knight, S.W.; Bass, B.L. A role for the RNase III enzyme DCR-1 in RNA interference and germ line development in Caenorhabditis elegans. *Science* **2001**, *293*, 2269–2271. [CrossRef]

48. Yang, J.S.; Lai, E.C. Alternative miRNA biogenesis pathways and the interpretation of core miRNA pathway mutants. *Mol. Cell* **2011**, *43*, 892–903. [CrossRef]

49. Lai, E.C. Micro RNAs are complementary to 3′UTR sequence motifs that mediate negative post-transcriptional regulation. *Nat. Genet.* **2002**, *30*, 363–364. [CrossRef]

50. Bernstein, E.; Caudy, A.A.; Hammond, S.M.; Hannon, G.J. Role for a bidentate ribonuclease in the initiation step of RNA interference. *Nature* **2001**, *409*, 363–366. [CrossRef]

51. Wang, Y.; Medvid, C.; Melton, R.; Jaenisch, R.; Blelloch, R. DGCR8 is essential for microRNA biogenesis and silencing of embryonic stem cell self-renewal. *Nat. Genet.* **2007**, *39*, 380–385. [CrossRef] [PubMed]

52. Förstemann, K.; Tomari, Y.; Du, T.; Vagin, V.V.; Denli, A.M.; Bratu, D.P.; Klattenhoff, C.; Theurkauf, W.E.; Zamore, P.D. Normal microRNA maturation and germ-line stem cell maintenance requires Loquacious, a double stranded RNA-binding domain protein. *PLoS Biol.* **2005**, *3*, e236. [CrossRef] [PubMed]

53. Huang, Y.; Shen, X.J.; Zou, Q.; Wang, S.P.; Tang, S.M.; Zhang, G.Z. Biological functions of microRNAs: A review. *J. Physiol. Biochem.* **2011**, *67*, 129–139. [CrossRef] [PubMed]

54. Melton, C.; Judson, R.L.; Blelloch, R. Opposing micro-RNA families regulate self-renewal in mouse embryonic stem cells. *Nature* **2010**, *463*, 621–626. [CrossRef] [PubMed]

55. Brennecke, J.; Hipfner, D.R.; Stark, A.; Russell, R.B.; Cohen, S.M. bantam encodes a developmentally regulated microRNA that controls cell proliferation and regulates the proapoptotic gene hid in Drosophila. *Cell* **2003**, *113*, 25–36. [CrossRef]

56. Hipfner, D.R.; Weigmann, K.; Cohen, S.M. The bantam gene regulates Drosophila growth. *Genetics* **2002**, *161*, 1527–1537. [PubMed]

57. Esau, C.; Davis, S.; Murray, S.F.; Yu, X.X.; Pandey, S.K.; Pear, M.; Watts, L.; Booten, S.L.; Graham, M.; McKay, R.; et al. miR-122 regulation of lipid metabolism revealed by in vivo antisense targeting. *Cell Metab.* **2006**, *3*, 87–98. [CrossRef]

58. Krützfeldt, J.; Rajewsky, N.; Braich, R.; Rajeev, K.G.; Tuschl, T.; Manoharan, M.; Stoffel, M. Silencing of microRNAs in vivo with ′antagomirs′. *Nature* **2005**, *438*, 685–689. [CrossRef]

59. Miska, E.A. How microRNAs control cell division, differentiation and death. *Curr. Opin. Genet. Dev.* **2005**, *15*, 563–568. [CrossRef]

60. Chen, C.Z.; Li, L.; Lodish, L.F.; Bartel, D. MicroRNAs modulate hematopoietic lineage differentiation. *Science* **2004**, *303*, 83–86. [CrossRef]

61. Tay, Y.; Rinn, J.; Pandolfi, P.P. The multilayered complexity of ceRNA crosstalk and competition. *Nature* **2014**, *505*, 344–352. [CrossRef] [PubMed]

62. Alvarez-Garcia, I.; Miska, E.A. MicroRNA functions in animal development and human disease. *Development* **2005**, *132*, 4653–4662. [CrossRef] [PubMed]

63. Gennari, L.; Bianciardi, S.; Merlotti, D. MicroRNAs in bone diseases. *Osteoporos. Int.* **2017**, *28*, 1191–1213. [CrossRef] [PubMed]

64. Clark, E.; Kalomoiris, S.; Nolta, J.; Fierro, F. Concise Review: MicroRNA Function in Multipotent Mesenchymal Stromal Cells. *Stem Cells* **2014**, *32*, 1074–1082. [CrossRef]

65. Peng, S.; Gao, D.; Gao, C.; Wei, P.; Niu, M.; Shuai, C. MicroRNAs regulate signaling pathways in osteogenic differentiation of mesenchymal stem cells (Review). *Mol. Med. Rep.* **2016**, *14*, 623–629. [CrossRef] [PubMed]

66. Fang, S.; Deng, Y.; Gu, P.; Fan, X. MicroRNAs Regulate Bone Development and Regeneration. *Int. J. Mol. Sci.* **2015**, *16*, 8227–8253. [CrossRef]

67. Ji, X.; Chen, X.; Yu, X. MicroRNAs in Osteoclastogenesis and Function: Potential Therapeutic Targets for Osteoporosis. *Int. J. Mol. Sci.* **2016**, *17*, 349. [CrossRef]

68. Dong, S.; Yang, B.; Guo, H.; Kang, F. MicroRNAs regulate osteogenesis and chondrogenesis. *Biochem. Biophys. Res. Commun.* **2012**, *418*, 587–591. [CrossRef]

69. Kiga, K.; Mimuro, H.; Suzuki, M.; Shinozaki-Ushiku, A.; Kobayashi, T.; Sanada, T.; Kim, M.; Ogawa, M.; Iwasaki, Y.W.; Kayo, H.; et al. Epigenetic silencing of miR-210 increases the proliferation of gastric epithelium during chronic Helicobacter pylori infection. *Nat. Commun.* **2014**, *5*, 4497. [CrossRef]

70. Chen, J.; Qiu, M.; Dou, C.; Cao, Z.; Dong, S. MicroRNAs in Bone Balance and Osteoporosis. *Drug Dev. Res.* **2015**, *76*, 235–245. [CrossRef]

71. Nugent, M. MicroRNAs and Fracture Healing. *Calcif. Tissue Int.* **2017**, *101*, 355–361. [CrossRef] [PubMed]

72. Wu, C.; Tian, B.O.; Qu, X.; Liu, F.; Tang, T.; Qin, A.N.; Zhu, Z.; Dai, K. MicroRNAs play a role in chondrogenesis and osteoarthritis (Review). *Int. J. Mol. Med.* **2014**, *34*, 13–23. [CrossRef] [PubMed]

73. Min, Z.; Zhang, R.; Yao, J.; Jiang, C.; Guo, Y.; Cong, F.; Wang, W.; Tian, J.; Zhong, N.; Sun, J.; et al. MicroRNAs associated with osteoarthritis differently expressed in bone matrix gelatin (BMG) rat model. *Int. J. Clin. Exp. Med.* **2015**, *8*, 1009–1017. [PubMed]

74. Seeliger, C.; Balmayor, E.; van Griensven, M. miRNAs Related to Skeletal Diseases. *Stem Cells Dev.* **2016**, *25*, 1261–1281. [CrossRef] [PubMed]

75. Anand, S.; Cheresh, D.A. Emerging Role of Micro-RNAs in the Regulation of Angiogenesis. *Genes Cancer* **2011**, *2*, 1134–1138. [CrossRef] [PubMed]

76. Wang, S.; Olson, E.N. AngiomiRs—Key Regulators of Angiogenesis. *Curr. Opin. Genet. Dev.* **2009**, *19*, 205–211. [CrossRef] [PubMed]

77. Small, E.M.; Olson, E.N. Pervasive roles of microRNAs in cardiovascular biology. *Nature* **2011**, *469*, 336–342. [CrossRef]

78. Weis, S.M.; Caheresh, D.A. Tumor angiogenesis: Molecular pathways and therapeutic targets. *Nat. Med.* **2011**, *17*, 1359–1370. [CrossRef]

79. Salinas-Vera, Y.; Marchat, L.; Gallardo-Rincon, D.; Ruiz-Garcia, E.; Astudillo- De La Vega, H.; Echavarria-Zepeda, R.; Lopez-Camarillo, C. AngiomiRs: MicroRNAs driving angiogenesis in cancer (Review). *Int. J. Mol. Med.* **2018**, *2018*. [CrossRef]

80. Suarez, Y.; Sessa, W.C. MicroRNAs As Novel Regulators of Angiogenesis. *Circ. Res.* **2009**, *104*, 442–454. [CrossRef]

81. Yang, W.; Yang, D.; Na, S.; Sandusky, G.; Zhang, Q.; Zhao, G. Dicer is required for embryonic angiogenesis during mouse development. *J. Biol. Chem.* **2005**, *280*, 9330–9335. [CrossRef] [PubMed]

82. Albinsson, S.; Suarez, Y.; Skoura, A.; Offermann, S.; Miano, J.M.; Sessa, W.C. MicroRNAs are necessary for vascular smooth muscle growth, differentiation, and function. *Arterioscler. Thromb. Vasc. Biol.* **2010**, *30*, 1118–1126. [CrossRef] [PubMed]

83. Wang, S.; Aurora, A.B.; Johnson, B.A.; Qi, X.; McAnnaly, J.; Hill, J.A.; Richardson, J.A.; Bassel-Duby, R.; Olson, E.N. The endothelial-specific microRNA miR-126 governs vascular integrity and angiogenesis. *Dev. Cell* **2008**, *15*, 261–271. [CrossRef] [PubMed]

84. Suarez, Y.; Wang, C.; Manes, T.D.; Pober, J.S. Cutting edge: TNF-induced microRNAs regulate TNF-induced expression of E-selectin and intercellular adhesion molecule-1 on human endothelial cells: Feedback control of inflammation. *J. Immunol.* **2010**, *184*, 21–25. [CrossRef] [PubMed]

85. Suarez, Y.; Fernandez-Hernando, C.; Pober, J.; Sessa, W. Dicer dependent microRNAs regulate gene expression and functions in human endothelial cells. *Circ. Res.* **2007**, *100*, 1164–1173. [CrossRef] [PubMed]

86. Greco, S.; Gaetano, C.; Martelli, F. HypoxamiR Regulation and Function in Ischemic Cardiovascular Diseases. *Antioxid. Redox Signal.* **2014**, *21*, 1202–1219. [CrossRef] [PubMed]

87. Samanta, S.; Balasubramanian, S.; Rajasingh, S.; Patel, U.; Dhanasekaran, A.; Dawn, B.; Rajasingh, J. MicroRNA: A new therapeutic strategy for cardiovascular diseases. *Trends Cardiovasc. Med.* **2016**, *26*, 407–419. [CrossRef]

88. Suarez, Y.; Fernandez-Hernando, C.; Yu, J.; Gerber, S.A.; Harrison, K.D.; Pober, J.S.; Iruela-Arispe, M.L.; Merkenschlager, M.; Sessa, W.C. Dicer-dependent endothelial microRNAs are necessary for postnatal angiogenesis. *Proc. Natl. Acad. Sci. USA* **2008**, *105*, 14082–14087. [CrossRef]

89. Würdinger, T.; Tannous, B.A.; Saydam, O.; Skog, J.; Grau, S.; Soutschek, J.; Weissleder, R.; Breakefield, X.O.; Krichevsky, A.M. miR-296 regulates growth factor receptor overexpression in angiogenic endothelial cells. *Cancer Cell* **2008**, *14*, 382–393. [CrossRef]

90. Anand, S.; Cheresh, D.A. MicroRNA-mediated Regulation of the Angiogenic Switch. *Curr. Opin. Hematol.* **2011**, *18*, 171–176. [CrossRef]

91. Nallamshetty, S.; Chan, S.Y.; Loscalzo, J. Hypoxia: A master regulator of microRNA biogenesis and activity. *Free Radic. Biol. Med.* **2013**, *64*, 20–30. [CrossRef] [PubMed]

92. Madanecki, P.; Kapoor, N.; Bebok, Z.; Ochocka, R.; Collawn, J.F.; Bartoszewski, R. Regulation of angiogenesis by hypoxia: The role of microRNA. *Cell. Mol. Biol. Lett.* **2013**, *18*, 47–57. [CrossRef] [PubMed]

93. el Azzouzi, H.; Leptidis, S.; Doevendans, P.A.; De Windt, L.J. HypoxamiRs: Regulators of cardiac hypoxia and energy metabolism. *Trends Endocrinol. Metab.* **2015**, *26*, 502–508. [CrossRef] [PubMed]

94. Greco, S.; Martelli, F. MicroRNAs in Hypoxia Response. *Antioxid. Redox Signal.* **2014**, *21*, 1164–1166. [CrossRef] [PubMed]

95. Collet, G.; Skrzypek, K.; Grillon, C.; Matejuk, A.; El Hafni-Rahbi, B.; Fayel, N.L.; Kieda, C. Hypoxia control to normalize pathologic angiogenesis: Potential role for endothelial precursor cells and miRNAs regulation. *Vascul. Pharmacol.* **2012**, *56*, 252–261. [CrossRef] [PubMed]

96. Bertero, T.; Rezzonico, R.; Pottier, N.; Mari, B. Impact of MicroRNAs in the Cellular Response to Hypoxia. *Int. Rev. Cell Mol. Biol.* **2017**, *333*, 91–158. [CrossRef] [PubMed]

97. Hua, Z.; Lv, Q.; Ye, W.; Wong, A.C.-K.; Cai, G.; Gu, D.; Ji, Y.; Zhao, C.; Wang, J.; Yang, B.B.; et al. MiRNA-Directed Regulation of VEGF and Other Angiogenic Factors under Hypoxia. *PloS ONE* **2006**, *1*, e116. [CrossRef]

98. Loscalzo, J. The cellular response to hypoxia: Tuning the system with microRNAs. *J. Clin. Invest.* **2010**, *120*, 3815–3817. [CrossRef]

99. Devlin, C.; Greco, S.; Martelli, F.; Ivan, M. MiR-210: More than a silent player in hypoxia. *IUBMB Life* **2011**, *63*, 94–100. [CrossRef]

100. Chan, S.; Loscalzo, J. MicroRNA-210: A unique and pleiotropic hypoxamir. *Cell Cycle* **2010**, *9*, 1072–1083. [CrossRef]

101. Ghosh, G.; Subramanian, I.V.; Adhikari, N.; Zhang, X.; Joshi, H.P.; Basi, D.; Chandrashekhar, Y.S.; Hall, J.L.; Roy, S.; Zeng, Y.; et al. Hypoxia-induced microRNA-424 expression in human endothelial cells regulates HIF-α isoforms and promotes angiogenesis. *J. Clin. Invest.* **2010**, *120*, 4141–4154. [CrossRef] [PubMed]

102. Taguchi, A.; Yanagisawa, K.; Tanaka, M.; Cao, K.; Matsuyama, Y.; Goto, H.; Takahashi, T. Identification of hypoxia-inducible factor-1alpha as a novel target for miR-17-92 microRNA cluster. *Cancer Res.* **2008**, *68*, 5540–5545. [CrossRef] [PubMed]

103. Liu, C.; Tsai, M.; Hung, P.; Kao, S.; Liu, T.; Wu, K.; Chiou, S.; Lin, S.; Chang, K. miR31 ablates expression of the HIF regulatory factor FIH to activate the HIF pathway in head and neck carcinoma. *Cancer Res.* **2010**, *70*, 1635–1644. [CrossRef] [PubMed]

104. Yuva-Aydemir, Y.; Simkin, A.; Gascon, E.; Gao, F.-B. MicroRNA-9: Functional evolution of a conserved small regulatory RNA. *RNA Biol.* **2011**, *8*, 557–564. [CrossRef] [PubMed]

105. Han, R.; Kan, Q.; Sun, Y.; Wang, S.; Zhang, G.; Peng, T.; Jia, Y. MiR-9 promotes the neural differentiation of mouse bone marrow mesenchymal stem cells via targeting zinc finger protein 521. *Neurosci. Lett.* **2012**, *515*, 147–152. [CrossRef] [PubMed]

106. Qu, J.; Lu, D.; Guo, H.; Miao, W.; Wu, G. MicroRNA-9 regulates osteoblast differentiation and angiogenesis via the AMPK signaling pathway. *Mol. Cell Biochem.* **2016**, *411*, 23–33. [CrossRef] [PubMed]

107. Liu, X.; Xu, H.; Kou, J.; Wang, Q.; Zheng, X.; Yu, T. MiR-9 promotes osteoblast differentiation of mesenchymal stem cells by inhibiting DKK1 gene expression. *Mol. Biol. Rep.* **2016**, *43*, 939–946. [CrossRef]

108. Zhuang, G.; Wu, X.; Jiang, Z.; Kasman, I.; Yao, J.; Guan, Y.; Oeh, J.; Modrusan, Z.; Bais, C.; Sampath, D.; et al. Tumour-secreted miR-9 promotes endothelial cell migration and angiogenesis by activating the JAK-STAT pathway. *EMBO J.* **2012**, *31*, 3513–3523. [CrossRef]

109. Jones, S.; Watkins, G.; Le Good, N.; Roberts, S.; Murphy, C.; Brockbank, S.; Needham, M.; Read, S.; Newham, P. The identification of differentially expressed microRNA in osteoarthritic tissue that modulate the production of TNF-alpha and MMP13. *Osteoarthr. Cartil.* **2009**, *17*, 464–472. [CrossRef]

110. Wang, S.; Tang, C.; Zhang, Q.; Chen, W. Reduced miR-9 and miR-181a expression down-regulates Bim concentration and promote osteoclasts survival. *Int. J. Clin. Exp. Pathol.* **2014**, *7*, 2209–2218.

111. Li, J.; Zhang, Y.; Zhao, Q.; Wang, J.; He, X. MicroRNA-10a Influences Osteoblast Differentiation and Angiogenesis by Regulating ß-Catenin Expression. *Cell. Physiol. Biochem.* **2015**, *37*, 2194–2208. [CrossRef]

112. Day, T.F.; Guo, X.; Garrett-Beal, L.; Yang, Y. Wnt/beta-catenin signaling in mesenchymal progenitors controls osteoblast and chondrocyte differentiation during vertebrate skeletogenesis. *Dev. Cell* **2005**, *8*, 739–750. [CrossRef] [PubMed]

113. Zhu, S.; Deng, S.; Ma, Q.; Zhang, T.; Jia, C.; Zhuo, D.; Yang, F.; Wei, J.; Wang, L.; Dykxhoorn, D.M.; et al. MicroRNA-10A* and MicroRNA-21 Modulate Endothelial Progenitor Cell Senescence Via Suppressing High-Mobility Group A2. *Circ. Res.* **2013**, *112*, 152–164. [CrossRef]

114. Hassel, D.; Cheng, P.; White, M.P.; Ivey, K.N.; Kroll, J.; Augustin, H.G.; Katus, H.A.; Stainier, D.Y.R.; Srivastava, D. MicroRNA-10 Regulates the Angiogenic Behavior of Zebrafish and Human Endothelial Cells by Promoting Vascular Endothelial Growth Factor Signaling. *Circ. Res.* **2012**, *111*, 1421–1433. [CrossRef]

115. Wang, X.; Ling, C.C.; Li, L.; Qin, Y.; Qi, J.; Liu, X.; You, B.; Shi, Y.; Zhang, J.; Xu, Q.J.H.; et al. MicroRNA-10a/10b represses a novel target gene mib1 to regulate angiogenesis. *Cardiovasc. Res.* **2016**, *110*, 140–150. [CrossRef]

116. Zhang, J.; Fu, W.; He, M.; Xie, W.; Lv, Q.; Li, G.; Wang, H.; Lu, G.; Hu, X.; Jiang, S.; et al. MiRNA-20a promotes osteogenic differentiation of human mesenchymal stem cells by co-regulating BMP signaling. *RNA Biol.* **2011**, *8*, 829–838. [CrossRef] [PubMed]

117. Doebele, C.; Bonauer, A.; Fischer, A.; Scholz, A.; Ress, Y.; Urbich, C.; Hofmann, W.-K.; Zeiher, A.M.; Dimmeler, S. Members of the microRNA-17-92 cluster exhibit a cell-intrinsic antiangiogenic function in endothelial cells. *Blood* **2010**, *115*, 4944–4950. [CrossRef]

118. Deng, H.-T.; Liu, H.-L.; Zhai, B.-B.; Zhang, K.; Xu, G.-C.; Peng, X.-M. Vascular endothelial growth factor suppresses TNFSF15 production in endothelial cells by stimulating miR-31 and miR-20a expression via activation of Akt and Erk signals. *FEBS Open Bio.* **2017**, *7*, 108–117. [CrossRef] [PubMed]

119. Luzi, E.; Marini, F.; Sala, S.C.; Tognarini, I.; Galli, G.; Brandi, M.L. Osteogenic Differentiation of Human Adipose Tissue–Derived Stem Cells Is Modulated by the miR-26a Targeting of the SMAD1 Transcription Factor. *J. Bone Miner. Res.* **2008**, *23*, 287–295. [CrossRef] [PubMed]

120. Luzi, E.; Marini, F.; Tognarini, I.; Galli, G.; Falchetti, A.; Brandi, M.L. The regulatory network menin-microRNA 26a as a possible target for RNA-based therapy of bone diseases. *Nucleic. Acid Ther.* **2012**, *22*, 103–108. [CrossRef]

121. Su, X.; Liao, L.; Shuai, Y.; Jing, H.; Liu, S.; Zhou, H.; Liu, Y.; Jin, Y. MiR-26a functions oppositely in osteogenic differentiation of BMSCs and ADSCs depending on distinct activation and roles of Wnt and BMP signaling pathway. *Cell Death Dis.* **2015**, *6*, e1851. [CrossRef] [PubMed]

122. Trompeter, H.-I.; Dreesen, J.; Hermann, E.; Iwaniuk, K.M.; Hafner, M.; Renwick, N.; Tuschl, T.; Wernet, P. MicroRNAs miR-26a, miR-26b, and miR-29b accelerate osteogenic differentiation of unrestricted somatic stem cells from human cord blood. *BMC Genom.* **2013**, *14*, 1–13. [CrossRef] [PubMed]

123. Icli, B.; Wara, A.K.M.; Moslehi, J.; Sun, X.; Plovie, E.; Cahill, M.; Marchini, J.F.; Schissler, A.; Padera, R.F.; Shi, J.; et al. MicroRNA-26a regulates pathological and physiological angiogenesis by targeting BMP/SMAD1 signaling. *Circ. Res.* **2013**, *113*, 1231–1241. [CrossRef] [PubMed]

124. Li, Z.; Hassan, M.Q.; Jafferji, M.; Aqeilan, R.I.; Garzon, R.; Croce, C.M.; Van Wijnen, A.J.; Stein, J.L.; Stein, G.S.; Lian, J.B. Biological Functions of miR-29b Contribute to Positive Regulation of Osteoblast Differentiation. *J. Biol. Chem.* **2009**, *284*, 15676–15684. [CrossRef] [PubMed]

125. Rossi, M.; Pitari, M.R.; Amodio, N.; Di Martino, T.M.; Conforti, F.; Leone, E.; Botta, C.; Paolino, F.M.; Giudice, T.D.E.L.; Iuliano, E.; et al. miR-29b Negatively Regulates Human Osteoclastic Cell Differentiation and Function: Implications for the Treatment of Multiple Myeloma-Related Bone Disease. *J. Cell. Physiol.* **2013**, *228*, 1506–1515. [CrossRef] [PubMed]

126. Zhang, K.; Cai, H.-X.; Gao, S.; Yang, G.-L.; Deng, H.-T.; Xu, G.-C.; Han, J.; Zhang, Q.-Z.; Li, L.-Y. TNSF15 suppresses VEGF production in endothelial cells by stimulating miR-29b expression via activation of JNK-GATA3 Signals. *Oncotarget* **2016**, *7*, 69436–69449. [CrossRef] [PubMed]

127. Li, Y.; Cai, B.; Shen, L.; Dong, Y.; Lu, Q.; Sun, S.; Liu, S.; Ma, S.; Ma, P.X.; Chen, J. MiRNA-29b suppresses tumor growth through simultaneously inhibiting angiogenesis and tumorigenesis by targeting Akt3. *Cancer Lett.* **2017**, *397*, 111–119. [CrossRef]

128. Granchi, D.; Ochoa, G.; Leonardi, E.; Devescovi, V.; Baglìo, S.R.; Osaba, L.; Baldini, N.; Ciapetti, G. Gene expression patterns related to osteogenic differentiation of bone marrow-derived mesenchymal stem cells during ex vivo expansion. *Tissue Eng. Part. C Methods* **2010**, *16*, 511–523. [CrossRef]

129. Baglìo, S.R.; Devescovi, V.; Granchi, D.; Baldini, N. MicroRNA expression profiling of human bone marrow mesenchymal stem cells during osteogenic differentiation reveals Osterix regulation by miR-31. *Gene* **2013**, *527*, 321–331. [CrossRef]

130. Deng, Y.; Wu, S.; Zhou, H.; Bi, X.; Wang, Y.; Hu, Y.; Gu, P.; Fan, X. Effects of a miR-31, Runx2, and Satb2 regulatory loop on the osteogenic differentiation of bone mesenchymal stem cells. *Stem Cells Dev.* **2013**, *22*, 2278–2286. [CrossRef]

131. Luo, J.; Lin, J.; Paranya, G.; Bischoff, J. Angiostatin Upregulates E-Selectin in Proliferating Endothelial Cells. *Biochem. Biophys. Res. Commun.* **1998**, *911*, 906–911. [CrossRef] [PubMed]

132. Chen, L.; Holmstrom, K.; Qiu, W.; Ditzel, N.; Shi, K.; Hokland, L.; Kassem, M. MicroRNA-34a Inhibits Osteoblast Differentiation and In Vivo Bone Formation of Human Stromal Stem Cells. *Stem Cells* **2014**, *32*, 902–912. [CrossRef] [PubMed]

133. Kang, H.; Chen, H.; Huang, P.; Qi, J.; Qian, N.; Deng, L.; Guo, L. Glucocorticoids impair bone formation of bone marrow stromal stem cells by reciprocally regulating microRNA-34a-5p. *Osteoporos. Int.* **2016**, *27*, 1493–1505. [CrossRef] [PubMed]

134. Fan, C.; Jia, L.; Zheng, Y.; Jin, C.; Liu, Y.; Liu, H.; Zhou, Y. Mir-34a Promotes Osteogenic Differentiation of Human Adipose-Derived Stem Cells via the RBP2/NOTCH I/CYCLIN DI Coregulatory Network. *Stem Cell Rep.* **2016**, *7*, 236–248. [CrossRef] [PubMed]

135. Engin, F.; Yao, Z.; Yang, T.; Zhou, G.; Bertin, T.; Jiang, M.M.; Chen, Y.; Wang, L.; Zheng, H.; Sutton, R.E.; et al. Dimorphic effects of Notch signaling in bone homeostasis. *Nat. Med.* **2008**, *14*, 299–305. [CrossRef] [PubMed]

136. Galindo, M.; Pratap, J.; Young, D.W.; Hovhannisyan, H.; Im, H.J.; Choi, J.Y.; Lian, J.B.; Stein, J.L.; Stein, G.S.; van Wijnen, A.J. The bone-specific expression of Runx2 oscillates during the cell cycle to support a G1-related antiproliferative function in osteoblasts. *J. Biol. Chem.* **2005**, *280*, 20274–20285. [CrossRef] [PubMed]

137. Zha, X.; Sun, B.; Zhang, R.; Li, C.; Yan, Z.; Chen, J. Regulatory Effect of MicroRNA-34a on Osteogenesis and Angiogenesis in Glucocorticoid-Induced Osteonecrosis of the Femoral Head. *J. Orthop. Res.* **2018**, *36*, 417–424. [CrossRef]

138. Zhao, T.; Li, J.; Chen, A.F. MicroRNA-34a induces endothelial progenitor cell senescence and impedes its angiogenesis via suppressing silent information regulator 1. *Am. J. Endocrinol. Metab.* **2010**, *299*, E110–E116. [CrossRef]

139. Mattagajasingh, I.; Kim, C.; Naqvi, A.; Yamamori, T.; Hoffman, T.; Jung, S.; DeRicco, J.; Kasuno, K.; Irani, K. SIRT1 promotes endothelium-dependent vascular relaxation by activating endothelial nitric oxide synthase. *Proc. Natl. Acad. Sci. USA* **2007**, *104*, 14855–14860. [CrossRef]

140. Kumar, B.; Yadav, A.; Lang, J.; Teknos, T.N.; Kumar, P. Dysregulation of MicroRNA-34a Expression in Head and Neck Squamous Cell Carcinoma Promotes Tumor Growth and Tumor Angiogenesis. *PLoS ONE* **2012**, *7*, e37601. [CrossRef]

141. Chai, Z.T.; Kong, J.; Zhu, X.D.; Zhang, Y.Y.; Lu, L.; Zhou, J.M.; Wang, L.R.; Zhang, K.Z.; Zhang, Q.B.; Ao, J.Y.; et al. MicroRNA-26a Inhibits Angiogenesis by Down-Regulating VEGFA through the PIK3C2α/Akt/HIF-1α Pathway in Hepatocellular Carcinoma. *PLoS ONE* **2013**, *8*, 1–12. [CrossRef] [PubMed]

142. Murata, K.; Ito, H.; Yoshitomi, H.; Yamamoto, K.; Fukuda, A.; Yoshikawa, J.; Furu, M.; Ishikawa, M.; Shibuya, H.; Matsuda, S. Inhibition of miR-92a enhances fracture healing via promoting angiogenesis in a model of stabilized fracture in young mice. *J. Bone Miner. Res.* **2014**, *29*, 316–326. [CrossRef] [PubMed]

143. Kalinina, N.; Klink, G.; Glukhanyuk, E.; Lopatina, T.; Anastassia, E.; Akopyan, Z.; Tkachuk, V. miR-92a regulates angiogenic activity of adipose-derived mesenchymal stromal cells. *Exp. Cell Res.* **2015**, *339*, 61–66. [CrossRef] [PubMed]

144. Daniel, J.-M.; Penzkofer, D.; Teske, R.; Dutzmann, J.; Koch, A.; Bielenberg, W.; Bonauer, A.; Boon, R.A.; Fischer, A.; Bauersachs, J.; et al. Inhibition of miR-92a improves re-endothelialization and prevents neointima formation following vascular injury. *Cardiovasc. Res.* **2014**, *103*, 564–572. [CrossRef] [PubMed]

145. Goettsch, C.; Rauner, M.; Pacyna, N.; Hempel, U.; Bornstein, S.R.; Hofbauer, L.C. MiR-125b regulates calcification of vascular smooth muscle cells. *Am. J. Pathol.* **2011**, *179*, 1594–1600. [CrossRef] [PubMed]

146. Mizuno, Y.; Yagi, K.; Tokuzawa, Y.; Kanesaki-Yatsuka, Y.; Suda, T.; Katagiri, T.; Fukuda, T.; Maruyama, M.; Okuda, A.; Amemiya, T.; et al. miR-125b inhibits osteoblastic differentiation by down-regulation of cell proliferation. *Biochem. Biophys. Res. Commun.* **2008**, *368*, 267–272. [CrossRef] [PubMed]

147. Zhou, S.; Zhang, P.; Liang, P.; Huang, X. The expression of miR-125b regulates angiogenesis during the recovery of heat-denatured HUVECs. *Burns* **2015**, *41*, 803–811. [CrossRef]

148. Huang, K.; Fu, J.; Zhou, W.; Li, W.; Dong, S.; Yu, S.; Hu, Z.; Wang, H.; Xie, Z. MicroRNA-125b regulates osteogenic differentiation of mesenchymal stem cells by targeting Cbfb in vitro. *Biochimie* **2014**, *102*, 47–55. [CrossRef]

149. Xihong, L.U.; Min, D.; Honghui, H.E.; Dehui, Z.; Wei, Z. miR-125b regulates osteogenic differentiation of human bone marrow mesenchymal stem cells by targeting Smad4. *J. Cent. South. Univ. (Med. Sci.)* **2013**, *38*, 341–346. [CrossRef]

150. Muramatsu, F.; Kidoya, H.; Naito, H.; Sakimoto, S.; Takakura, N. microRNA-125b inhibits tube formation of blood vessels through translational suppression of VE-cadherin. *Oncogene* **2013**, *32*, 414–421. [CrossRef]

151. Schaap-Oziemlak, A.M.; Raymakers, R.A.; Bergevoet, S.M.; Gilissen, C.; Jansen, B.J.H.; Adema, G.J.; Kögler, G.; le Sage, C.; Agami, R.; van der Reijden, B.A.; et al. MicroRNA hsa-miR-135b Regulates Mineralization in Osteogenic Differentiation of Human Unrestricted Somatic Stem Cells. *Stem Cells Dev.* **2010**, *19*, 877–885. [CrossRef] [PubMed]

152. Umezu, T.; Tadokoro, H.; Azuma, K.; Yoshizawa, S.; Ohyashiki, K.; Ohyashiki, J.H. Exosomal miR-135b shed from hypoxic multiple myeloma cells enhances angiogenesis by targeting factor-inhibiting HIF-1. *Blood* **2014**, *124*, 3748–3757. [CrossRef] [PubMed]

153. Xu, S.; Santini, G.C.; De Veirman, K.; Broek, I.V.; Leleu, X.; De, A.; Van Camp, B.; Vanderkerken, K.; Van Riet, I. Upregulation of miR-135b Is Involved in the Impaired Osteogenic Differentiation of Mesenchymal Stem Cells Derived from Multiple Myeloma Patients. *PLoS ONE* **2013**, *8*, e79752. [CrossRef]

154. Sumiyoshi, K.; Kubota, S.; Ohgawara, T.; Kawata, K.; Abd El Kader, T.; Nishida, T.; Ikeda, N.; Shimo, T.; Yamashiro, T.; Takigawa, M. Novel Role of miR-181a in Cartilage Metabolism. *J. Cell. Biochem.* **2013**, *114*, 2094–2100. [CrossRef] [PubMed]

155. Gabler, J.; Ruetze, M.; Kynast, K.L.; Grossner, T.; Diederichs, S.; Richter, W. Stage-Specific miRs in Chondrocyte Maturation: Differentiation-Dependent and Hypertrophy-Related miR Clusters and the miR-181 Family. *Tissue Eng. Part. A* **2015**, *21*, 2840–2851. [CrossRef] [PubMed]

156. Sun, X.; Charbonneau, C.; Wei, L.; Chen, Q.; Terek, R.M. miR-181a Targets RGS16 to Promote Chondrosarcoma Growth, Angiogenesis, and Metastasis. *Mol. Cancer Res.* **2015**, *13*, 1347–1357. [CrossRef] [PubMed]

157. Sun, X.; Wei, L.; Chen, Q.; Terek, R.M. MicroRNA Regulates Vascular Endothelial Growth Factor Expression in Chondrosarcoma Cells. *Clin. Orthop. Relat. Res.* **2015**, *473*, 907–913. [CrossRef] [PubMed]

158. Almeida, M.I.; Silva, A.M.; Vasconcelos, D.M.; Almeida, C.R.; Caires, H.; Pinto, M.T.; Calin, A.; Santos, S.G.; Barbosa, M.A. miR-195 in human primary mesenchymal stromal/stem cells regulates proliferation, osteogenesis and paracrine effect on angiogenesis. *Oncotarget* **2015**, *7*, 7–22. [CrossRef] [PubMed]

159. Wang, R.; Zhao, N.; Li, S.; Fang, J.; Chen, M.; Yang, J.; Jia, W.; Yuan, Y.; Zhuang, S. MicroRNA-195 Suppresses Angiogenesis and Metastasis of Hepatocellular Carcinoma by Inhibiting the Expression of VEGF, VAV2, and CDC42. *Hepatology* **2013**, *58*, 642–653. [CrossRef] [PubMed]

160. Li, Y.; Zeng, C.; Tu, M.; Jiang, W.; Dai, Z.; Hu, Y.; Deng, Z.; Xiao, W. MicroRNA-200b acts as a tumor suppressor in osteosarcoma via targeting ZEB1. *Onco Targets Ther.* **2016**, *9*, 3101–3111.

161. Fan, X.; Teng, Y.; Ye, Z.; Zhou, Y.; Tan, W.-S. The effect of gap junction-mediated transfer of miR-200b on osteogenesis and angiogenesis in a co-culture of MSCs and HUVECs. *J. Cell Sci.* **2018**, *131*, jcs216135. [CrossRef] [PubMed]

162. Choi, Y.; Yoon, S.; Jeong, Y.; Yoon, J.; Baek, K. Regulation of Vascular Endothelial Growth Factor Signaling by miR-200b. *Mol. Cells* **2011**, *32*, 77–82. [CrossRef] [PubMed]

163. Mizuno, Y.; Tokuzawa, Y.; Ninomiya, Y.; Yagi, K.; Yatsuka-Kanesaki, Y.; Suda, T.; Fukuda, T.; Katagiri, T.; Kondoh, Y.; Amemiya, T.; et al. miR-210 promotes osteoblastic differentiation through inhibition of AcvR1b. *FEBS Lett.* **2009**, *583*, 2263–2268. [CrossRef] [PubMed]

164. Liu, X.-D.; Cai, F.; Liu, L.; Zhang, Y.; Yang, A.-L. microRNA-210 is involved in the regulation of postmenopausal osteoporosis through promotion of VEGF expression and osteoblast differentiation. *Biol. Chem.* **2015**, *396*, 339–347. [CrossRef] [PubMed]

165. Fasanaro, P.; D'Alessandra, Y.; Di Stefano, V.; Melchionna, R.; Romani, S.; Pompilio, G.; Capogrossi, M.C.; Martelli, F. MicroRNA-210 modulates endothelial cell response to hypoxia and inhibits the receptor tyrosine kinase ligand ephrin-A3. *J. Biol. Chem.* **2008**, *283*, 15878–15883. [CrossRef] [PubMed]

166. Yan, J.; Guo, D.; Yang, S.; Sun, H.; Wu, B.; Zhou, D. Inhibition of miR-222-3p activity promoted osteogenic differentiation of hBMSCs by regulating Smad5-RUNX2 signal axis. *Biochem. Biophys. Res. Commun.* **2016**, *470*, 498–503. [CrossRef] [PubMed]

167. Takigawa, S.; Chen, A.; Wan, Q.; Na, S.; Sudo, A.; Yokota, H.; Hamamura, K. Role of miR-222-3p in c-Src-Mediated Regulation of Osteoclastogenesis. *Int. J. Mol. Sci.* **2016**, *17*, 240. [CrossRef] [PubMed]

168. Poliseno, L.; Tuccoli, A.; Mariani, L.; Evangelista, M.; Citti, L.; Woods, K.; Mercatanti, A.; Hammond, S.; Rainaldi, G. MicroRNAs modulate the angiogenic properties of HUVECs. *Blood* **2006**, *108*, 3068–3071. [CrossRef]

169. Gao, J.; Yang, T.; Han, J.; Yan, K.; Qiu, X.; Zhou, Y.; Fan, Q.; Ma, B. MicroRNA Expression During Osteogenic Differentiation of Human Multipotent Mesenchymal Stromal Cells From Bone Marrow. *J. Cell. Biochem.* **2011**, *112*, 1844–1856. [CrossRef]

170. Vimalraj, S.; Selvamurugan, N. MicroRNAs expression and their regulatory networks during mesenchymal stem cells differentiation toward osteoblasts. *Int. J. Biol. Macromol.* **2014**, *66*, 194–202. [CrossRef]

171. Li, L.; Qi, Q.; Luo, J.; Huang, S.; Ling, Z.; Gao, M.; Zhou, Z.; Stiehler, M.; Zou, X. FOXO1-suppressed miR-424 regulates the proliferation and osteogenic differentiation of MSCs by targeting FGF2 under oxidative stress. *Sci. Rep.* **2017**, *7*, 1–12. [CrossRef]

172. de Pontual, L.; Yao, E.; Callier, P.; Faivre, L.; Drouin, V.; Cariou, S.; Van Haeringen, A.; Geneviève, D.; Goldenberg, A.; Oufadem, M.; Manouvrier, S.; Munnich, A.; et al. Germline deletion of the miR-17 ~ 92 cluster causes skeletal and growth defects in humans. *Nat. Genet.* **2011**, *43*, 1026–1030. [CrossRef] [PubMed]

173. Penzkofer, D.; Bonauer, A.; Fischer, A.; Tups, A.; Brandes, R.P.; Zeiher, A.M.; Dimmeler, S. Phenotypic Characterization of miR-92a - /- Mice Reveals an Important Function of miR-92a in Skeletal Development. *PLoS ONE* **2014**, *9*, e101153. [CrossRef] [PubMed]

174. Zhou, M.; Ma, J.; Chen, S.; Chen, X.; Yu, X. MicroRNA-17-92 cluster regulates osteoblast proliferation and differentiation. *Endocrine* **2014**, *45*, 302–310. [CrossRef] [PubMed]

175. Mao, G.; Wu, P.; Zhang, Z.; Zhang, Z.; Liao, W.; Li, Y.; Kang, Y. MicroRNA-92a-3p Regulates Aggrecanase-1 and Aggrecanase-2 Expression in Chondrogenesis and IL-1β- Induced Catabolism in Human Articular Chondrocytes. *Cell. Physiol. Biochem.* **2017**, *44*, 38–52. [CrossRef]

176. Zhang, Z.; Kang, Y.; Zhang, Z.; Zhang, H.; Duan, X.; Liu, J.; Li, X.; Liao, W. Expression of microRNAs during chondrogenesis of human adipose-derived stem cells. *Osteoarthr. Cartil.* **2012**, *20*, 1638–1646. [CrossRef] [PubMed]

177. Mao, G.; Zhang, Z.; Huang, Z.; Chen, W.; Huang, G.; Meng, F.; Zhang, Z.; Kang, Y. MicroRNA-92a-3p regulates the expression of cartilage-specific genes by directly targeting histone deacetylase 2 in chondrogenesis and degradation. *Osteoarthr. Cartil.* **2017**, *25*, 521–532. [CrossRef] [PubMed]

178. Bonauer, A.; Carmona, G.; Iwasaki, M.; Mione, M.; Koyanagi, M.; Fischer, A.; Burchfield, J.; Fox, H.; Doebele, C.; Ohtani, K.; et al. MicroRNA-92a Controls Angiogenesis and Functional Recovery of Ischemic Tissues in Mice. *Science* **2009**, *324*, 1710–1713. [CrossRef] [PubMed]

179. Elmén, J.; Lindow, M.; Schütz, S.; Lawrence, M.; Petri, A.; Obad, S.; Lindholm, M.; Hedtjärn, M.; Hansen, H.; Berger, U.; et al. LNA-mediated microRNA silencing in non-human primates. *Nature* **2008**, *452*, 896–899. [CrossRef] [PubMed]

180. Zhang, L.; Zhou, M.; Qin, G.; Weintraub, N.L.; Tang, Y. MiR-92a regulates viability and angiogenesis of endothelial cells under oxidative stress. *Biochem. Biophys. Res. Commun.* **2015**, *446*, 952–958. [CrossRef] [PubMed]

181. Rippe, C.; Blimline, M.; Magerko, K.A.; Lawson, B.R.; LaRocca, T.; Donato, A.; Seals, D.R. MicroRNA Changes in Human Arterial Endothelial Cells with Senescence: Relation to Apoptosis, eNOS and Inflammation Catarina. *Exp. Gerontol.* **2012**, *47*, 45–51. [CrossRef] [PubMed]

182. Lee, Y.S.; Kim, H.K.; Chung, S.; Kim, K.S.; Dutta, A. Depletion of human micro-RNA miR-125b reveals that it is critical for the proliferation of differentiated cells but not for the down-regulation of putative targets during differentiation. *J. Biol. Chem.* **2005**, *280*, 16635–16641. [CrossRef] [PubMed]

183. Xu, J.-F.; Yang, G.-H.; Pan, X.-H.; Zhang, S.-J.; Zhao, C.; Qiu, B.-S.; Gu, H.-F.; Hong, J.-F.; Cao, L.; Chen, Y.; et al. Altered MicroRNA Expression Profile in Exosomes during Osteogenic Differentiation of Human Bone Marrow- Derived Mesenchymal Stem Cells. *PLoS ONE* **2014**, *9*, e114627. [CrossRef] [PubMed]

184. Fan, G. Hypoxic exosomes promote angiogenesis Platelets: Balancing the septic triad. *Blood* **2014**, *124*, 3669–3670. [CrossRef] [PubMed]

185. Zhang, L.; Sun, Z.-J.; Bian, Y.; Kulkarni, A.B. MicroRNA-135b acts as a tumor promoter by targeting the hypoxia-inducible factor pathway in genetically defined mouse model of head and neck squamous cell carcinoma. *Cancer Lett.* **2013**, *331*, 230–238. [CrossRef] [PubMed]

186. Kazenwadel, J.; Michael, M.Z.; Harvey, N.L. Prox1 expression is negatively regulated by miR-181 in endothelial cells. *Blood* **2010**, *116*, 2395–2401. [CrossRef]

187. Naguibneva, I.; Ameyar-Zazoua, M.; Polesskaya, A.; Ait-Si-Ali, S.; Groisman, R.; Souidi, M.; Cuvellier, S.; Harel-Bellan, A. The microRNA miR-181 targets the homeobox protein Hox-A11 during mammalian myoblast differentiation. *Nat. Cell Biol.* **2006**, *8*, 278–284. [CrossRef]

188. Akiyama, T.; Bouillet, P.; Miyazaki, T.; Kadono, Y.; Chikuda, H.; Chung, U.; Fukuda, A.; Hikita, A.; Seto, H.; Okada, T.; et al. Regulation of osteoclast apoptosis by ubiquitination of proapoptotic BH3-only Bcl-2 family member Bim. *EMBO J.* **2003**, *22*, 6653–6664. [CrossRef]

189. Palmieri, A.; Pezzetti, F.; Brunelli, G.; Zollino, I.; Scapoli, L.; Martinelli, M.; Arlotti, M.; Carinci, F. Differences in osteoblast miRNA induced by cell binding domain of collagen and silicate-based synthetic bone. *J. Biomed. Sci.* **2007**, *14*, 777–782. [CrossRef]

190. Christoffersen, N.R.; Silahtaroglu, A.; Ørom, U.L.F.A.; Kauppinen, S.; Lund, A.H. miR-200b mediates post-transcriptional repression of ZFHX1B. *RNA* **2007**, *13*, 1172–1178. [CrossRef]

191. Baglio, S.R.; Rooijers, K.; Koppers-lalic, D.; Verweij, F.J.; Lanzón, M.P.; Zini, N.; Naaijkens, B.; Perut, F.; Niessen, H.W.M.; Baldini, N.; et al. Human bone marrow- and adipose- mesenchymal stem cells secrete exosomes enriched in distinctive miRNA and tRNA species. *Stem Cell Res. Ther.* **2015**, *6*, 1–20. [CrossRef] [PubMed]

192. Zong, L.; Zhu, Y.; Liang, R.; Zhao, H.-B. Gap junction mediated miRNA intercellular transfer and gene regulation: A novel mechanism for intercellular genetic communication. *Sci. Rep.* **2016**, *6*, 19884. [CrossRef] [PubMed]

193. Liu, G.-T.; Chen, H.-T.; Tsou, H.-K.; Tan, T.-W.; Fong, Y.-C.; Chen, P.-C.; Yang, W.-H.; Wang, S.-W.; Chen, J.-C.; Tang, C.-H. CCL5 promotes VEGF-dependent angiogenesis by downregulating miR-200b through PI3K/Akt signaling pathway in human chondrosarcoma cells. *Oncotarget* **2014**, *5*, 10718–10731. [CrossRef] [PubMed]

194. Chan, Y.C.; Khanna, S.; Roy, S.; Sen, C.K. miR-200b targets Ets-1 and is down-regulated by hypoxia to induce angiogenic response of endothelial cells. *J. Biol. Chem.* **2011**, *286*, 2047–2056. [CrossRef] [PubMed]

195. Lou, Y.-L.; Guo, F.; Liu, F.; Gao, F.-L.; Zhang, P.-Q.; Niu, X.; Guo, S.-C.; Yin, J.-H.; Wang, Y.; Deng, Z.-F. miR-210 activates notch signaling pathway in angiogenesis induced by cerebral ischemia. *Mol. Cell Biochem.* **2012**, *370*, 45–51. [CrossRef] [PubMed]

196. Fasanaro, P.; Greco, S.; Lorenzi, M.; Pescatori, M.; Brioschi, M.; Kulshreshta, R.; Banfi, C.; Stubbs, A.; Calin, G.A.; Ivan, M.; et al. An integrated approach for experimental target identification of hypoxia-induced miR-210. *J. Biol. Chem.* **2009**, *284*, 35134–35143. [CrossRef] [PubMed]

197. Ivan, M.; Huang, X. miR-210: Fine-Tuning the Hypoxic Response. *Adv. Exp. Med. Biol.* **2014**, *772*, 205–227. [CrossRef]

198. Ivan, M.; Harris, A.L.; Martelli, F.; Kulshreshtha, R. Hypoxia response and microRNAs: No longer two separate worlds. *J. Cell Mol. Med.* **2008**, *12*, 1426–1431. [CrossRef]

199. Kuijper, S.; Turner, C.J.; Adams, R.H. Regulation of angiogenesis by Eph-ephrin interactions. *Trends Cardiovasc. Med.* **2007**, *17*, 145–151. [CrossRef]

200. Pandey, A.; Shao, H.; Marks, R.M.; Polverini, P.J.; Dixit, V.M. Role of B61, the ligand for the Eck receptor tyrosine kinase, in TNF-alpha-induced angiogenesis. *Science* **1995**, *268*, 567–569. [CrossRef]

201. Matsui, J.; Wakabayashi, T.; Asada, M.; Yoshimatsu, K. Stem cell factor/c-kit signaling promotes the survival, migration, and capillary tube formation of human umbilical vein endothelial cells. *J. Biol. Chem.* **2004**, *279*, 18600–18607. [CrossRef] [PubMed]

202. Jung, S.; Kleinheinz, J. Angiogenesis—The Key to Regeneration. In *Tissue Engineering and Regenerative Medicine*; Andrades, J.A., Ed.; InTechOpen: London, UK, 2013; pp. 453–473.

203. Kanczler, J.M.; Oreffo, R.O.C. Osteogenesis and angiogenesis: The potential for engineering bone. *Eur. Cells Mater.* **2008**, *15*, 100–114. [CrossRef]

204. Hou, H.; Zhang, X.; Tang, T.; Dai, K.; Ge, R. Enhancement of bone formation by genetically-engineered bone marrow stromal cells expressing BMP-2, VEGF and angiopoietin-1. *Biotechnol. Lett.* **2009**, *31*, 1183–1189. [CrossRef] [PubMed]

205. Zhang, F.; Qiu, T.; Wu, X.; Wan, C.; Shi, W.; Wang, Y.; Chen, J.; Wan, M.; Clemens, T.L.; Cao, X. Sustained BMP Signaling in Osteoblasts Stimulates Bone Formation by Promoting Angiogenesis and Osteoblast Differentiation. *J. Bone Miner. Res.* **2009**, *24*, 1224–1233. [CrossRef] [PubMed]

206. Shi, Z.; Wang, K. Effects of recombinant adeno-associated viral vectors on angiopoiesis and osteogenesis in cultured rabbit bone marrow stem cells via co-expressing hVEGF and hBMP genes: A preliminary study in vitro. *Tissue Cell* **2010**, *42*, 314–321. [CrossRef] [PubMed]

207. Hon, L.S.; Zhang, Z. The roles of binding site arrangement and combinatorial targeting in microRNA repression of gene expression. *Genome Biol.* **2007**, *8*, R166. [CrossRef]

208. Grimson, A.; Farh, K.K.; Johnston, W.K.; Garrett-Engele, P.; Lim, L.P.; Bartel, D.P. MicroRNA targeting specificity in mammals: Determinants beyond seed pairing. *Mol. Cell* **2007**, *27*, 91–105. [CrossRef] [PubMed]

209. Betel, D.; Wilson, M.; Gabow, A.; Marks, D.S.; Sander, C. The microRNA.org resource: Targets and expression. *Nucleic Acids Res.* **2008**, *36*, D149–D153. [CrossRef]

210. Krek, A.; Grün, D.; Poy, M.N.; Wolf, R.; Rosenberg, L.; Epstein, E.J.; MacMenamin, P.; da Piedade, I.; Gunsalus, K.C.; Stoffel, M.; et al. Combinatorial microRNA target predictions. *Nat. Genet.* **2005**, *37*, 495–500. [CrossRef]

211. Brennecke, J.; Stark, A.; Russell, R.; Cohen, S. Principles of microRNA-target recognition. *PLoS Biol.* **2005**, *3*, e85. [CrossRef]

212. Tsang, J.; Zhu, J.; van Oudenaarden, A. MicroRNAmediated feedback and feedforward loops are recurrent network motifs in mammals. *Mol. Cell* **2007**, *26*, 753–767. [CrossRef] [PubMed]

213. Lewis, B.; Burge, C.; Bartel, D. Conserved seed pairing, often flanked by adenosines, indicates that thousands of human genes are microRNA targets. *Cell* **2005**, *120*, 15–20. [CrossRef] [PubMed]

214. Li, H.; Xie, H.; Liu, W.; Hu, R.; Huang, B.; Tan, Y.; Xu, K.; Sheng, Z.; Zhou, H.; Wu, X.; et al. A novel microRNA targeting HDAC5 regulates osteoblast differentiation in mice and contributes to primary osteoporosis in humans. *J. Clin. Invest.* **2009**, *119*, 3666–3677. [CrossRef] [PubMed]

215. Mariner, P.; Johannesen, E.; Anseth, K. Manipulation of miRNA activity accelerates osteogenic differentiation of hMSCs in engineered 3D scaffolds. *J. Tissue Eng. Regen Med.* **2012**, *6*, 314–324. [CrossRef] [PubMed]

216. Lee, W.Y.; Li, N.; Lin, S.; Wang, B.; Lan, H.Y.; Li, G. miRNA-29b improves bone healing in mouse fracture model. *Mol. Cell. Endocrinol.* **2016**, *430*, 97–107. [CrossRef] [PubMed]

217. Li, Y.; Fan, L.; Liu, S.; Liu, W.; Zhang, H.; Zhou, T.; Wu, D.; Yang, P.; Shen, L.; Chen, J.; et al. The promotion of bone regeneration through positive regulation of angiogenic-osteogenic coupling using microRNA-26a. *Biomaterials* **2013**, *34*, 5048–5058. [CrossRef] [PubMed]

218. Yoshizuka, M.; Nakasa, T.; Kawanishi, Y.; Hachisuka, S.; Furuta, T.; Miyaki, S.; Adachi, N.; Ochi, M. Inhibition of microRNA-222 expression accelerates bone healing with enhancement of osteogenesis, chondrogenesis, and angiogenesis in a rat refractory fracture model. *J. Orthop. Sci.* **2016**, *21*, 852–858. [CrossRef] [PubMed]

219. Ørom, U.A.; Kauppinen, S.; Lund, A.H. LNA-modified oligonucleotides mediate specific inhibition of microRNA function. *Gene* **2006**, *372*, 137–141. [CrossRef] [PubMed]

220. Adams, B.D.; Parsons, C.; Walker, L.; Zhang, W.C.; Slack, F.J. Targeting noncoding RNAs in disease. *J. Clin. Investig.* **2017**, *127*, 761–771. [CrossRef]

221. Simonson, B.; Das, S. MicroRNA Therapeutics: The Next Magic Bullet? *Mini Rev. Med. Chem.* **2016**, *15*, 467–474. [CrossRef]

222. Esquela-Kerscher, A.; Slack, F.J. Oncomirs-microRNAs with a role in cancer. *Nat. Rev. Cancer* **2006**, *6*, 259–269. [CrossRef] [PubMed]

223. López-Camarillo, C.; Marchat, L.A.; Aréchaga-Ocampo, E.; Azuara-Liceaga, E.; Pérez-Plasencia, C.; Fuentes-Mera, L.; Fonseca-Sánchez, M.A.; Flores-Pérez, A. *Functional Roles of microRNAs in Cancer: microRNomes and oncomiRs Connection*; Oncogenomi.; In Tech Open Science: London, UK, 2013.

224. Senanayake, U.; Das, S.; Vesely, P.; Alzoughbi, W.; Fröhlich, L.F.; Chowdhury, P.; Leuschner, I.; Hoefler, G.; Guertl, B. miR-192, miR-194, miR-215, miR-200c and miR-141 are downregulated and their common target ACVR2B is strongly expressed in renal childhood neoplasms. *Carcinogenesis* **2012**, *33*, 1014–1021. [CrossRef] [PubMed]

12

Biology and Function of miR159 in Plants

Anthony A. Millar *, Allan Lohe and Gigi Wong

Division of Plant Science, Research School of Biology, The Australian National University,
Canberra ACT 2601, Australia
* Correspondence: tony.millar@anu.edu.au

Abstract: MicroR159 (miR159) is ancient, being present in the majority of land plants where it targets a class of regulatory genes called *GAMYB* or *GAMYB-like* via highly conserved miR159-binding sites. These *GAMYB* genes encode R2R3 MYB domain transcription factors that transduce the gibberellin (GA) signal in the seed aleurone and the anther tapetum. Here, *GAMYB* plays a conserved role in promoting the programmed cell death of these tissues, where miR159 function appears weak. By contrast, *GAMYB* is not involved in GA-signaling in vegetative tissues, but rather its expression is deleterious, leading to the inhibition of growth and development. Here, the major function of miR159 is to mediate strong silencing of *GAMYB* to enable normal growth. Highlighting this requirement of strong silencing are conserved RNA secondary structures associated with the miR159-binding site in *GAMYB* mRNA that promotes miR159-mediated repression. Although the miR159-*GAMYB* pathway in vegetative tissues has been implicated in a number of different functions, presently no conserved role for this pathway has emerged. We will review the current knowledge of the different proposed functions of miR159, and how this ancient pathway has been used as a model to help form our understanding of miRNA biology in plants.

Keywords: miR159; *GAMYB*; programmed cell death; aleurone; tapetum; vegetative growth; flowering

1. Introduction

Associated with the emergence and diversification of land plants, is a core set of conserved microRNA (miRNA) families that arose early in terrestrial plant evolution and which are conserved in modern day plant species [1]. This conservation implies these endogenous gene regulators are fundamental to plant biology and have been indispensable for the conquest of plant life on land. One such core family is microR159 (miR159), which has now been extensively studied in multiple, diverse plant species. In this review, we will highlight the major functions identified for miR159, its use as a model for gaining greater insights into miRNA biology in general, and finally highlight the many outstanding questions surrounding this ancient gene regulator.

2. MiR159 is Strongly Conserved and Highly Abundant Throughout the Plant Kingdom

Surveys of the many deep sequencing experiments on the small RNA fractions of plants, find miR159 ubiquitously present as a 21 nucleotide (nt) miRNA in all eudicots and monocotyledonous plants examined [2], and present in the majority of basal angiosperms, gymnosperms, ferns, and lycopods examined [3]. It has either been classified as a Class I (ubiquitous) or a Class II (present in most taxonomic groups) miRNA [2,3]. There is some uncertainty regarding whether miR159 is present in Bryophytes, where it is generally regarded as being absent [4], but it has been reported in a liverwort [5]. However, the reported miR159 sequence was only 18 nt long, suggesting it may not be a genuine miR159 homologue, so further analyses will be required to resolve this. Nevertheless, it is apparent that miR159 arose early in basal land plants and has been strongly conserved henceforth.

MiR159 fits the curious observation that the stronger the miRNA is conserved, the greater its expression or abundance [2]. This was derived from a multitude of small RNA-sequencing experiments from a wide diversity of plant species, where miR159 is often among the most abundant small RNA species (e.g., [6–10]). Additionally, highly similar miR159 isoforms are present in most land plants (Figure 1; [11]), so the sequence of this canonical miR159 appears to have remained fixed for 100s of millions of years. Nevertheless, like most miRNA families, considerable variation exists within small RNAs defined as miR159, with most plant species containing multiple family members that encode identical or highly similar isoforms, or "isomiRs", that differ by one to several nucleotides. For example, maize has 11 different *MIR159* loci, encoding four different miR159 isoforms [12]. For the most part, nucleotide variation occurs at the extremities of the miRNA, at positions considered less important for its specificity [13]. This is the case for the three different miR159 isoforms found in Arabidopsis that vary by 1–2 nucleotides; however, as these isoforms appear functionally redundant, this variation unlikely impacts which genes they target for repression [14,15]. Some species have even more variant miR159 isoforms (e.g., poplar; grape, soybean, and maize with 3–5 sequence variations [16]), so whether these miR159 variants have sub-functionalized to regulate different targets remains a possibility. Indeed, the ancient miRNA miR319 is closely related to miR159. In Arabidopsis, these two families are identical at 17 of 21 nucleotide positions, but have distinct target genes, demonstrating their sub-functionalization [17]. Their similarity extends to their primary-*MIRNA* precursors, where *pri-MIR159* and *pri-MIR319* are both unusually long fold-back structures that are processed in a non-canonical loop-to-base direction [18]. Phylogenetic analysis of primary-*MIRNA* precursor sequences of these families supports a common origin of miR319 and miR159, with the likelihood that miR159 has arisen and specialized from miR319 in basal land plants [19].

3. GAMYB and GAMYB-like Genes are the Only Conserved Targets of miR159

Core to understanding the function of a miRNA is the identification of the genes that they target. A clear and recurrent theme is that miR159 targets a family of genes encoding R2R3 MYB transcription factors referred to as "GAMYB" or "GAMYB-like". Similar to the conservation of miR159, *GAMYB*-homologues with a highly conserved miR159 binding site are found in most lineages of land plants (Figure 1). This extends to basal plants such as lycopods (e.g., *Selaginella moellendorffii*), moss (e.g., *Physcomitrella patens*), and the liverwort *Marchantia polymorpha* [20,21]. However, in *Marchantia* it appears that the *GAMYB* homologue is not regulated by miR159, but rather by miR319 [21,22]. Even in Arabidopsis, miR319 can regulate the *GAMYB* targets [17]. However, as miR319 in Arabidopsis is narrowly and weakly expressed compared to the widely and abundantly expressed miR159, miR319-mediated regulation of the *GAMYB-like* genes is insignificant relative to miR159-mediated regulation [17]. This makes miR159 functionally specific for the *GAMYB-like* targets, whereas miR319 is functionally specific for genes encoding another class of transcription factors, the *TCP* family, which miR159 is unable to regulate [17,23]. Therefore, it appears likely the more specific miR159 has arisen from miR319 in basal land plants, sub-functionalizing to become specific for the *GAMYB-like* genes. Although there is sequence variation in both the miR159 and its binding site within the *GAMYB-like* genes, both have appeared to have become fixed, arguing that this ancient miR159-*GAMYB* target relationship is critical for the life of land plants [24].

Strong experimental evidence supports the prediction of conserved miR159-mediated regulation of *GAMYB*. Firstly, degradome analysis from multiple diverse species confidently identifies *GAMYB* homologues are being regulated via a miR159-mediated cleavage mechanism. Although this analysis only detects targets regulated by the miRNA-guided cleavage mechanism (and not the translational- repression mechanism), functionally important targets appear to be preferentially detected [25]. Degradome experiments have been mainly performed on higher plants, including eudicots such as Arabidopsis [25], soybean [26], cotton [27], tomato [28], orchids [29], and peach [30]; also monocots such as wheat [31], rice [32], and barley [33], among many others, all of which experimentally validate *GAMYB* homologues as targets of miR159. Although many of these degradome experiments also pick up other target genes (e.g., [27,31]), these other targets are diverse in their identity and do not appear to be broadly conserved miR159 targets; i.e., they are not identified in degradomes from multiple diverse species. This argues that although miR159 may regulate additional targets, this does not appear to be at the expense of its main target, *GAMYB*. For instance, in tomato, miR159 has acquired a novel target, a gene that encodes a protein with a NOZZLE-like domain, and this miR159-mediated regulation is important for tomato development [34]. However, miR159 still regulates *GAMYB-like* genes in tomato, which is important for fruit development [35].

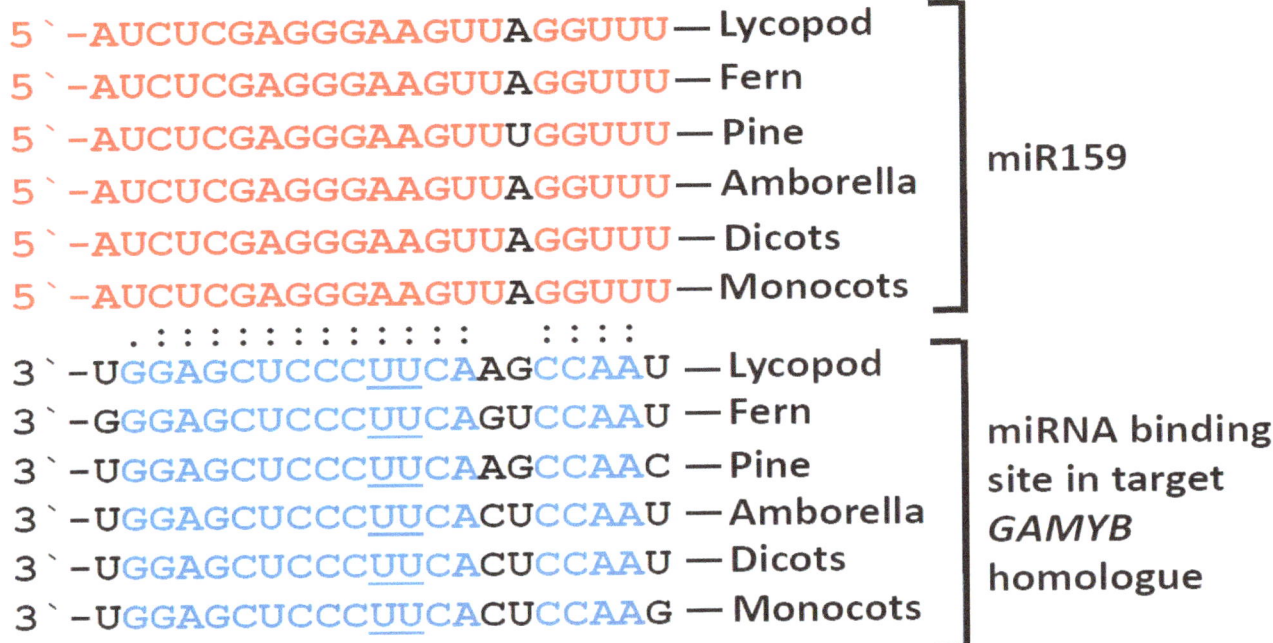

Figure 1. The microR159 (miR159)-*GAMYB* regulatory pathway appears highly conserved in land plants. Similar/identical miR159 isomiRs (shown in red) are found in most plant linages, including Lycopods (*Selaginella uncinata*; [3]) ferns (*Salvinia cucullata*; [3]), pine (*Pinus densata*; [36]), Amborella [37], dicots, and monocots (miRbase; [11]). Highly similar and complementary miR159 binding sites (shown in blue) are found in *GAMYB* homologues from lycopods (*Selaginella moellendorffii*; [20]), ferns (*Salvinia cucullata*; [11]), pine (*Larix kaempferi*; [38]), Amborella and many different monocots and dicots. Variant nucleotide positions are shown in black. However, throughout the plant kingdom, variation is not limited to these positions; for example, see [39].

4. The miR159-*GAMYB* Pathway in Arabidopsis

The Arabidopsis miR159 family has been extensively studied as a model for plant miRNA-mediated gene regulation (Figure 2). Arabidopsis has three *MIR159* genes (*MIR159a*, *MIR159b*, and *MIR159c*), each encoding a distinct isoform that differ from one another by 1–2 nucleotides [8]. Examination of their expression domains with promoter: GUS constructs found *MIR159a: GUS* and *MIR159b: GUS* had highly similar expression patterns, being broadly expressed throughout the plant, but strongest in shoot and root meristematic regions [14]. By contrast, the expression domain of a *MIR159c: GUS* reporter gene was much narrower, being restricted mainly to anthers and the shoot apical region [15], suggesting sub-functionalization. Regarding their level of expression, both deep sequencing and qPCR has found miR159a to be the most abundant family member, with miR159c being very weakly expressed [8,15]. To investigate their function, T-DNA loss-of-function mutant alleles were generated for each gene, however, none of the single *mir159* mutant plants displayed any phenotypic defects [14,15]. However, consistent with the highly similar expression domains, miR159a and miR159b were demonstrated to be functionally redundant, as a double *mir159a.mir159b* (*mir159ab*) mutant displayed severe growth and developmental defects, most notably a smaller rosette with upwardly curled leaves (Figure 2; [14]).

As a triple *mir159abc* mutant appeared indistinguishable from *mir159ab*, this and other data suggested miR159c in Arabidopsis has little to no activity and possibly corresponds to a pseudogene [15,17]. This is one of the few instances in Arabidopsis where T-DNA mutants have been identified and combined for all members of a miRNA family, and the *mir159ab* and *mir159abc* mutants have been used extensively in the functional characterization of miR159.

A bioinformatic search of miR159 targets in Arabidopsis using the standard target prediction program psRNATarget, identifies almost 100 potential miR159 targets with four or less mismatches [40]. The top twenty targets are shown in Table 1, which includes eight *MYB* genes with highly conserved miR159 binding sites [15]. By contrast, the non-*MYB* genes are highly diverse and their miR159-binding sites do not appear conserved [15]. Of the conserved *MYB* targets, seven are *GAMYB-like* genes (*MYB33, MYB65, MYB81, MYB97, MYB101, MYB104,* and *MYB120*), and the other is non-*GAMYB-like* gene, *DUO1 (DUO POLLEN1)*, which has a conserved miR159 binding site at a position distinct from the *GAMYB-like* genes [17]. Despite the fact that miR159-mediated cleavage products can be isolated for many of these predicted targets (Table 1), transcript profiling of the *mir159ab* mutant only identified two genes that appeared strongly de-regulated, the *GAMYB-like* targets, *MYB33* and *MYB65* (Table 1). This de-regulation resulted in *MYB33* and *MYB65* being strongly expressed throughout the plant [14,41]. Consistently, the only genes detected in multiple degradome analyses were *MYB33* and *MYB65* (Table 1) [25,42]. Eliminating the expression of these genes via the introduction of *myb33* and *myb65* loss-of-function alleles, suppressed all vegetative phenotypic defects of *mir159ab*, as a *mir159ab.myb33.myb65* quadruple mutant appeared indistinguishable from wild-type, other than male sterility [14]. The phenotype of male sterility is the only apparent defect of *myb33.myb65* plants, as *MYB33* and *MYB65* are two redundant genes that facilitate anther development (Figure 2) [43].

These genetic experiments demonstrated the major role of miR159 in Arabidopsis as being the widespread suppression of *MYB33* and *MYB65* expression, whose activity has severe deleterious impacts on plant growth and development, including stunted growth and curled leaves (Figure 2). The experiments also defined the functional specificity of miR159 in Arabidopsis as being *MYB33* and *MYB65* [14]. Supporting this is the expression of either a miR159-resistant *MYB33* or miR159-resistant *MYB65* transgene, both of which can phenocopy the *mir159ab* mutant [14,23,44]. This much narrower functional specificity compared to the bioinformatic prediction of many more targets is a common theme in miRNA biology, where both in animal and plants, pleiotropic defects of miRNA mutants can be suppressed via the repression of one-two target genes, despite bioinformatic programs predicting many targets with conserved miRNA binding sites [24,45]. Partially explaining this phenomenon for miR159, many of the bioinformatically predicted miR159 targets appear to have transcriptional domains that are mutually exclusive to that of miR159. Hence, the miRNA and targets are physically separated spatially and/or temporally preventing interaction (Figure 2) [14,15].

Despite their deleterious impact on vegetative growth, *MYB33* and *MYB65* appear ubiquitously transcribed throughout the plant, but only to be strongly and ubiquitously silenced, other than in seeds and anthers (Figure 2) [41,46]. There are multiple lines of evidence supporting this claim; (1) the vegetative phenotype of *myb33.myb65* appears indistinguishable from wild-type; (2) the transcriptome profiles of shoot apical regions of wild-type versus *myb33.myb65* appear indistinguishable; (3) the expression of a *MYB33: GUS* transgene is undetectable in GUS-stained vegetative tissues, but a miR159 resistant version of the reporter gene, *mMYB33: GUS*, is widely and strongly expressed (Figure 2) [41]. The efficiency of this silencing is highlighted by the *mir159a* single mutant; although deep sequencing demonstrates miR159a is the predominant isoform (e.g., miR159a–6621 reads, miR159b–982 reads [8]), the *mir159a* mutant appears indistinguishable from wild-type [14], implying strong reductions in miR159 levels do not impact the silencing of *MYB33/MYB65*. Conversely, overexpression of a wild-type *MYB33* gene fails to result in any phenotypic defects [47]. Although these *MYB33* overexpressing Arabidopsis plants have high *MYB33* mRNA levels, they do not exhibit any phenotypic defects, indicating miR159 also represses expression of *MYB33/MYB65* mRNA via a translational repression mechanism [47]. The importance of this mechanism was shown via the complementation of *mir159ab* with a mutated miR159 variant that had two mismatches to *MYB33/MYB65* at the cleavage site; although this attenuated cleavage, this miR159 variant could still potently silence *MYB33/MYB65* [47]. Therefore, these combined silencing mechanisms ensure *MYB33/MYB65* are strongly repressed in vegetative tissues.

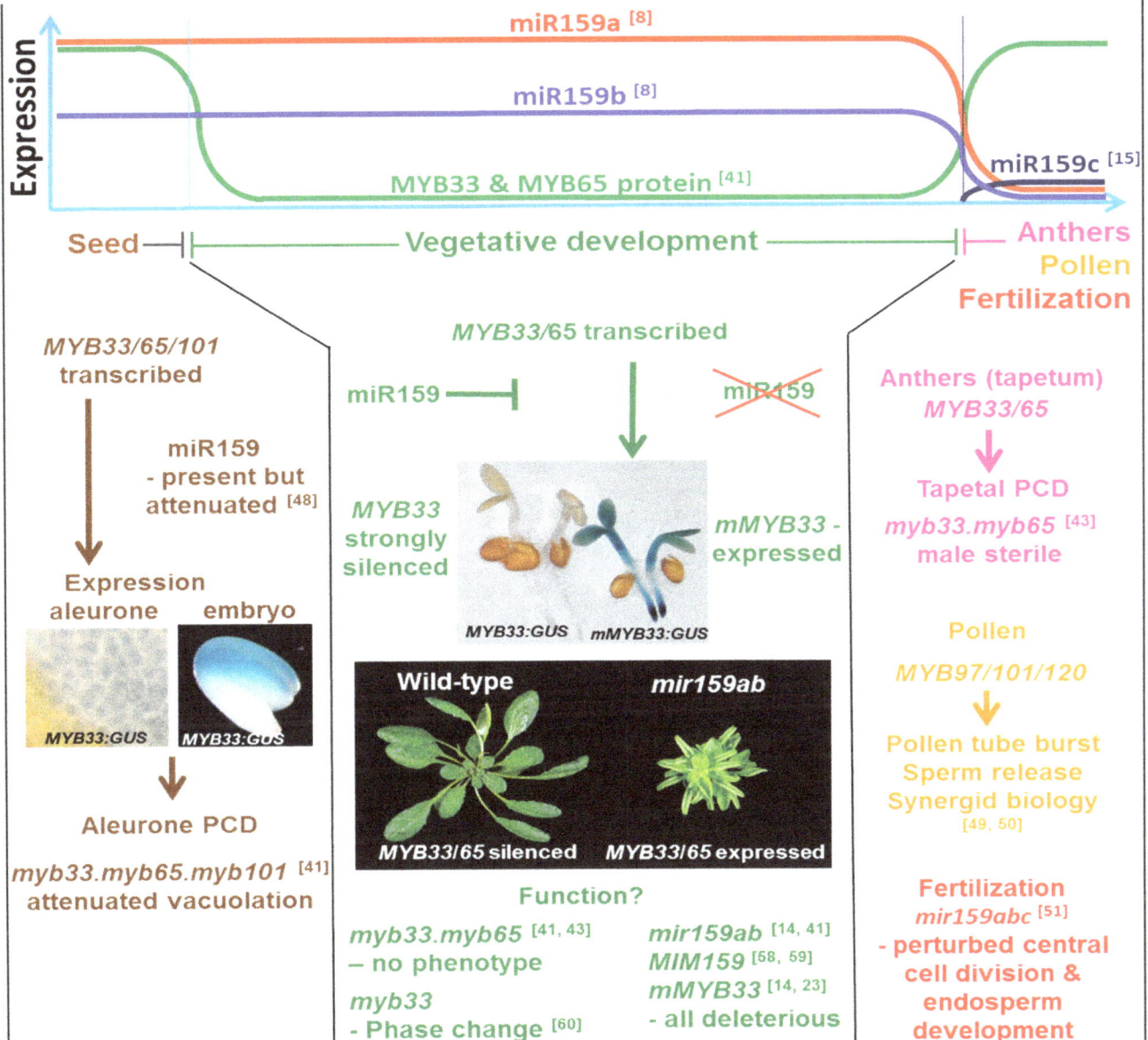

Figure 2. The *miR159-GAMYB-like* pathway in Arabidopsis. miR159a is the predominant family member, being expressed in seed and throughout plant development at a constantly high level, but it is absent in anthers [8,14,46]. miR159b is expressed at a lower level than miR159a [8,15,46], but its expression pattern appears highly similar to miR159a [14]. miR159c, is weakly expressed and appears mainly confined to anthers [15]. In seeds, miR159 efficacy appears attenuated [48], enabling *GAMYB-like* gene expression which promotes PCD of the aleurone [41]. In contrast, throughout vegetative development, miR159 efficacy is strong, and *MYB33/65* expression is strongly silenced. Only via inhibition of miR159, or mutation of the miR159 binding site within *MYB33* or *MYB65*, will expression occur, which leads to strong deleterious outcomes, such as stunted growth and curled leaves [14,41]. Although the function of the pathway has been suggested to be involved in flowering-time and phase change, the purpose of this pathway in vegetative development is still unclear. In anthers, miR159 activity is low. Here, *MYB33* and *MYB65* are expressed to promote PCD in the tapetum [43]. *MYB97/101/120* expression is required for pollen function [49,50]. Finally, miR159 is required for fertilization [51].

Table 1. miR159 targets in Arabidopsis as determined by different approaches. The top 20 miR159a targets in Arabidopsis as identified by the bioinformatic program psRNATarget with standard search parameters [40]; the number of mismatches is indicated by the score. Confirming this prediction, 5'-RACE analysis can detect miR159-guided cleavage products for at least nine of these genes. In contrast, the more quantitative degradome analysis only identifies three of these genes, with only *MYB33* and *MYB65* being frequently detected in multiple degradome analyses [25,42]. Overexpression of miR159 could detect down-regulation of multiple targets [13,52]. However, genetic analysis using a loss-of-function *mir159ab* mutant identify *MYB33* and *MYB65* as the major important targets [14,41].

	At Number	Score	Name	5'-RACE	Degradome	miR159 OE	*miR159ab*
1	AT4G37770	1.5	*ACS8*	[53]		[13]	
2	AT2G32460	2	*MYB101*	[15,17,53]		[13]	
3	AT3G60460	2	*DUO1*	[15,17,53]			
4	AT2G26950	2	*MYB104*				
5	AT4G26930	2	*MYB97*				
6	AT5G06100	2.5	*MYB33*	[15,23]	[25,42]	[52]	[14,41]
7	AT3G11440	2.5	*MYB65*	[23]	[25,42]		[14,41]
8	AT2G34010	2.5	*MRG1*	[53]	[42]		
9	AT2G21600	2.5	*RER1B*				
10	AT5G55020	2.5	*MYB120*	[15]		[13]	
11	AT4G27330	2.5	*SPL*				
12	AT5G27395	2.5	*Tim44-related*				
13	AT3G61740	3	*SDG14, ATX3*				
14	AT1G29010	3	*MRG-LIKE*				
15	AT4G31240	3	*NRX2*				
16	AT2G26960	3	*MYB81*	[15]			
17	AT2G22810	3	*ACS4*				
18	AT3G08850	3	*RAPTOR1B*				
19	AT5G55930	3.5	*OPT1*	[13]		[13]	[41]
20	AT2G44450	3.5	*beta gluc 15*				

5. Conserved RNA Secondary Structures in *MYB33/65* Promote miR159-Mediated Silencing

Highlighting this efficient silencing were miR159 efficacy assays performed on the various Arabidopsis *MYB* targets [44]. Here, it was demonstrated that *MYB33* and *MYB65* were very sensitive targets of miR159, being strongly silenced. In contrast, the other *MYB* genes (*MYB81*, *MYB97*, *MYB101*, *MYB104*, and *DUO1*), were poorly silenced by miR159. As all these *MYB* targets had highly complementary miR159 binding sites, it implies factors other than complementarity must be contributing to this differential miR159-mediated silencing [44]. Correlating with this difference, is a predicted RNA secondary structure that abuts the miR159 binding site of *MYB33* and *MYB65*, but which is absent in the poorly regulated targets (Figure 3; also see [44] for RNA secondary structures of the various Arabidopsis *GAMYB-like* genes). To determine the significance of this in silico predicted RNA structure, a structure/function analysis was performed. Mutation of this structure within the *MYB33* context attenuates silencing, whereas the restoration of the structure, although with a different primary nucleotide sequence, restores strong silencing of *MYB33* [44]. Therefore, this demonstrates that this RNA secondary structure facilitates *MYB33* and *MYB65* silencing, earmarking them as functional targets of miR159. It argues that a fully functional miR159 target site of *MYB33/MYB65* encompasses nucleotides beyond that of the binding site.

Further evidence of the importance of this RNA secondary structure is its strong conservation in *GAMYB-like* homologues throughout the plant kingdom (Figure 3), as the nucleotides that correspond to the stems of the RNA secondary structures are conserved in *GAMYB* homologues of eudicots, monocots, and basal angiosperms, such as Amborella [44]. This indicates this structure is part of the miR159-*GAMYB* regulatory relationship and that the mechanism of regulation is likely more complex than miRNA-binding site complementarity alone. Given that so many miRNA-target relationships are ancient, it will be interesting to investigate how many other miRNA targets have conserved RNA elements associated with their miRNA binding sites, as these ancient regulatory relationships have had 100s of millions of years to evolve greater regulatory complexity.

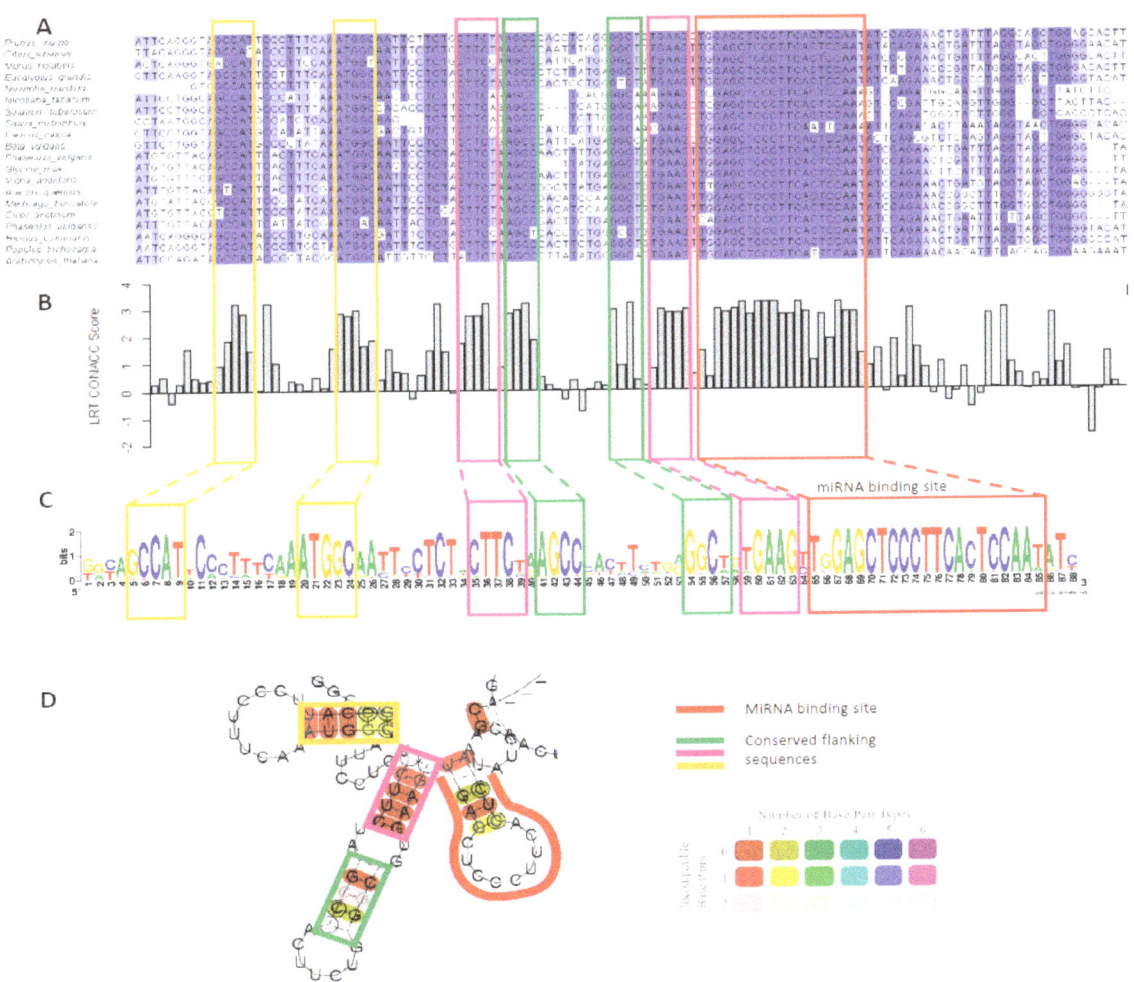

Figure 3. (**A**) Multiple alignment of *MYB33* homologues from different plants species. The binding site is boxed in red, and the conserved flanking sequences in yellow, pink, and green throughout the Figure. (**B**) phyloP score of the multiple sequence alignment of *MYB33* sequences. A positive score denotes evolutionary conservation, whereas, a negative score denotes acceleration [54]. A likelihood ratio test (LRT) was used as the method to detect non-neutral substitution rates. Scores were generated using rPHAST [55]. (**C**) Sequence logo of the binding site and conserved flanking sequences generated using WebLogo [56]. (**D**) RNA secondary structure prediction of the consensus sequence from the multiple alignment in A. generated using RNAalifold [57] at 22 °C and default parameters. Colours represent the number of base pairs types (i.e., AU, UA, CG, GC, UG, GU), and hue the number of non-conserved nucleotides at that position.

In Arabidopsis, not only do the poorly regulated *GAMYB-like* genes lack this conserved RNA structure, but they have highly specific transcriptional domains, predominantly in seeds and anthers, where miR159 activity appears attenuated or absent [14,15,49,50]. Therefore, strong selection of this RNA secondary structural element may have occurred for *GAMYB* homologues that are transcribed in vegetative tissues and require strong miR159-mediated silencing to prevent deleterious outcomes [44]. Investigating whether this also applies to other species with multiple *GAMYB* homologues will be interesting to follow up. Finally, the efficacy of miR159-mediated silencing of *MYB33* in Arabidopsis varies between tissues, being strong in the rosette, but weak in the seed [48]. As RNA secondary structures are dynamic in vivo, they may be operating as a riboswitch, with certain formations facilitating silencing, and others attenuating silencing. It will be interesting to determine whether the conformation of the RNA secondary structure changes between tissues, controlling the ability of miR159 to silence *MYB33*.

6. The Function of miR159-*MYB* Pathway in Plant Development

The functional role of the miR159-*GAMYB* pathway has been studied in numerous plant species, and this is summarized in Table 2.

Table 2. Functional analyses of the miR159-*GAMYB* pathway in plants.

Species	Approach	Phenotype	Ref.
Arabidopsis	T-DNA *mir159ab* mutant	Pleiotropic defects, stunted growth, curled leaves, reduced apical dominance	[14]
Arabidopsis	T-DNA *mir159c* mutant	none	[15]
Arabidopsis	T-DNA *mir159abc* mutant	Perturbed fertilization	[51]
Arabidopsis	*MIM159* mimic–loss-of-function.	Pleiotropic defects, stunted growth, curled leaves, defective sepals, petals and anthers	[58,59]
Arabidopsis (Col-0)	miR159a overexpression	Male sterility	[13]
Arabidopsis (Ler)	miR159a overexpression	Male sterility, delayed flowering-time	[52]
Arabidopsis	T-DNA *myb33.myb65* mutant	Male sterile	[43]
Arabidopsis	T-DNA *myb33* mutant	Altered phase change	[60]
Rice	*STTM159* mimic–loss-of-function	Stunted growth, curled leaves, smaller seeds	[61,62]
Rice	miR159 overexpression	Delayed heading, shorten internode I, malformed flowers, male sterility	[63]
Rice	*gamyb-1* insertion mutant	Male sterility	[64]
Barley	miR159 overexpression	Male sterility	[65]
Wheat	miR159 overexpression	Delayed heading, male sterility, increased tillering	[66]
Gloxinia	*MIM159* mimic loss-of-function, miR159 over-expression (OE)	Accelerated flowering (*MIM159*) or delayed flowering (miR159 OE)	[67]
Tomato	miR159 overexpression	Fruit set, parthenocarpy, ovule development, seedless fruits	[35]
Cucumber	RNAi against *GAMYB*	Altered ratio of male to female flowers	[68]
Strawberry	RNAi against *GAMYB*	Inhibition of receptacle ripening	[69]

6.1. A Role in Male Reproductive Development

The GAMYB/GAMYB-like family of transcription factors is found throughout the plant kingdom, where they share high sequence similarity in their R2R3 DNA-binding domains located towards the N-terminal region, but are much more diverse in their C-terminal regions [20]. Nevertheless, the functions of these *GAMYB* homologues appear to be highly conserved, as *GAMYB* homologues from *Lycopods* or Bryophytes can partially complement a *gamyb-2* rice mutant [20], or a cucumber *GAMYB* can complement the male sterile phenotype of the Arabidopsis *myb33.myb65* mutant [68]. Hence, despite the sequence diversity of the C-terminal regions, this complementation of distant species argues the biochemical function of GAMYB has been strongly conserved.

To date, a role in male reproductive development appears the clearest function for *GAMYB* [70]. Inhibition of its activity perturbs male development, whether in basal plants such as lycopods (*Selaginella moellendorffii*) or bryophytes (moss-*Physcomitrella patens*) [20], or in higher flowering plants, such as rice [64] or Arabidopsis [43]. Moreover, GAMYB was shown to positively regulate the *CYP703* gene, which is required for male development in both basal and higher plants [20,71]. It appears this *GAMYB-CYP703* pathway arose very early in land plant development, and then has come under the control of gibberellin (GA) in lycopods, likely explaining why male reproductive development in plants is under the control of GA [20,71]. It was speculated that the GA regulation of the *GAMYB-CYP703* pathway was a step in the evolution of the sporophyte-dominated life cycle, which requires greater regulatory control for its more complex reproductive system [20].

Consistently, there have been many reports of plants with multiple *GAMYB* homologues for which at least one is strongly transcribed in anthers (e.g., *CsGAMYB1* in cucumber, [68]; *TaGAMYB1* in wheat, [66]; *HvGAMYB* in barley, [65]), or is anther-specific (e.g., *PtrMYB012* in poplar, [16]; *MYB97*, *MYB120* in Arabidopsis, [49]), many of which are positively regulated by GA. Inhibition of GAMYB activity perturbs programmed cell death (PCD) in the anther tapetum [71], where in both a rice *gamyb* mutant, and the Arabidopsis *myb33.myb65* mutant, the tapetum fails to degenerate, resulting in hypertrophy, leading to male sterility [43,64,71]. Additionally, *MYB33* and *MYB65* in Arabidopsis are also required for the formation of the radial microtubule array surrounding nuclei immediately following meiosis II [72]. In the *myb33.myb65* mutant, the resulting defects in male meiotic cytokinesis produce diploid pollen with a defective pollen wall morphology [72]. However, the role that miR159 plays in regulating GAMYB expression in male development is unclear, where it may be fine-tuning expression or preventing expression occurring in particular cell layers, but for most species, this is yet to be resolved. In rice anthers, miR159 and *GAMYB* are co-expressed, suggesting potential fine-tuning of *GAMYB* expression [63].

In Arabidopsis, the role of the miR159-*GAMYB* pathway and its interaction with the miR319-*TCP* pathway in flower maturation has been investigated using miRNA loss-of-function *MIM159* and *MIM319* transgenic plants, both of which display multiple pleiotropic defects [59]. In sepals, petals and anthers of these plants, it was found that the GAMYB and TCP proteins are expressed and directly interact to regulate another miRNA, miR167, which creates a miR159-miR319-miR167 network. It is proposed that the function of miR159/miR319 is to dampen MYB/TCP expression, resulting in low miR167, and hence enabling strong ARF6/8 expression, which in turn regulates many genes required for floral development including that of auxin signaling [59]. However, in wild-type plants, it appears *MYB33/65* expression in flowers is restricted to anthers, and the *myb33.myb65* mutant only displays anther defects [43]. Therefore, in wild-type, it appears the role of miR159 in flowers is to strongly repress the *MYB* genes in sepals and petals to prevent strong expression of miR167 to ultimately enable strong ARF6/8 expression.

MiR159 is present in pollen where it has a crucial role in fertility [51]. It has been known for some time that sperm cell entry alone triggers central cell division, suggesting that the male genome and/or unknown factors transmitted by the sperm control the initiation of endosperm development. Unexpectedly, the central cell usually fails to initiate division after pollination by *mir159abc* mutants, or stops dividing after one or two divisions, resulting in reduced seed set. It was found that both *MYB33* and *MYB65* are highly expressed in the central cell of the embryo sac before fertilization, but after fertilization, both transcripts are rapidly cleared from the central cell and the endosperm initiates development. It was observed that *MYB33* and *MYB65* transcripts are not cleared in pollinations with *mir159abc* pollen, suggesting that miR159 in pollen is transmitted to the central cell by fertilization where it degrades *MYB33* and *MYB65* transcripts [51]. Thus, miR159 has a paternal effect on seed development: miR159 carried in pollen abolishes central cell repression after fertilization permitting endosperm nuclear divisions [51]. Loss of maternal miR159 also results in seed defects but these defects are less severe on seed set than loss of paternal miR159, and the mechanism of this maternal effect is unknown.

6.2. A Role in Seed Development

GAMYB was first identified as a GA signaling component in the barley aleurone, hence giving these MYB genes their name "GA"MYB [73]. Here, GAMYB positively transduces the GA signal to activate expression of α-amylase and other hydrolytic enzymes [74], as well as promote PCD in the aleurone [75]. This latter function appears conserved in Arabidopsis, as a *myb33.myb65.myb101* triple mutant has attenuated vacuolation in aleurone cells, a PCD-mediated process that is positively regulated by GA [41]. Therefore, a conserved role in PCD in the aleurone and tapetum in both monocots and dicots is currently a unifying function for these GAMYB transcription factors. Curiously, both these tissues are single cell layers that provide nutrients upon death to the embryo (aleurone) or pollen (tapetum). It is possible other *GAMYB-like* genes may play similar roles in terms of inhibiting growth and promoting cell death. For example, *MYB97*, *MYB101*, and *MYB120* are all expressed in the pollen tube, and in a *myb97.myb101.myb120* mutant, the pollen tube fails to undergo growth arrest and then fails to degenerate in order to release the sperm cells to the ovules [49,50].

Many downstream targets of the miR159-*GAMYB* pathway in Arabidopsis support a role in PCD. Micro-array analysis on the shoot apical region of *mir159ab* plants found that of the 166 up-regulated genes, many appeared aleurone related [41]. Many of these aleurone related genes were also down-regulated in *myb33.myb65.myb101* seeds, making them strong candidates of being downstream of *GAMYB* activity [41]. This includes the most up-regulated gene in *mir159ab*, *CYSTEINE PROTEINASE 1* (*CP1*), whose expression appears tightly correlated with *GAMYB* expression [41,46], and corresponds to a class of enzymes which have been associated with PCD and cell lysis. Similarly, inhibition of miR159 in transgenic rice results in up-regulation of pathways associated with PCD, suggesting that *GAMYB* promotes these pathways in the rice grain [61]. Again, what role miR159 plays in regulating GAMYB activity in the seed is unclear. In Arabidopsis germinating seeds, miR159 and *MYB33* are co-transcribed in the aleurone and embryo, however, MYB33 protein is expressed, which suggests miR159 may only be fine-tuning the expression of *MYB33* in this tissue [48]. Nevertheless, *mir159ab* plants produce malformed seeds [14], implying miR159 is required for proper seed development.

6.3. The Role of miR159-GAMYB Pathway in Vegetative Tissues

In Arabidopsis, the widespread transcription of *MYB33/MYB65*, only to be strongly silenced by miR159 raises the question of what is the purpose of this seemingly futile regulatory pathway. Although miR159 is sometimes associated with leaf development due to the smaller, upwardly curled leaves of the *mir159ab* mutant (Figure 2), this phenotype appears more a consequence of the deleterious impact of *MYB33/MYB65* expression rather than the alteration of a developmental program [41]. In general, de-regulated expression of GAMYB in leaves results in strong perturbation of growth. This was shown in *mir159ab*, as well as transgenic Arabidopsis expressing miRNA decoys to inhibit miR159 function, either *MIMIC159* (*MIM159*), Short target tandem *MIMIC159* (*STTM159*), or *SPONGE159* (*SP159*) [58,76], or with *STTM159* rice [61,62], which all result in the similar phenotypic defect of stunted growth. For instance, *STTM159* rice plants are shorter than wild-type rice, with decreased cell numbers, and the most down-regulated genes in *STTM159* rice are associated with cell division. Therefore, the main role of rice miR159 is to suppress *GAMYB* expression to enable cell proliferation [61]. Likewise, the expression of miR159-resistant *GAMYB* transgenes in Arabidopsis [14,16,23,44], lead to the same phenotypic defects. Therefore, it is clear that these *GAMYB* genes encode a class of transcription factors that when expressed inhibit growth, a phenotype contrary to a role in promoting the GA signal for which they were originally identified. Supporting this, GA treatments do not alter the RNA levels of *MYB33*, *MYB65* or miR159 in Arabidopsis rosettes, and the response of *myb33.myb65* plants to GA is not perturbed in vegetative tissues [41]. Therefore, the role the *GAMYB* in transducing the GA signal appears to be tissue dependent, where it is involved in transducing the GA signal in seeds and anthers [70,73,77], but not in vegetative tissues [41]. Supporting this, the Arabidopsis *myb33.myb65* or rice *gamyb* mutants do not appear to have any obvious phenotypic defects at the vegetative stage [43,64].

Contrary to the growth inhibition phenotype of the leaves, the roots of *mir159ab* Arabidopsis are longer than wild-type and have a larger apical meristem zone. Thus, in roots, *GAMYB* expression appears to enhance cell cycle progression, leading to extended roots. However, the root lengths of *myb33.myb65* or *myb33.myb65.myb101* plants were unchanged compared to wild-type, again indicating that these *GAMYB-like* genes are likely fully silenced in roots, again raising the question of what is the role of this pathway in this vegetative tissue [78].

6.4. A Role of miR159 in Controlling GA-Mediated Flowering-Time and Growth?

This clear role in inhibiting growth appears at odds with a role often ascribed to *GAMYB* in promoting flowering-time [79,80]. This idea arose from the fact that GA promotes flowering-time, and that the *GAMYB* or *GAMYB-like* genes were thought to be positive regulators of GA throughout the plant [81]. Supporting this idea was the finding that the *LEAFY* gene, a central regulator of flowering, contained a MYB-binding site within its promoter, and this binding site appeared critical in transducing the GA-signal [82]. Subsequently, it was shown that *MYB33* transcription was induced at the shoot apical region upon the induction of flowering, either through GA-application or exposure to long-day conditions, and that the MYB33 protein could bind the *LEAFY* promoter in in vitro gel shift assays [81]. Then, overexpression of miR159 in Arabidopsis [ecotype Landsberg *erecta*, (L*er*)] resulted in down-regulation of *MYB33* expression, which correlated with a decrease in *LEAFY* expression and a delayed flowering-time under short-day conditions [52]. Supporting this is the manipulation of miR159 levels in other plant species that result in altered flowering-times. This includes overexpression of miR159 in rice and wheat which lead to a reduced heading-time [63,66]. Additionally, in the ornamental flowering plant *Gloxinia*, the over-expression of miR159 delayed flowering, whereas the inhibition of miR159 with a *MIM159* transgene accelerated flowering-time [67]. Unlike *MIM159* Arabidopsis or *STTM159* rice [58,59,61,62,76], *MIM159 Gloxinia* did not exhibit any defects in vegetative growth or development [67].

Such evidence argues for a clear and conserved role for miR159 in flowering-time, and that *GAMYB* is likely promoting the GA-signal with regard to flowering. However, overexpression of miR159 in Arabidopsis (ecotype Columbia) did not affect flowering-time [13]. So, although overexpression of miR159 in both ecotypes (L*er* and Columbia) resulted in male sterility due the requirement of GAMYB activity in anthers, there was a differential response with regard to flowering-time. Moreover, a *myb33.myb65* mutant (ecotype Columbia) did not have a delayed flowering-time, and the *mir159ab* mutant (greater GAMYB activity) displayed a late flowering-time under short-day conditions, implying greater GAMYB activity was inhibiting flowering [41]. Given the severe pleiotropic defects of *mir159ab*, it is uncertain whether delayed flowering is a direct result of greater GAMYB activity, or a secondary effect of the severe growth and developmental defects [41].

In addition to delayed flowering, the *mir159ab* Arabidopsis mutant was found to have a strong delay in vegetative phase change (VPC), with the first leaf with abaxial trichomes being *leaf* 16.0 as opposed to *leaf* 7.9 for wild-type, and this was tightly correlated with the increased levels of miR156, one of the key determinants of VPC [60]. In a complex regulatory mechanism, it was found that MYB33 activated transcription of both the *MIR156* gene and its target, *SPL9*, via direct interaction with their promoters [60]. Conversely, Arabidopsis plants overexpressing miR159 (*leaf* 7.3) or the *myb33* mutant (*leaf* 7.1) only had slight increases in VPC compared to wild-type (*leaf* 7.9). This argued that MYB33 protein is expressed to some extent in the Arabidopsis rosette. However, the VPC of a *myb65* mutant (*leaf* 8.1) was unchanged from wild-type, implying *MYB65* did not appear to impact this pathway [60]. Given the subtle changes to vegetative phase change in the *myb33* mutant, the miR159-*GAMYB* pathway was considered a modifier of VPC, where miR159 promotes VPC by preventing MYB33 expression which negatively regulates VPC [60]. It will be interesting to see what role the miR159-*GAMYB* pathway is found to have in this process in other plant species.

Therefore, regarding the miR159-*GAMYB* pathway in growth and flowering, there is strong conflicting evidence. Although the difference in Arabidopsis is possibly due to ecotype variation, a role for GAMYB in either promoting flowering, or alternatively, deleteriously inhibiting growth, will need further experimentation for clarification of how such diametrically opposed outcomes can arise.

6.5. Fruit and Reproductive Development

There is growing evidence that the miR159-*GAMYB* pathway plays a role in fruit development. In strawberries, fruit development is GA-regulated, and miR159 is strongly expressed in the fruit's receptacle tissue and appears to regulate *GAMYB*, as miR159 and *GAMYB* expression is reciprocal [83]. *GAMYB* is a key regulator of strawberry fruit development, as repression of *GAMYB* via RNAi inhibits receptacle ripening and color formation [69]. In tomato, the miR159-*GAMYB* pathway is present in ovules, and overexpression of miR159 resulted in abnormal ovule development, precocious fruit initiation and seedless fruits [35]. Similarly, in grapes, the pathway appears active in the fruits, and under the control of GA [84]. In the monoecious plant cucumber, inhibition of *GAMYB* activity via RNAi altered the ratio of male to female flowers, decreasing the number of nodes with male flowers [68]. Therefore, it appears this pathway is involved in many different functions of the reproductive process in different plant species.

7. The Function of the miR159-*MYB* Pathway in Plant Stress

7.1. Abiotic Stress

Given the ubiquity and abundance of miR159 throughout the plant kingdom, it is not surprising that numerous studies have implicated miR159 in a wide range of stresses from many different plant species (for review see [85]). In Arabidopsis, miR159 levels increase under salinity [86], and in germinating seeds, miR159 has been found to accumulate in response to the stress hormone ABA as well as to drought [87]. MiR159 also accumulates to higher levels in response to drought in maize, wheat and barley [85]. Such results suggests that increased levels of miR159 may result in greater stress tolerance. However, in some species, miR159 levels decrease in response to drought or salinity [85], and overexpression of miR159 in rice resulted in increased sensitivity to heat-stress [66]. In potato, in which the drought tolerant gene *cap-binding 80* protein has been down-regulated, miR159 levels were decreased and mRNA levels of *GAMYB-like* homologues were higher [88]. Therefore, these studies have found no consistent or unified role for miR159 in plant stress response.

The functional role of the Arabidopsis miR159-*GAMYB* pathway to abiotic stress was investigated by comparing the response a mutant lacking this entire pathway, the *mir159ab.myb33.myb65* quadruple mutant, to that of wild-type plants [46]. Two-week old plants were exposed to three weeks of treatments with either ABA, high temperature, high light, drought or cold. However, no differential response between the *mir159ab.myb33.myb65* mutant and wild-type plants were identified. As it was demonstrated that miR159 fully represses *MYB33* and *MYB65* in vegetative tissues of Arabidopsis plants [41], it was rationalized that miR159 levels would need to decrease in Arabidopsis to enable activation of these two *GAMYB-like* genes [46]. However, none of the treatments appeared to repress miR159 to levels in which would allow *MYB33* and *MYB65* expression, and this was supported by assaying the downstream marker gene *CP1*, whose levels appeared completely repressed [46]. Based on this, no clear role for this pathway was identified, and it remains uncertain what role it plays in stress response in Arabidopsis.

7.2. Biotic Stress

Similarly, the levels of miR159 respond to many different biotic stresses. Recently it was shown that cotton and Arabidopsis accumulate elevated levels of miR159 in response to the fungus, *Verticillium dahlia* [89]. MiR159 was exported into the fungal hyphae, where it targeted the gene encoding isotrichodermin C-15 hydroxylase (HiC-15), which is critical for hyphal growth. As expression of a miR159-resistant HiC-15 gene in *V. dahlia* resulted in greater virulence, it was concluded that exporting miR159 from the plant was conferring greater pathogenic resistance. Given that the miR159-binding site is highly conserved in HiC-15, it was hypothesized that this has evolved to dampen HiC-15 expression as to avoid rapid death of the host, which then enables establishment of the fungus on the plant [89]. Currently, this is the only clear role for miR159 in pathogen response.

MiR159 also accumulates to higher levels in Arabidopsis root galls that form in response to root knot nematodes (RKN). The *MYB33* gene appears dynamically expressed during gall formation, as a *MYB33: GUS* reporter was expressed during early gall development, but not at later stages. Functional evidence for the involvement of the miR159-*GAMYB* pathway is that an Arabidopsis *mir159abc* triple mutant has greater resistance to root knot nematodes (RKN) [90]. Further investigation will be needed to understand the precise role of the pathway in gall formation and the response pathway to RKN infection.

8. Conclusions and Some Unresolved Questions

The miR159-*GAMYB* pathway appears nearly ubiquitous in terrestrial plants, implying it has played an important role in plant's conquest of the land. Although its role in some tissues now appear to be relatively clear, this is far from the case in others. Below are some of the questions we believe still need to be resolved.

1. Why are *MYB33* and *MYB65* transcribed in vegetative tissues where failure to fully repress them results in a detrimental effect? What selective advantage does this give the plant?

a.   One hypothesis is that if miR159 is inhibited by a certain trigger, and strong *MYB33/65* expression occurs, growth inhibition (or another unknown process) may result in a beneficial outcome (e.g., drought conditions to slow growth). However, currently, no triggers to inhibit miR159 to enable strong *MYB* expression are known.

b.   A second hypothesis would be that *MYB33/65* are not silenced in all vegetative tissues, but in certain cells they are expressed where they confer a selective advantage. Some evidence suggests *GAMYB* is involved in the transition to flowering, and VPC in Arabidopsis. But currently there is much conflicting data. For instance in Arabidopsis, overexpressing miR159 represses flowering-time, and inhibition of miR159 represses VPC. Other studies have found no role for miR159 in flowering. More work is needed here to clarify these roles, and how conserved they are across species.

2.   Why is expression of *GAMYB* in vegetative tissues deleterious and how does it inhibit growth? What down-stream events are these genes triggering? Although some studies have started to address this, more work is needed for a clearer understanding.

3.   Is *GAMYB* function related to the way it is regulated, i.e., strongly transcribed, only to then be strongly silenced by miR159? Does miR159 have a role in stress response? Again, many studies have identified changes to miR159 levels in response to a host of different biotic/abiotic stresses, but currently there is no clearly defined role for this miR159 concerning stress tolerance/response.

4.   How does the conserved RNA secondary structure associated with the miR159-binding sites of *GAMYB* genes promote their silencing by miR159? Can this structure facilitate a complex regulatory mechanism, enabling strong silencing in some tissues, but poor silencing in others, depending on a dynamic secondary structure configuration? i.e., acting like a riboswitch concerning silencing.

5.   What is the role of miR159-mediate regulation on non-*GAMYB* targets? For example, *DUO1* has a conserved miR159-binding site, but the role of miR159 in controlling the expression of this gene remains unclear.

6.   What is the role of miR159 in female fertility? Why are Arabidopsis *mir159ab* seeds small and misshapen (likewise rice STTM159 grains are small)? Why does the central cell still divide in some *mir159abc* ovules? How can a seed still form (from *mir159abc* pollen) with a viable embryo when the endosperm divisions stop apparently so early?

Author Contributions: All authors contributed to the writing of the review.

Acknowledgments: G.W. was supported by an Australian Government Research Training Program RTP Scholarship.

References

1.   Axtell, M.J.; Bartel, D.P. Antiquity of microRNAs and their targets in land plants. *Plant Cell* **2005**, *17*, 1658–1673. [CrossRef] [PubMed]

2.   Chávez Montes, R.A.; de Fátima Rosas-Cárdenas, F.; De Paoli, E.; Accerbi, M.; Rymarquis, L.A.; Mahalingam, G.; Marsch-Martínez, N.; Meyers, B.C.; Green, P.J.; de Folter, S. Sample sequencing of vascular plants demonstrates widespread conservation and divergence of microRNAs. *Nat. Commun.* **2014**, *5*, 3722. [CrossRef] [PubMed]

3.   You, C.; Cui, J.; Wang, H.; Qi, X.; Kuo, L.Y.; Ma, H.; Gao, L.; Mo, B.; Chen, X. Conservation and divergence of small RNA pathways and microRNAs in land plants. *Genome Biol.* **2017**, *18*, 158. [CrossRef] [PubMed]

4.   Axtell, M.J.; Meyers, B.C. Revisiting criteria for plant microRNA annotation in the era of big data. *Plant Cell* **2018**, *30*, 272–284. [CrossRef] [PubMed]

5.   Alaba, S.; Piszczalka, P.; Pietrykowska, H.; Pacak, A.M.; Sierocka, I.; Nuc, P.W.; Singh, K.; Plewka, P.; Sulkowska, A.; Jarmolowski, A.; et al. The liverwort Pellia endiviifolia shares microtranscriptomic traits that are common to green algae and land plants. *Plant J.* **2014**, *80*, 331–344.

6.   Fahlgren, N.; Howell, M.D.; Kasschau, K.D.; Chapman, E.J.; Sullivan, C.M.; Cumbie, J.S.; Givan, S.A.; Law, T.F.; Grant, S.R.; Dang, J.L.; et al. High-throughput sequencing of Arabidopsis microRNAs: Evidence for frequent birth and death of *MIRNA* genes. *PLoS ONE* **2007**, *2*, e219. [CrossRef] [PubMed]

7.   Jeong, D.H.; Park, S.; Zhai, J.; Gurazada, S.G.R.; De Paoli, E.; Meyers, B.C.; Green, P.J. Massive analysis of rice small RNAs: Mechanistic implications of regulated microRNAs and variants for differential target RNA cleavage. *Plant Cell* **2011**, *23*, 4185–4207. [CrossRef]

8.   Rajagopalan, R.; Vaucheret, H.; Trejo, J.; Bartel, D.P. A diverse and evolutionarily fluid set of microRNAs in *Arabidopsis thaliana*. *Genes Dev.* **2006**, *20*, 3407–3425. [CrossRef] [PubMed]

9.   Szittya, G.; Moxon, S.; Santos, D.M.; Jing, R.; Fevereiro, M.P.; Moulton, V.; Dalmay, T. High-throughput sequencing of Medicago truncatula short RNAs identifies eight new miRNA families. *BMC Genom.* **2008**, *9*, 593. [CrossRef]

10.   Mao, W.; Li, Z.; Xia, X.; Li, Y.; Yu, J. A combined approach of high-throughput sequencing and degradome analysis reveals tissue specific expression of microRNAs and their targets in cucumber. *PLoS ONE* **2012**, *7*, e33040. [CrossRef]

11.   Kozomara, A.; Birgaoanu, M.; Griffiths-Jones, S. miRBase: From microRNA sequences to function. *Nucleic Acids Res.* **2019**, *47*, D155–D162. [CrossRef] [PubMed]

12.   Zhang, L.; Chia, J.M.; Kumari, S.; Stein, J.C.; Liu, Z.; Narechania, A.; Maher, C.A.; Guill, K.; McMullen, M.D.; Ware, D. A genome-wide characterization of microRNA genes in maize. *PLoS Genet.* **2009**, *5*, e1000716. [CrossRef] [PubMed]

13.   Schwab, R.; Palatnik, J.F.; Riester, M.; Schommer, C.; Schmid, M.; Weigel, D. Specific effects of microRNAs on the plant transcriptome. *Dev. Cell* **2005**, *8*, 517–527. [CrossRef] [PubMed]

14.   Allen, R.S.; Li, J.; Stahle, M.I.; Dubroué, A.; Gubler, F.; Millar, A.A. Genetic analysis reveals functional redundancy and the major target genes of the Arabidopsis miR159 family. *Proc. Natl. Acad. Sci. USA* **2007**, *104*, 16371–16376. [CrossRef] [PubMed]

15.   Allen, R.S.; Li, J.; Alonso-Peral, M.M.; White, R.G.; Gubler, F.; Millar, A.A. MicroR159 regulation of most conserved targets in Arabidopsis has negligible phenotypic effects. *Silence* **2010**, *1*, 18. [CrossRef] [PubMed]

16.   Kim, M.H.; Cho, J.S.; Lee, J.H.; Bae, S.Y.; Choi, Y.I.; Park, E.J.; Lee, H.; Ko, J.H. Poplar MYB transcription factor PtrMYB012 and its Arabidopsis *AtGAMYB* orthologs are differentially repressed by the Arabidopsis miR159 family. *Tree Physiol.* **2018**, *38*, 801–812. [CrossRef] [PubMed]

17.   Palatnik, J.F.; Wollmann, H.; Schommer, C.; Schwab, R.; Boisbouvier, J.; Rodriguez, R.; Warthmann, N.; Allen, E.; Dezulian, T.; Huson, D.; et al. Sequence and expression differences underlie functional specialization of Arabidopsis microRNAs miR159 and miR319. *Dev. Cell* **2007**, *13*, 115–125. [CrossRef]

18.   Bologna, N.G.; Mateos, J.L.; Bresso, E.G.; Palatnik, J.F. A loop-to-base processing mechanism underlies the biogenesis of plant microRNAs miR319 and miR159. *EMBO J.* **2009**, *28*, 3646–3656. [CrossRef]

19.   Li, Y.; Li, C.; Ding, G.; Jin, Y. Evolution of *MIR159/319* microRNA genes and their post-transcriptional regulatory link to siRNA pathways. *BMC Evol. Biol.* **2011**, *11*, 122. [CrossRef]

20.   Aya, K.; Hiwatashi, Y.; Kojima, M.; Sakakibara, H.; Ueguchi-Tanaka, M.; Hasebe, M.; Matsuoka, M. The Gibberellin perception system evolved to regulate a pre-existing GAMYB-mediated system during land plant evolution. *Nat. Commun.* **2011**, *2*, 544. [CrossRef]

21.   Tsuzuki, M.; Nishihama, R.; Ishizaki, K.; Kurihara, Y.; Matsui, M.; Bowman, J.L.; Kohchi, T.; Hamada, T.; Watanabe, Y. Profiling and characterization of small RNAs in the Liverwort, *Marchantia polymorpha*, belonging to the first diverged land plants. *Plant Cell Physiol.* **2016**, *57*, 359–372. [CrossRef] [PubMed]

22.   Lin, S.S.; Bowman, J.L. MicroRNAs in *Marchantia polymorpha*. *New Phytol.* **2018**, *220*, 409–416. [CrossRef] [PubMed]

23.   Palatnik, J.F.; Allen, E.; Wu, X.; Schommer, C.; Schwab, R.; Carrington, J.C.; Weigel, D. Control of leaf morphogenesis by microRNAs. *Nature* **2003**, *425*, 257–263. [CrossRef] [PubMed]

24.   Li, J.; Reichel, M.; Li, Y.; Millar, A.A. The functional scope of plant microRNA-mediated silencing. *Trends Plant Sci.* **2014**, *19*, 750–756. [CrossRef] [PubMed]

25.   Addo-Quaye, C.; Eshoo, T.W.; Bartel, D.P.; Axtell, M.J. Endogenous siRNA and miRNA targets identified by sequencing of the Arabidopsis degradome. *Curr. Biol.* **2008**, *18*, 758–762. [CrossRef] [PubMed]

26.   Song, Q.X.; Liu, Y.F.; Hu, X.Y.; Zhang, W.K.; Ma, B.; Chen, S.Y.; Zhang, J.S. Identification of miRNAs and their target genes in developing soybean seeds by deep sequencing. *BMC Plant Biol.* **2011**, *11*, 5. [CrossRef] [PubMed]

27.   Liu, N.; Tu, L.; Tang, W.; Gao, W.; Lindsey, K.; Zhang, X. Small RNA and degradome profiling reveals a role for miRNAs and their targets in the developing fibers of *Gossypium barbadense*. *New Phytol.* **2015**, *206*, 352–367.

28. Zhang, J.; Zeng, R.; Chen, J.; Liu, X.; Liao, Q. Identification of conserved microRNAs and their targets from *Solanum lycopersicum* Mill. *Gene* **2008**, *423*, 1–7. [CrossRef]

29. An, F.M.; Chan, M.T. Transcriptome-wide characterization of miRNA-directed and non-miRNA-directed endonucleolytic cleavage using degradome analysis under low ambient temperature in *Phalaenopsis aphrodite* subsp. formosana. *Plant Cell Physiol.* **2012**, *53*, 1737–1750. [CrossRef]

30. Luo, X.; Gao, Z.; Shi, T.; Cheng, Z.; Zhang, Z.; Ni, Z. Identification of miRNAs and their target genes in peach (*Prunus persica* L.) using high-throughput sequencing and degradome analysis. *PLoS ONE* **2013**, *8*, e79090. [CrossRef]

31. Sun, F.; Guo, G.; Du, J.; Guo, W.; Peng, H.; Ni, Z.; Sun, Q.; Yao, Y. Whole-genome discovery of miRNAs and their targets in wheat (*Triticum aestivum* L.). *BMC Plant Biol.* **2014**, *14*, 142. [CrossRef] [PubMed]

32. Li, Y.F.; Zheng, Y.; Addo-Quaye, C.; Zhang, L.; Saini, A.; Jagadeeswaran, G.; Axtell, M.J.; Zhang, W.; Sunkar, R. Transcriptome-wide identification of microRNA targets in rice. *Plant J.* **2010**, *62*, 742–759. [CrossRef]

33. Curaba, J.; Spriggs, A.; Taylor, J.; Li, Z.; Helliwell, C. miRNA regulation in the early development of barley seed. *BMC Plant Biol.* **2012**, *12*, 120. [CrossRef] [PubMed]

34. Buxdorf, K.; Hendelman, A.; Stav, R.; Lapidot, M.; Ori, N.; Arazi, T. Identification and characterization of a novel miR159 target not related to *MYB* in tomato. *Planta* **2010**, *232*, 1009–1022. [CrossRef] [PubMed]

35. Da Silva, E.M.; Silva, G.F.F.E.; Bidoia, D.B.; da Silva Azevedo, M.; de Jesus, F.A.; Pino, L.E.; Peres, L.E.P.; Carrera, E.; López-Díaz, I.; Nogueira, F.T.S. microRNA159-targeted *SlGAMYB* transcription factors are required for fruit set in tomato. *Plant J.* **2017**, *92*, 95–109. [CrossRef] [PubMed]

36. Wan, L.C.; Zhang, H.; Lu, S.; Zhang, L.; Qiu, Z.; Zhao, Y.; Zeng, Q.Y.; Lin, J. Transcriptome-wide identification and characterization of miRNAs from *Pinus densata*. *BMC Genom.* **2012**, *13*, 132. [CrossRef] [PubMed]

37. Amborella Genome Project. The Amborella genome and the evolution of flowering plants. *Science* **2013**, *342*, 1241089. [CrossRef]

38. Li, W.F.; Zhang, S.G.; Han, S.Y.; Wu, T.; Zhang, J.H.; Qi, L.W. Regulation of *LaMYB33* by miR159 during maintenance of embryogenic potential and somatic embryo maturation in *Larix kaempferi* (Lamb.) Carr. *Plant Cell Tissue Organ Cult.* **2013**, *113*, 131–136. [CrossRef]

39. Pappas, M.D.C.R.; Pappas, G.J.; Grattapaglia, D. Genome-wide discovery and validation of *Eucalyptus* small RNAs reveals variable patterns of conservation and diversity across species of *Myrtaceae*. *BMC Genom.* **2015**, *16*, 1113. [CrossRef]

40. Dai, X.; Zhuang, Z.; Zhao, P.X. psRNATarget: A plant small RNA target analysis server (2017 release). *Nucleic Acids Res.* **2018**, *46*, W49–W54. [CrossRef]

41. Alonso-Peral, M.M.; Li, J.; Li, Y.; Allen, R.S.; Schnippenkoetter, W.; Ohms, S.; White, R.G.; Millar, A.A. The microRNA159-regulated *GAMYB-like* genes inhibit growth and promote programmed cell death in Arabidopsis. *Plant Physiol.* **2010**, *154*, 757–771. [CrossRef] [PubMed]

42. German, M.A.; Pillay, M.; Jeong, D.H.; Hetawal, A.; Luo, S.; Janardhanan, P.; Kannan, V.; Rymarquis, L.A.; Nobuta, K.; German, R.; et al. Global identification of microRNA–target RNA pairs by parallel analysis of RNA ends. *Nat. Biotech.* **2008**, *26*, 941. [CrossRef] [PubMed]

43. Millar, A.A.; Gubler, F. The Arabidopsis *GAMYB-like* genes, *MYB33* and *MYB65*, are microRNA-regulated genes that redundantly facilitate anther development. *Plant Cell* **2005**, *17*, 705–721. [CrossRef] [PubMed]

44. Zheng, Z.; Reichel, M.; Deveson, I.; Wong, G.; Li, J.; Millar, A.A. Target RNA secondary structure is a major determinant of miR159 efficacy. *Plant Physiol.* **2017**, *174*, 1764–1778. [CrossRef] [PubMed]

45. Seitz, H. Redefining microRNA targets. *Curr. Biol.* **2009**, *19*, 870–873. [CrossRef] [PubMed]

46. Li, Y.; Alonso-Peral, M.; Wong, G.; Wang, M.B.; Millar, A.A. Ubiquitous miR159 repression of *MYB33/65* in Arabidopsis rosettes is robust and is not perturbed by a wide range of stresses. *BMC Plant Biol.* **2016**, *16*, 179. [CrossRef] [PubMed]

47. Li, J.; Reichel, M.; Millar, A.A. Determinants beyond both complementarity and cleavage govern microR159 efficacy in Arabidopsis. *PLoS Genet.* **2014**, *10*, e1004232. [CrossRef] [PubMed]

48. Alonso-Peral, M.M.; Sun, C.; Millar, A.A. MicroRNA159 can act as a switch or tuning microRNA independently of its abundance in Arabidopsis. *PLoS ONE* **2012**, *7*, e34751. [CrossRef] [PubMed]

49. Leydon, A.R.; Beale, K.M.; Woroniecka, K.; Castner, E.; Chen, J.; Horgan, C.; Palanivelu, R.; Johnson, M.A. Three MYB transcription factors control pollen tube differentiation required for sperm release. *Curr. Biol.* **2013**, *23*, 1209–1214. [CrossRef] [PubMed]

50. Liang, Y.; Tan, Z.M.; Zhu, L.; Niu, Q.K.; Zhou, J.J.; Li, M.; Chen, L.Q.; Zhang, X.Q.; Ye, D. *MYB97*, *MYB101* and *MYB120* function as male factors that control pollen tube-synergid interaction in *Arabidopsis thaliana* fertilization. *PLoS Genet.* **2013**, *9*, e1003933. [CrossRef]

51. Zhao, Y.; Wang, S.; Wu, W.; Li, L.; Jiang, T.; Zheng, B. Clearance of maternal barriers by paternal miR159 to initiate endosperm nuclear division in Arabidopsis. *Nat. Commun.* **2018**, *9*, 5011. [CrossRef] [PubMed]

52. Achard, P.; Herr, A.; Baulcombe, D.C.; Harberd, N.P. Modulation of floral development by a gibberellin-regulated microRNA. *Development* **2004**, *131*, 3357–3365. [CrossRef] [PubMed]

53. Alves-Junior, L.; Niemeier, S.; Hauenschild, A.; Rehmsmeier, M.; Merkle, T. Comprehensive prediction of novel microRNA targets in *Arabidopsis thaliana*. *Nucleic Acids Res.* **2009**, *37*, 4010–4021. [CrossRef] [PubMed]

54. Pollard, K.S.; Hubisz, M.J.; Rosenbloom, K.R.; Siepel, A. Detection of nonneutral substitution rates on mammalian phylogenies. *Genome Res.* **2010**, *20*, 110–121. [CrossRef]

55. Hubisz, M.J.; Pollard, K.S.; Siepel, A. PHAST and RPHAST: Phylogenetic analysis with space/time models. *Brief Bioinform.* **2011**, *12*, 41–51. [CrossRef] [PubMed]

56. Crooks, G.; Hon, G.; Chandonia, J.; Brenner, S. WebLogo: A sequence logo generator. *Genome Res.* **2004**, *14*, 1188–1190. [CrossRef] [PubMed]

57. Bernhart, S.H.; Hofacker, I.L.; Will, S.; Gruber, A.R.; Stadler, P.F. RNAalifold: Improved consensus structure prediction for RNA alignments. *BMC Bioinform.* **2008**, *9*, 1–13. [CrossRef]

58. Todesco, M.; Rubio-Somoza, I.; Paz-Ares, J.; Weigel, D. A collection of target mimics for comprehensive analysis of microRNA function in *Arabidopsis thaliana*. *PLoS Genet.* **2010**, *6*, e1001031. [CrossRef]

59. Rubio-Somoza, I.; Weigel, D. Coordination of flower maturation by a regulatory circuit of three microRNAs. *PLoS Genet.* **2013**, *9*, e1003374. [CrossRef]

60. Guo, C.; Xu, Y.; Shi, M.; Lai, Y.; Wu, X.; Wang, H.; Zhu, Z.; Poethig, R.S.; Wu, G. Repression of miR156 by miR159 regulates the timing of the juvenile-to-adult transition in Arabidopsis. *Plant Cell* **2017**, *29*, 1293–1304. [CrossRef]

61. Zhao, Y.; Wen, H.; Teotia, S.; Du, Y.; Zhang, J.; Li, J.; Sun, H.; Tang, G.; Peng, T.; Zhao, Q. Suppression of microRNA159 impacts multiple agronomic traits in rice (*Oryza sativa* L.). *BMC Plant Biol.* **2017**, *17*, 215. [CrossRef] [PubMed]

62. Zhang, H.; Zhang, J.; Yan, J.; Gou, F.; Mao, Y.; Tang, G.; Botella, J.R.; Zhu, J.K. Short tandem target mimic rice lines uncover functions of miRNAs in regulating important agronomic traits. *Proc. Natl. Acad. Sci. USA* **2017**, *114*, 5277–5282. [CrossRef] [PubMed]

63. Tsuji, H.; Aya, K.; Ueguchi-Tanaka, M.; Shimada, Y.; Nakazono, M.; Watanabe, R.; Nishizawa, N.K.; Gomi, K.; Shimada, A.; Kitano, H.; et al. *GAMYB* controls different sets of genes and is differentially regulated by microRNA in aleurone cells and anthers. *Plant J.* **2006**, *47*, 427–444. [CrossRef] [PubMed]

64. Kaneko, M.; Inukai, Y.; Ueguchi-Tanaka, M.; Itoh, H.; Izawa, T.; Kobayashi, Y.; Hattori, T.; Miyao, A.; Hirochika, H.; Ashikari, M.; et al. Loss-of-function mutations of the rice *GAMYB* gene impair α-amylase expression in aleurone and flower development. *Plant Cell* **2004**, *16*, 33–44. [CrossRef] [PubMed]

65. Murray, F.; Kalla, R.; Jacobsen, J.; Gubler, F. A role for *HvGAMYB* in anther development. *Plant J.* **2003**, *33*, 481–491. [CrossRef] [PubMed]

66. Wang, Y.; Sun, F.; Cao, H.; Peng, H.; Ni, Z.; Sun, Q.; Yao, Y. TamiR159 directed wheat *TaGAMYB* cleavage and its involvement in anther development and heat response. *PLoS ONE* **2012**, *7*, e48445. [CrossRef]

67. Li, X.; Bian, H.; Song, D.; Ma, S.; Han, N.; Wang, J.; Zhu, M. Flowering time control in ornamental gloxinia (*Sinningia speciosa*) by manipulation of miR159 expression. *Ann. Bot.* **2013**, *111*, 791–799. [CrossRef]

68. Zhang, Y.; Zhang, X.; Liu, B.; Wang, W.; Liu, X.; Chen, C.; Liu, X.; Yang, S.; Ren, H. A *GAMYB* homologue *CsGAMYB1* regulates sex expression of cucumber via an ethylene-independent pathway. *J. Exp. Bot.* **2014**, *65*, 3201–3213. [CrossRef]

69. Vallarino, J.G.; Osorio, S.; Bombarely, A.; Casañal, A.; Cruz-Rus, E.; Sánchez-Sevilla, J.F.; Amaya, I.; Giavalisco, P.; Fernie, A.R.; Botella, M.A.; et al. Central role of FaGAMYB in the transition of the strawberry receptacle from development to ripening. *New Phytol.* **2015**, *208*, 482–496. [CrossRef]

70. Plackett, A.R.; Thomas, S.G.; Wilson, Z.A.; Hedden, P. Gibberellin control of stamen development: A fertile field. *Trends Plant Sci.* **2011**, *16*, 568–578. [CrossRef]

71. Aya, K.; Ueguchi-Tanaka, M.; Kondo, M.; Hamada, K.; Yano, K.; Nishimura, M.; Matsuoka, M. Gibberellin modulates anther development in rice via the transcriptional regulation of *GAMYB Plant Cell* **2009** *21* 1453–1472. [CrossRef] [PubMed]

72. Liu, B.; De Storme, N.; Geelen, D. Gibberellin induces diploid pollen formation by interfering with meiotic cytokinesis. *Plant Physiol.* **2017**, *173*, 338–353. [CrossRef] [PubMed]

73. Gubler, F.; Kalla, R.; Roberts, J.K.; Jacobsen, J.V. Gibberellin-regulated expression of a myb gene in barley aleurone cells: Evidence for Myb transactivation of a high-pI alpha-amylase gene promoter. *Plant Cell* **1995**, *7*, 1879–1891. [CrossRef] [PubMed]

74. Gubler, F.; Raventos, D.; Keys, M.; Watts, R.; Mundy, J.; Jacobsen, J.V. Target genes and regulatory domains of the GAMYB transcriptional activator in cereal aleurone. *Plant J.* **1999**, *17*, 1–9. [CrossRef] [PubMed]

75. Guo, W.J.; Ho, T.H.D. An abscisic acid-induced protein; HVA22, inhibits gibberellin-mediated programmed cell death in cereal aleurone cells. *Plant Physiol.* **2008**, *147*, 1710–1722. [CrossRef] [PubMed]

76. Reichel, M.; Li, Y.; Li, J.; Millar, A.A. Inhibiting plant microRNA activity: Molecular *SPONGEs*, target *MIMICs* and STTMs all display variable efficacies against target microRNAs. *Plant Biotech. J.* **2015**, *13*, 915–926. [CrossRef] [PubMed]

77. Gong, X.; Bewley, D.J. A *GAMYB-like* gene in tomato and its expression during seed germination. *Planta* **2008**, *228*, 563–572. [CrossRef] [PubMed]

78. Xue, T.; Liu, Z.; Dai, X.; Xiang, F. Primary root growth in Arabidopsis thaliana is inhibited by the miR159 mediated repression of *MYB33, MYB65* and *MYB101*. *Plant Sci.* **2017**, *262*, 182–189. [CrossRef]

79. Spanudakis, E.; Jackson, S. The role of microRNAs in the control of flowering time. *J. Exp. Bot.* **2014**, *65*, 365–380. [CrossRef]

80. Conti, L. Hormonal control of the floral transition: Can one catch them all? *Dev. Biol.* **2017**, *430*, 288–301. [CrossRef]

81. Gocal, G.F.; Sheldon, C.C.; Gubler, F.; Moritz, T.; Bagnall, D.J.; MacMillan, C.P.; Li, S.F.; Parish, R.W.; Dennis, E.S.; Weigel, D.; et al. *GAMYB-like* genes, flowering, and gibberellin signaling in Arabidopsis. *Plant Physiol.* **2001**, *127*, 1682–1693. [CrossRef] [PubMed]

82. Blazquez, M.A.; Green, R.; Nilsson, O.; Sussman, M.R.; Weigel, D. Gibberellins promote flowering of arabidopsis by activating the *LEAFY* promoter. *Plant Cell* **1998**, *10*, 791–800. [CrossRef]

83. Csukasi, F.; Donaire, L.; Casañal, A.; Martínez-Priego, L.; Botella, M.A.; Medina-Escobar, N.; Llave, C.; Valpuesta, V. Two strawberry miR159 family members display developmental-specific expression patterns in the fruit receptacle and cooperatively regulate *Fa-GAMYB*. *New Phytol.* **2012**, *195*, 47–57. [CrossRef] [PubMed]

84. Wang, C.; Jogaiah, S.; Zhang, W.; Abdelrahman, M.; Fang, J.G. Spatio-temporal expression of miRNA159 family members and their *GAMYB* target gene during the modulation of gibberellin-induced grapevine parthenocarpy. *J. Exp. Bot.* **2018**, *69*, 3639–3650. [CrossRef] [PubMed]

85. Zhang, B. MicroRNA: A new target for improving plant tolerance to abiotic stress. *J. Exp. Bot.* **2015**, *66*, 1749–1761. [CrossRef] [PubMed]

86. Liu, H.H.; Tian, X.; Li, Y.J.; Wu, C.A.; Zheng, C.C. Microarray-based analysis of stress-regulated microRNAs in Arabidopsis thaliana. *RNA* **2008**, *14*, 836–843. [CrossRef] [PubMed]

87. Reyes, J.L.; Chua, N.H. ABA induction of miR159 controls transcript levels of two *MYB* factors during Arabidopsis seed germination. *Plant J.* **2007**, *49*, 592–606. [CrossRef]

88. Pieczynski, M.; Marczewski, W.; Hennig, J.; Dolata, J.; Bielewicz, D.; Piontek, P.; Wyrzykowska, A.; Krusiewicz, D.; Strzelczyk-Zyta, D.; Konopka-Postupolska, D.; et al. Down-regulation of *CBP80* gene expression as a strategy to engineer a drought-tolerant potato. *Plant Biotechnol. J.* **2013**, *11*, 459–469. [CrossRef]

89. Zhang, T.; Zhao, Y.L.; Zhao, J.H.; Wang, S.; Jin, Y.; Chen, Z.Q.; Fang, Y.Y.; Hua, C.L.; Ding, S.W.; Guo, H.S. Cotton plants export microRNAs to inhibit virulence gene expression in a fungal pathogen. *Nat. Plants* **2016**, *2*, 16153. [CrossRef]

90. Medina, C.; da Rocha, M.; Magliano, M.; Ratpopoulo, A.; Revel, B.; Marteu, N.; Magnone, V.; Lebrigand, K.; Cabrera, J.; Barcala, M.; et al. Characterization of microRNAs from Arabidopsis galls highlights a role for miR159 in the plant response to the root-knot nematode *Meloidogyne incognita*. *New Phytol.* **2017**, *216*, 882–896. [CrossRef]

13

A Novel Circular RNA Generated by FGFR2 Gene Promotes Myoblast Proliferation and Differentiation by Sponging miR-133a-5p and miR-29b-1-5p

Xiaolan Chen [1,2], **Hongjia Ouyang** [1,3], **Zhijun Wang** [1,2], **Biao Chen** [1,2] and **Qinghua Nie** [1,2,*]

[1] Department of Animal Genetics, Breeding and Reproduction, College of Animal Science, South China Agricultural University, Guangzhou 510642, China; xiaolanchen@stu.scau.edu.cn (X.C.); oyolive@stu.scau.edu.cn (H.O.); zhijunwang@stu.scau.edu.cn (Z.W.); biaochen@stu.scau.edu.cn (B.C.)

[2] National-Local Joint Engineering Research Center for Livestock Breeding, Guangdong Provincial Key Lab of Agro-Animal Genomics and Molecular Breeding, and the Key Lab of Chicken Genetics, Breeding and Reproduction, Ministry of Agriculture, Guangzhou 510642, China

[3] College of Animal Science & Technology, Zhongkai University of Agriculture and Engineering, Guangzhou 510225, China

* Correspondence: nqinghua@scau.edu.cn;

Abstract: It is well known that fibroblast growth factor receptor 2 (*FGFR2*) interacts with its ligand of fibroblast growth factor (*FGF*) therefore exerting biological functions on cell proliferation and differentiation. In this study, we first reported that the *FGFR2* gene could generate a circular RNA of circFGFR2, which regulates skeletal muscle development by sponging miRNA. In our previous study of circular RNA sequencing, we found that circFGFR2, generated by exon 3–6 of *FGFR2* gene, differentially expressed during chicken embryo skeletal muscle development. The purpose of this study was to reveal the real mechanism of how circFGFR2 affects skeletal muscle development in chicken. In this study, cell proliferation was analyzed by both flow cytometry analysis of the cell cycle and 5-ethynyl-2′-deoxyuridine (EdU) assays. Cell differentiation was determined by analysis of the expression of the differentiation marker gene and Myosin heavy chain (MyHC) immunofluorescence. The results of flow cytometry analysis of the cell cycle and EdU assays showed that, overexpression of circFGFR2 accelerated the proliferation of myoblast and QM-7 cells, whereas knockdown of circFGFR2 with siRNA reduced the proliferation of both cells. Meanwhile, overexpression of circFGFR2 accelerated the expression of myogenic differentiation 1 (*MYOD*), myogenin (*MYOG*) and the formation of myotubes, and knockdown of circFGFR2 showed contrary effects in myoblasts. Results of luciferase reporter assay and biotin-coupled miRNA pull down assay further showed that circFGFR2 could directly target two binding sites of miR-133a-5p and one binding site of miR-29b-1-5p, and further inhibited the expression and activity of these two miRNAs. In addition, we demonstrated that both miR-133a-5p and miR-29b-1-5p inhibited myoblast proliferation and differentiation, while circFGFR2 could eliminate the inhibition effects of the two miRNAs as indicated by rescue experiments. Altogether, our data revealed that a novel circular RNA of circFGFR2 could promote skeletal muscle proliferation and differentiation by sponging miR-133a-5p and miR-29b-1-5p.

Keywords: circular RNA; circFGFR2; *FGFR2*; miR-133a-5p; miR-29b-1-5p; skeletal muscle; proliferation; differentiation

1. Introduction

Circular RNA is a large class of endogenous RNA with a covalently closed loop. It was actually discovered in plants, mouse, and yeast twenty years ago [1–3]. However, it has been regarded as unvalued mis-splicing product of mRNA in the last decades as a few kinds and a small quantity of circular RNAs have been found [4]. In addition, circular RNA has no 5′ caps and 3′ tails, and it could be easily abandoned by traditional sequencing technology [5]. Fortunately, with the rapid development of high throughput sequencing technology, the mysterious veil of circular RNA was revealed step by step [6]. Large amounts of circular RNAs were discovered in many species, including human [7], monkey [8], and pig [9].

Nowadays, circular RNA is considered as an up-rising star in the small RNAs interaction network with regulatory potency [10]. The diverse functions of circular RNA act as miRNA sponge, participating in regulating the expression of its own linear RNA in different ways [10,11], sequestering proteins [12,13], coding protein in vitro [14–16], and deriving pseudogenes [17]. Acting as miRNA sponge is a well-studied function of circular RNA, also known as a competing endogenous RNA mechanism (ceRNA). The CeRNA mechanism is that messenger RNAs, transcribed pseudogenes, and long noncoding RNAs competitively combine with the same miRNA response elements (MREs), and then eliminate the inhibition of miRNA on their target genes. Circular RNA interacted with miRNA are ubiquitous in a variety of tissues. A well-known example is that ciRS-7 has more than 70 highly conserved target sites of miR-7 and can extremely repress the activity of miR-7 [18]. This is the strongest evidence for a circRNA function as the miRNA sponge has thrust circRNAs into the spotlight and spurred a multitude of studies searching for functional circRNA sponges [19–21].

In previous work [22], we used leg muscle tissues of two female XingHua chickens from each at days E11, E16, and P1 for circRNA sequencing to comprehensively identify stably expressed circRNAs during skeletal muscle development at the embryonic stage. As a result, 13,377 potential circRNAs were identified and abundantly expressed among different development stages. Furthermore, the differentially expressed genes (DEGs) analysis showed 462 of them were differentially expressed at different development stages. CircFGFR2 was one of the DEcircRNAs with high expression during skeletal muscle development. Through divergent reverse-transcription PCR and RNase R treatment, in previous work [22], we confirmed that circFGFR2 was a stable exonic circular RNA formed by 3–6 exons of fibroblast growth factor receptor 2 (*FGFR2*), with a length of 636 bp. As a member of *FGFRs* family, *FGFR2* interacts with fibroblast growth factor (*FGF*) to exert biological effects on cell proliferation and differentiation as well as skeletal development [23]. The different expression level of circRNAs implied that they could potentially regulate skeletal muscle development. We previously revealed that circRBFOX2 could interact with miR-206 to regulate skeletal muscle cell proliferation and differentiation [22]. Considering all of that, we assumed that circFGFR2 was another candidate circRNA that probably affects skeletal muscle development.

In comparison to circular RNA, miRNAs are extremely well studied non-coding RNAs that suppress protein expression by targeting the 3′-UTR (Untranslated Region) of their mRNA with Argonaute effector protein [24,25]. The MiR-133 family has two members of miR-133a and miR-133b, which are found to specifically express in skeletal muscle and cardiac [26]. MiR-133a-5p belongs to the miR-133a cluster. Many studies have shown that miR-133 families are involved in regulating the proliferation and differentiation of various kinds of skeletal muscle cells [27,28]. However, the role of miR-133a-5p on skeletal muscle development has not been reported in poultry. MiR-29b-1-5p is a mature miRNA and belongs to the miR-29b cluster of the miR-29 family. This family has other clusters of miR-29a and miR-29c [29]. In chicken, the gga-miR-29b cluster contains gga-miR-29b-1-5p, gga-miR-29b-2-5p, and gga-miR-1701. MiR-29s are efficient regulators in the process of cell proliferation [30], differentiation [31], apoptosis [32–34] as well as DNA methylation [35,36] in different cell types. In skeletal muscle, miR-29s could participate in regulating skeletal myogenesis through different pathways. In mouse C2C12 cells, they could down-regulate *Rybp* (Ring1 and YY1-binding Protein) [37], AKT serine/threonine kinase 3 (*AKT3*) [38], and histone deacetylase 4 (*HDAC4*) [39] to regulate the differentiation of skeletal muscle cell. In addition, miR-29s were also related to some muscle diseases, including muscle atrophy [40], dystrophic muscle pathogenesis [41], and Duchenne muscular dystrophy [42]. Obviously, miR-29s play important roles in muscle development.

In this study, we aim to investigate the effects of circFGFR2 on skeletal muscle cell development, and to reveal its regulatory mechanism by interacting miR-133a-5p and miR-29b-1-5p.

2. Materials and Methods

2.1. Ethics Statement

This study was carried out in accordance with the principles of the Basel Declaration and recommendations of the Statute on the Administration of Laboratory Animals, the South China Agriculture University Institutional Animal Care and Use Committee. The protocol was approved by the South China Agriculture University Institutional Animal Care and Use Committee (approval, 19 November 2017, ID: 2017046).

2.2. Primers

All primers used in this study were designed by Premier Primer 5.0 software (Premier Bio-soft International, Palo Alto, CA, USA) and synthesized by Sangon (Sangon Biotech, Shanghai, China). The detailed information of all primers is listed in Table 1.

Table 1. Primers used in this study.

Name	Nucleotide Sequences (5′→3′)	Annealing Temperature (°C)	Size	Application
circFGFR2	F: ACATCGTATTGGCGGCTAT R: ACCCCATCCTTAGTCCAAC	60	267	qRT-PCR for circFGFR2
FGFR2-1	F: GTCCGCTGTATGTGATTGTAG R: TGAATGTCATCTGCTCCTCT	56	129	qRT-PCR for FGFR2 gene
FGFR2-2	F: AGCCGCCAACCAAATACCAAATR: CGACAACATCGAGATGGTAAGT	56	636	Amplification of the whole linear sequence of circFGFR2
MYOD	F: GCTACTACACGGAATCACCAAAT R: CTGGGCTCCACTGTCACTCA	58	200	qRT-PCR
MYOG	F: CGGAGGCTGAAGAAGGTGAA R: CGGTCCTCTGCCTGGTCAT	60	320	qRT-PCR
β-actin	F: ACCACAGGACTCCATACCCAAGAAAG R: GCCGAGAGAGAAATTGTGCGTGAC	52–60	146	qRT-PCR

2.3. RNA Extraction, cDNA Synthesis and Quantitative Real-Time PCR

The total RNA was extracted from cells by using RNAiso reagent (TaKaRa, Otsu, Japan). The quality and concentration of all obtained RNA samples were determined by 1.5% agarose gel electrophoresis and evaluated for optical density 260/280 ratio by Nanodrop 2000 spectrophotometer (Thermo, Waltham, MA, USA). For mRNA and circFGFR2 expression analysis, cDNA synthesis for mRNA was performed using a PrimeScript RT Reagent Kit (Perfect Real Time) (TaKaRa, Otsu, Japan). The β-actin gene was used as an internal control for quantitative real-time PCR (qRT-PCR) analysis. The reverse transcription reaction for miRNA was performed using ReverTra Ace qPCR RT Kit (Toyobo, Osaka, Japan). The specific Bulge-loop miRNA qRT-PCR Primer for miR-133a-5p, miR-29b-1-5p and U6 were designed by RiboBio (RiboBio, Guangzhou, China). qRT-PCR was performed on a Bio-Rad CFX96 Real-Time Detection System (Bio-Rad, Hercules, CA, USA) using iTaq™ Universal SYBR® Green Supermix Kit (Bio-Rad, Hercules, CA, USA). Each sample was assayed in triplicate, following the manufacturer's instructions. The specificity of the product was evaluated by the melting curve, and the quantitative values were obtained from the threshold PCR cycle number (Ct) at which the increase in signal is associated with an exponential growth at which the PCR product starts to be detected. The relative mRNA level in each sample was indicated by $2^{-\Delta\Delta Ct}$.

2.4. RNA Oligonucleotides and Plasmids Construction

The gga-miR-133a-5p mimic, gga-miR-29b-1-5p mimic and mimic control duplexes, the 3' end biotinylated gga-miR-133a-5p, gga-miR-29b-1-5p and mimic control duplexes, siRNA target against circFGFR2 (si-circFGFR2, 5'-CGATGTTGTCGAGCCGCCA-3') and non-specific siRNA negative control were synthesized by RiboBio (Guangzhou, China). For circFGFR2 overexpression plasmids construction, the linear sequences of circFGFR2 was amplified by PCR with primer *FGFR2-2*, and the cDNA template was synthesized from the RNA of chicken primary myoblast by RT-PCR. Then, the obtained linear sequences were cloned into *KpnI* and *BamHI* restriction sites of a circular expression vector-the pCD2.1-ciR vector (Geneseed Biotech, Guangzhou, China) according to the manufacturer's protocol, so as to generate the pCD2.1-circFGFR2 overexpression vector. For pmirGLO dual-luciferase reporter construction: the whole linear sequences of circFGFR2 were cloned into *XhoI and SalI* restriction sites of pmirGLO vector to generate the wild reporter vector (PGLO-WT reporter vector), which includes the predicted binding sites of miR-133a-5p and miR-29b-1-5p. PGLO-MT1 and PGLO-MT2 were two mutational reporter vectors of miR-133a-5p which were cloned into *XhoI and SalI* restriction sites of pmirGLO vector by PCR mutagenesis. We changed one of miR-133a-5p binding seed sequences from "CCAG" to "TTGA" in PGLO-MT1, while in PGLO-MT2 we changed another miR-133a-5p binding seed sequence (which included the binding site of miR-29b-1-5p) from "CCAG" to "GTTG". All luciferase reporters were constructed by Hongxun Biotech (Suzhou, China).

2.5. Cell Culture

Chicken embryo fibroblast cell line (DF-1) cells were cultured in high-glucose Dulbecco's modified Eagle's medium (Gibico, Grand Island, NY, USA) with 10% (*v/v*) fetal bovine serum (FBS) (Gibco, Grand Island, NY, USA) and 0.2% penicillin/streptomycin (Invitrogen, Carlsbad, CA, USA). Quail muscle cell line (QM-7) cells were cultured in high-glucose M199 medium (Gibco, USA) with 10% (*v/v*) FBS, 10% tryptose phosphate broth solution (Sigma, Louis, MO, USA) and 0.2% penicillin/streptomycin (Invitrogen, Carlsbad, CA, USA). Chicken primary myoblasts were isolated from the leg muscles of 11-day embryo age (E11) chickens. Leg tissues were collected from E11 chickens by completely removing skin and bones. Leg muscle was minced into sections of approximately 1 mm with scissors and then digested with 0.25% trypsin (Gibco, Grand Island, NY, USA) at 37 °C in a shaking water bath (90 oscillations/min) for 20 min. Digestions were terminated by adding equal values of complete medium-(RPMI)1640 medium with 20% FBS, 1% nonessential amino acids and 0.2% penicillin/streptomycin (Invitrogen, Carlsbad, CA, USA). The mixture was filtered through a nylon mesh with 70 mm pores (BD Falcon, Greiner, Germany). The filtered cells were centrifuged at $500 \times g$ for 5 min, and maintained in complete medium at 37 °C in a 5% CO_2, humidified atmosphere. Serial plating was performed to enrich myoblasts and to remove fibroblasts.

2.6. Transfections

Transfections were performed with Lipofectamine 3000 reagent (Invitrogen, Carlsbad, CA, USA) according to the manufacturer's instruction. Nucleic acids were diluted in OPTI-MEM Medium (Gibco, Grand Island, NY, USA).

2.7. 5-Ethynyl-2'-Deoxyuridine (EdU) Assays

After cells were transfected for 48 h, myoblasts were exposed to 50 µM 5-ethynyl-2'-deoxyuridine (EdU) (RiboBio, Guangzhou, China) for 2 h at 37 °C. Next, the cells were fixed in 4% paraformaldehyde (PFA) for 30 min and 2 mg/mL glycine solution was used to neutralize the 4% PFA. Cells were, then, permeabilized with 0.5% Triton X-100. Subsequently, $1 \times$ Apollo reaction cocktail (RiboBio, Guangzhou, China) was added to the cells and incubated for 30 min. The cells were stained with Hoechst 33342 for 30 min for DNA content analysis. Finally, the EdU-stained cells were visualized under a fluorescence microscope (Nikon, Tokyo, Japan or Leica, Wetzlar, Germany). The analysis of myoblast proliferation (ratio of EdU+ to all myoblasts) was performed using images of randomly selected fields obtained on the fluorescence microscope.

2.8. Flow Cytometry Analysis of the Cell Cycle

Myoblast cultures in growth medium (GM) were collected after a 48 h or 36 h-transfection and then fixed in 70% ethanol overnight at $-20\,°C$. After incubation in 50 μg/mL propidium iodide (PI) (Sigma, Louis, MO, USA) containing 10 μg/mL RNase A (TaKaRa, Otsu, Japan) and 0.2% (v/v) Triton X-100 (Sigma, Louis, MO, USA) for 30 min at 4 °C, the cells were analyzed by using a BD AccuriC6 flow cytometer (BD Biosciences, San Jose, CA, USA) and FlowJo7.6 software (Treestar Incorporated, Ashland, OR, USA).

2.9. Immunofluorescence

For immunofluorescence, cells were seeded in 24-well plates. After transfection for 48 h, cells were fixed in 4% formaldehyde for 20 min then washed three times with PBS for 5 min. Subsequently, the cells were permeabilized by adding 0.1% Triton X-100 for 5 min and blocked with goat serum for 30 min. After incubation with MyHC (B103; DSHB, Iowa City, IA, USA; 0.5 μg/mL) at 37 °C for 2 h, the Fluorescein (FITC)-conjugated AffiniPure Goat Anti-Mouse IgG (H + L) (Bioworld, Minneapolis, MN, USA; 1:200) or FITC (Bioworld, Minneapolis, MN, USA; 1:50) was added and the cells were incubated at room temperature for 1 h. The cell nuclei were stained with 4',6-diamidino-2-phenylindole (DAPI, Beyotime, Shanghai, China; 1:50) for 5 min. Images were obtained with a fluorescence microscope (Leica, Wetzlar, Germany). The area of cells labeled with anti-MyHC was measured by using ImageJ software (National Institutes of Health, Bethesda, MD, USA), and the total myotube area was calculated as a percentage of the total image area covered by myotubes.

2.10. Luciferase Reporter Assay

To investigate the binding sites of circFGFR2 with miR-133a-5p/miR-29b-1-5p, DF-1 cells were seeded in 96-well plates and then co-transfected with 100 ng of PGLO-WT reporter vector or mutant vectors PGLO-MT1 or PGLO-MT2, and 50 nM of miR-133a-5p/miR-29b-1-5p mimics or mimic control duplexes by using Lipofectamine 3000 reagent (Invitrogen, Carlsbad, CA, USA). To investigate whether circFGFR2 could inhibit the activity of miR-133a-5p/miR-29b-1-5p, DF-1 cells were seeded in 96-well plates and then co-transfected with 100 ng of PGLO-WT reporter vector or circFGFR2 overexpression vector, and 50 nM of miR-133a-5p/miR-29b-1-5p mimics or mimic control duplexes by using Lipofectamine 3000 reagent (Invitrogen, Carlsbad, CA, USA). After 48 h post transfection, luciferase activity analysis was performed using a Fluorescence/Multi-Detection Microplate Reader (BioTek, Winooski, VT, USA) and a Dual-GLO® Luciferase Assay System Kit (Promega, Madison, WI, USA). Firefly luciferase activities were normalized to Renilla luminescence in each well.

2.11. Biotin-Coupled miRNA Pull Down Assay

Transfection procedure: the 100 nM of 3' end biotinylated miR-133a-5p, miR-29b-1-5p or mimic NC (RiboBio, Guangzhou, China) were transfected into QM-7 cells along with 30 μg circFGFR2 expression vector in T75 cell culture bottle. At 24 h after transfection, the cells were harvested and washed in PBS, then lysed in lysis buffer. A total of 100 μL washed streptavidin magnetic beads were blocked for 2 h and then added to each reaction tube to pull down the biotin-coupled RNA complex. All the tubes were incubated for 4 h on a rotator at a low speed (10 r/min). The beads were washed with lysis buffer five times and RNAiso reagent (TaKaRa, Otsu, Japan) was used to recover RNAs specifically interacting with miRNA. The abundance of circFGFR2 in bound fractions was evaluated by qRT-PCR analysis.

2.12. Statistical Analysis

In all panels, results are expressed as the mean ± S.E.M. of three independent experiments. For two group comparison analysis, statistical significance of differences between means was analyzed by unpaired Student's t-test. For multiple comparison analysis, data were analyzed by one-way ANOVA followed by both least significant difference (LSD) and Duncan test through Statistical Package for the Social Sciences software (SPSS 17.0, Chicago, IL, USA). We considered $p < 0.05$ to be statistically significant. $^{*} p < 0.05$; $^{**} p < 0.01$. NC, negative control.

3. Results

3.1. CircFGFR2 Promotes Myoblast Proliferation

To investigate the role of circFGFR2 in skeletal muscle cell proliferation, we conducted overexpression and knocked down experiments by transfecting circFGFR2 overexpression vector and siRNAs (pCD2.1-circFGFR2 and si-circFGFR2) into chicken primary myoblast and QM-7 cell. The relative expression of circFGFR2 was detected after 48 h post transfection by qRT-PCR. Result showed that both the effect of overexpression and knockdown had reached a significant level in both myoblast and QM-7 cell (Figure 1A–D), and si-circFGFR2 specifically downregulated the expression of circFGFR2 but not linear mRNA of FGFR2 (Figure 2B). Furthermore, we detected the proliferation process of both chicken primary myoblast and QM-7 cell by flow cytometry for cell cycle analysis and 5-ethynyl-2'-deoxyuridine (EdU) incorporation assays after transfecting with pCD2.1-circFGFR2/pCD2.1-ciR, or si-circFGFR2/control. Cell cycle analysis showed that overexpression of circFGFR2 increased the cell population in S phase and decreased the cell population in G1/0 and G2/M phases (Figure 1E) while knockdown of circFGFR2 decreased the cell population in S phase and increased the cell population in G1/0 phase, as observed in chicken primary myoblast (Figure 1F). Meanwhile, the result of EdU strain assay showed that there were significantly more cells in the pCD2.1-circFGFR2 transfected group than in the control group (Figure 1G,H), whereas knockdown of circFGFR2 significantly decreased the numbers of EdU strained cells (Figure 1G,I). These results indicated that circFGFR2 could promote the proliferation rate of chicken primary myoblast. As expected, we obtained similar results in QM-7 cell (Figure 1J–N). These results suggested that circFGFR2 could significantly promote the proliferation of myoblast and QM-7 cell.

3.2. CircFGFR2 Promotes Myoblast Differentiation

Myogenesis is a complex process including myoblast proliferation, differentiation and myotube formation and is controlled by myogenic regulatory factors (MRFs), MYOD, MYOG, myogenic factor 5 (Myf5), and myogenic factor 6 (Myf6, also known as myogenic regulatory factor 4, MRF4). These factors activate muscle-specific genes to coordinate myoblasts to terminally withdraw from the cell cycle and subsequently fuse into multinucleated myotubes [43]. Following proliferation, the initiation of terminal differentiation and fusion begins with the expression of myogenin, which together with MYOD, activates the muscle specific structural and contractile genes to stimulate myoblast differentiation [44]. To address the potential role of circFGFR2 in primary myoblast differentiation, the expression of differentiation marker genes, including MYOG and MYOD were analyzed by qRT-PCR after transfecting with pCD2.1-circFGFR2/pCD2.1-ciR, or si-circFGFR2/control. Result showed that overexpression of circFGFR2 significantly promoted the expression of MYOD and MYOG while knockdown of circFGFR2 significantly inhibited the expression of MYOD and MYOG (Figure 2A,B). It indicated that circFGFR2 may promote chicken primary myoblast differentiation. Subsequently, we induced chicken primary myoblast differentiation in vitro, as soon as the muscle cells started to differentiate into myotubes (the first day of differentiation, DM1), we transfected them with pCD2.1-circFGFR2/pCD2.1-ciR. MyHC immunofluorescence staining was carried out on the differentiated myoblasts after 36 h post transfection (DM3). According to the immunofluorescence staining, we found that the areas of myotubes of pCD2.1-circFGFR2 transfected group were prominently greater than that of the control group (Figure 2C,D). Conversely, the areas of myotubes in the si-circFGFR2 transfected group were lower than that of the control group (Figure 2E,F). The result showed that circFGFR2 could promote the formation of myotubes and promote the early differentiation of chicken primary myoblast.

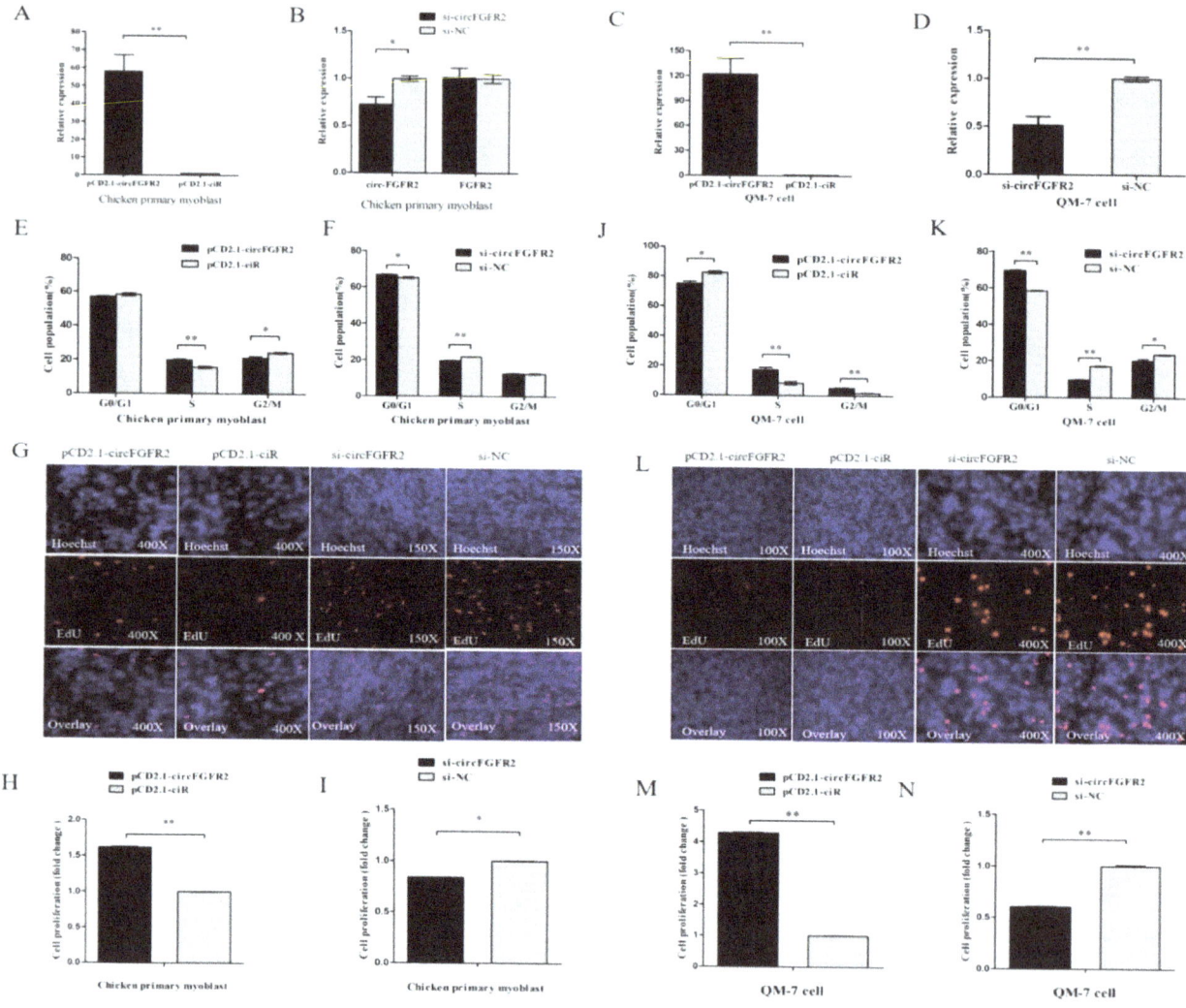

Figure 1. CircFGFR2 promotes myoblast proliferation. (**A,B**) The relative expression of circFGFR2 after transfected chicken primary myoblasts with 1 µg pCD2.1-circFGFR2 or 50 nM si-circFGFR2 for 48 h. (**C,D**) The relative expression of circFGFR2 after transfected QM-7 cells with 1 µg pCD2.1-circFGFR2 or 50nM si-circFGFR2 for 48 h. (**E,F**) Cell cycle analysis of chicken primary myoblasts transfected with 1 µg circFGFR2 pCD2.1-circFGFR2 or 50 nM si-circFGFR2 for 36 h. (**G**) 5-ethynyl-2'-deoxyuridine (EdU) assays for chicken primary myoblasts transfected with 1 µg circFGFR2 pCD2.1-circFGFR2 or 50 nM si-circFGFR2 for 36 h. (**H,I**) The percentage of EdU-stained chicken primary myoblasts after overexpression or knockdown of circFGFR2 for 36 h. (**J,K**) Cell cycle analysis of QM-7 cells transfected with 1 µg circFGFR2 pCD2.1-circFGFR2 or 50 nM si-circFGFR2 for 48 h. (**L**) EdU assays for QM-7 cells transfected with 1 µg circFGFR2 pCD2.1-circFGFR2 or 50 nM si-circFGFR2 for 48 h. (**M,N**) The percentage of EdU-stained chicken primary myoblasts after overexpression or knockdown of circFGFR2 for 48 h. In all panels, the results are shown as mean ± S.E.M., and the data are represented by three independent assays. Statistical significance of differences between means was assessed using an unpaired Student's t-test (* $p < 0.05$; ** $p < 0.01$).

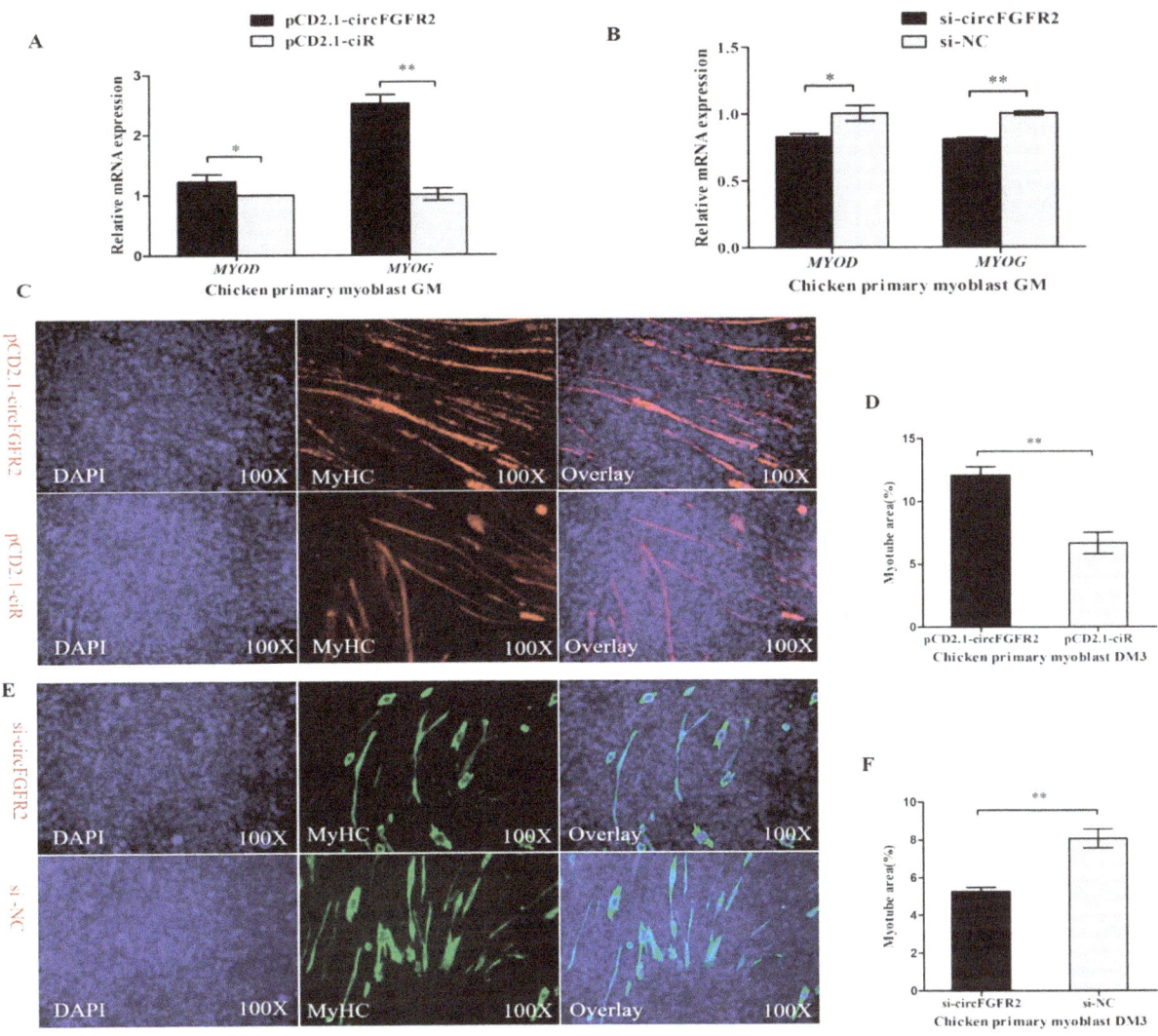

Figure 2. CircFGFR2 promotes myoblast differentiation. (**A**) Overexpression of circFGFR2 promotes mRNA expression of MYOD and MYOG. (**B**) Knockdown of circFGFR2 inhibits the mRNA expression of MYOD and MYOG. (**C,D**) Overexpression of circFGFR2 facilitates the formation of myotubes. (**E,F**) Down-regulation of circFGFR2 suppresses the formation of myotubes. In all panels, data are presented as mean $\pm$ S.E.M. of three biological replicates. Statistical significance of differences between means was assessed using an unpaired Student's t-test (* $p < 0.05$; ** $p < 0.01$).

3.3. CircFGFR2 Interacts with miR-133a-5p and miR-29b-1-5p, and Inhibits the Expression of miR-133a-5p and miR-29b-1-5p in Myoblast

Circular RNA has been shown to act as miRNA sponge and circFGFR2 could promote myoblast proliferation and differentiation. We hypothesized that circFGFR2 exerts function by acting as miRNA sponge as well as regulating the expression of miRNA. To screen potential miRNAs that bind to circFGFR2, we used RNAhybrid to conduct the putative combination site between circFGFR2 and miR-133a-5p/miR-29b-1-5p. Interestingly, we found that circFGFR2 has two potential miR-133a-5p binding sites (binding site 1 and binding site 2) and one potential miR-29b-1-5p binding site. The potential miR-29b-1-5p binding site shares six of seven nucleotides with the binding site 2 of miR-133a-5p. The mature sequence of miR-133a-5p/miR-29b-1-5p and the predicted binding sites of these two miRNAs are shown in Figure 3A–D.

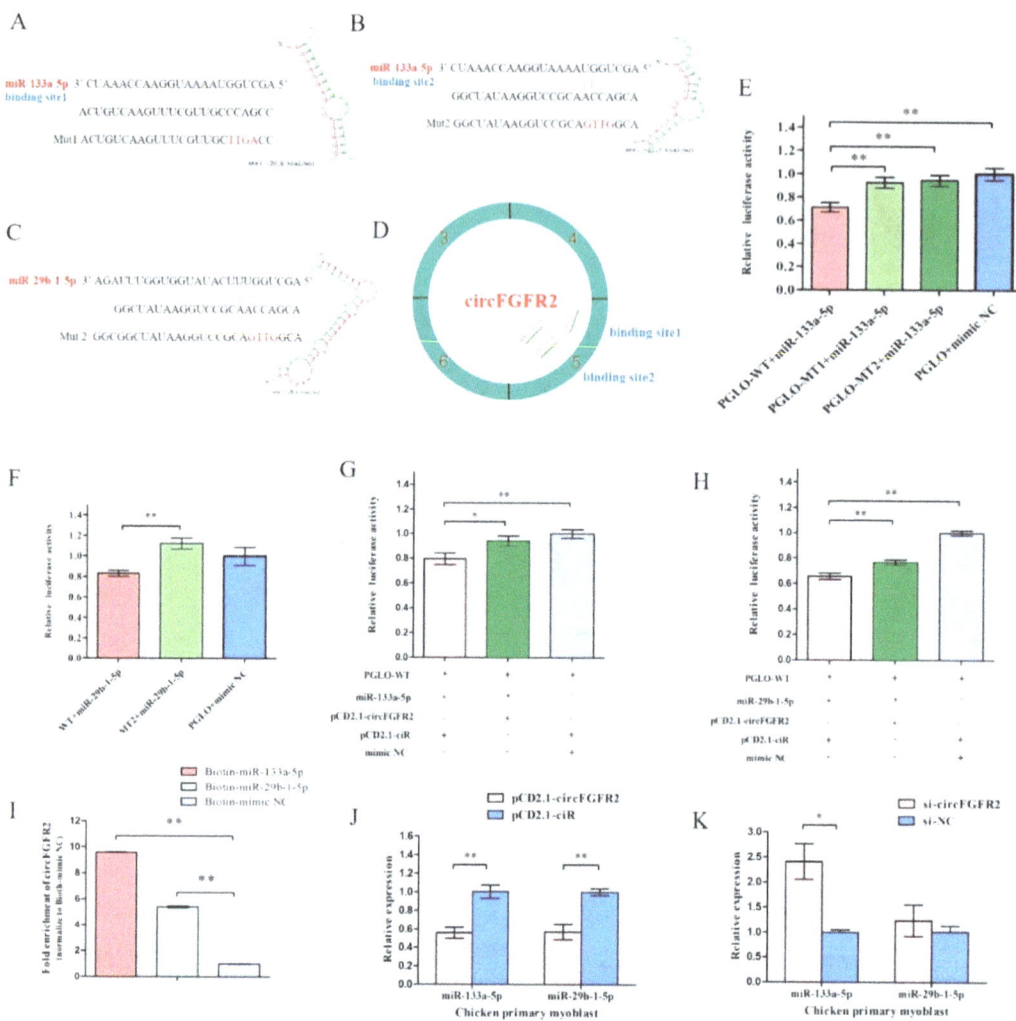

Figure 3. CircFGFR2 sponges with miR-133a-5p and miR-29b-1-5p, and inhibits the expression of miR-133a-5p and miR-29b-1-5p in myoblast. (**A–C**) The potential binding sites of miR-133a-5p and miR-29b-1-5p in circFGFR2. The mutant sequences in binding sites are highlighted in red. (**D**) A schematic drawing showing the putative binding sites of miR-133a-5p/miR-29b-1-5p associated with circFGFR2. (**E,F**) Luciferase assay was conducted by co-transfecting wild type or mutant linear sequence of circFGFR2 with miR-133a-5p/miR-29b-1-5p mimic or mimic-NC in DF-1 cells. (**G,H**) Luciferase assay was conducted by co-transfecting wild type circFGFR2 linear sequence and miR-133a-5p/miR-29b-1-5p mimic or mimic-NC and with circFGFR2 overexpression vector (pCD2.1-circFGFR2) or empty vector (pCD2.1-ciR). (**I**) Biotin-coupled miRNA pull down assay from the myoblast lysates after transfection with 3′ end biotinylated miR-133a-5p, miR-29b-1-5p or mimic NC. The expression level of circFGFR2 was quantified by qRT–PCR, and fold enrichment in the streptavidin captured fractions are plotted. (**J,K**) qRT–PCR analysis of the relative expression of miR-133a-5p and miR-29b-1-5p after overexpression or inhibition of circFGFR2. In all panels, results are expressed as the mean ± S.E.M. of three independent experiments. For two group comparison analysis, statistical significance of differences between means was analyzed by unpaired Student's t-test. For multiple comparison analysis, data were analyzed by one-way ANOVA followed by both least significant difference (LSD) and Duncan test through SPSS software. We considered $p < 0.05$ to be statistically significant. * $p < 0.05$; ** $p < 0.01$. NC, negative control.

To investigate the binding site of circFGFR2 with miR-133a-5p/miR-29b-1-5p, we constructed a dual-luciferase reporter by inserting the wild type (WT) or mutant (MT) linear sequence of circFGFR2 into the 3' end of *firefly* luciferase of pmirGLO (PGLO) luciferase vector to generate a wild type reporter (PGLO-WT) and two mutant reporters (PGLO-MT1 and PGLO-MT2). PGLO-MT1 vector contains the mutated seed sequences for the binding site 1 of mir-133a-5p, and PGLO-MT2 contains the mutated seed sequence for miR-133a-5p binding site 2 and miR-29b-1-5p binding site. The mutant sequences are shown in Figure 3A–C. Then DF-1 cells were co-transfected with PGLO-WT, PGLO-MT1/PGLO-MT2/PGLO luciferase vector and co-transfected with miR-133a-5p/miR-29b-1-5p mimic/control duplexes, respectively. The relative luciferase activity in DF-1 cell line was significantly decreased when miR-133a-5p/miR-29b-1-5p mimic were co-transfected with PGLO-WT reporter (Figure 3E,F) compared with the miR-133a-5p/miR-29b-1-5p mimic and their correspondent mutant reporter co-transfected group. This result demonstrated that miR-133a-5p and miR-29b-1-5p could really combine with the predicted binding sites and miR-133a-5p could combine with both binding site 1 and site 2.

To study the effect of circFGFR2 on the activity of miR-133a-5p/miR-29b-1-5p, we conducted another luciferase reporter assay by co-transfected pCD2.1-circFGFR2 (circFGFR2 overexpression vector)/pCD2.1-ciR (the empty overexpression vector), miR-133a-5p/miR-29b-1-5p/mimic NC with PGLO-WT reporter vector. Luciferase reporter assay showed that the relative luciferase activity was significantly decreased when cells were co-transfected miR-133a-5p/miR-29b-1-5p mimic with PGLO-WT reporter, while the relative luciferase activity was significantly increased when cells were co-transfected the miR-133a-5p/miR-29b-1-5p mimic with pCD2.1-circFGFR2 (Figure 3G,H). It suggested that circFGFR2 could combine with exogenetic miR-133a-5p and miR-29b-1-5p and eliminate the activity of both miRNAs.

Subsequently, we also conducted biotin-coupled miRNA pull down assay to further confirm the interaction between circFGFR2 and miR-133a-5p/miR-29b-1-5p by using biotin-coupled miR-133a-5p and miR-29b-1-5p mimics. Compared with the negative control, we observed more than 8-fold enrichment of circFGFR2 in miR-133a-5p-captured fraction and more than 5-fold enrichment of circFGFR2 in miR-29b-1-5p-captured fraction (Figure 3I), which demonstrated that circFGFR2 could directly sponge miR-133a-5p and miR-29b-1-5p. The greater enrichment observed in miR-133a-5p-captured fraction is probably due to the fact that circFGFR2 contained two binding sites for miR-133a-5p but only one for miR-29b-1-5p.

In addition, the qRT-PCR result showed that overexpression of circFGFR2 could significantly decrease the expression of miR-133a-5p and miR-29b-1-5p (Figure 3J), while knockdown of circFGFR2 could up-regulate the expression of miR-133a-5p and miR-29b-1-5p in chicken primary myoblast (Figure 3K).

3.4. MiR-133a-5p and miR-29b-1-5p Inhibit Myoblast Proliferation

As circFGFR2 had an effect on myoblast proliferation, we also confirmed that circFGFR2 could inhibit the expression and activity of miR-133a-5p and miR-29b-1-5p. We speculated that miR-133a-5p and miR-29b-1-5p had a potential effect on myoblast proliferation. To confirm our hypothesis, we synthesized miR-133a-5p and miR-29b-1-5p mimic. In chicken primary myoblast, we detected the expression of miR-133a-5p and miR-29b-1-5p after transfected chicken primary myoblast with 50 nM miR-133a-5p or miR-29b-1-5p mimic for 48 h. The expression of miR-133a-5p or miR-29b-1-5p was significantly increased by mimic (Figure 4A,B). Subsequently, in chicken primary myoblast, flow cytometry analysis showed that overexpression of miR-133a-5p or miR-29b-1-5p could prominently increase the numbers of cells that progressed to G0/G1 and reduced the numbers of S phase cells (Figure 4C,D). Meanwhile, we found similar results in QM-7 cells as indicated by cycle analysis. MiR-133a-5p or miR-29b-1-5p overexpression significantly increased the number of QM-7 cells that progressed to G0/G1 and reduced the number of S phase cells (Figure 4E,F). Furthermore, the EdU assay demonstrated that overexpression of miR-133a-5p and miR-29b-1-5p dramatically decreased the numbers of EdU strained cells (Figure 4G–J) in both chicken primary myoblast and QM-7 cell, which indicated that miR-133a-5p and miR-29b-1-5p could inhibit the proliferation rate of skeletal muscle cells. These results revealed that miR-133a-5p and miR-29b-1-5p could suppress myoblast proliferation.

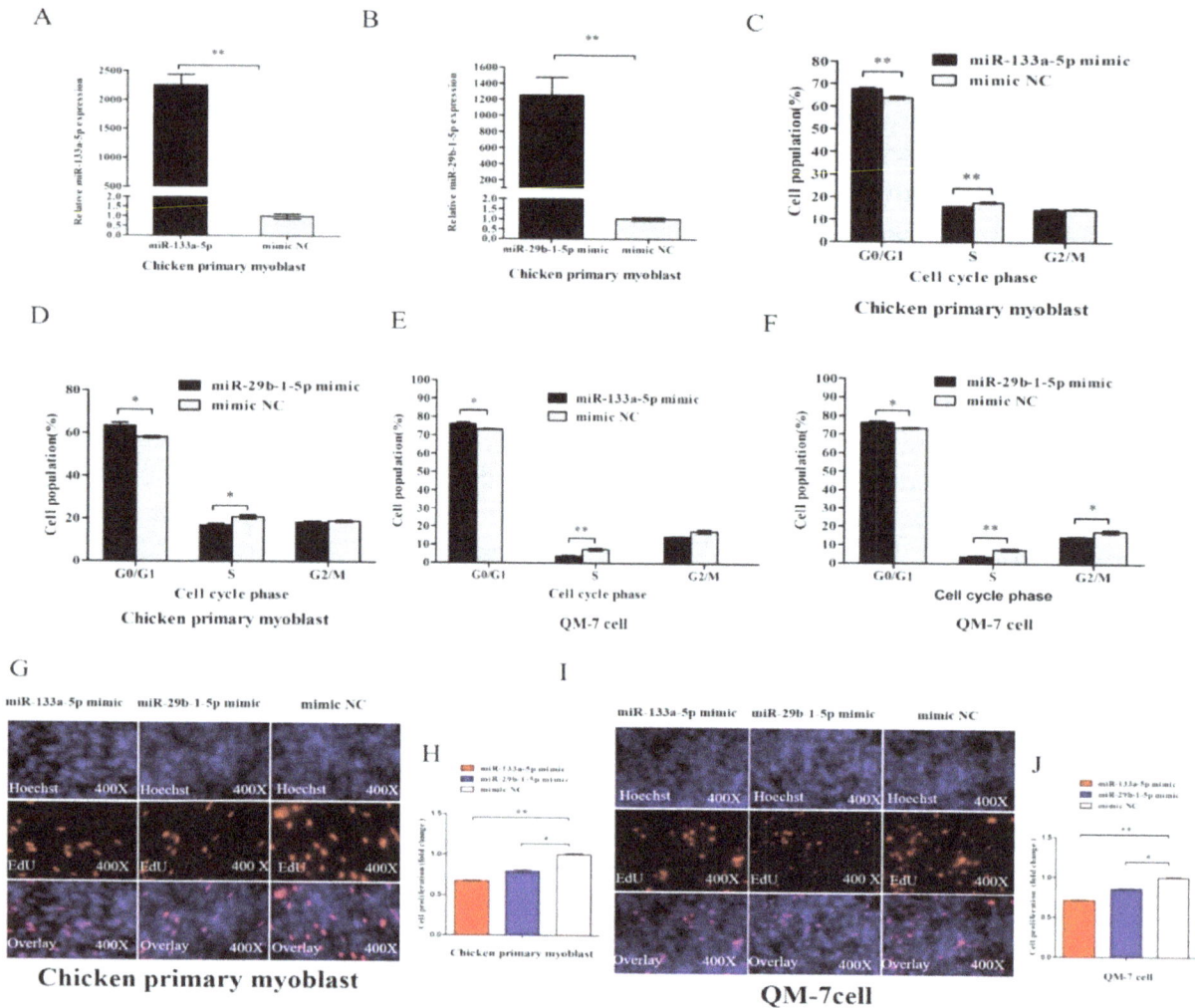

Figure 4. miR-133a-5p and miR-29b-1-5p inhibit myoblast proliferation. (**A,B**) The relative expression of miR-133a-5p and miR-29b-1-5p after transfected chicken primary myoblast with 50 nM miR-133a-5p and miR-29b-1-5p mimic for 48 h. (**C,D**) Cell cycle analysis of chicken primary myoblasts transfected with 50 nM miR-133a-5p and miR-29b-1-5p mimic for 36 h. (**E,F**) Cell cycle analysis of QM-7 cell transfected with 50 nM miR-133a-5p and miR-29b-1-5p mimic for 48 h. (**G,H**) EdU assay of chicken primary myoblasts transfected with 50 nM miR-133a-5p or miR-29b-1-5p mimic for 36 h. (**I,J**) EdU assay of QM-7 cell transfected with 50 nM miR-133a-5p or miR-29b-1-5p mimic for 48 h. In all panels, results are expressed as the mean ± S.E.M. of three independent experiments, and statistical significance of differences between means was assessed using an unpaired Student's t-test (* $p < 0.05$; ** $p < 0.01$). NC, negative control.

3.5. CircFGFR2 Eliminates the Inhibition Effect of miR-133a-5p and miR-29b-1-5p on Myoblast Proliferation

Considering the interaction between circFGFR2 and miR-133a-5p/miR-29b-1-5p, rescue experiments were conducted by co-transfecting circFGFR2 with miR-133a-5p/miR-29b-1-5p mimics to assess whether the inhibition on proliferation of two miRNAs could be blocked by circFGFR2 overexpression. As expected, flow cytometry analysis and EdU assay confirmed that circFGFR2 could eliminate the inhibition from overexpressed miR-133a-5p (Figure 5A–F) or miR-29b-1-5p on the proliferation of both chicken primary myoblast and QM-7 cell (Figure 5G–L).

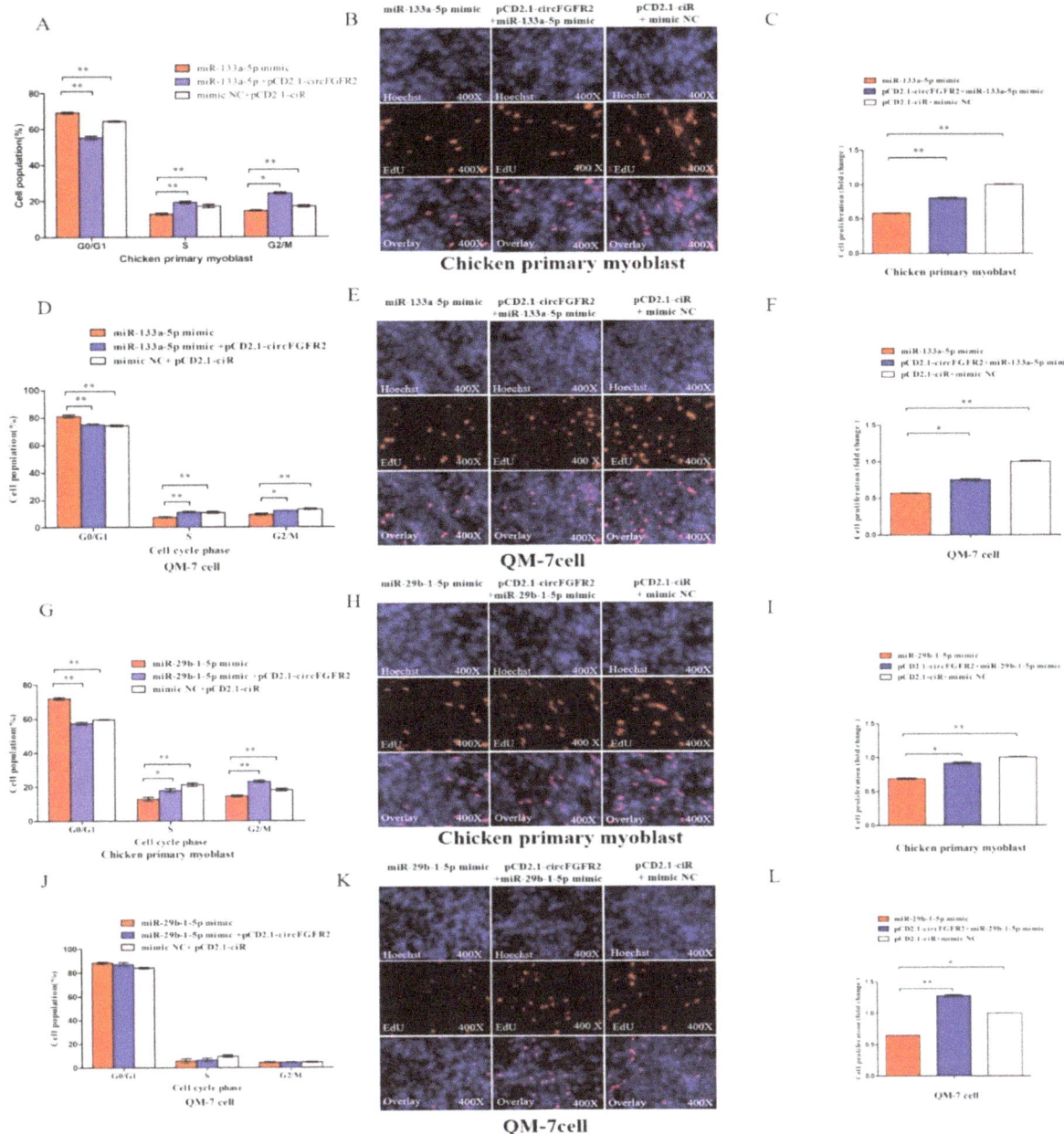

Figure 5. CircFGFR2 eliminates the inhibition effect of miR-133a-5p and miR-29b-1-5p on myoblast proliferation. (**A**) Cell cycle analysis of chicken primary myoblasts after co-transfection with the listed nucleic acids (miR-133a-5p, circFGFR2 overexpression vector and miR-133a-5p, empty overexpression vector and mimic NC, respectively) for 36 h. (**B,C**) EdU assays of chicken primary myoblasts after co-transfection with the listed nucleic acids (miR-133a-5p, circFGFR2 overexpression vector and miR-133a-5p, empty overexpression vector and mimic NC, respectively) for 36 h. (**D**) Cell cycle analysis of QM-7 cells after co-transfection with the listed nucleic acids (miR-133a-5p, circFGFR2 overexpression vector and miR-133a-5p, empty overexpression vector and mimic NC, respectively) for 48 h. (**E,F**) EdU assays of QM-7 cells after co-transfection with the listed nucleic acids (miR-133a-5p, circFGFR2 overexpression vector and miR-133a-5p, empty overexpression vector and mimic NC, respectively) for 48 h. (**G**) Cell cycle analysis of chicken primary myoblasts after co-transfection with the listed nucleic acids (miR-29b-1-5p, circFGFR2 overexpression vector and miR-29b-1-5p, empty overexpression vector and mimic NC, respectively) for 36 h. (**H,I**) EdU assays of chicken primary myoblasts after co-transfection with the listed nucleic acids (miR-29b-1-5p, circFGFR2 overexpression vector and miR-29b-1-5p, empty overexpression vector and mimic NC, respectively) for 36 h. (**J**) Cell cycle analysis

of QM-7 cells after co-transfection with the listed nucleic acids (miR-29b-1-5p, circFGFR2 overexpression vector and miR-29b-1-5p, empty overexpression vector and mimic NC, respectively) for 48 h. (**K,L**) EdU assays of QM-7 cells after co-transfection with the listed nucleic acids (miR-29b-1-5p, circFGFR2 overexpression vector and miR-29b-1-5p, empty overexpression vector and mimic NC, respectively) for 48 h. In all panels, results are expressed as the mean ± S.E.M. of three independent experiments. For two group comparison analysis, statistical significance of differences between means was analyzed by unpaired Student's t-test. For multiple comparison analysis, data were analyzed by one-way ANOVA followed by both least significant difference (LSD) and Duncan test through SPSS software. We considered $p < 0.05$ to be statistically significant. * $p < 0.05$; ** $p < 0.01$. NC, negative control.

3.6. miR-133a-5p and miR-29b-1-5p Repress Myoblast Differentiation

To unveil the potential roles of miR-133a-5p and miR-29b-1-5p in chicken primary myoblast differentiation, the expression of the myoblast differentiation marker genes including *MYOG* and *MYOD* were evaluated by qRT-PCR in myoblast transfected with miR-133a-5p or miR-29b-1-5p. Overexpression of miR-133a-5p notably inhibited the expression of *MYOD* and *MYOG*, and overexpression of miR-29b-1-5p could also inhibit the expression of *MYOD* and *MYOG* (Figure 6A,B). Furthermore, we synthesized miR-133a-5p and miR-29b-1-5p inhibitor to down-regulate the expression of miR-133a-5p or miR-29b-1-5p, and we found that down-regulation of miR-133a-5p or miR-29b-1-5p accelerated the expression of *MYOD* and *MYOG* (Figure 6C,D). Subsequently, we induced chicken primary myoblast differentiation in vitro, and we transfected them with miR-133a-5p or miR-29b-1-5p mimic/inhibitor at DM1. MyHC immunofluorescence staining was carried out on the transfected differentiated myoblasts at DM3. According to immunofluorescence staining, we found that the total areas of myotubes of miR-133a-5p or miR-29b-1-5p mimic transfected group were prominently less than that of the control group (Figure 6E,F). On the contrary, the areas of myotubes in miR-133a-5p or miR-29b-1-5p (Figure 6G,H) inhibitor transfected group were more than that of the control group. The results demonstrated that miR-133a-5p and miR-29b-1-5p could repress chicken primary myoblast differentiation.

3.7. CircFGFR2 Eliminates the Inhibition Effect of miR-133a-5p and miR-29b-1-5p on Myoblast Differentiation

We further performed a rescue experiment to investigate whether the suppressing effects of miR-133a-5p and miR-29b-1-5p on myoblast differentiation could be eliminated by circFGFR2 overexpression. As shown in Figure 7A, the expressions of *MYOD* and *MYOG* in miR-133a-5p and circFGFR2 co-transfected group were dramatically elevated compared with the miR-133a-5p transfected group. For miR-29b-1-5p, circFGFR2 also eliminated its repression effect on *MYOD* and *MYOG* (Figure 7B). Further MyHC immunofluorescence showed that overexpression of circFGFR2 eliminated the inhibition on myotube formation at DM3 caused by either miR-133a-5p or miR-29b-1-5p (Figure 7C–F). Taken together, these results demonstrated that circFGFR2 could eliminate the inhibition effect of miR-133a-5p and miR-29b-1-5p on myoblast differentiation.

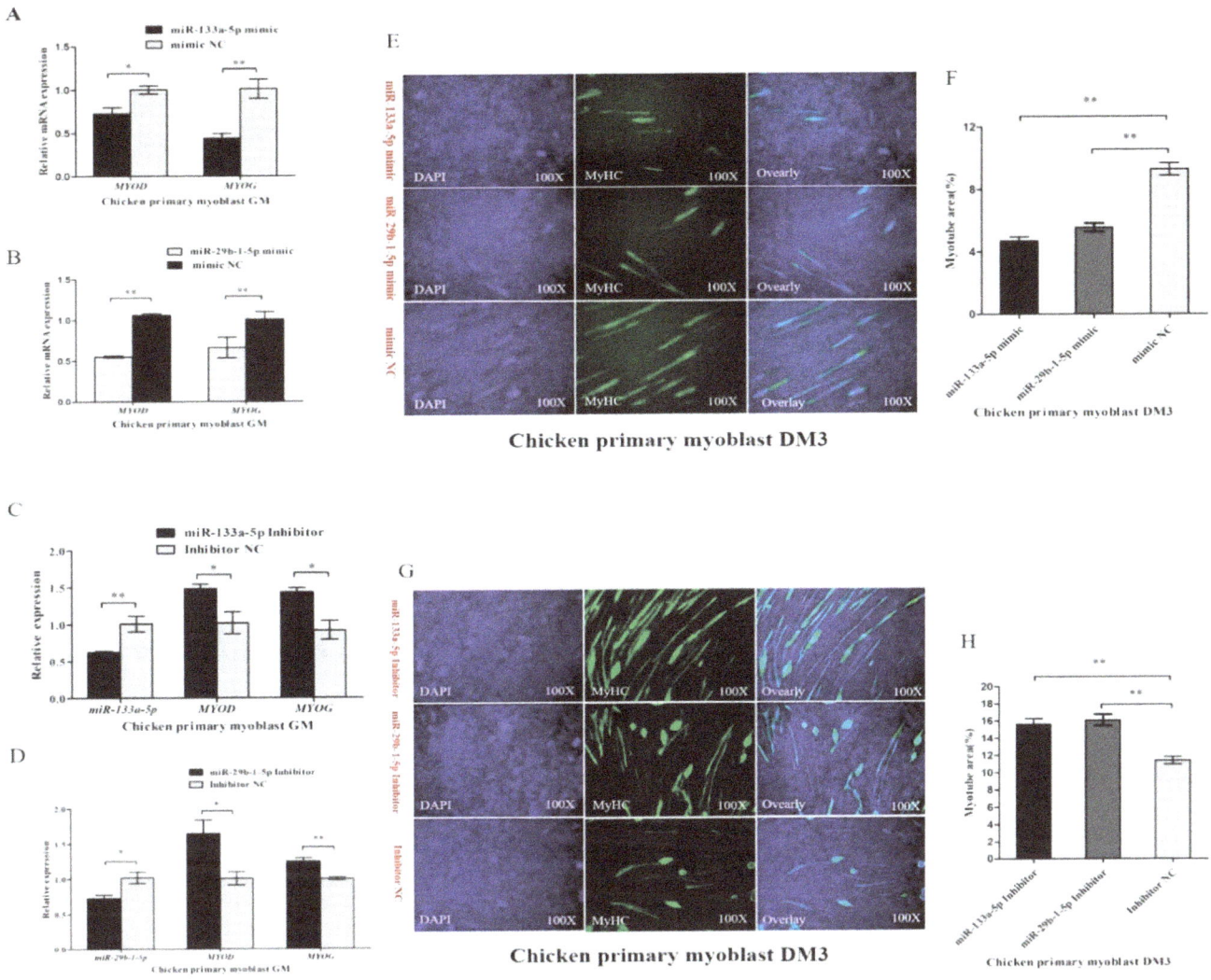

Figure 6. miR-133a-5p and miR-29b-1-5p repress myoblast differentiation. (**A,B**) Overexpression of miR-133a-5p and miR-29b-1-5p reduced the expression of *MYOD* and *MYOG*. (**C,D**) Inhibition of miR-133a-5p and miR-29b-1-5p accelerated the expression of *MYOD* and *MYOG*. (**E,F**) Immunofluorescence analysis of MyHC-staining cells after overexpression miR-133a-5p or miR-29b-1-5p. (**G,H**) Immunofluorescence analysis of MyHC-staining cells after down-regulation of miR-133a-5p or miR-29b-1-5p. In all panels, results are expressed as the mean ± S.E.M. of three independent experiments, and statistical significance of differences between means was assessed using an unpaired Student's *t*-test (* *p* < 0.05; ** *p* < 0.01). NC, negative control.

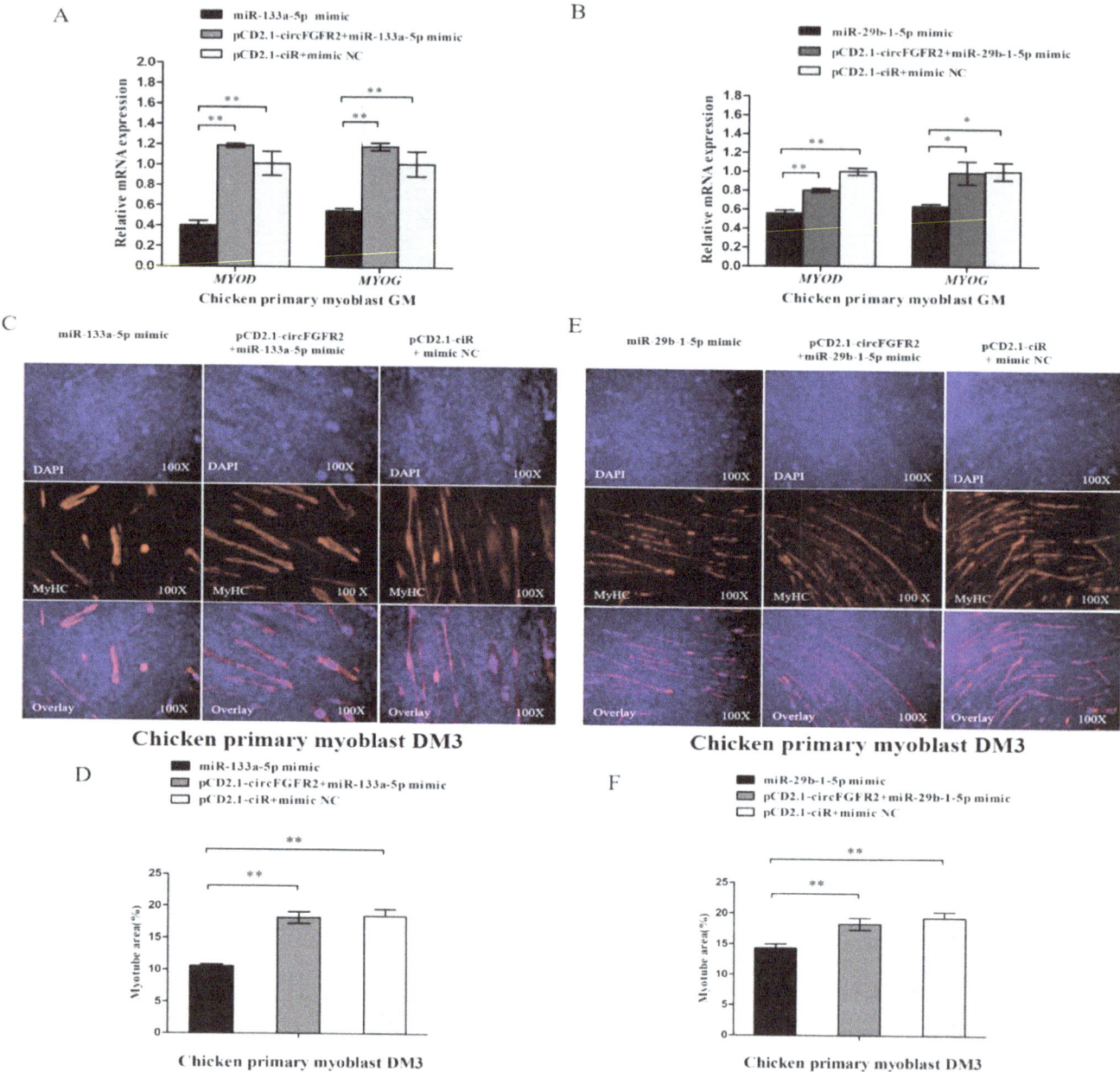

Figure 7. CircFGFR2 eliminates the inhibition effect of miR-133a-5p and miR-29b-1-5p on myoblast differentiation. (**A**) The mRNA expression of *MYOD* and *MYOG* of chicken primary myoblasts after co-transfection with the listed nucleic acids (miR-133a-5p, circFGFR2 overexpression vector and miR-133a-5p, empty overexpression vector and mimic NC, respectively). (**B**) The mRNA expression of *MYOD* and *MYOG* of chicken primary myoblasts after co-transfection with the listed nucleic acids (miR-29b-1-5p, circFGFR2 overexpression vector and miR-29b-1-5p, empty overexpression vector and mimic NC, respectively). (**C,D**) The myotube area of chicken primary myoblasts after co-transfection with the listed nucleic acids (miR-133a-5p, circFGFR2 overexpression vector and miR-133a-5p, empty overexpression vector and mimic NC, respectively). (**E,F**) The myotubes area of chicken primary myoblasts after co-transfection with the listed nucleic acids (miR-29b-1-5p, circFGFR2 overexpression vector, and miR-29b-1-5p, empty overexpression vector and mimic NC, respectively). In all panels, results are expressed as the mean ± S.E.M. of three independent experiments, and statistical significance of differences between means were analyzed by one-way ANOVA followed by both least significant difference (LSD) and Duncan test through SPSS software. We considered $p < 0.05$ to be statistically significant. * $p < 0.05$; * $p < 0.01$. NC, negative control.

4. Discussion

In recent years, circular RNAs have been successfully identified in various cell types across different species [7,9]. They have shown features of dynamic and tissue-specific expression, which indicate a distinct function in diverse tissues [45,46]. CircFGFR2 is a highly expressed DGcircRNA among millions of circRNAs during embryonic muscle development according to our previous circRNA sequencing results [22], which indicates that it has a potential effect in regulating skeletal muscle development. Here we primarily confirmed that circFGFR2 has a crucial function on skeletal muscle development. In both chicken primary myoblast and QM-7 cell, cell cycle analysis demonstrated that overexpression of circFGFR2 could significantly increase the cell numbers in S phase and reduce the cell numbers in G0/G1 phase, while downregulation of circFGFR2 showed the opposite effects. In addition, EdU incorporation assay confirmed that circFGFR2 elevated the cell proliferation rate as shown by overexpression and knockdown of circFGFR2. The results strongly supported that circFGFR2 could promote skeletal muscle cell proliferation. Skeletal myogenesis comes after cell cycle termination, which is coordinated by various regulatory transcription factors, including *MYOD*, *MYOG*, myogenic factor 5 (*Mrf5*), the muscle regulatory factor 4 (*Mrf4*), and myocyte enhancer factor-2 (*Mef2*) families [47,48]. *MYOD* and *MYOG* can regulate most myogenesis-related genes thus facilitating myoblast differentiation into myotubes [49,50]. *MyHC* is a differentiation marker gene of muscle and forms the backbone of the sarcomere thick filaments [51]. The circFGFR2 exerts a function in skeletal muscle cell proliferation, we detected whether circFGFR2 was also involved in skeletal muscle cell differentiation by monitoring the impact of circFGFR2 on the expression of *MYOD* and *MYOG*. As expected, circFGFR2 could promote the expression of *MYOD* and *MYOG*. MyHC immunofluorescence suggested that circFGFR2 accelerated the formation of myotubes, which confirmed another important role of circFGFR2 in skeletal muscle cell, i.e., it can facilitate myoblast differentiation.

Circular RNA is known to be a functional molecule transcribed from protein-encoding genes which contain MREs like other mRNAs or lncRNAs [52]. However, circular RNA was capable of escaping from degradation as it has no poly A tail could not be recognized by exonuclease compared with mRNAs or lncRNAs [5]. In addition, the expression level of some circular RNAs were not lower than their linear mRNAs [53]. Based on that advantage, they are efficient to act as ceRNA, which are enriched for stable miRNA binding sites and regulate the activity of miRNA. Bioinformatics technology is universally applicable for the analysis of the binding relationship of ceRNA and miRNA [54]. In this study, using the bioinformatics program RNAhybrid, we found that circFGFR2 had two possible binding sites for miR-133a-5p and one site for miR-29b-1-5p. Subsequently, we confirmed that miR-133a-5p and miR-29b-1-5p were actually combined with the predicated sites of circFGFR2 but not with *FGFR2* mRNA as indicated by two dual-luciferase reporter assays. Biotin-coupled miRNA pull down is an efficient method to verify the combined relationship between circular RNA and miRNA [18,19,55]. In this study, biotin-miR-133a-5p and biotin-miR-29b-1-5p were efficient in enriching circFGFR2, and overexpression of circFGFR2 significantly inhibits the expression of miR-133a-5p and miR-29b-1-5p which confirm the interacted relationship between circFGFR2 and miR-133a-5p/miR-29b-1-5p.

miR-133a-5p and miR-29b-1-5p belong to two miRNA families, miR-133 and miR-29, respectively. These two families have been well-studied miRNAs, and found to be involved in skeletal muscle cell proliferation and differentiation [27,38,56]. In mouse C2C12 cell line, miR-133 which contain a seed sequence of "UUGGUCC" could promote myoblast differentiation and inhibit cell proliferation, and miR-29 which contains a seed sequence of "AGCACCA" could reduce proliferation and facilitate differentiation [28,56]. The roles of miR-133a-5p and miR-29b-1-5p in avian skeletal muscle development still remain unclear. Here we first reported that miR-133a-5p and miR-29b-1-5p could repress the proliferation and differentiation of skeletal muscle cell. The roles of these two miRNAs were different from the studied miR-133 and miR-29 in mouse. We compared the sequence of miR-133a-5p and miR-29b-1-5p with other miR-133s and miR-29s in both chicken and mouse, and found that the mature sequences of gga-miR-133a-5p and gga-miR-29b-1-5p were different from the studied miR-133 and miR-29. Since the seed sequence was different, and miRNA exerts function by targeting the 3′-UTR of their target genes, it is possibly that the function of gga-miR-133a-5p and gga-miR-29b-1-5p was different from the miR-133 and miR-29 which have been studied in mouse. On the other hand, the roles of gga-miR-133a-5p or gga-miR-29b-1-5p were opposite to the effect of circFGFR2 in myoblast. It is therefore reasonable that circFGFR2 could act as a molecular sponge for miR-133a-5p and miR-29b-1-5p. To confirm this, we further performed rescue experiments and found that circFGFR2 eliminated the inhibition effect of miR-133a-5p and miR-29b-1-5p on myoblast proliferation and differentiation. Considering all of this, we declared that circFGFR2 regulates skeletal muscle cell proliferation and differentiation by inhibiting the expression and activity of miR-133a-5p and miR-29b-1-5p in poultry.

5. Conclusions

In conclusion, we found that a novel circular RNA of circFGFR2, generated by the FGFR2 gene, could regulate myoblast proliferation and differentiation by acting as a sponge of miR-133a-5p and miR-29b-1-5p in poultry.

Author Contributions: X.C. conceived the study, carried out all experiments, analyzed data, and wrote the paper. H.O. provided essential logistical help. Z.W. and B.C. participated in partial experiments. Q.N. conceived the study, and participated in its design and coordination.

Acknowledgments: We thank the Chicken Breeding Farm of South China Agricultural University for providing the eggs for hatching chickens. We thank Endashaw Jebessa for his edit of the manuscript.

References

1.   Sanger, H.L.; Klotz, G.; Riesner, D.; Gross, H.J. Kleinschmidt, A.K. Viroids are single-stranded covalently closed circular RNA molecules existing as highly base-paired rod-like structures. *Proc. Natl. Acad. Sci. USA* **1976**, *73*, 3852–3856. [CrossRef] [PubMed]

2.   Capel, B.; Swain, A.; Nicolis, S.; Hacker, A.; Walter, M.; Koopman, P.; Goodfellow, P.; Lovell-Badge, R. Circular transcripts of the testis-determining gene Sry in adult mouse testis. *Cell* **1993**, *73*, 1019–1030. [CrossRef]

3.   Arnberg, A.C.; Van Ommen, G.J.; Grivell, L.A.; Van Bruggen, E.F.; Borst, P. Some yeast mitochondrial RNAs are circular. *Cell* **1980**, *19*, 313–319. [CrossRef]

4.   Cocquerelle, C.; Mascrez, B.; Hetuin, D.; Bailleul, B. Mis-splicing yields circular RNA molecules. *FASEB J.* **1993**, *7*, 155–160. [CrossRef] [PubMed]

5.   Jeck, W.R.; Sharpless, N.E. Detecting and characterizing circular RNAs. *Nat. Biotechnol.* **2014**, *32*, 453–461. [CrossRef] [PubMed]

6.   Zhang, Z.; Qi, S.; Tang, N.; Zhang, X.; Chen, S.; Zhu, P.; Ma, L.; Cheng, J.; Xu, Y.; Lu, M.; et al. Discovery of replicating circular RNAs by RNA-seq and computational algorithms. *PLoS Pathog.* **2014**, *10*, e1004553. [CrossRef] [PubMed]

7.   Salzman, J.; Gawad, C.; Wang, P.L.; Lacayo, N.; Brown, P.O. Circular RNAs are the predominant transcript isoform from hundreds of human genes in diverse cell types. *PLoS ONE* **2012**, *7*, e30733. [CrossRef] [PubMed]

8.   Abdelmohsen, K.; Panda, A.C.; De, S.; Grammatikakis, I.; Kim, J.; Ding, J.; Noh, J.H.; Kim, K.M.; Mattison, J.A.; de Cabo, R.; et al. Circular RNAs in monkey muscle: Age-dependent changes. *Aging* **2015**, *7*, 903–910. [CrossRef] [PubMed]

9.   Veno, M.T.; Hansen, T.B.; Veno, S.T.; Clausen, B.H.; Grebing, M.; Finsen, B.; Holm, I.E.; Kjems, J. Spatio-temporal regulation of circular RNA expression during porcine embryonic brain development. *Genome Biol.* **2015**, *16*, 245. [CrossRef] [PubMed]

10.   Qu, S.; Yang, X.; Li, X.; Wang, J.; Gao, Y.; Shang, R.; Sun, W.; Dou, K.; Li, H. Circular RNA: A new star of noncoding RNAs. *Cancer Lett.* **2015**, *365*, 141–148. [CrossRef] [PubMed]

11.   Chen, L.L. The biogenesis and emerging roles of circular RNAs. *Nat. Rev. Mol. Cell Biol.* **2016**, *17*, 205–211. [CrossRef] [PubMed]

12. Du, W.W.; Yang, W.; Chen, Y.; Wu, Z.K.; Foster, F.S.; Yang, Z.; Li, X.; Yang, B.B. Foxo3 circular RNA promotes cardiac senescence by modulating multiple factors associated with stress and senescence responses. *Eur. Heart J.* **2017**, *38*, 1402–1412. [CrossRef] [PubMed]

13. Du, W.W.; Yang, W.; Liu, E.; Yang, Z.; Dhaliwal, P.; Yang, B.B. Foxo3 circular RNA retards cell cycle progression via forming ternary complexes with p21 and CDK2. *Nucleic Acids Res.* **2016**, *44*, 2846–2858. [CrossRef] [PubMed]

14. Chen, C.Y.; Sarnow, P. Initiation of protein synthesis by the eukaryotic translational apparatus on circular RNAs. *Science* **1995**, *268*, 415–417. [CrossRef] [PubMed]

15. Wang, Y.; Wang, Z. Efficient backsplicing produces translatable circular mRNAs. *RNA* **2015**, *21*, 172–179. [CrossRef] [PubMed]

16. Abe, N.; Matsumoto, K.; Nishihara, M.; Nakano, Y.; Shibata, A.; Maruyama, H.; Shuto, S.; Matsuda, A.; Yoshida, M.; Ito, Y.; et al. Rolling Circle Translation of Circular RNA in Living Human Cells. *Sci. Rep.* **2015**, *5*, 16435. [CrossRef] [PubMed]

17. Dong, R.; Zhang, X.O.; Zhang, Y.; Ma, X.K.; Chen, L.L.; Yang, L. CircRNA-derived pseudogenes. *Cell Res.* **2016**, *26*, 747–750. [CrossRef] [PubMed]

18. Hansen, T.B.; Jensen, T.I.; Clausen, B.H.; Bramsen, J.B.; Finsen, B.; Damgaard, C.K.; Kjems, J. Natural RNA circles function as efficient microRNA sponges. *Nature* **2013**, *495*, 384–388. [CrossRef] [PubMed]

19. Wang, K.; Long, B.; Liu, F.; Wang, J.X.; Liu, C.Y.; Zhao, B.; Zhou, L.Y.; Sun, T.; Wang, M.; Yu, T.; et al. A circular RNA protects the heart from pathological hypertrophy and heart failure by targeting miR-223. *Eur. Heart J.* **2016**, *37*, 2602–2611. [CrossRef] [PubMed]

20. Yang, C.; Yuan, W.; Yang, X.; Li, P.; Wang, J.; Han, J.; Tao, J.; Li, P.; Yang, H.; Lv, Q.; et al. Circular RNA circ-ITCH inhibits bladder cancer progression by sponging miR-17/miR-224 and regulating p21, PTEN expression. *Mol. Cancer* **2018**, *17*, 19. [CrossRef] [PubMed]

21. Zheng, Q.; Bao, C.; Guo, W.; Li, S.; Chen, J.; Chen, B.; Luo, Y.; Lyu, D.; Li, Y.; Shi, G.; et al. Circular RNA profiling reveals an abundant circHIPK3 that regulates cell growth by sponging multiple miRNAs. *Nat. Commun.* **2016**, *7*, 11215. [CrossRef] [PubMed]

22. Ouyang, H.; Chen, X.; Wang, Z.; Yu, J.; Jia, X.; Li, Z.; Luo, W.; Abdalla, B.A.; Jebessa, E.; Nie, Q.; et al. Circular RNAs are abundant and dynamically expressed during embryonic muscle development in chickens. *DNA Res.* **2017**. [CrossRef] [PubMed]

23. Ornitz, D.M.; Marie, P.J. Fibroblast growth factor signaling in skeletal development and disease. *Genes Dev.* **2015**, *29*, 1463–1486. [CrossRef] [PubMed]

24. Ambros, V. The functions of animal microRNAs. *Nature* **2004**, *431*, 350–355. [CrossRef] [PubMed]

25. Baek, D.; Villen, J.; Shin, C.; Camargo, F.D.; Gygi, S.P.; Bartel, D.P. The impact of microRNAs on protein output. *Nature* **2008**, *455*, 64–71. [CrossRef] [PubMed]

26. Chen, J.F.; Mandel, E.M.; Thomson, J.M.; Wu, Q.; Callis, T.E.; Hammond, S.M.; Conlon, F.L.; Wang, D.Z. The role of microRNA-1 and microRNA-133 in skeletal muscle proliferation and differentiation. *Nat. Genet.* **2006**, *38*, 228–233. [CrossRef] [PubMed]

27. Luo, Y.; Wu, X.; Ling, Z.; Yuan, L.; Cheng, Y.; Chen, J.; Xiang, C. microRNA133a targets Foxl2 and promotes differentiation of C2C12 into myogenic progenitor cells. *DNA Cell Biol.* **2015**, *34*, 29–36. [CrossRef] [PubMed]

28. Mishima, Y.; Abreu-Goodger, C.; Staton, A.A.; Stahlhut, C.; Shou, C.; Cheng, C.; Gerstein, M.; Enright, A.J.; Giraldez, A.J. Zebrafish miR-1 and miR-133 shape muscle gene expression and regulate sarcomeric actin organization. *Genes Dev.* **2009**, *23*, 619–632. [CrossRef] [PubMed]

29. Kriegel, A.J.; Liu, Y.; Fang, Y.; Ding, X.; Liang, M. The miR-29 family: Genomics, cell biology, and relevance to renal and cardiovascular injury. *Physiol. Genom.* **2012**, *44*, 237–244. [CrossRef] [PubMed]

30. Lee, J.; Lim, S.; Song, B.W.; Cha, M.J.; Ham, O.; Lee, S.Y.; Lee, C.; Park, J.H.; Bae, Y.; Seo, H.H.; et al. MicroRNA-29b inhibits migration and proliferation of vascular smooth muscle cells in neointimal formation. *J. Cell Biochem.* **2015**, *116*, 598–608. [CrossRef] [PubMed]

31. Li, Z.; Hassan, M.Q.; Jafferji, M.; Aqeilan, R.I.; Garzon, R.; Croce, C.M.; van Wijnen, A.J.; Stein, J.L.; Stein, G.S.; Lian, J.B. Biological functions of miR-29b contribute to positive regulation of osteoblast differentiation. *J. Biol. Chem.* **2009**, *284*, 15676–15684. [CrossRef] [PubMed]

32. Fu, Q.; Shi, H.; Shi, M.; Meng, L.; Zhang, H.; Ren, Y.; Guo, F.; Jia, B.; Wang, P.; Ni, W.; et al. bta-miR-29b

attenuates apoptosis by directly targeting caspase-7 and NAIF1 and suppresses bovine viral diarrhea virus replication in MDBK cells. *Can. J. Microbiol.* **2014**, *60*, 455–460. [CrossRef] [PubMed]

33. Shen, L.; Song, Y.; Fu, Y.; Li, P. MiR-29b mimics promotes cell apoptosis of smooth muscle cells via targeting on MMP-2. *Cytotechnology* **2018**, *70*, 351–359. [CrossRef] [PubMed]

34. Mott, J.L.; Kobayashi, S.; Bronk, S.F.; Gores, G.J. mir-29 regulates Mcl-1 protein expression and apoptosis. *Oncogene* **2007**, *26*, 6133–6140. [CrossRef] [PubMed]

35. Fabbri, M.; Garzon, R.; Cimmino, A.; Liu, Z.; Zanesi, N.; Callegari, E.; Liu, S.; Alder, H.; Costinean, S.; Fernandez-Cymering, C.; et al. MicroRNA-29 family reverts aberrant methylation in lung cancer by targeting DNA methyltransferases 3A and 3B. *Proc. Natl. Acad. Sci. USA* **2007**, *104*, 15805–15810. [CrossRef] [PubMed]

36. Fu, Q.; Shi, H.; Chen, C. Roles of bta-miR-29b promoter regions DNA methylation in regulating miR-29b expression and bovine viral diarrhea virus NADL replication in MDBK cells. *Arch. Virol.* **2017**, *162*, 401–408. [CrossRef] [PubMed]

37. Zhou, L.; Wang, L.; Lu, L.; Jiang, P.; Sun, H.; Wang, H. A novel target of microRNA-29, Ring1 and YY1-binding protein (Rybp), negatively regulates skeletal myogenesis. *J. Biol. Chem.* **2012**, *287*, 25255–25265. [CrossRef] [PubMed]

38. Wei, W.; He, H.B.; Zhang, W.Y.; Zhang, H.X.; Bai, J.B.; Liu, H.Z.; Cao, J.H.; Chang, K.C.; Li, X.Y.; Zhao, S.H. miR-29 targets Akt3 to reduce proliferation and facilitate differentiation of myoblasts in skeletal muscle development. *Cell Death Dis.* **2013**, *4*, e668. [CrossRef] [PubMed]

39. Winbanks, C.E.; Wang, B.; Beyer, C.; Koh, P.; White, L.; Kantharidis, P.; Gregorevic, P. TGF-beta regulates miR-206 and miR-29 to control myogenic differentiation through regulation of HDAC4. *J. Biol. Chem.* **2011**, *286*, 13805–13814. [CrossRef] [PubMed]

40. Li, J.; Chan, M.C.; Yu, Y.; Bei, Y.; Chen, P.; Zhou, Q.; Cheng, L.; Chen, L.; Ziegler, O.; Rowe, G.C.; et al. miR-29b contributes to multiple types of muscle atrophy. *Nat. Commun.* **2017**, *8*, 15201. [CrossRef] [PubMed]

41. Wang, L.; Zhou, L.; Jiang, P.; Lu, L.; Chen, X.; Lan, H.; Guttridge, D.C.; Sun, H.; Wang, H. Loss of miR-29 in myoblasts contributes to dystrophic muscle pathogenesis. *Mol. Ther.* **2012**, *20*, 1222–1233. [CrossRef] [PubMed]

42. Zanotti, S.; Gibertini, S.; Curcio, M.; Savadori, P.; Pasanisi, B.; Morandi, L.; Cornelio, F.; Mantegazza, R.; Mora, M. Opposing roles of miR-21 and miR-29 in the progression of fibrosis in Duchenne muscular dystrophy. *Biochim. Biophys. Acta* **2015**, *1852*, 1451–1464. [CrossRef] [PubMed]

43. Abmayr, S.M.; Pavlath, G.K. Myoblast fusion: Lessons from flies and mice. *Development* **2012**, *139*, 641–656. [CrossRef] [PubMed]

44. Sassoon, D.A. Myogenic regulatory factors: Dissecting their role and regulation during vertebrate embryogenesis. *Dev. Biol.* **1993**, *156*, 11–23. [CrossRef] [PubMed]

45. Salzman, J.; Chen, R.E.; Olsen, M.N.; Wang, P.L.; Brown, P.O. Cell-type specific features of circular RNA expression. *PLoS Genet.* **2013**, *9*, e1003777. [CrossRef]

46. Westholm, J.O.; Miura, P.; Olson, S.; Shenker, S.; Joseph, B.; Sanfilippo, P.; Celniker, S.E.; Graveley, B.R.; Lai, E.C. Genome-wide analysis of drosophila circular RNAs reveals their structural and sequence properties and age-dependent neural accumulation. *Cell Rep.* **2014**, *9*, 1966–1980. [CrossRef] [PubMed]

47. Dodou, E.; Xu, S.M.; Black, B.L. mef2c is activated directly by myogenic basic helix-loop-helix proteins during skeletal muscle development in vivo. *Mech. Dev.* **2003**, *120*, 1021–1032. [CrossRef]

48. Blum, R.; Vethantham, V.; Bowman, C.; Rudnicki, M.; Dynlacht, B.D. Genome-wide identification of enhancers in skeletal muscle: The role of MYOD1. *Genes Dev.* **2012**, *26*, 2763–2779. [CrossRef] [PubMed]

49. Berkes, C.A.; Tapscott, S.J. MYOD and the transcriptional control of myogenesis. *Semin. Cell Dev. Biol.* **2005**, *16*, 585–595. [CrossRef] [PubMed]

50. Cao, Y.; Kumar, R.M.; Penn, B.H.; Berkes, C.A.; Kooperberg, C.; Boyer, L.A.; Young, R.A.; Tapscott, S.J. Global and gene-specific analyses show distinct roles for MYOD and Myog at a common set of promoters. *EMBO J.* **2006**, *25*, 502–511. [CrossRef] [PubMed]

51. Tajsharghi, H.; Oldfors, A. Myosinopathies: Pathology and mechanisms. *Acta Neuropathol.* **2013**, *125*, 3–18. [CrossRef] [PubMed]

52. Ashwal-Fluss, R.; Meyer, M.; Pamudurti, N.R.; Ivanov, A.; Bartok, O.; Hanan, M.; Evantal, N.; Memczak, S.; Rajewsky, N.; Kadener, S. circRNA biogenesis competes with pre-mRNA splicing. *Mol. Cell* **2014**, *56*, 55–66. [CrossRef] [PubMed]

53. Jeck, W.R.; Sorrentino, J.A.; Wang, K.; Slevin, M.K.; Burd, C.E.; Liu, J.; Marzluff, W.F.; Sharpless, N.E.

Circular RNAs are abundant, conserved, and associated with ALU repeats. *RNA* **2013** *19*, 141–157. [CrossRef] [PubMed]

54. Dudekula, D.B.; Panda, A.C.; Grammatikakis, I.; De, S.; Abdelmohsen, K.; Gorospe, M. CircInteractome: A web tool for exploring circular RNAs and their interacting proteins and microRNAs. *RNA Biol.* **2016**, *13*, 34–42. [CrossRef] [PubMed]

55. Lal, A.; Thomas, M.P.; Altschuler, G.; Navarro, F.; O'Day, E.; Li, X.L.; Concepcion, C.; Han, Y.C.; Thiery, J.; Rajani, D.K.; et al. Capture of microRNA-bound mRNAs identifies the tumor suppressor miR-34a as a regulator of growth factor signaling. *PLoS Genet.* **2011**, *7*, e1002363. [CrossRef] [PubMed]

56. Zhou, L.; Wang, L.; Lu, L.; Jiang, P.; Sun, H.; Wang, H. Inhibition of miR-29 by TGF-beta-Smad3 signaling through dual mechanisms promotes transdifferentiation of mouse myoblasts into myofibroblasts. *PLoS ONE* **2012**, *7*, e33766. [CrossRef] [PubMed]

Permissions

The contributors of this book come from diverse backgrounds, making this book a truly international effort. This book will bring forth new frontiers with its revolutionizing research information and detailed analysis of the nascent developments around the world.

We would like to thank all the contributing authors for lending their expertise to make the book truly unique. They have played a crucial role in the development of this book. Without their invaluable contributions this book wouldn't have been possible. They have made vital efforts to compile up to date information on the varied aspects of this subject to make this book a valuable addition to the collection of many professionals and students.

This book was conceptualized with the vision of imparting up-to-date information and advanced data in this field. To ensure the same, a matchless editorial board was set up. Every individual on the board went through rigorous rounds of assessment to prove their worth. After which they invested a large part of their time researching and compiling the most relevant data for our readers.

The editorial board has been involved in producing this book since its inception. They have spent rigorous hours researching and exploring the diverse topics which have resulted in the successful publishing of this book. They have passed on their knowledge of decades through this book. To expedite this challenging task, the publisher supported the team at every step. A small team of assistant editors was also appointed to further simplify the editing procedure and attain best results for the readers.

Apart from the editorial board, the designing team has also invested a significant amount of their time in understanding the subject and creating the most relevant covers. They scrutinized every image to scout for the most suitable representation of the subject and create an appropriate cover for the book.

The publishing team has been an ardent support to the editorial, designing and production team. Their endless efforts to recruit the best for this project, has resulted in the accomplishment of this book. They are a veteran in the field of academics and their pool of knowledge is as vast as their experience in printing. Their expertise and guidance has proved useful at every step. Their uncompromising quality standards have made this book an exceptional effort. Their encouragement from time to time has been an inspiration for everyone.

The publisher and the editorial board hope that this book will prove to be a valuable piece of knowledge for researchers, students, practitioners and scholars across the globe.

List of Contributors

Duy N. Do
Agriculture and Agri-Food Canada, Sherbrooke Research and Development Centre, Sherbrooke, QC J1M 0C8, Canada
Department of Animal Science, McGill University, Ste-Anne-de-Bellevue, QC H9X 3V9, Canada

Pier-Luc Dudemaine and Eveline M. Ibeagha-Awemu
Agriculture and Agri-Food Canada, Sherbrooke Research and Development Centre, Sherbrooke, QC J1M 0C8, Canada

Bridget E. Fomenky
Agriculture and Agri-Food Canada, Sherbrooke Research and Development Centre, Sherbrooke, QC J1M 0C8, Canada
Département de Sciences Animale, Université Laval, Quebec, QC G1V 0A6, Canada

Claudia Ricci, Carlotta Marzocchi and Stefania Battistini
Department of Medical, Surgical and Neurological Sciences, University of Siena, 53100 Siena, Italy

María José López-Galiano, Inmaculada García-Robles, M. Dolores Real and Carolina Rausell
Department of Genetics, University of Valencia, Burjassot, 46100 Valencia, Spain

Ana I. González-Hernández, Gemma Camañes and Begonya Vicedo
Plant Physiology Area, Biochemistry and Biotechnology Group, Department CAMN, University Jaume I, 12071 Castellón, Spain

Adele Vivacqua, Anna Sebastiani, Damiano Cosimo Rigiracciolo, Francesca Cirillo, Giulia Raffaella Galli, Marianna Talia, Maria Francesca Santolla, Rosamaria Lappano, Francesca Giordano, Maria Luisa Panno and Marcello Maggiolini
Department of Pharmacy, Health and Nutritional Sciences, University of Calabria, 87036 Rende, Italy

Anna Maria Miglietta
Regional HospitalCosenza, 87100 Cosenza, Italy

Michael Kravchik, Ran Stav, Eduard Belausov and Tzahi Arazi
Institute of Plant Sciences, Agricultural Research Organization, Volcani Center, Bet Dagan 50250, Israel

James Jabalee and Rebecca Towle
Department of Integrative Oncology, British Columbia Cancer Research Center, Vancouver V5Z 1L3, BC, Canada

Cathie Garnis
Department of Integrative Oncology, British Columbia Cancer Research Center, Vancouver V5Z 1L3, BC, Canada
Division of Otolaryngology, Department of Surgery, University of British Columbia, Vancouver V6T 1Z4, BC, Canada

Xin Shu, Xinyuan Zang, Xiaoshuang Liu, Jie Yang and Jin Wang
The State Key Laboratory of Pharmaceutical Biotechnology and Jiangsu Engineering Research Center for MicroRNA Biology and Biotechnology, NJU Advanced Institute for Life Sciences (NAILS), School of Life Science, Nanjing University, Nanjing 210023, China

Joseph L. Pegler, Jackson M. J. Oultram, Christopher P. L. Grof and Andrew L. Eamens
Centre for Plant Science, School of Environmental and Life Sciences, Faculty of Science, University of Newcastle, Callaghan 2308, New South Wales, Australia

Felipe Fenselau de Felippes
Science and Engineering Faculty, Queensland University of Technology, Brisbane, Australia

Shuyuan Wang, Wencan Wang, Qianqian Meng, Xueyan Ma, Xu Zhou, Hui Liu and Xiaowen Chen
College of Bioinformatics Science and Technology, Harbin Medical University, Harbin 150081, China

Shunheng Zhou, Haizhou Liu and Wei Jiang
College of Automation Engineering, Nanjing University of Aeronautics and Astronautics, Nanjing 211106, China

Leopold F. Fröhlich
Department of Cranio-Maxillofacial Surgery, University of Münster, Albert-Schweitzer-Campus 1, 48149 Münster, Germany

Anthony A. Millar, Allan Lohe and Gigi Wong
Division of Plant Science, Research School of Biology, The Australian National University, Canberra ACT 2601, Australia

Xiaolan Chen, Zhijun Wang, Biao Chen and Qinghua Nie
Department of Animal Genetics, Breeding and Reproduction, College of Animal Science, South China Agricultural University, Guangzhou 510642, China National-Local Joint Engineering Research Center for Livestock Breeding, Guangdong Provincial Key Lab of Agro-Animal Genomics and Molecular Breeding, and the Key Lab of Chicken Genetics, Breeding and Reproduction, Ministry of Agriculture, Guangzhou 510642, China

Hongjia Ouyang
Department of Animal Genetics, Breeding and Reproduction, College of Animal Science, South China Agricultural University, Guangzhou 510642, China College of Animal Science & Technology, Zhongkai University of Agriculture and Engineering, Guangzhou 510225, China

Index

www.ingramcontent.com/pod-product-compliance
Lightning Source LLC
Chambersburg PA
CBHW080411190526
45161CB00003B/200